Hamas Jihad

All rights reserved. No part of this publication may be reproduced, stored in a retrieval system, or transmitted in any form or by any means, without the permission of the copyright holder.

Copyright ©2016 Yisrael Ne'eman
Cover & Interior Design by White Hart Publications

Short excerpts and quotes may be used to advance research, debate and discussion, as long as citation credit is given to the author. For other copyright permissions, contact Yisrael Ne'eman at www.HamasJihad.com, or via email at author@hamasjihad.com.

Library of Congress Cataloging-in-Publication Data
Ne'eman, Yisrael 1956
Hamas Jihad: Antisemitism, Islamic World Conquest and Manipulation of Palestinian Nationalism

Library of Congress Control Number: 2016936543

ISBN: 978-1-942923-14-5 (paperback)
978-1-942923-15-2 (electronic)

Hamas Jihad

Antisemitism, Islamic World Conquest & Manipulation of Palestinian Nationalism

Yisrael Ne'eman

White Hart Publications, USA | Ne'eman Books, Israel

Dedication

To all those who defend against
the dictates and ravages of Jihad

Acknowledgments

Some fifteen years ago, I realized the need to expose the dangers of a rising Jihadist tide and antisemitism, not only in the Middle East but the West as well. Exposing Hamas was the first order of business. Many encouraged me to do so and to those who assisted me in this venture I express my thanks.

I am grateful to Professor David Patterson, who from the outset offered his encouragement, insights, and time in reviewing the bulk of the material covered in this volume. Likewise, my long-term friend and colleague, Elliot Chodoff from my hometown of Eshchar, offered advice and constructive criticism as he read through the manuscript over the years. He helped me more clearly express certain perspectives and conclusions appearing in this work. To both, I offer my heartfelt thanks.

I would like to thank Rachael Hartman, the publisher, editor, and owner of White Hart Publications. She labored long and hard over the past year, at times assisted by an associate, Kent Curry, to rework an often complicated text to make it comprehensible to the non-professional reader. Rachael made sure this book would see the light of day, and to her, I extend my great appreciation.

There were many "readers," none professionals in the field, who suggested an insight on one topic or another. They were quite helpful and adjustments were made as a result of their comments. On the professional side, I would like to thank Dr. Yigal Carmon of Memri for providing access to the English translation of *The Hamas Covenant*.

And of course, there is my wife, Susan, to whom I am most appreciative. Manuscripts can be quite challenging. She was always there to offer support and advice while reading through this entire work. To her, I express my continual gratitude.

Contents

Introduction   **The Pivotal Importance of Understanding**
11   ***The Hamas Covenant* and Its Ramifications**
       A Personal Perspective 18

Chapter I   **Negative Image of the Jew in the Arab Muslim World**
23   Historical Review up to the Mid Twentieth Century

   Part I   Judeo-Islamic Relationship: History and Theology
       Arabian Origins of Islam 24
       Negative Stereotypes of the Jew in the Koran. 27
       Christians and Jews Condemned Together in the Koran . . .32
       Jews in the Hadith. 35
       The Charter of Omar and Dhimmi Status. 36

   Part II   Historic Survey: Islamic Attitudes/Policies Toward the Jews
       The Middle Ages through the Nineteenth Century 41
       Mid Twentieth Century Onward. 45

   Part III   From Ottoman Islamic Rule to Secular Nationalism
       Jews and Muslims in the Land of Israel/Palestine. 49
       Rollback of the 1000 Year Jihad: Christendom Defeats
          Ottoman Imperialism. 52
       Confluence of Interests and Theology: Christian-Islamic
          Antisemitism 55
       Islam Incorporates Czarist/Nazi Indictments Against Jews 61

Chapter II   **Arab Islamic Ideologues - Jihad Past and Present**
71   Hasan al-Banna, Sayyid Qutb, and Abdullah Azzam

   Part I   Hasan al-Banna: Founding the Muslim Brotherhood,
       Activating Jihad
       Jihad and Arabization. 72
       Jihad Today the Battle Against the West 77

Contents

Part II **Sayyid Qutb: Theological Father, *The Hamas Covenant***
Jahiliyyah and Jihad 79
The Jahili in Other Cultures: East, West, Judaism,
  and Christianity 80
The Jahili Infected Muslim World. 84
Islam in Contrast to Secular Arab Nationalism
  and the PLO/Fatah 85
Women 87
Jihad in the Battle Against Jahiliyyah 89
Qutb's Continuing Antisemitic Influence 92

Part III **Abdullah Azzam: Jihadi Scholar and Warrior**
Afghani Jihad and World Conquest 98
Afghanistan and Palestine. 103

Part IV **Al-Banna, Qutb, and Azzam**
Comparisons and Impact on Hamas. 106

Chapter III Zionism
117 Jewish National Liberation Catalyzes Islamic Antisemitism to New Extremes

Turkish Policy Toward Zionism 1882-1918 118
Islamic Fundamentalism and Palestinian Nationalism
  in the British Mandate 124
Haj Amin el-Husseini's Alliance with the Nazis 130
The Muslim Brotherhood and the Palestine Mandate:
  1945-48 131
Jihad and the Dhimma: 1948 and Beyond 135
The PLO and PA Adopt the Hamas Jewish Stereotype ... 143

Chapter IV Development of the Palestinian Muslim
161 Brotherhood/Hamas 1948 to 2000

The 1948 War to the 1987 Rise of Independent Hamas .. 162
Hamas and the Intifada 1987-93 173
Oslo Accords and the PA Mini-State 1993-2000 181
Decisive Crossroads: Summer 2000 192

Chapter V — 199
Hamas Ideological Victory, Resistance, and Pragmatism 2000 to 2016

The Low Intensity Conflict (LIC) of 2000 to 2004,
 The Second or "Al-Aksa" Intifada. 201
End of the LIC: Hamas Gains Power Despite Fatah, Israeli,
 and Western Opposition 215
The "Cast Lead" Gaza War and Repercussions 227
Sharia Law, Hudnas, and the 2011 Islamic Awakening. . .235

Chapter VI — 255
The Hamas Covenant Analysis
Islamism, Jihad, and Antisemitism

Covenant of the Islamic Resistance Movement–Hamas . . 257
Preamble 259
Chapter One: Introduction to the Movement Ideological
 Premises 262
Chapter Two: Causes and Goals 270
Chapter Three: Strategy and Means 271
Chapter Four: Our Positions On
   A. The Islamic Movements. 295
   B. Nationalist Movements in the
      Palestinian Arena 297
   C. Palestine Liberation Organization 299
   D. Arab and Islamic States and Governments 302
Chapter Five: Historical Evidence throughout the Generations
 Regarding Confrontation with Aggressors 314
Conclusion: The Islamic Resistance Movement Soldiers . . 317
Summary and Conclusion 319

Chapter VII — 323
A Comparative Analysis
The Palestinian National Charter and *The Hamas Covenant*

The Palestinian National Charter 323
Conclusion. 351

Chapter VIII — Czarist-Nazi Integration into Palestinian Islamist Jihad — 355

The Grand Mufti Haj Amin el-Husseini and the Nazis... 355
Jewish Involvement in WWI, WWII,
    and the Bolshevik Revolution 361
Czarist/Nazi Antisemitism in *The Hamas Covenant*...... 363

Chapter IX — Conflict Resolution in the Shadow of Islamic Abrogation — 375
Constructive Perspectives Toward Jews, Christians, and Others

Islamic Endorsement of the Israelite Covenant at Sinai... 380
Israel and the Promised Land in the Koran............ 385
Changing Islamic Attitudes Toward Christians,
    and Christianity 389
Non-Western, non-Abrahamic Faiths 393
Decision Making in Political Islam and Hamas 395
Western Tolerance of Jihadi Islam.................. 396

Chapter X — Summary and Conclusions — 405
Analysis of Hamas Revolution and World Islamism

Discontent, Revolutionary Stages, and Hamas 406
Hamas Revolution Disrupted by Outside Forces 410
Islamic World Domain, Background, Message,
    and Ramifications 417
Hamas Future within the Palestinian and Arab/Islamic
    World 427

List of Maps 10
Glossary 433
Bibliography......................................455
Index...475
About the Author 492

List of Maps

Map of the Arab World 22

Ottoman Empire at the Height of Expansion in 1683 22

Map of the Caliphate as demanded by the Islamic State of Iraq
    and the Levant 70

1947 Partition Plan (replaces British Palestine Mandate) 116

1949-1967 Armistice Lines 160

Israel 2016 198

Gaza Strip 254

West Bank 322

Introduction

The Pivotal Importance of Understanding *The Hamas Covenant* and Its Ramifications

We are fortunate the Palestinian wing of the Muslim Brotherhood known as "Hamas" chose to write down its *Covenant* in 1988. *The Hamas Covenant* is a revealing window into the heart and soul of today's Islamic extremists. Hamas providing the written document was a great unintentional service by an Islamic fundamentalist group, benefitting secular Muslims, Christians, non-believers and especially Jews. Sixty years after Hasan al-Banna established the radical Muslim Brotherhood in Egypt Hamas compiled the concise ideological document explaining the religio-philosophical pillars of the movement. The *Covenant* was also presented as a Palestinian "nationalist" platform, but removing the facade reveals the current demands on followers and the future world as extremist political Islamism envisions it. Now translated to English, the *Covenant* is accessible to hundreds of millions of Westerners. The central pillar of the ideology is commitment to Islam through Jihad, with Jews presented as the manifest and hidden enemy ultimately destined for annihilation. The document is clearly antisemitic, but other people groups such as "Crusaders" (Christians) and "Tatars" (Far Easterners) are also labeled as enemies of Islam. Social ideals and economic systems such as capitalism, socialism, communism, and with a bit of research and interpretation, democracy as well, are all considered anti-Islamic ways of life, and in essence pagan and/or Jewish plots (Koran 5:82). The root of each clause in the *Covenant* stretches back decades, at times even as far back to the rise of Islam in the seventh century CE.

Hamas is not a Palestinian liberation movement in the secular sense of the term. It advocates fundamentalist, revolutionary Islamic initiatives no less extreme than the Muslim Brotherhood dominated Sudan, Khomeinist Iran, or the Taliban dictatorship in Afghanistan prior to 2001. All are "diocentric," or theocratic regimes; Allah or God is the center of all law, worship and action. In the year 2011, we saw these ideals in action, revolution throughout the Arab/Muslim world and the increase of political Islamification. Sometimes conquest occurred in a moderate fashion, such as in Tunisia, but more often it was radical, as in Egypt where we also witnessed the rise of the Salafist movement which made the Muslim Brotherhood victory look flexible. To this day the Western media refers to the Arab uprisings of 2011 as the "Arab Spring," inferring protestors sought liberty, democracy and individual human rights. This may have been true at the very outset. However, in actuality, the "Arab Spring" was really an "Islamic Awakening." Hamas fits the pattern well; its objectives are Islamic, demanding *Sharia* law. Secular nationalism seeking cultural liberation and individual rights is its enemy, not its objective. The secular Palestinian nationalist group Fatah was evicted from Gaza in 2007. In essence, Fatah's exit made Hamas among the first victors in the "Islamic Awakening." Palestine is only one small geographic region in the overall effort for the "liberation" of the entire Muslim world from any secular or non-Islamist way of life.

Based on "*fatwas*," or religious judgments, the physical conquering Holy War, or Jihad, is said to be imperative among believers. Islam's Holy War has two steps. First, there is what is termed "defensive" Jihad, meaning the recapture of lands held by Muslims at some point in history such as the Iberian Peninsula, the Balkans, India and Israel. The second, or "offensive" step, involves the remainder of the planet, including North and South America, Russia, etc. Believing otherwise is either a misread of *The Hamas Covenant* and the texts comprising the foundation upon which it rests, or groundless optimism. One other option remains; whereby Hamas leadership, its followers and Muslim Brotherhood activists worldwide were not telling us the truth and developed a harsh fundamentalist document as an inspiration for Jihad and world conquest, simply as a bargaining chip in what some refer to as "posturing." This last option appears highly unlikely, and demands feeding outright lies to believers and enthusiastic supporters. The *Covenant* was written as a Divinely inspired work in the name of Allah for the Hamas faithful, more than for anyone else. *The Hamas Covenant* does not seek to deceive non-believers; it is clear in its message. Those who do not take it at face value deceive themselves and endanger all concerned.

Although the heart of this work rests on an appreciation of *The Hamas Covenant* itself, most have never read the seminal document, and among those who have, few relate to it seriously. Readers fail to take the text at face value despite the fact it declares itself as a modern day religious document, also known as "The Charter of Allah." A look at history recalls most "thinking" people never cared to indulge in reading the texts presented by Lenin, Mao or Hitler. Despite the dangers such ideals continue to embody for today's global society, the writings of Ayatollah Khomeini are ignored, as are those of his Sunni ideological partner, Sayyid Qutb, who was the major inspirational source for Hamas and considered the "Father of 9/11." *The Hamas Covenant* gives an overall picture of Muslim Brotherhood thought. The same perspective is held by its radical offshoots, such as the Salafists, and its two most infamous children, al-Qaeda and the Islamic State. Without the Muslim Brotherhood it would be difficult to imagine Osama bin Laden. Understanding the *Covenant's* contents is crucial for those who value freedom, the democratic lifestyle and Western society as a whole. Jews, more than all others as the proclaimed ultimate perfidious enemy of Hamas and the Muslim Brotherhood, need to be aware of the rabid antisemitism contained in Hamas doctrines. Jews are to suffer death as their "rightly earned" fate to pay for the evil deeds Hamas and their universal extension, the Muslim Brotherhood, attributed to them. A few may save their own lives, provided they subject themselves to the constraints of Sharia or Islamic law.

One weighty counter point is clear; the Muslim Brotherhood is an Islamist political organization not representing all Muslims. Many estimate that of the one and a half billion Muslims worldwide, only a 15 percent minority supports the Brotherhood. This percentage may no longer hold true after 2011 and therefore a reassessment may be in order. This leads others to believe that we may be speaking of possibly 30-40 percent of the worldwide Islamic population supporting the Brotherhood, or numerically half a billion Muslims or so, quite a large minority.

This book concentrates on what the *Covenant* advocates and how those ideals played out historically, especially concerning Jews. This discussion will deal with Muslim-Jewish relations as well as the attitude in the *Covenant* toward Jewish nationalism, which culminated in the modern State of Israel. The *Covenant* was written less than three decades ago and reflects anti-Jewish attitudes, which permeated Arab/Muslim society in previous centuries and are influential to this day. *The Hamas Covenant* is the ultimate integration of these Islamist ideals with traditional Christian European antisemitism. Committed to print, the combination became holy writ.

Whether the Muslim Brotherhood and Hamas are more or less popular at any given moment is less of a concern. In early 2016 Hamas remains on the defensive after the overthrow of the Morsi regime in Egypt and its own unsuccessful military encounter with Israel known as the Protective Edge Operation. Jihadists are involved in the continuing turmoil in Libya, Syria and Iraq but have not been able to declare victory. Lately they are facing the gains of the more liberal, secular nationalist democratic trends as we note in Tunisia. Yet Jihadists often retake the offensive, evidenced by the rise of the Islamic State and its allies. Regardless, Hamas ideals are integral to the Islamist political mind in the Arab world, and are not about to vanish. Due to temporary set-backs as a result of conflict with Israel, the Egyptian blockade and the resulting economic nosedive, Hamas was negotiating entry into a short-term national unity government with Fatah as part of the Palestinian Authority framework said to be arranging general elections in both the Gaza Strip and the West Bank. Win or lose, Hamas ideals will certainly gain traction, and the Islamic wheel will continue to turn. Hamas captured Gaza by force and may eventually take the Fatah-dominated West Bank through the ballot box, should elections take place at all. At worst, Hamas will view itself as delayed in achieving its ultimate goals, but it certainly has not conceded the battle for Israel's destruction and Islamic rule from the Jordan River to the Mediterranean Sea.

Internationally, the Muslim Brotherhood and their more radical offshoots are a grave challenge to the West, especially if their ideals become main stream in the Arab world. The Islamists declare their plans outright and implement them on a practical level. These are not unbalanced or insane individuals advocating extremist Islamic ideologies, but rather highly intelligent political leaders and clerics. Too often we hear Western commentators, whether liberal or conservative, declaring extreme Islamic perspectives as "insane." The people touting Hamas doctrine are no more mentally unbalanced or illogical than the Czarists, Bolsheviks, Nazis, Maoists, Khomeinists, or those who support the democratically elected regimes in the West. It is a danger if the West declares Islamist zealous types "insane," and thereby not responsible for their actions. Alternatively, understanding *The Hamas Covenant* ideology can be the first step in containing the real and potential threat of Jihadi Islam. Failing to do so may extract an unbearably heavy price. Successful Jihad must be well thought out; the *Covenant* shows us that Jihad is a plan, not just a whim.

For advocates of democracy and conflict resolution, no room exists for compromise as far as the Islamists are concerned. When Arabic-speaking Islamists explain their intent to conquer the world, they mean it literally. Hamas preaches "the truth" to their own people, garnering solid support for their unshakable religious values and actions. At times Hamas and other Islamists will engage in "pragmatic" policies such as a "*hudna*," or temporary Islamic cease-fire, to ensure organizational self-preservation. A *hudna* is not a cease-fire dedicated to mutual recognition and the peace process, but rather for Islamists to gain breathing space to retrain, re-arm, and go on the offensive as a Divine command when the time is right. Their mindset is very different from a secular politician calling a cease-fire for negotiations. In the Hamas world, tactical moderation is a prudent tool during difficult times. The Islamist leadership, whether political or military, acts in the name of Allah. The Israeli/Palestinian conflict represents the local military front for Hamas, but global Jihad is the overall strategy, and Islamic victory is the ultimate objective.

To understand Islamists, specifically Hamas, one needs a historic and ideological perspective which spans generations and emphasizes the last one hundred and fifty years. This book focuses on that timeframe, and particularly emphasizes the impact the establishment of the State of Israel has had on Islamist/Jihadi activism. In Islam, the negative image of Jews began from the time of the first clash between Jews and Muslims in the Arabian Peninsula some fourteen hundred years ago. The Koran records anti-Jewish attitudes in numerous clauses which continue to be used in promoting Islamist Judeophobia to this day. At first, the Jews were a vicious enemy. Later they became a conquered people living in the subservient "*dhimma*" status under the restrictive Charter of Omar statutes developed during the seventh and eighth centuries CE. Throughout Islamic history the dominant Jewish stereotype was of a scheming coward who would never succeed in his plots. It is crucial for anyone discussing Hamas and its Islamist allies to understand that discriminatory Islamic anti-Jewish attitudes began more than a millennium ago, prior to the rise of modern political Zionism. The advent of Jewish nationalism triggered Islamist reactions to the Jewish struggle for equality and independence. The stereotype of the Jew shifted from a scheming coward, to one of a rising and oppressive evil at the outset of the twentieth century. By the time of Israel's independence, Islamic antisemitism, as represented by the Muslim Brotherhood, augmented itself with many Czarist/Nazi stereotypes developed during the previous century. Traditional Islamic antisemitism provided fertile ground for the integration of the two traditions, particularly in light of the successful battle for survival of the State of Israel in the geographical heart of the Arab Islamic world.

One often forgets it is the intellectuals who set the pace of thought in a community, whether in the name of peace or hatred. The power of Arab/Muslim scholars, most notably Hasan al-Banna, Sayyid Qutb and Abdullah Azzam set the tone by bringing Jihad, antisemitism, and increasing opposition to all secular ideals including Arab nationalism. These perspectives are at the forefront of modern Islamist movements in the Arab world today. The above activist ideologues were the conduit to fanaticism among Arab Islamists, most notably in their attitude toward Israel and world Jewry. Hamas sees its primary battle against Jewish national legitimacy and denies the continuation of the Jewish connection to the ancient homeland, the Land of Israel. In line with the Islamist-Jihadi understanding, the Jews broke their covenant with

Allah, were banished from the Land twice and were never to return. Until the twentieth century, this theological understanding made perfect sense when looked at through the historical record of the Jewish People's expulsions, wanderings and persecutions. In the mind of Islamists, the Jewish bond to the Land was annulled for eternity. Jihadi Islam, as expressed by Hamas, adopted the worst Czarist/Nazi antisemitic stereotypes and demanded Jewish destruction, whether in Israel or the Diaspora. Still, the Islamic faithful lived and live the contradiction of a re-established independent Jewish nation in the Land of the Covenant as expounded upon in the *Tanakh*, also called the Hebrew Scriptures, or Old Testament. For Jihadi activists, battling the State of Israel became and remained the immediate primary front. It was the theological anomaly of Jewish independence, which demanded correction. The global enemy, Diaspora Jewry, was next in line.

The foundation of Palestinian Jihadi Islam pre-dates 1948, but underwent serious repression by secular Arab nationalism in the 1950s. Ironically, in the 1970s under Israeli occupation of the West Bank and Gaza Strip the Palestinian Muslim Brotherhood flourished again. Their regaining of power was primarily due to the Israeli focus on destroying the nationalist Palestine Liberation Organization or PLO led by Yasir Arafat. Israel unwittingly empowered the Islamists. Both secular PLO ideals and Hamas doctrines developed between the 1960s and 1980s. The secular PLO wrote and revised its ideals in the 1960s and 1970s. Hamas doctrines were committed to print in the late 1980s. There were and are cross-influences between *The Palestinian National Charter* and *The Hamas Covenant,* especially concerning policies calling for the destruction of the Jewish State. The two documents differ in their approach to the Jewish People on a global scale. Hamas seeks wide-range Jewish destruction, while the PLO denies all national memory and identity pertaining to the Jews. The PLO approach defines Judaism as a religious congregation devoid of peoplehood and homeland, discounting large portions of the *Tanakh* and Jewish historical consciousness.

An analysis of the interaction between Jewish nationalism and Palestinian Islamism is required in order to understand the unique characteristics Hamas embodies. Study must include the period well before the 1948 War and continue through 2000. The Palestinian Muslim Brotherhood, calling itself "Hamas" by late 1987, solidified its base despite serious disabilities, through trial and error, failure and success vis-à-vis Israel and the PLO/Fatah. From 2000 onward, Hamas won popular acclaim both through elections and the military conquest of the Gaza Strip. To date, we have over fifteen years of history revealing Hamas ideological triumph intertwined with their pragmatic approach, which is necessary to guarantee survival of the group. Most notably, Jewish nationalism and Islamic fundamentalism clashed on physical, ideological and religious levels. The conflict continues to this day.

By and large, Hamas plays a significant militant function in the worldwide Islamist movement, despite the perception by many who believe its true role is limited to the Palestine national struggle. Hamas is, in fact, one small actor in the global Islamic offensive. Along with other Islamists, Hamas, which is only nominally identified as Palestinian, takes great pride in the 2011 Islamic Awakening. Hamas made serious inroads in its struggle against Palestinian Arab secular nationalism and in its battle with the Jewish State and the West. Still, the Hamas Revolution remains incomplete as secular Palestinian national aspirations persist. Nevertheless, despite difficulties, the world Islamic Revolution including Hamas continues to move forward to attain their final objectives.

Such circumstances lead us to the overall question as to whether there can be peace with Hamas and by extension the Muslim Brotherhood. The Koran makes both positive and negative statements about Jews, Christians and others. Over the years, Muslim jurists have abrogated, or abolished, the idea of reconciling positive and negative contradictions in the Koran. They disregard positive statements and emphasize demands for discriminatory *dhimma* status regulations and Jihad against adversaries. Specifically, abrogation is fully invoked by denying God's covenant with the Israelite/Jewish People and their rights to the Land of Israel, precepts clearly stated in the Koran. Abrogation may be the key to conflict resolution, not only with Hamas, but with the entire Muslim world. As Muslim jurists reserve the right to nullify positive comments about non-Muslims in the Koran, they also have the power to reinstate such beliefs, thus canceling calls for Jihad and the destruction and dominance over Jews, Christians and others. Global Islamic leadership can use their own discretion in acting within the theological bounds of responsibility and using abrogation for a pluralistic interpretation of the Koran, as opposed to encouraging Jihad and erasing possibilities for peace. This suggestion may sound naïve, yet when we consider the alternative of continuing Jihad perpetuated against the entire world in an era of weapons of mass destruction, "reverse abrogation" may be the only answer that can alter Islamist thinking at its core. Attempting to impose Western ideals on Islamists offers virtually no possibility of curtailing conflict.

A Personal Perspective

To conclude this introduction, I offer my personal perspective on these issues. I originally moved to Israel, or made *"aliya,"* in the 1970s. I followed Labor Zionist ideals advocating secular Jewish nationalism and universal socialist humanist values, believing all people could find their place on earth. Overall, this meant I believed in compromise concerning the Land of Israel and the establishment of a Palestinian State, provided the Palestinians accepted Israel's right to exist. Thus, I supported the Rabin-Peres initiative and the Oslo Accords in the 1990s. To me it seemed clear that Israel was established to ensure the continued existence of the Jewish People. In order to end the decades-old conflict with the Palestinian Arabs, there needed to be a compromise and the re-establishment of the two-state solution rejected by the Palestinians and Arab world in 1947-48. An agreed upon "End of Conflict" through the establishment of an independent Palestinian State would ensure both groups' existence and allow each to preserve their national and religious identities through nation state sovereignty and joint security arrangements. Neither side was obligated to accept the other side's narrative as the defining truth, but they would commit themselves to understanding the other side's narrative was the absolute truth for him or her despite years of antagonism. One would look to the future in building a better tomorrow with two national entities living side by side.

From the Israeli perspective, *The Palestinian National Charter* needed to be revised by the secular PLO/Fatah. A two-thirds majority as proclaimed by the PLO itself could accomplish this change, particularly since we are not speaking of a sanctified document, but rather one admittedly written by men. Altering the *The Palestinian National Charter,* as was agreed to in late 1998, would prove Palestinian goodwill and pave the way for conflict resolution. In comparison, *The Hamas Covenant* cannot be altered; its contents are considered the word of Allah. The *The Palestinian National Charter* was never changed and the Oslo peace process broke down when it seemed the sides were very close to agreement. It appeared to me the Hamas ideological impact was much greater than previously imagined and should be credited with heavily influencing secular Palestinian and Fatah/PLO thinking, yet for years I kept those thoughts on hold.

In the 1990s as a result of the Oslo Accords there was great hope for conflict resolution. That hope was deferred indefinitely in 2000 by what is often referred to as the "Second Intifada," but in reality was a war, or "Low Intensity Conflict." From that point on, Fatah and Hamas appeared unified in their anti-Israel approach. Even in North America, most of the pro-Palestinian groups on college campuses rarely argued over the terms of the Oslo Accords and the conditions for a two-state solution, but rather they sought to influence the average student not only to question Israel's legitimacy, but to demand its destruction. The idea for this book came about from my own and my colleagues' experiences on college campuses engaging with the general student body during lecture tours pertaining to Arab-Israel issues of war and peace. In particular, discussions with Zionist student activists, both Jewish and non-Jewish, proved most informative. The pro-Israel students were constantly under attack in the never-ending debate concerning Palestinian and Israeli matters. Even advocating the two-state solution, as most pro-Israel students did, was usually not good enough. The issues were less about borders, settlements, security or even "Palestinian refugees," but rather the continued questioning of Israel's right to exist. The distinct tinge of antisemitism was quite noticeable, especially when confronting the unholy alliance of certain Muslim and Arab activists working alongside the extreme leftist fringe-anarchists. Such attitudes manifested with questions and debates laced with inferred accusations of Jewish influence, especially concerning finances and the media, thereby forcing US support for Israel against American and Western interests.

Simultaneously, a very different local political event was unwinding on the hilltop where I reside in central Galilee. I live in the small town of Eshchar, which at the time had 76 families. It is a community advocating pluralism concerning religious, secular and traditional Jewish lifestyle. In 1998, we were faced with questions concerning our neighbors living in the unrecognized Bedouin village of Arab al-Nai'm. They lived in corrugated tin shacks and were expected to move at the government's behest to one of the neighboring Arab or Bedouin villages, a plan they refused to accept. Some of my more liberal neighbors decided to help the Bedouin tribe in various ways, often putting themselves at loggerheads with the regional council and by extension the authorities in Jerusalem. With enough issues on my plate as the volunteer local council chairman of our community, the Bedouin subject was not my priority, though demands to take action to help Arab al-Na'im mounted. External forces also began to add pressure, including the arrival of certain left-wing activists from outside organizations—whose motives I often questioned, inquiries by the press, and warnings that the calm relations between our two communities could be disturbed. It seemed imperative to meet with the Bedouin local council. I was forewarned there would be a litany of complaints from their representatives, which proved accurate.

That same day I met them in June 1998, we agreed upon a reasonable geographical border between our two communities. Within a few weeks our general assembly approved advancement of the border policy. We convinced the ministry of interior and the regional council to allow for the development and establishment of a permanent Bedouin village. It was made clear to us that such an initiative by two communities was basically unheard of, and that we would need to bear responsibility for the consequences of our decisions. At that point, the government, regional council and local volunteers took over. Personally, I was never involved again except for dealing with a technical planning issue or two.

The community border agreement led to what would become a short-lived exaggerated optimism on my part. I believed Jews and Arabs could work together locally and that the Oslo Accords would succeed in ending the clash between Israel and the Palestinians. After the failed talks between Ehud Barak and Yasir Arafat in the summer of 2000, Arab riots exploded in Galilee. On our hill though, despite certain attempts to incite our neighbors against us, relations remained cordial.

Once again I thought in terms of Hamas as the spoiler. No doubt other forces were at work in the Galilee, but certainly the Hamas inspired "Islamic Front" in Israel had an influence either directly, or through more secular aspects of Palestinian nationalism. Hamas doctrine and antisemitism were now in our backyard. Just five years previously it appeared peace was around the corner. On the other hand, Arab al-Na'im moved ahead with plans for permanent establishment, although not all went smoothly due to drawn out negotiations between residents and the government concerning town planning, lands, and financing. In 2013, infrastructure and housing construction commenced.

Such divergent experiences made me ponder the possibilities for peace. Whatever understandings existed on our hill would not work with more extreme elements. Between developments with Arab al-Na'im, the Galilee riots, my acquaintance with student issues in America, and the overall situation in this corner of the Middle East, I realized the pressing necessity for conflict resolution. Islam's doctrines and underlying attitudes had to be considered for their impact on all Muslims, including those with more secular viewpoints. I began living a contradiction as I witnessed the almost daily terror attacks and suicide-homicide bombings in the early 2000s. The need to investigate Islamic extremism became more paramount as did finding non-military solutions, if any existed. Hamas won the Palestinian parliamentary elections in early 2006, overthrew the PA government in Gaza in June 2007, and essentially established an Islamic mini-state, clashing with Israel in the Cast Lead Operation of 2008-09.

Together, the popular support for and rise of Hamas forced my realization that the roots of this renewed anti-Zionism and antisemitism were far deeper than I previously imagined. Liberal democratic peace-making solutions inspired by Enlightenment ideas would be of little help. Would the future hold never-ending conflict, or was there hope? I needed to explore some very disturbing realities, and only then look for answers. In retrospect, two major truths emerged: *hudnas* do not lead to peace, and Islamist ideals will rebound and therefore need to be fundamentally altered as relates to non-Muslims, most notably Jews.

At that time, I made the decision to scrutinize *The Hamas Covenant* and to write a short booklet to get to the core of the matter. Of course, as we all know, once delving into a specific realm of research there are far too many topics to cover and questions to answer to limit writing to only a short booklet. This work is the result of my studies. Although reading this book may not be the most pleasant at times, my hope is that people will become more aware and better educated upon doing so. After all, it is a matter of self-preservation, not only for the State of Israel and Jews wherever their domicile, but for anyone who refuses to accept the dictates of Jihadi Islam.

Yisrael Ne'eman
March 2016

Map of the Arab World

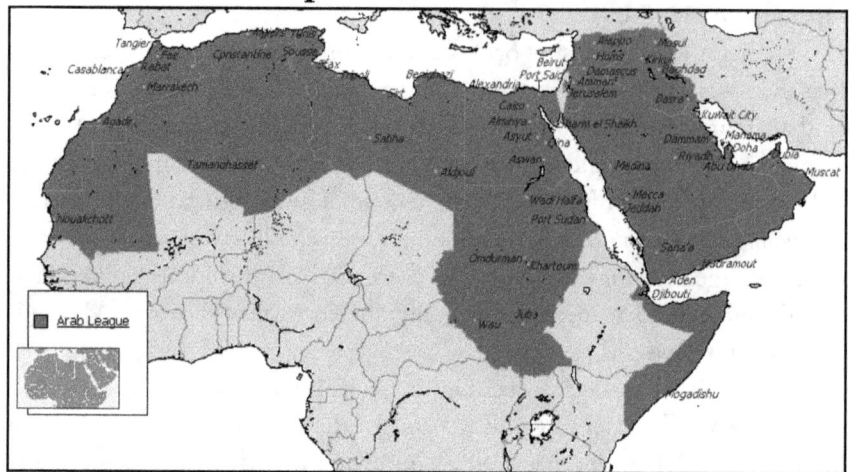

Credit: Arab Hafez, Public domain, Wikimedia Commons

Ottoman Empire at the Height of Expansion in 1683

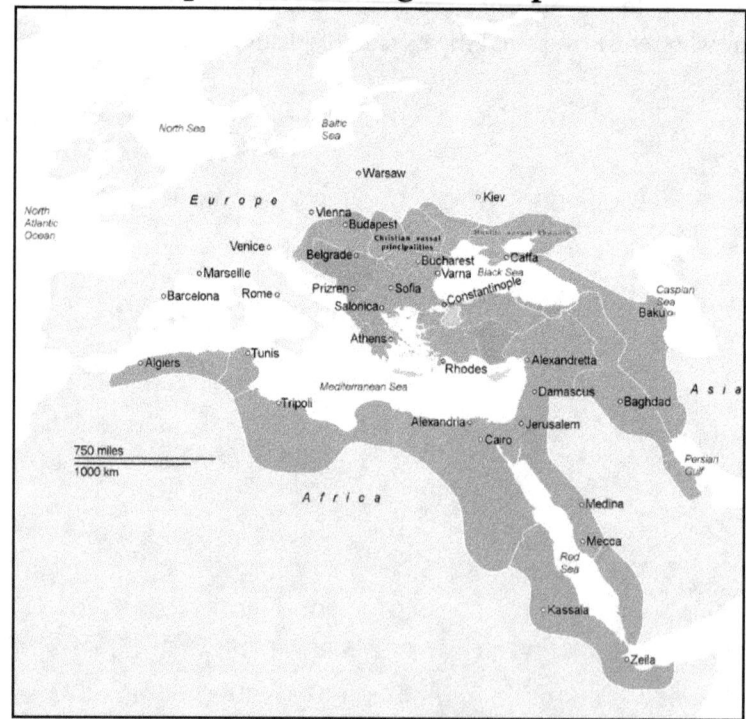

Credit: Modified from "Ottoman Empire in 1683," Wikimedia Commons

1

Negative Image of the Jew in the Arab Muslim World

Historical Review up to the Mid Twentieth Century

Overview

Vicious antisemitism is currently sweeping the Islamic world, but it is not a new trend. There is a direct line from antiquity to the present. Muslim dislike for the Jews began in Islam's formative years during the seventh century in the Arabian Peninsula and continues until the present day. Currently Hamas, or the Palestinian Muslim Brotherhood, is in the forefront of anti-Jewish activities as expressed through its *Covenant*, leading to both words and deeds taken in an effort to destroy the Jews. The Palestinian Islamist fundamentalist organization Hamas was not created in a vacuum, hence the need for an overall historical, theological and legal review of the Jewish-Islamic relationship over the past 1400 years to acquaint the reader with these deeply ingrained negative Muslim attitudes toward the Jews. To simplify the major concepts under discussion, this initial chapter is divided into three parts:

Part I
Development of the Judeo-Islamic Relationship: History and Theology

Presented is a review of the early development of Islam during the seventh and eighth centuries and the accompanying attitudes toward the Jews. As expressed in the Islamic holy text, the Koran, the Jews were defined as an adversary and will find themselves suffering under a second class legal status known as the *dhimma*, a result of the Charter of Omar dictates. Simultaneously the Hadith literature, which is a collection of testimonies purporting to quote the Prophet Mohammed as concerns numerous topics and second in importance only to the Koran, reinforced negative Jewish stereotypes.

Part II
Historic Survey: Islamic Attitudes/Policies Toward the Jews

Here we have a very brief historic overview spanning more than a millennia examining Islamic behavior toward the Jews throughout the Middle East and North Africa. Muslim behavior is in line with a "spirit of the time" as determined theologically by the Koran, the Hadith and the legally binding

Charter of Omar. Permanent anti-Jewish attitudes became a way of life in the Islamic world.

Part III
From Ottoman Islamic Rule to Secular Nationalism

Leading into the modern period we will survey the decline of the Ottoman Empire commencing with the rollback of the one thousand year Jihad and the rise of regional secular identities, particularly Arab nationalism. The Islamist backlash against secularism and equal rights for minorities boded ill for both the Jews and Christians. In particular the rise of Jewish nationalism or Zionism infuriated the Islamists. By the twentieth century Islamic anti-Jewish attitudes were augmented by European models, specifically those of the Czarist and Nazi variety demonizing the Jews. The links in the chain were complete – the Palestinian Muslim Brotherhood, commonly known as Hamas, fully integrated traditional Islamic, Czarist Russian Christian Orthodox and the Nazi pseudo-scientific racial hatred for the Jews into its *Covenant* and today are at the cutting edge of antisemitism in the Arab/Muslim world.

Part I
Judeo-Islamic Relationship: History and Theology

Arabian Origins of Islam

Recounting seventh century Arabian history is necessary for insight into the roots of the Jewish-Muslim clash. Prior to the birth of Mohammed and the rise of Islam, Judaism was the dominant religious group in the Arabian Peninsula. Jewish tribes traversed the Hejaz western mountain range after Rome destroyed the Second Temple in 70 CE and the Bar Kokhva rebellion failed in Judea sixty-five years later. In the sixth century, Yusuf Dhu Nuwas established a Jewish kingdom in Yemen. He had a two-pronged policy of solidifying his kingdom against the Abyssinian Ethiopian Christian threat, and of ensuring trade route links with the Land of Israel to the north. Medina, Khaybar and other cities on the caravan route had fair-sized Jewish populations, or were Jewish strongholds, possible links in the expanding chain of influence for Dhu Nuwas' Himyarite Kingdom. In the end, Dhu Nuwas lost to Christian forces and his kingdom was destroyed, yet Jewish influence continued to permeate throughout Arabia.[1]

Mohammed entered the Arabian scene at a period of intense competition and clash between the pro-Persian forces as represented by the Jews, and the

Byzantine alliances with Christian powers such as Abyssinia. The Byzantines temporarily gained the upper hand, but the Jews still played a major role in Arabian politics and society. The Muslim community's flight from Mecca to Medina in 622, called the "Hejira," took place against the background of the Prophet Mohammed's rejection in his former home-town. The Banu Qaylah pagan Arab tribe, which consisted of the Banu Aus and Banu Khazraj tribes, some of whom may have previously converted to Islam, invited Mohammed and his followers to Medina. Together they worked to erode the power of the Jewish majority. All the Medina tribes were involved in conflict, especially over land. Alliances were in constant flux as treaties were signed and violated time and again.

Mohammed became an arbiter, or a judge. He established the *Constitution of Medina*, whereby he solidified a determining role in daily decision-making processes of the region. Jewish and Muslim tribes were expected to follow the Prophet's directives and were to receive aid, "equality and shall not be wronged." Disagreements were referred to Mohammed, Allah's Messenger. All those adhering to the *Constitution* were expected to believe in Allah and the Last Day. They understood that any violation would incur Divine wrath on the Day of Resurrection, whereby the offending party would be found guilty. The Jews were perceived as allies of the Believers (Muslims) and were expected to help with defense costs while reaping benefits of the alliance. Jews and Muslims were each seen as a community or "*umma*" and were together accredited with having their own courts of law. When looking in retrospect at the religious-historical record, Muslims viewed any Jewish failure to follow Mohammed's directives as a betrayal.[2] Many scholars question as to whether there is proof that Jews were signatories to the *Constitution*, or even shared the Muslim interpretation. From the Muslim perspective, it is believed the Jews signed and agreed to the *Constitution*.

In these early days, Jews and Muslims had much in common. There was belief in one God, Jerusalem as the "*qibla*," or prayer direction, and a familiarity with the same Biblical stories. For a few short years there was a relationship involving a fair amount of cooperation. This is evidenced by the early positive statements pertaining to Jews in the Islamic Holy Scriptures, the Koran. It is said many Muslims studied the Torah alongside the Koran. On the other hand, it appears Jews and Muslims competed for converts even if the Jews were subtle about proselytizing efforts. While Mohammed saw himself as the continuation of Jewish prophecy, the vast majority of Jews rejected him. Considering that politics and religion were one, Jewish religious opposition to Mohammed became a political and theological affront. The clash began around the time of the Battle of Badr against the pagan Meccans when Mohammed began to suspect Jewish disloyalty. This theological turning point in 624 led to an intense Jewish-Muslim rivalry. Mecca became the new *qibla* along with reverence for the Ka'aba, the ancient pagan Arabian holy site. Hence, Mohammed began consolidating Islam into a religious community completely separate from the Jews.[3]

First, Mohammed expelled the Banu Qaynuqa and Banu an-Nadir Jewish tribes from Medina, and, after the Battle of the Trench in 627, the Banu Qurayza tribe was destroyed. The Banu Qurayza Jews were expected to go to battle alongside Mohammed against the Meccans and not adopt a neutral position by lending only minimal aid to the Muslims. Their inaction became etched in the collective Islamic memory as a Jewish refusal to adhere to the *Medina Constitution*, and therefore a betrayal. The Jews were charged with betraying Mohammed in the heat of battle. Mohammed appointed a judge by the name of Sa'd, who was known as a confident of his. Sa'd "ordered that the men should be killed [some 400-900], their property divided, and the women and children taken as captives." On the defensive, the Jews suffered from the stigma of disloyalty and now suffered persecution by these first Muslims. The pagan Meccans were weakened, but they temporarily came to terms with Mohammed after the Battle of Hudaybia. Shortly afterward, Islam captured the Jewish farming town of Kaybar in the north, and by 630 CE Mohammed made his victorious entry into Mecca. From the earliest times, Jews and Christians were able to retain their religious beliefs as second-class *dhimmis,* paying the discriminatory *jizya* tax. Tradition says Mohammed's final command was for the expulsion of the Jews from the Arabian Peninsula. He explained the inadmissibility of the two religions existing side by side in the Islamic heartland.[4] In addition the fact that Jews were not pro-active in expanding monotheism and retained friendly relations with the pagan Arabs in their midst,[5] caused them to be viewed as an implacable enemy.[6]

Abu Bakr became the first caliph after the death of Mohammed. He not only reconsolidated his power, but further weakened Jewish influence. Omar followed, expropriating Jewish lands and expelling the Jews from Kaybar and Najran. In the Middle Ages there were reports of Jewish tribes in Arabia, in particular the fierce Rechabites. It appears Jewish existence in northern Arabia came to an end by the mid-eighteenth century coinciding with the violent rise of the modern Wahhabist movement in the same region.[7]

This early period is of overriding significance because devout Muslims view the legacy of Jewish-Islamic relations through the prism of the seventh century, even today 1400 years later. Theologically frozen in time for believers, this

Jewish adversarial position is eternalized through the Islamic Holy Scriptures - the Koran, the later Hadith writings and the Charter of Omar discriminatory legal strictures. Early on anti-Jewish attitudes became intrinsic in Islam. For a fuller understanding of Islam's animosity toward the Jews we will review below the roots of early Islamic antisemitism contained in holy writ.

Negative Stereotypes of the Jew in the Koran[8]

For a true understanding of Muslim negative attitudes toward Jews, we need to return to the foundational belief document, the Koran, for a brief review of the root causes. According to historians the Koran was compiled in the late seventh and into the eighth century CE recounting much of the life of the Prophet Mohammed and the activities of his early followers. For believers it is understood that the Koran contains the revelations of Allah recited through Mohammed and recorded by his scribes. For devout Muslims the Koran is the infallible pillar upon which Islam rests and from where Sharia (Islamic) Law is derived. (For positive attitudes about Jews and their possible ramifications see Chapter IX "Islamic Abrogation.")

The Jewish perspective concerning the authenticity of Mohammed's mission was and is quite different. One must remember Mohammed presented himself as the Moses of the new Islamic community,[9] continuing the revelations of Allah to all humankind. In the Jewish mind, Mohammed is often seen as uninformed and confused when recounting events from the *Tanakh*, also known as the Hebrew Scriptures or Old Testament. Examples of such mix-ups are found in the Koran 28:38, when Mohammed recites the story of Pharaoh instructing his advisor Haman to build him a tower to the God of Moses, seemingly a Tower of Babel.[10] Furthermore, Ezra the Scribe is said to be the Son of God, a belief not contained in Judaism.[11] These are not the only cases of such confusion, and hence Mohammed incurred ridicule and was rejected by the Jews; thus the Jews became the implacable enemy of Islam and often incurred the wrath of Koranic condemnation for their rejection of Allah's Divinely revealed faith. Below are quotes censuring the Jewish People.

Islam condemns Israelite rebelliousness, most notably in the building of the Golden Calf during the Sinai Exodus (Koran 4:153-155):

> The People of the Book ask you to bring down for them a book from heaven. Of Moses they demanded a harder thing than that. They said to him: 'Show us Allah distinctly.' And for their wickedness a thunderbolt smote them. They worshipped the calf after

> We had revealed to them Our signs; yet We forgave them that, and bestowed on Moses clear authority.
>
> When We made a covenant with them We raised the Mount above them and said: 'Enter the gates in adoration. Do not break the Sabbath.' We took from them a solemn covenant. But they broke the covenant, denied the revelations of Allah, and killed their prophets unjustly. They said 'Our hearts are sealed.'
>
> (It is Allah who has sealed their hearts, on account of their unbelief. They have no faith, except a few of them.)

It should be noted that worship of the Golden Calf is first condemned in the *Tanakh*[12] and is in partial agreement with the Koranic verse above. The Koran insinuates that virtually all the People of the Book, meaning the Israelites in this instance, were and are responsible for the breaking of the covenant, denying Allah's revelations and engaging in the all encompassing murder of their prophets. The Israelites were beyond salvation once they declared, "Our hearts are sealed." The understanding is clear: there can be no repentance and, Allah won't forgive. Accusations of lack of faith and belief sealed the Jewish fate, especially when Allah took the step of hardening their hearts. There were a few exceptions, but from the perspective of the Koran the Jews as a people chose to defy the Divine.

As for the killing of prophets, the *Tanakhic* recounting is quite different. Korach the rebel and his cohorts Dathan and Aviram died in an earthquake. The 250 disobedient chieftains burned in a fire and the people suffered from a plague as a result of their rebellion.[13] There were further challenges and conflicts recounted in the Exodus story, but those who were not righteous faced punishment, and in the end the Children of Israel entered the "Promised Land" under Joshua's leadership, meaning they must have found favor with Allah. The Koran admits as much even if Joshua is not mentioned by name (see Chapter IX "Abrogation"). However, this negative image continues to be reinforced, for example in Koran 5:12-13.

> Allah made a covenant with the Israelites and raised among them twelve chieftans. He said: 'I shall be with you. If you attend to your prayers and pay the alms-tax; if you believe in My apostles and assist them and give Allah a generous loan, I shall

forgive you your sins and admit you to gardens watered by running streams. But he that hereafter denies Me shall stray from the right path.' But because they broke their covenant We laid on them Our curse and hardened their hearts. They have perverted the words of the Scriptures and forgotten much of what they were enjoined. You will ever find them deceitful, except for a few of them. But pardon them and bear with them, Allah loves the righteous.

A few Jews were righteous while others needed to be pardoned; however, deceitfulness was now added to disobedience and the defiance of Allah. The Israelites were cursed for warping the words of their own Divinely revealed Scriptures. Beyond the Torah there were those Israelites who challenged God and were punished in the end. Verses 5:12-13 mention the righteous few, while transgressors may be pardoned. Jihadists and antisemites emphasize the curse placed on the Israelites while ignoring Allah's clemency.

From here, the emphasis is on those who continue to disobey, fully conscious of their evils. Not only do these Jews deny Allah, they insidiously declare they know the truth of Allah's message, yet insist on denying Him as shown in Koran 4:46.

> Some Jews take words out of their context and say to the Apostle: 'We hear, but disobey. May you be bereft of hearing! Listen to us!' – thus distorting the phrase with their tongues and reviling the true faith. But if they said: 'We hear and obey: Hear us and Look upon us', it would be better and more proper for them. Allah has cursed them in their unbelief. They have no faith, except a few of them.

Koran 5:62-64 paints the picture that the Jews continue in their "sin and wickedness," achieving "unlawful gain" while their rabbis acquiesce by doing nothing.

> You see many of them vie with one another in sin and wickedness and eat the fruits of unlawful gain. Evil is what they do. Why do their rabbis and divines not forbid them to blaspheme or to practice what is unlawful? Evil indeed are their doings. The Jews say: 'Allah's hand is chained.' May their own hands be chained!

> May they be cursed for what they say! By no means. His hands are both outstretched: He bestows as He will. That which Allah has revealed to you will surely increase the wickedness and unbelief of many of them. We have stirred among them enmity and hatred, which will endure till the Day of Resurrection. Whenever they kindle the fire of war, Allah puts it out. They spread evil in the land, but Allah does not love the evil-doers.

Instead of competing for righteousness and mending their ways, the Jews compete in the ultimate evil through the defiance of Allah. They declare, "Allah's hand is chained," attesting to a Divine impotence. The Jews are cursed once again for challenging God, seemingly aligning themselves with the devil. At the end of the quote in Koran 5:64, believers in Islam are told that as more is revealed to them, the Jews will counterbalance such good tidings through increased "wickedness," thereby stirring among themselves "enmity and hatred" to continue until "the Day of Resurrection." The Jews are the eternal enemy as "they kindle the fire of war" only to be extinguished by Allah.

The once righteous Jews, who have since broken their covenant with Allah, manifest their evil through their everyday practice of money lending, a forbidden profession in Islam, and one that leads to cheating humanity out of their property as noted in Koran 4:160-162.

> Because of their iniquity we forbade the Jews good things which were formerly allowed them; because time after time they have debarred others from the path of Allah; because they practice usury – although they were forbidden it – and cheat others of their possessions. We have prepared a stern chastisement for those of them that disbelieve. But those of them that have deep learning and those that truly believe in what has been revealed to you and to other prophets before you; who attend to their prayers and pay the alms-tax and have faith in Allah and the Last Day – these shall be richly rewarded.

In following the last verse of the above quote, we understand those who pay the alms tax and are of "deep learning and truly believe" having "faith in Allah and the Last Day" will receive their just reward. Overall, even in the most damning Koranic verses there are exceptions made for those Jews who are righteous (Koran 4:46, 4:155, 5:13). This allows for a curtailed, non-equal

Jewish existence under Islamic sovereignty in this world such as ascribed to in the *dhimmi* status, while recognizing Allah makes the final judgment.

Koran 4:54-56 states when the Jews, descendants of Abraham, do not believe in Mohammed's mission they are to burn in Hell with their skins devoured over and over again.

> Do they envy others what Allah has of His bounty given Them? We gave Abraham's descendants scriptures and prophethood, and an illustrious kingdom. Some believe in him [Mohammed], but others reject him. Sufficient scourge is the fire of Hell. Those that deny Our revelations We will burn in Hell-fire. No sooner will their skins be consumed than We shall give them other skins, so that they may truly taste Our scourge. Allah is mighty and wise.

On earth, the Koran states rebellious Jews were turned into apes as punishment for breaking the Sabbath (Koran 2:65 and 7:166). Believers are urged to associate with other righteous Muslims (5:57-58) and not to befriend the People of the Book, in this case the reference is understood to be directed at the Jews, who mock Islam and even laugh at their own religion. These will be turned into "apes and swine" as stated in Koran 5:59-60.

> Say: 'People of the Book, do you hate us for any reason other than that we believe in Allah and in what has been revealed to us and to others before us, and that most of you are evil doers?' Say: 'Shall I tell you who will receive the worse reward from Allah? Those on whom Allah has laid His curse and with whom he has been angry, transforming them into apes and swine, and those who worship false gods. Worse is the plight of these, and they have strayed farther from the right path.'

The issue here is whether such allegory is taken literally, something of which is quite prevalent in the Arab Muslim world. Nowadays, Islamic dignitaries and scholars use the literal interpretation to prove the Jews to be deniers of Allah and therefore inferior. Such modern day vilification is dangerous.[14] First, the Jews were turned into apes and swine and then were condemned to continual incineration in Hell, which was all justified by the Jewish rejection of their own prophets, as stated in Koran 5:70-71.

> We made a covenant with the Israelites and sent forth apostles among them. But whenever an apostle came to them with a message that did not suit their fancies they either rejected him or slew him. They thought no harm would come to them: They were blind and deaf. Allah turned to them in mercy, but many of them again became blind and deaf. Allah is ever watching over their actions.

The Israelites were accused of slaughtering any apostle or messenger of their own who brought tidings not to their liking, behaving as if they were deaf and blind. This is a clear denial of Allah's revelations. The innuendo contained within the accusation is not quite at the level of deicide, yet Jewish perfidy is clearly understood in light of the covenant made between God and the Israelites. It appears these accusations were taken from the Christian Gospels where the Jews are condemned for the killing of Jesus and persecuting his followers. Islam borrowed these antisemitic accusations from Christianity and would do so again with institutionalized social and legal prejudice through the *dhimma* statutes.[15]

Christians and Jews condemned together in the Koran

Koran 5:51 limits Muslims from befriending Jews and Christians, since a believer must never be converted to an inferior faith. "Believers, take neither Jews nor Christians for your friends. They are friends with one another. Whoever of you seeks their friendship shall become one of their number. Allah does not guide the wrongdoers."

Jews and Christians are said to be friends, and cooperate as tricksters and connivers leading faithful Muslims astray (Koran 3:71-73). True Muslims will know better and not fall into the trap of accepting either religion. Koran 2:135-137 states:

> They say: 'Accept the Jewish or Christian faith and you shall be rightly guided.' Say: 'We believe in Allah and that which is revealed to us; we believe in what was revealed to Abraham, Ishmael, Isaac, Jacob, and the tribes; to Moses and Jesus and the other prophets. We make no distinction between any of them, and to Allah we have surrendered ourselves.' If they accept your faith they shall be rightly guided; if they reject it they shall surely be in schism. Against them Allah is your all-sufficient defender. He hears all and knows all.

On an economic level, both Jews and Christians were to be treated equally when Muslims implemented the *jizya* or humiliating head tax. Here we find a contradiction. Jews and Christians were denying Allah, which was very much a capital offense in Islam, yet if they pay the head tax and accept submission they were allowed to survive. As materialistic thieves, their punishment is physical branding by their own wealth in the fires of Hell. Their denial of Islam cannot be all encompassing because to do so would lead to death. Submission and financial tribute remained the socio-religious and economic basis for allowing the People of the Book to exist under Islamic overlords. The following quote, from Koran 9:29-35, is extensive but of great consequence, setting the *jizya* tax. Unrelated, but notable in the following passage, there is no basis in Judaism for the Koran's assertion that "Jews say Ezra is the son of Allah."

> Fight against such of those to whom the Scriptures were given as believe neither in Allah nor the Last Day, who do not forbid what Allah and His apostle have forbidden, and do not embrace the true faith, until they pay tribute out of hand [jizya tax] and are utterly subdued.
>
> The Jews say Ezra is the son of Allah, while the Christians say the Messiah is the son of Allah. Such are the assertions, by which they imitate the infidels of old. Allah confound them! How perverse they are!
>
> They worship their rabbis and their monks, and the Messiah son of Mary, as gods besides Allah; though they were ordered to serve one God only. There is no god but Him. Exalted be He above those whom they deify beside Him!
>
> They would extinguish the light of Allah with their mouths; but Allah seeks only to perfect His light, though the infidels abhor it.
>
> It is He who has sent forth His apostle with guidance and the true faith to make it triumphant over all religions, however much the idolaters may dislike it.
>
> Believers, many are the rabbis and the monks who defraud men of their possessions and debar them from the path of Allah. Pro-

claim a woeful punishment to those that hoard up gold and silver and do not spend it in Allah's cause. The day will surely come when their treasures shall be heated in the fire of Hell, and their foreheads, sides, and backs branded with them. Their tormentors will say to them: 'These are the riches which you horded. Taste then the punishment which is your due.'

Although the People of the Book are condemned together, they are not all considered equal. The hierarchy is shown in the following Koranic quote. Koran 5:82 states, "You will find that the most implacable of men in their enmity to the faithful are the Jews and the pagans, and that the nearest in affection to them are those who say; 'We are Christians.' That is because there are priests and monks among them; and because they are free from pride." In this passage, Jews and pagans are equally evil, and Christians are somewhat better. At least theoretically, and very often in practice as well, Christians are a rung higher than Jews on the social ladder. In modern times, Koran 5:82 serves as the basis for accusations against the Jews for adhering to atheist doctrines advocating a non-Divine socio-economic system (see Chapter II "Ideologues").

Beyond the Jewish-Christian friendship is the Jewish-pagan alliance in 5:82 which Islamists see as the depths of evil, especially because the Israelites violated their covenant with Allah. The verse states Christians have a better demeanor than Jews, because they are free of arrogance. Muslims see the Jews as having debased themselves not only through their constant exploitation of others and materialistic behavior, but as true infidels working alongside pagans who rejected Allah's existence. The Jews knew and know the truth, and not only lie as recounted in Koran 58:14-19, but actively work toward the destruction of Allah's message as passed on through Mohammed and the Koran. The Muslim mind understands that once the Jews have broken their own covenant with Allah, they will prevent all others from ever doing the work of the Divine in the future. Inferred is an inborn nefarious Jewish nature, since logically why would anyone willingly break the covenant with Allah?

The answer is found in the supposedly innate Jewish characteristics of betrayal, materialism, usury, cheating one's neighbor, murder of holy men and the rejection of Allah's final messenger, Mohammed. Jews are characterized as devoid of spirituality and pursuing narrow self-interests, which exclude righteousness. Yet there are a few who behave correctly and being that Allah judges all at the End of Days, He will decide their eternal fate. In the meantime, Muslims who have surrendered all to Allah must suffer the Jews continued existence, relegating them to a tolerated religious community bereft of any national claims to land and subject to Islamic or Sharia law. The Islam-imposed punishment began with the *jizya* tax (9:29) and evolved throughout the centuries until Jews existed in second-class community status with restrictions as stipulated under the Charter of Omar. Implementation of such discriminatory regulations is dependent on the whim of the ruling clique or the leader himself when imposing the limits of the *dhimmi* status.

Muslims see themselves as the best nation of all humankind, as expressed in Koran 3:110-111 and quoted in *The Hamas Covenant* Preamble. Believers in Islam are righteous and just, firmly at the top of the ladder closest to Allah. To them, Christians do not believe as they should, but their personal behavior is better and their motives less dishonorable than the Jews. The Jewish-pagan alliance is purposed to destroy Islam and Allah's universe. In today's world, traditional paganism barely exists and therefore it is the Jews who are accused of carrying on Satan's work,[16] placing them almost or completely beyond redemption. They occupy the lowest rung, or fall off the ladder altogether. While Muslims preach faith in Allah, the Jews strive for the anti-Allah. In the modern world, paganism is replaced by atheist ideologies, in particular communism. To this one can add socialism, capitalism, liberalism and democracy, all global ideologies and socio-economic systems. By the twentieth century, Muslims viewed these political ideologies as the Judeo-pagan alliance.

Physically the "necessary" step of cleansing the Arabian Peninsula of anyone who was not Muslim was left to Omar b. al-Khattab (Caliph Omar I), who drove the Jews of Kaybar from their homes after Mohammed had subjugated them. He exclaimed, "If Allah prolongs my life, I shall certainly chase all the Jews and Christians from Arabia and will leave only Muslims."[17] Today, except for a handful of Jews in Yemen, his goal is achieved.

Jews in the Hadith

Narratives quoting the Prophet Mohammed as recorded by different scribes and passed down through the generations form the collection of Islamic religious literature known as the Hadith, a work second in authority only to the Koran itself. The Hadith accounts were completed well after the Prophet's death. Here too, writers place the Jews in a negative light. The scholar Georges Vajda summarized the Muslim attitude as: the Jews are simply not liked.[18] Vajda said the Hadith often reminds one of the Jewish defeats at Kaybar and the Jewish refusal to convert to Islam. Jews are accused of worshiping Ezra the Scribe as the son of God, an idea for which there is no support in any Jewish texts. Most vilifying is the Hadith claim that a Jewish woman was responsible for poisoning the Prophet Mohammed. Muslims are called upon not only to curse Jews and Christians, but in particular are urged to kill the Jews.[19] This demand is made clear in *The Hamas Covenant*, Article 7.

Vajda goes on to explain that Muslims do their best not to adopt Jewish customs, whether dealing with hairstyle, ritual purity, sexual relations or funeral arrangements. Jews and Christians are said to worship at the graves of prophets and are condemned in such behavior. Vajda says Islam asserts that because Jews are wealthier, they see themselves as superior to Muslims while out of malicious hatred and jealousy they knowingly deny Mohammed is the Prophet. To cover their deceit, the Jews and Christians are accused of erasing all references to Mohammed in their holy texts. Jewish denial is expressed through ridicule of the prophet and his followers, which is in particular done by testing Mohammed with questions to trip him up in matters of ritual law and specific beliefs.[20]

Muslims accuse Jews of not enforcing the laws of the Torah, especially when it comes to punishments for transgressions. They say the rabbis have become overly lenient. In other words, the Jews are no longer following the laws of God, but rather are bypassing Divine instructions for man-made interpretations. Overall, the Muslim attitude is one of supposed tolerance, but in reality is one of discrimination and oppression. It is out of this mentality Judaism is subjugated to Islamic superiority and Jews are distrusted and humiliated in everyday life, even though converting a Jew or Christian is considered a major accomplishment in Islam.[21]

The most damning Hadith accusation toward Jews is the story of a Jewess poisoning Mohammed to death, and the traditional belief of a Jew being the Dajjal, Anti-Christ, anti-Messiah or anti-Allah devil, who meets his destruction in the End Time. Thus, here is another imperative for Muslims to kill Jews. On Judgment Day, Jews and Christians will be sent to hell, but Muslims will achieve salvation.[22] The Jew stands accused of taking action to ensure the reign of evil and the destruction of Islam. According to these views, the Jew is at the depths of debasement, and his behavior undermines Allah's laws and all that is good. In sum, Islam views Jews as the most evil of all because of their capricious denial of Mohammed's prophecy, the jealousy of Islam, hatred of truth and malicious falsification of their own scriptures.

The Charter of Omar and Dhimmi Status
The Charter of Omar is the legalization of the *dhimmi* status, which is the protected second-class sufferance granted to Jews and Christians living under

Islamic rule. These statutes are attributed to Omar I, the second caliph in the mid-seventh century. Many believe the Charter of Omar was more likely the work of Omar ibn Abdel Aziz (Omar II) in the early-eighth century. Although not mentioned directly in the Charter, the poll tax or *jizya* was an extra, and often heavy, tax payment required of *dhimmi* communities. As stated previously, the tax requirement comes from the Koranic verse 9:29. "Fight against such of those to whom the Scriptures were given as believe neither in Allah nor the Last Day, who do not forbid what Allah and His apostle have forbidden, and do not embrace the true faith, until they pay tribute out of hand and are utterly subdued." The payment conditions were humiliating. Islamic overlords often struck a Jew or Christian during a public ceremony to prove the latter's submission. The precedent for such discriminatory taxes and behavior originate in the fifth and sixth centuries in the Byzantine Empire.[23] Islam expanded the limitations of the *dhimmi* status. Below are the stipulations from the Charter of Omar.

Dhimmis

- Must not build new churches or synagogues
- Are not to learn the Koran, since a Muslim would need to be employed to teach them, which was forbidden
- Cannot shelter anyone considered hostile to the regime or considered a spy
- Cannot buy Muslim slaves, nor could they buy a slave previously owned by a Muslim
- Cannot sell liquor or animals which were not slaughtered by correct Islamic ritual practices
- Cannot employ Muslims
- Must stand in the presence of a Muslim and are not allowed to deceive, or hit a Muslim, even in self-defense
- Must host Muslim travelers for up to three days
- Must wear distinctive clothing; Jews often wore yellow and Christians blue
- Footwear had to be different from that worn by Muslims
- Are disallowed Muslim names

- Must wear a special sign around their necks when using the bathhouses.
- Are forbidden to have weapons
- Cannot ride a horse or a mule, but only a donkey.
- Must ride side-saddle without a riding saddle, only a pack-saddle.
- Homes must be lower than Muslim homes
- Tombs must be lower than Muslim tombs
- Are not to raise their voices in houses of worship
- Christians could not be seen publicly with crosses
- Are not to have a government position giving them authority over Muslims
- Would lose their inheritance to the Islamic authorities upon the death of a relative until they could prove their right of inheritance of the family property in question[24]

In addition to all of those regulations, a *dhimmi's* testimony against a Muslim was inadmissible in court since the oath of a Muslim was accepted and that of a *dhimmi* was not. Theoretically, should such a clash take place the Muslim could swear the *dhimmi* had cursed or insulted the prophet Mohammed and capital punishment was inflicted. Obviously the charge need not be true. Hence Jewish and Christian rights were severely curtailed under Islamic rule.[25]

The *jizya* was required in addition to the "*karaj,*" or land tax,[26] and "*avariz,*" or irregular tax. These taxes were applied most arbitrarily toward the *dhimmi* populations.[27] Failure to pay resulted in the death penalty since non-payment was a violation of the *dhimmi* contractual status of protection given by the Muslim rulers. It must be noted that such a contract allowed for autonomy in the *dhimmi* communities. Usually a local leader had access to the Islamic authorities and played the part of the "court Jew" similar to the arrangement in Christian Europe. As opposed to pagans, the People of the Book had a certain freedom of religion, a right to existence and the ownership of property. Pagan infidels refusing conversion to Islam faced slavery or death.

Enforcement of the Charter of Omar *dhimmi* status was at the discretion of the local leaders. Implementation could be fairly lenient, as was the case with certain Ottoman administrations and specific rulers in Muslim Spain.

I Negative Image of the Jew in the Arab/Muslim World

On the other hand, extreme enforcement occurred with the Almohads of North Africa in the 1100s and the persecutions under the Fatimid Caliph al-Hakim at the turn of the eleventh century, though most Fatimid rulers were considered fairly liberal. In the twentieth century, outbreaks of extremism plagued Jewish communities in Persia (Iran), North Africa, and especially Yemen where the *dhimmi* status was harshly enforced through the middle of the twentieth century, causing the vast majority of Yemenite Jews to seek refuge in the newly born State of Israel.

In its historic context, the Charter of Omar came on the heels of institutionalized Byzantine discrimination against the Jewish community. Some consider it a fairly liberal document allowing for semi-autonomy of subject Jewish and Christian communities. In its early years it was particularly liberal compared to the atmosphere of intolerance against the Jews reigning in Christian Byzantium prior to the Arab Muslim invasion. In the Middle Ages, the Jewish community initially welcomed the Arab Muslim arrival; however it appears the Jewish world was unaware of the destruction of the Arabian Jewish community, most notably at Medina and Khaybar. The Arab Muslim "lenient" attitude apparently lasted until the rule of Omar II in 717. For the first eighty years, Islam was dependent on *dhimmi* populations living in the former Byzantine Empire for administrative help. Once the Arab overlords felt secure, Byzantine-style discrimination and worse was re-instituted under the Charter of Omar.

Under the Charter of Omar, all *dhimmis* were officially excluded from public office, since they were deemed unqualified to rule over Muslims in any way. Despite that, there was a breach. Muslim leadership needed *dhimmi* expertise in administration, finance, tax collection, language skills and commercial initiatives, in particular with Europe. Such times of leniency were invariably followed by harsh anti-*dhimmi* persecutions, including massacres, once the liberal Muslim ruler was replaced. Successors faced popular demands to return *dhimmi* communities to their proper oppressed status in Islamic society. Such public pressure could force a change of policy on a reigning monarch, and abruptly shift the culture into full persecution.

Jihad is and was the Divine instrument for subjugating the entire world to Islam, as expressed in Koran 8:39-40. "Make war on them until idolatry is no more and Allah's religion reigns supreme. If they desist Allah is cognizant of all their actions; but if they give no heed, know then that Allah will protect you." All material goods belong to Allah, and Muslims obtain them as victory spoils.[28] Koran 9:29 demands *dhimmi* populations be taxed, humiliated and physically forced into submissive degrading contracts, such as expressed in the

Charter of Omar. All unequal behavior toward subjugated *dhimmi* communities was and is honorable when done in the name of Islam, which is the true victorious religion Jews and Christians consciously reject for no other reason than stubbornness and conceit. *Dhimmi* rejection of Islam is seen as evil and a challenge to Allah. Through Muslim graciousness, Jewish and Christian communities are allowed to survive until the End of Days when Islamic truth will prevail. The People of the Book are all *dhimmis,* but a time lag exists for the fulfillment of total forced submission of all. In the meantime, Muslims collected the *jizya* tax while simultaneously progressing with the Holy War. In theory there were exemptions from the *jizya,* but in practice it was not unusual for a warrior society to demand taxes from orphans, women, children and even the dead. Forced to pay the *jizya* or face imprisonment, the *dhimmi* was further expected to pay double levies on everything else.[29] A *dhimmi* was always suspected, loathed and abused. Muslim overlords used *dhimmi* taxes to fund Jihad; taxes and war were the inseparable two sides of the same coin.[30]

In addition, the "*fay*" and "*karaj*" taxes provided for the good of the Islamic nation, or "*umma.*" The *fay* tax has its origins in Koran 59:6-10, and the *karaj* in agreements made between Mohammed and Jewish farmers subject to his rule. *Karaj* taxes were often collected in the most brutal fashion, with Muslim overlords forcing the *dhimmi* to abandon his land. *Fay* taxes came from infidel territories now properties of the state, and could be confiscated by Muslim lords as logic dictated that all property belonged to Allah and those in his service—Muslims.[31]

The *dhimmi* had no "basic right to life." The *dhimmi* was and is a quasi legal entity, safest when adopted as property by a Muslim sovereign, in essence accepting a status similar to slavery. He was forced to behave in the most humble manner, forbidden to defend himself, and obligated to prostrate himself before the powers of Islam proving his gratefulness at their permitting him to live. The *dhimmi* lived on sufferance, not toleration.[32] Toleration infers a certain pluralism or equality, which was non-existent in the Muslim-*dhimmi* relationship. Should a *dhimmi* violate a rule, the penalty could be death. A *dhimmi* could not testify against a Muslim, yet a Muslim could kill a *dhimmi* without worry of any repercussion, much less capital punishment. Muslims were equal to each other, while *dhimmis* were inferior, suffering constant compulsory degradation as a religious duty. The *dhimmi* community survived living under the protection, as property or chattel, of the Islamic sovereign.[33] In everyday existence Jews lived in fear of violence and even death, a result of being Jewish.

Jews could not even escape through conversion. A convert to Islam was still viewed as Jewish.[34] They were always suspected of conspiracies to undermine Islam. The Jewish philosopher Ibn Aqnin, himself a convert to Islam who lived under Muslim rule in twelfth century Spain, described not only the continued suffering of Jews who converted to Islam, but the never ending abuses heaped on their descendants over a century later. He said suffering was a result of Jews abandoning their ancient faith for personal gain. In the seventeenth century the Persian monarch Abbas II pursued a similar policy with devastating results for the Jewish community. Forced conversions were a matter of course, yet Jewish converts were not trusted and had to constantly assert their Islamic identity. They were referred to as "New Muslims," similar to "New Christians" in Spain. As a result, they often secretly practiced Judaism, similar to the Spanish Marranos who were forced to convert under the Inquisition.[35]

Part II
Historic Survey: Islamic Attitudes/Policies Toward the Jews

The Middle Ages through the Nineteenth Century

Before we begin a survey of Jewish persecution throughout the Muslim world illustrating the intensity and destructiveness of Islamic antisemitism let it be made clear that there were places and specific periods of time such as the "Golden Age" in Spain or Turkish rule in the Land of Israel during the Kabbalistic period of the sixteenth century where Jews were treated fairly well and the community flourished. However from the Middle Ages until the present, the traditional Jewish stereotype was and is still one of a corrupting, evil, degenerate,[36] debased, criminal, wicked people. Jews were considered more evil than polytheists and allied with the Dajjal or anti-Allah devil. Certain Islamic thinkers such as the Islamic-Indian writer Mohamad Yasin Owadally, claimed the Dajjal is himself Jewish. The Egyptian Sayyid Ayyub, as well as others, declared the evil Dajjal Jew was already functioning on earth.[37] The most famous rabbi of the Middle Ages Rambam (Maimonides) had this to say in a letter of empathy to his suffering Yemenite co-religionists, while he also counseled faith in God. "God has hurled us in the midst of this people, the Arabs, who have persecuted us severely, and passed baneful and discriminatory legislation against us . . . Never did a nation molest, degrade, debase and hate us as much as they . . . No matter how we suffer and elect to remain at peace with them, they stir up strife and sedition."[38] Below follows a very brief review by region and/or country of harassment, abuse, discrimination and persecution suffered by Jews under the Charter of Omar throughout the Muslim Middle East for over a millennium.

Persia/Iran - Muslims considered *dhimmi* communities defiled, especially in Persia (Iran) where even being touched by a Jew was regarded an offense.[39] Within the general domain of persecution and humiliation, Jews suffered forced conversions and pogroms, or massacres, throughout Persian history. In the sixteenth and seventeenth centuries, Iranian Jewry was almost wiped out by Shiite fanaticism whose dearly held theology insisted Jews were unclean and polluted. Such persecutions reached similar heights of frenzy not long ago, specifically in Meshhed in 1839 and Shiraz in 1910.[40] Less intense outbreaks of violence and discrimination were reported in Hamadan in 1875 and Tehran in 1897.[41]

In the late nineteenth century, Shah Nasr-ad-Din re-issued the following dictates: Jews could not leave their homes during rain or snowstorms to prevent impurities being passed on to Muslims. Any water that touched a Jew was considered polluted. Jewish women were forbidden from covering their faces in public, which made them similar to prostitutes in Iranian Shiite culture. Jews were forced to wear special colored clothes and mismatching shoes. Jews were to be subservient to Muslims in all matters and were even forbidden to travel in the countryside. Violation of the edict resulted in forced conversion or death.[42] Although the *dhimma* status was canceled in the early twentieth century, Uri Lubrani, an Israeli official who spent many years in Iran during the rule of the Shah in the 1970s, recently recalled the fact that the average religious Shiite refused to accept anything from his hand since he was Jewish and therefore "polluted."[43]

Yemen - Yemen was no better and even worse if one considers that the *dhimma* status was harshly enforced until the Jewish community arrived in Israel in the mid twentieth century. Due to their unequal status, *dhimmi* communities were forced to quarter soldiers either at their residence or in the community house of worship. The *dhimmi* were forbidden to enter a mosque, display overt public religious symbols or sounds, and were assigned to public works or degrading labor at the will of the local despot. One of the worst abuses occurred from 1646-1950, when Jews were forced to clean public latrines in Yemen.[44] It required constant allocation of manpower to remove feces from public toilets. This edict went into effect in Yemen in 1646 and continued for centuries. It was made "incumbent upon the Jews to cleanse the country and the public latrines from all excrement and loathsome materials. With these [excrements] they will heat the public baths. They shall also remove the carcasses of dogs, donkeys, horses, and camels."[45] Jews were associated with feces and its stench and were treated in a commensurate manner, meaning their lives were close to worthless.

The Messianic fervor of 1666, stirred by the false Messiah Shabtai Tzvi ignited persecution. Rabbi Sulayman al-Aqta, the great Sana'a scholar and Kabbalist, approached the king of Yemen to request permission to allow Jews to travel to the Land of Israel. According to Muslim records, he informed the king of Yemen that his earthly rule had come to an end. The rabbi was jailed, tortured and beheaded. His body was displayed naked at the city gate. The Jews paid a ransom for his burial and for the release of other tortured community leaders from imprisonment.[46] In 1679, the greatest disaster befell the Jewish community with the expulsion to Mauza. Jews were offered Islam or death. Choosing expulsion, the community was almost annihilated. Muslims confiscated property and destroyed synagogues and holy books. Once the decree was finally annulled, it was too late. The community was devastated, and three quarters perished within a year.[47]

In 1905, the Imam Yahya issued an edict reinforcing the *jizya* poll tax for all male Jews above the age of thirteen, so they would "be assured of their existence." He then enforced fourteen limitations on Jews.[48] The most humiliating was a return to the somewhat lapsed forced labor and latrines cleaning edict. Up to the moment they left for Israel, Jews wore distinctive clothing and always showed deference to Muslims. Jews were arbitrarily beaten and killed, while Jewish orphans under the age of eighteen were forced to convert to Islam. Most lived a frightened, impoverished existence.[49] Only Jewish flight to the State of Israel in 1949-50 brought freedom from *dhimmi* servitude.

Iraq - The life of Iraqi Jews, or what is often referred to as the Babylonian Jewish community, was full of uncertainty, instability and persecution throughout much of their existence under Islam. Islamic fanaticism began with the above-mentioned Omar II, and continued through Harun al-Rashid, Mutawakkil and al-Muqtadi during the Middle Ages. In particular, they imposed humiliating dress codes and crushing taxes. From the end of the 1200s and for the next century, instability played havoc with Jewish fortunes. Muslims blamed society's ills on the Jews, which quite often led to Jewish casualties. Over the years and into the modern period Jews fled, especially under Da'ud Pasha's oppressive rule from 1817 to 1831. Upon entering the twentieth century, Iraqi Jewish existence vacillated between persecution and new freedoms, all dependent on the whim of the ruling authorities.[50] Jews were finally deemed fully equal in Iraq after the British victory in WWI. Fifteen years later, Iraq moved toward independence and once again Jews suffered discrimination, particularly in employment under the Iraqi State. Nazi influenced Iraqi antisemitism was particularly vicious during and after WWII (see Chapter III "Jewish National Liberation").

Egypt - Living under the Charter of Omar was problematic enough, but sudden brutal attacks added to Jewish woes. Such was life under the Fatimid Egyptian Caliph al-Hakim in 1012 where Jews found themselves persecuted and their community life on the verge of destruction for no other reason than popular demand.[51] Egyptian Muslims in particular reviled the Jews far more than Christians, as noted by Edward Lane in the nineteenth century. Specifically, he emphasized Koran 5:82, which was used to justify beatings and executions of Jews on false charges. If a Muslim wanted to insult someone he called the person "a Jew." During the same period Moritz Luttke likewise reported hatred of Jews was unbounded, while deference was shown to the Christian population. Even the peasantry who had never met a Jew held him in contempt.[52]

North Africa/Magreb - North African Jewry did not fare any better under Islamic rule. Tolerance was shown toward *dhimmi* peoples at the time of the Islamic arrival during the second half of the 600s, but within a century oppression became the byword and Jews and Christians were forced to convert to Islam. The Emir Idris ravaged the countryside, laying waste to property and people alike. Historian Andre Chouraqui compared North African and European Jewish suffering. He said, "Under Islam, however, the stubborn Jews who clung to their ancestral faith were subjected to such repression, restriction and humiliation as to exceed anything in Europe." Rigid enforcement of the Charter of Omar heralded "the departure of virtually all Christian communities from North Africa, and subjected the Jews to the harshest conditions of inferiority even under the most benevolent rulers." Jews lived in special ghettoized quarters and "reached a state of indescribable misery and squalor," all the while being forced to wear either black or yellow to distinguish them from the Muslim population.[53]

Jews were blamed for famines, epidemics, regime change or any other unfortunate event and paid the price in blood, as victims to the raging Muslim hordes. Plunder, rape, massacre, the destruction of property and sacred writings including Torah scrolls, were often the communal misfortune. Almohads and Almoravids were devastating with bloody attacks against Jews in Fraa, Sijilmasa, Tlemcen, Marrakesh, Fez, Ceuta and Meknes in the twelfth and thirteenth centuries. Often, the choice was to flee or convert to Islam. Even though they tried to buy protection through payment of the *jizya* tax and generous gifts given to local officials, Jewish communities often found themselves at the mercy of frenzied mobs driven by lust and seeking plunder in times of instability. The authorities could not, or would not, protect them. At the outset of the French Protectorate in Morocco in 1912, the Jewish community suffered massacre and plunder, culminating in the sacking and burning of Jewish quarters.[54] For much of Morocco's history, Jews endured *dhimmi* slavery conditions.[55] Other disasters included the Jihadi slaughter in Fez in 1465[56] and again in 1820. The latter was accompanied by pillage, rape and mass destruction of the Jewish Quarter.[57] Similar attacks occurred in Algiers in the early nineteenth century.[58] Still, this list is only a tiny sampling of overall antisemitic outrages in the Muslim world throughout the centuries.

Mid Twentieth Century Onward

Such attitudes continued into World War II. In his famous book about the American war effort, *Crusade in Europe,* the Commander of European Allied forces and later to be President General Dwight D. Eisenhower spoke of the intractability of Arab prejudice against the Jews in North Africa. Arabs still refused Jews equal rights and saw them as inferior, even though the official *dhimmi* status was abolished. Eisenhower insisted that as long as the US army was responsible for civil affairs, all citizens were equal before the law. The American commander spoke of his clash with Muslim authorities as one of the most complicated civil administration issues he ever faced. The root cause of the problem was his refusal to agree to the spirit of discrimination expressed in the *dhimmi* status. Eisenhower understood that not giving in to antisemitic discrimination might jeopardize certain American objectives, yet he ignored threats to the US military administration while working to remove pro-Nazi officials and ensure equality despite warnings of possible pogroms.[59] He recounted, "The Arab population was then sympathetic to the Vichy French [pro-Nazi] regime, which had effectively eliminated Jewish rights in the region, and an Arab uprising against us, which the Germans were definitely trying to foment, would have been disastrous."[60] He explained:

> One complication in the Arab tangle was the age-old antagonism existing between the Arab and the Jew. Since the former outnumbered the latter by some forty to one in North Africa,

it had become local policy to placate the Arab at the expense of the Jew; repressive laws had resulted and the Arab population regarded any suggestion for amelioration of such laws as the beginning of an effort to establish a Jewish government, with consequent persecution of themselves. Remembering that for years the uneducated population had been subjected to intensive Nazi propaganda calculated to fan these prejudices, it is easy to understand that the situation called more for caution and evolution than it did for precipitate action and possible revolution. The country was ridden, almost ruled, by rumor. One rumor was to the effect that I was a Jew, sent into the country by the Jew, Roosevelt, to grind down the Arabs and turn over North Africa to Jewish rule.[61]

Eisenhower was unaware of the continual persecution of Jews into the twentieth century and the expectation of the furthering of the *dhimmi* status as a matter of course, regardless of who ruled. Vichy French persecution of Jews was a natural state of affairs, as opposed to American enforced equality for all. The Vichy French incarcerated thousands of Libyan and Tunisian Jews in concentration camps prior to the Allied arrival, without protest from Muslim authorities.[62] Libyans herded over two thousand Jews into the Giado concentration camp where over 500 died. Two months after WWII nation-wide pogroms broke out, killing well over one hundred more Jews.[63] During the war when Nazis rounded up Jews for forced labor, many found their erstwhile friends had turned against them and were actively supporting the pro-Nazi Vichy regimes persecutions.[64] By the mid-1960s, most North African Jews left their homes, the majority fleeing to Israel.

After WWII, antisemitism became unacceptable and was replaced with the veneer of anti-Zionism. Jewish community suffering continued unabated. Jewish support of Zionism was a violation of the Charter of Omar and the no longer "officially" existent *dhimmi* status. Jews could not support Zionism since they were to have no other loyalty except for their Islamic overlords, or the secular metamorphosis, Arab nationalism.

There were continuing anti-Jewish attacks in Cairo throughout the late 1940s and in Libya from the end of WWII until the early 1950s. Iraqi Jews suffered from the time of the "Farhud" massacre in 1941 when 180 Jews were killed until the community dwindled to just a few thousand by 1952. Yemenite and Aden Jewry fled in mass by 1951, especially in the wake of the persecutions of December 1947 where 82 people were killed. Between 6,000 and 7,000 Jews abandoned Aleppo the same month in order to escape the popular onslaught in Syria. Simultaneously, there were further riots in Egypt and Bahrain, while North African Jewry prepared to flee by early 1948. The pace of attacks quickened as a result of the complete Arab/Muslim world rejection of the UN Partition Plan on November 29, 1947. The resolution called for Jewish and Arab States to live side by side in peace. The Arab invasion and resulting clash in the Palestine Mandate served as a catalyst for pogroms against Jews and the looting of their properties. When Israel declared independence in May 1948, persecution intensified, resulting in the virtual liquidation of ancient Jewish communities in various places.[65] The vast majority of the above mentioned communities found sanctuary in the newly established State of Israel.

From 1950-67 Moroccan and Tunisian Jewry fled mostly to the Jewish State as well, while Algerian Jewry immigrated almost entirely to France. Jewish communities were no longer; any sign of their existence was obliterated and their memories erased, Libya serving as a case in point.[66] Egyptian President Nasser ordered the removal of all references to the Jewish community from Egyptian history books. For the few remaining helpless Jews, Israel's victory in the 1967 Six Day War only made the situation worse. As a manner of covering for their defeat, Arab regimes and populace took out their anger on the few Jews in their midst. Looting, beatings, imprisonment, torture, the occasional murder and accusations of betrayal, which for instance resulted in the execution of nine Jewish victims in Iraq 1969, became a recurring experience. The most viciously abused Iraqi and Syrian Jews were refused the right to leave, thus many fled illegally.[67] By the 1990s, there were only a few thousand Jews left in the entire Arab world. Little had changed since the Middle Ages.

Persecutions continued, even though there were almost no Jews left in Arab States. Such a blatant violation of human rights had little to do with Zionism, except for the fact that it is the modern nationalist extension of Jewish identity. This violence against a few thousand helpless Jews had everything to do with continued antisemitism. Those Jews remaining in Arab countries exhibited pro-Arab loyalties. They defied the majority and willingly decided to remain in the Arab world, choosing not to immigrate to Israel or the West. Yet Jew hatred reached new heights.

As illustrated throughout much of the Middle East, Jews often occupied the lowest rung on the social ladder both historically and at the moment of their immigration to the State of Israel. Situated in the heart of the Arab Middle East, Jewish sovereignty was an affront and humiliation to the Arab Muslim world. In 1948, the Arabs were disgraced on an international level when the newborn State of Israel survived a combined offensive by the five Arab armies of Egypt, Jordan, Iraq, Syria, Lebanon and the Arab Liberation Army in conjunction with Palestinian irregular forces. By 1967, the Middle East was swept up in "secular" revolutionary Arab nationalism and Egypt's President Nasser publicly led the drive for Israel's elimination, only to encounter defeat along with his Jordanian and Syrian allies. Worse yet was the capture of the Old City of Jerusalem and the Temple Mount, known to Muslims as the "Noble Sanctuary" and considered to be the third holiest site in Islam. The Arab Muslim world suffered from national and theological defeat at the hands of the inferior *dhimmi* Jews.

By the mid 1960s, close to 60 percent of Israel's Jews were of Asian and North African backgrounds, and either they or their ancestors had suffered the *dhimmi* existence under Muslim Arab overlords. European Jews dominated Israel's power elite and Western Diaspora Jewish communities came to Israel's aid. Still, the majority of Israel's fighting forces were made up of those same docile, "tolerated" and "protected" Jews who were *dhimmi* status just a few decades previous. Nothing could be more insulting to Arab Muslim pride than to suffer the physical loss of men, equipment and land to a despised, landless, exile nation, supposedly punished for all eternity for breaking their covenant with Allah. As Israelis they broke their *dhimma* "protection for servitude" contract, and to intensify the humiliation, defeated their former masters on the battlefield. Jews from Arab countries finally had a feeling of vindication.

Due to the sudden strength projected by this previously passive minority, the image of the Jew and of Israelis at large took on new, reinvigorated, negative stereotypes alongside those in the Koran and others acquired while living as a *dhimmi*. This third level of an anti-Jewish image needed to be integrated with the other two, despite the fact that at times they appeared contradictory. Emphasis was placed on Jewish cunning, plotting, cruelty and international Jewish efforts to defeat the Arab and Muslim world. Simultaneously, the Jew was accused of aligning himself with the anti-Messiah Dajjal, or devil, but all his schemes failed. By the late twentieth century, the Jewish stereotype took on monstrously evil proportions, similar to those projected by Nazism.

Sheikh Abdul-Hamid Attiyah Al-Dibani, the Rector of the Islamic University in Libya, had the objective to keep the Jew in his place through the sacred Islamic legal system. Not keeping with Divine jurisprudence led to Jewish empowerment and Muslim defeat. He declared at a Jihadi conference after the 1967 Six Day War, "Once Muslim jurisprudence had been discarded as a rule of life, the Jews could establish a State of their own in the heart of the Muslim world, to defy Muslims, and to gain victory over the Arabs in three consecutive battles. Hence, present-day Muslims should never treat with them for peace, since it has been proved beyond doubt that they [the Jews] are a mere gang of robbers and criminals, to whom trust, faith and conscience mean nothing."[68]

Sheikh Abdullah Ghoshah, the Chief Judge of the Hashemite Kingdom of Jordan, a territory where few Jews if any ever lived during the modern period, had this to say at the same event, "Treachery was the business of Jews throughout their ages and times as it was their instinct to break their covenant with others and resort to treachery as soon as they had any chance to betray others."[69]

Part III
From Ottoman Islamic Rule to Secular Nationalism

Jews and Muslims in the Land of Israel/Palestine

In the aftermath of the Bar Kokhva Revolt from 132-135 the Romans renamed the Land of Israel "Palestine" in honor of the Philistines who were of Greek origin. The Romans sought to obliterate any memory of the Jewish People's connection to the Land of the Covenant. Jews were reduced to a minority in their homeland. Two centuries after the Arab conquest of 638 CE, Muslims became the majority as Jews and Christians converted to Islam while others immigrated to friendlier lands. By the end of the eleventh century, few Jews remained in the Holy Land when it was conquered by the Christian Crusaders between 1099 and 1291. Later, Muslim Egyptian Mamluk and then Turkish rule (1517-1918) continued into the twentieth century. None of the above rulers brought much comfort to the Jews even if at times there were more liberal regimes.

Except for the pre-Crusader period, Arabs did not rule the Holy Land. Prior to the advent of modern political Zionism in the late 1800s, Jews continued in their attempts to return to the Land of Israel, most specifically to the four holy cities of Jerusalem, Safed, Tiberias and Hebron.[70] They experienced the traditional discrimination and abuse suffered as *dhimmis*, yet their unshakeable beliefs brought many to the ancient homeland undeterred. Especially in Galilee, the ancient Jewish imprint was noticeable. Many villages still bore Arabized Hebrew names originating in the Talmudic period.[71]

Palestine became a neglected region under Mamluk and Turkish rule. After Sultan Sulieman's fairly positive policy encouraging Jewish return to the Land of Israel,[72] Murad III declared a policy of expulsion of the Safed Jewish community to Cyprus, although it is believed the order was never carried out. Forced relocations known as "*surgun*" were implemented from the Balkan region to Istanbul and from Salonika to Rhodes in the late fifteenth and early sixteenth centuries. The Jews were then tied to new residences and had to pay double taxes.[73] Persecution in the form of extortionist taxation was carried out in Jerusalem, leading to the expropriation of the last remaining synagogue. Jerusalem Jews were heavily taxed while the Safed Jewish community continued to suffer anarchy, Bedouin raids, and other massacres at the hands of local Arabs during the nineteenth century.[74]

Taxes in Jerusalem remained a terrible burden as related by Gedaliah of Siemiatyce in the year 1700:

> In addition to the expenses in bribes destined to win the favor of the Muslims, each male was obliged to pay an annual poll tax of two pieces of gold to the sultan. The rich man was not obliged to give more, but the poor man could not give less. Every year, generally during the festival of Passover, an official from Constantinople would arrive in Jerusalem. He who did not have the means to pay the tax was thrown into prison and the Jewish community was obliged to redeem him. The official remained in Jerusalem for about two months and consequently, during that period, the poor people would hide where ever they could, but if ever they were caught they would be redeemed by community funds.[75]

Extortion was rampant in the nineteenth century. Jews paid hundreds of pounds to Muslim authorities to ensure access to the Western Wall in Jerusalem, for admittance to Rachel's tomb north of Bethlehem, for the right to use the Jaffa-Jerusalem road without harassment and to guarantee the ancient cemetery on the Mount of Olives would not be desecrated. Muslims denied Jews access to the Cave of the Machpela in Hebron where the Hebrew patriarchs and matriarchs are buried. Still, attacks continued against Jews in the four holy cities, particularly in the 1830s, with extortions, beatings and murders in Hebron and massacres in Safed. Despite constant appeals to British consular authorities for help, little was offered. Robbery, pillage, plunder, rape and murder perpetrated against the Jewish community were commonplace. If caught on the highway, a Jew could be not only robbed, but also forced to strip and left to continue on his way barefoot and naked. Any Jew attempting to give evidence against a Muslim assailant had his testimony nullified in accordance with the Charter of Omar—even after the Ottomans canceled the Charter in 1856. Local authorities continued to adhere to traditional Islamic ways, while Muslim attackers took vengeance on Jews if they dared complain to authorities.[76]

British Consul W.T. Young was beside himself with despair when he reported, "scarcely a day passes that I do not hear of some act of tyranny and oppression against a Jew." He continued in his analysis of the Jewish predicament:

> Like a miserable dog without an owner he is kicked by one because he crosses his path, and cuffed by another because he cried out - to seek redress he is afraid, lest it bring worse upon him; he thinks it better to endure than to live in the expectation of his complaint being revenged upon him. Brought up from infancy to look upon his civil disabilities everywhere as a mark of degradation, his heart becomes the cradle of fear and suspicion – he finds he is trusted by none – and therefore he lives himself without confidence in any.[77]

For Young any declaration of equality for Jews was useless and often only brought about more harm. While enumerating several reasons for such abuse he emphasized such behavior was attributable *"to the blind hatred and ignorant prejudices of a fanatical populace."*[78]

Several points need to be made regarding the above persecutions. In the midst of institutionalized religious discrimination against Jews, there were periods of stability. Second, there are many historians who claim the Jewish predicament was no worse than that of the average Muslim peasant. This is the same argument made concerning the Jews of Europe and the Christian peasantry yet there is an important difference. In both cases the peasant did not have to worry about forced conversion, expulsion or massacre by his overlords. He suffered as an individual, not as a "tolerated" out-group who at any moment could find his community at the mercy of blood-thirsty mobs. The peasantry suffered material deprivation in addition to whatever other maladies existed in society such as invasion, taxation or disease, yet the peasants were not singled out for punishment and destruction. The above examples are only a small sampling of the numerous persecutions that befell Jewry under Islamic rule in the Holy Land.

Over the centuries, comparisons were and are made between Christian and Islamic antisemitism. At the outset of modern research, many historians believed the Muslim world to have shown greater leniency and even favor toward Jews. A review of the sources proves that belief incorrect; however, the reverse is also not accurate. Christian persecution of Jews was horrific especially in Eastern Europe into the early modern period, even pre-dating the Holocaust. We can conclude our survey of this period by recalling the great Hebrew poet of the Middle Ages, Yehuda Halevy, who lived in both Christian and Muslim Spain. He deemed the evils of Christian and Islamic antisemitism to be equally brutal.[79]

Rollback of the 1000 Year Jihad
Christendom Defeats Ottoman Imperialism

To understand Islamic frustrations underlying extremist attitudes toward the West in the modern period we need a brief review of the fall of the Ottoman Empire. The last major Muslim advance into Europe reached its height when the Turks approached the gates of Vienna in 1683 and were rolled back by Polish, Hapsburg and German forces by 1699. Except for the Iberian Peninsula, the wildly successful, one thousand year Jihad was finally halted and reversed. The Christian Balkans went into revolt along church and national lines serving as an example to others. Once, Muslim imperialism in the form of Jihad flowed into Europe; now the Balkan "liberation wars" and European imperialism streamed in the opposite direction penetrating the Ottoman Empire by the eighteenth and nineteenth centuries. For the *dhimmi* peoples, in particular the Christians, there was hope. The unbreakable connection between Jihad and the *dhimmi* existence backfired when the Turks lost their previous conquests.

With the weakening of the Ottoman Empire, the European powers made steady inroads all over the map. France moved into North Africa, Russian influence coupled with indigenous Slavic populations in the Balkans, while Britain and France penetrated into the Levant of Syria, Lebanon and Palestine. European ideals advocating secular nationalism and equality swept into the Middle East through Christian communities, in particular those in Beirut and Damascus.[80] The Europeans worked for *dhimmi* emancipation and equality as a matter of principle but also to undermine and bring about the collapse of the Ottoman Empire. It was hoped the former *dhimmi* communities, in particular the Christians, would show appreciation and increasing loyalty to the Europeans for intervention to ensure their equal rights.

The confluence of material and community interests often brought about a European alliance with the *dhimmis*. Western pressure resulted in the abolishment of the *dhimmi* legal statute in 1856. The European-initiated Capitulation Treaties, designed for economic reasons in 1535, were expanded to award specific rights to particular *dhimmis,* either as individuals or as a group over the centuries. The full cancellation of institutionalized discrimination infuriated the "*ulema*," Muslim religious leadership, and stirred up the accumulated anger of the "*umma,*" or people, toward the *dhimmi* communities, in particular against the Christians. This was a major cause leading to popular outbreaks of violence and the massacre of some 20,000 Christians throughout the Levant by 1860.

When Christians asserted their equal rights they violated the stipulations of the traditional Islamic toleration, Muslims saw them as breaking the *dhimma* agreement despite its repeal. Thereby these Christians no longer deserved the protection of their Ottoman Muslim overlords. In addition, the Turks suffered territorial losses, which led to Muslim refugee problems particularly in the Balkans. In a defensive move, the Ottomans resettled refugees in Palestine and Armenia to solidify Turkish Muslim control over those areas where large non-Muslim populations asserted themselves nationally, religiously and increasingly in alliance with European interests. Such moves led to radicalization, further instability and clashes, which traditional Islamic thinking interpreted as religious wars.[81] One result of WWI was the Ottoman perpetrated Armenian slaughter, the first genocide of the twentieth century. Many Jews knew they could be next, but suffered "only" expulsions and persecutions in Palestine from 1915-17.

Overall the Jewish predicament was both better and worse. Equality led to economic gain, which was often connected to finding common cause with the British. Jews were not a threat on the political or diplomatic level, but they were considered the most servile of all groups, and therefore any equality obtained was far more insulting to a Muslim's honor than awarding Christians their rights. The Jews tread carefully; their enhanced status often stirred up Islamic resentment as well as Christian antisemitism, whether of local variety Eastern Orthodox roots, or those imported from Europe. All of this made Jews more vulnerable in many locales.

Europe put pressure on the Balkan, Russian, and Armenian borders and opened a domestic front by advocating equality for the *dhimmi* communities throughout the Ottoman Empire. It was Europe's attempt to gain *dhimmi* loyalties through liberation from Muslim disabilities, which often made the *dhimmi* or "*raya*" communities into pawns in the crucial power struggle between the Turks and Europe, Islam and Christendom. Attacks against the *dhimmis* and their property were not unusual as the average Muslim resented the equal status awarded to those who denied Allah's perfect belief system. In attempting to spread their influence, Europeans prodded the Ottomans toward modernization and a secular Turkish nationalism in return for help developing a modern military. This was partially realized when the "Young Turks" overthrew the sultan in 1908. Both Germany and Britain vied to manipulate the Ottomans into their sphere. By WWI the Germans succeeded, but were defeated alongside the Ottomans and their empire was shattered. Secular Turkish nationalism and European ideals triumphed over the previous Islamic identity and Sharia law. In the 1920s Mustafa Kemal Ataturk developed modern Turkey, a culmination of this process.[82]

Quite a few members of the former *dhimmi* groups, Jews included, advocated full loyalty to the Ottoman Empire, believing their dreams would be fulfilled in the aftermath of the Young Turk revolution. Such hopes continued into WWI, but for the most part they were dashed with the Armenian genocide, the general persecution of Christians and the Greco-Turkish clash of the early 1920s. Luckily, Jews suffered no mass murder, but they understood their place. A broader discussion of the development of modern Turkey and both Ottoman and Turkish policies toward minorities is beyond the scope of this work.

The Ottoman defeat in WWI brought about full European imperial intrusion into the Middle East. Traditional Muslims saw any alliance of *dhimmi* populations with Europe as a betrayal. Most notably, they saw the Lebanese Maronite Catholic and French alliance in this light. According to Bat Ye'or, Zionism could not develop in Middle Eastern Jewish communities the way it did in Europe due to overriding fears of massacre facing any *dhimmi* community daring to declare loyalty to an overlord other than Islam. Great massacres and persecutions of Greeks, Maronites, Armenians and others drove this message home to those Jewish communities still under the *dhimma*.[83]

As Jewish nationalism challenged the Arab world, Muslims saw Jews in Arab countries as easy prey. They made public threats in international arenas. For instance, the Syrian delegate to the United Nations, Faris el-Khouri, gave full expression to this line of thinking in a *New York Times* interview on February 19, 1947 when the international community began considering the Partition Plan. "Unless the Palestine problem is settled," he said, "we shall have difficulty in protecting and safeguarding the Jews in the Arab world."[84] His inference of "settled" meant no place for a Jewish national entity.

Confluence of Interests and Theology
Christian-Islamic Antisemitism

Christian and Muslim antisemites borrowed from each other in their antipathy toward the Jews. Despite being *dhimmi* themselves, Eastern Churches blamed the Jews for their sufferings under Muslim rule, in essence transferring their anger against the oppressor onto the Jews, whose social position was often far less than their own.[85] Although given equality, Middle Eastern Christians still suffered persecution, as did the Jews; however, with support from France and the Catholic Church their hope of solidifying equal status was understood to be within reach. Theologically, despite conflict with Islam, they could agree with the Muslim authorities to disenfranchise the Jews from any claims of finding favor with God, or in a more physical sense any claim to the Holy Land. Christians set off the Damascus Blood Libel of 1840 and the outbreak of violence against Jews in the area of the Church of the Holy Sepulcher in Jerusalem in the mid nineteenth century.[86] Antisemitism was indigenous to much of French nationalism and such tendencies were often passed on to Christian Arabs.[87]

French-influenced Christians initiated the rise of secular Arab nationalism in the nineteenth century in an effort to gain equality and a foothold in the teetering Ottoman Empire. Jews were not included, but instead disenfranchised here as well. The Arab nationalist formula meant identifying oneself as a Muslim Arab or Christian Arab, where the term "Arab" was the noun. All Arabs, Muslims or Christians, were to be equal in the future Arab national state living under secular law. However, when religion and Islamic Sharia law ruled, a person was either an Arab Christian or Arab Muslim; one's religion was the noun and determining factor of identity and law, meaning Christians would continue to suffer under the *dhimma* restrictions.

Borrowing from the European understanding, secular Arab nationalism made inroads not only among Christians but within elements of the majority Muslim population as well. At the conclusion of WWI in 1918 Britain and France divided the Middle East into regional interests and issued mandates with international League of Nations support. The British and French hoped to build economic ties established on the European secular model with the new Arab national entities no longer part of the Ottoman Empire as a result of WWI.

WWI and the Turkish defeat led to a Muslim backlash. The Middle East saw massacres of some 300,000 Assyrian Christians, including Nestorians and Chaldeans,[88] and one and a half million Armenians from the outset of WWI into the 1920s. The Christian dilemma was at its height. Lebanese Christians were divided between attempting independence with French support, or accepting a role within the secular Arab nationalist movement. Most chose the latter believing they could integrate fully into the new Middle Eastern realities achieving equality and security once religious differences were downplayed. In the Holy Land, Christians aligned themselves with the Muslims in the newly developing Palestinian Arab nationalism.[89] More than elsewhere in the twentieth century, antisemitism characterized these churches under the Palestine Mandate. Catholics in particular accused Jews of being "deicidal." During and after WWI, Christian refugee survivors of Jihadist massacres often fled to Jerusalem and the Levant in general where sizable Christian communities remained. Augmented by the newly Arabic translated Czarist *Protocols of the Learned Elders of Zion*[90] charging the Jews with seeking world conquest, these Christian refugees deflected hatred from their Muslim oppressors and projected it toward the Jews, accusing them of causing their misfortunes[91] (see Chapter III "Jewish Nationalism"). Some might consider their attitude "displaced aggression," or a defensive measure. Seeking and receiving acceptance from the majority Muslims while identifying the Jews as the common enemy could bring physical security to the Christian minority. Secular Arab nationalism promised to guarantee physical existence and equality for Christians by replacing the Sharia-driven Muslim societies.

One can ask, "Why did the Jews not adopt this new Arab nationalist identity?" In fact some tried, in particular in Iraq from the 1920s through the 1940s but the movement failed.[92] Religious Muslims supporting a form of secular Arab nationalism still viewed the Jews as an exiled national entity suffering punishment for deliberately breaking the covenant. Christians were viewed in a more positive light; they were understood to be misguided, but not malicious. In addition, the rise of the Jewish national movement or Zionism commencing in the 1890s was seen as a full rebellion against the *dhimma* status and clashed with deeply held theological beliefs among both Christians and Muslims that Jews had broken their covenant with God and lost any chance of redemption in the Promised Land. Zionism competed directly with secular Palestinian Arab nationalism, Jewish national success being an affront to God. Conversely Christians were seeking to merge with the Arab nationalist majority without claiming any specific Muslim *waqf* lands for an independent Christian entity.

Rejecting any national or separatist movement of their own, Christians felt it best to align with the rising Arab national movement. Most well known for this policy is the historian George Antonius during the British Mandate, PLO members George Habash and Nayef Hawatmeh after Israel's independence, and in recent years the academic Edward Said. This was the safest policy to achieve equality and survive in the Middle East, yet it demanded the concession of Christian identity and the acceptance of the Muslim Arab national narrative. The community nullified any previous Byzantine or Christian claims of primary distinctiveness. Points of unity within the Arab national identity found prominence. Canceling points of conflict, the Christian communities made common cause with Muslims through traditional staunch antisemitism awakened by the rise of Zionism and together denied the legitimacy of the Jewish national movement.[93]

Only a small European Christian minority viewed a Jewish return to the Promised Land as a harbinger of the End Time and the return of Jesus of Nazareth. Such was the case with certain British Protestant groups supporting the effort in the 1800s. Politically, Britain viewed an alliance with Zionism as working in its favor and issued the Balfour Declaration supporting a Jewish National Home, but no state, in November 1917 and later obtained League of Nations approval for the Palestine Mandate. Stitching together economic and military interests stretching from Iraq through Transjordan to Palestine and Egypt, London established friendly Arab regimes to ensure the free flow of oil from Iraq to Haifa for civilian and military needs.

Jews were not only behaving as equals, but Muslims saw them as audacious in their work toward national independence in the heart of the Arab Muslim world. Christians likewise were appalled; theological disaster was on the way should Jews gain sovereignty in the Holy Land and in particular over Jerusalem. A similar fate confronted Islam, should *waqf* land[94] fall under Jewish rule along with the Noble Sanctuary or Temple Mount, home to the Al-Aksa Mosque and Dome of the Rock. Christians, once downtrodden and humiliated through Muslim persecution, were now equals in the Arab national movement. Christian equality dealt religious Muslims a blow to their dignity, but common ground existed when confronting the theologically "dispossessed" Jews. Islam and Eastern Christianity intensified their antisemitic campaign against the Jews to deny them equal status as a nation, a people or as individuals. Eventually the venomous pitch of antisemitism reached levels of hatred only appeasable through Jewish destruction.

Muza Kazem el-Husseini, president of the Arab Palestinian Congress representing both Muslims and Christians, demanded the implementation of

Sharia Law and the abridgment of the Balfour Declaration, but he somehow excluded Christians from discriminatory statutes. At the same time Christians lobbied the British to accept the arguments in the *Protocols*. The Muslim Christian Association determined Jews had no historic-religious right to Palestine/Land of Israel, and they were to remain in the Diaspora. Muslims and Christians, unified as Arabs, were the only true owners of the land. Jews were satanic and evil by nature. The "Arabs were the creators of science and civilization" and all accusations made against the Jews in the *Protocols* were true. Theology also played an increasingly political role. Christian accusations of deicide and the emphasis on superseding the Jews as God's "chosen" accompanied the adaptation of the Muslim Jihadist ideal to be carried out jointly in the name of a unified Arab people against Jews.[95]

Theology served secular political purposes. Political anti-Zionism met on a platform of theological antisemitism. Both Christians and Muslims faced down the British in the Palestinian Arab demand for nullification of the Jewish National Home. Such visceral hatred reinforced the political clash and could be counted upon to stir up violence against the Jews. Anti-Jewish riots in 1920, 1921, 1929 and 1936-39 forced the British to reconsider the promise of a Jewish National Home. Reviling the Jews was expected to bring Christians acceptance in the Palestine Arab national movement, although even here the Islamist Izz a-Din al-Qassam saw Christians as an additional enemy. Theological condemnation of the Jews would not bring Christians any closer to Islam, but adopting a Palestinian Arab identity provided an acceptable secular interpretation. Christianity and Islam replaced the Jews who were now to be destroyed by a joint military front in the service of the collective Palestinian Arab people. Relegating the Jews to fossilization, Christian "Replacement Theology" reigned again, even though from the Muslim perspective Mohammed's revelations replaced both Christianity and Judaism. True believers accepted the eternal absolutist claim of Islamic existence and superiority commencing with the creation of the world. The *dhimmi* peoples were only an aberration to be "tolerated" or "suffered." Somehow the Eastern Churches thought they would become acceptable to the Muslim world by embracing secular Arab nationalism while battling against the common Jewish enemy. This policy failed as shown by Jihadi attacks against Middle Eastern Christians into the early twenty-first century.

Vicious antisemitism characterizing the *dhimmi* Jew and Czarist *Protocols* stereotype were prominent into the 1930s and the outbreak of WWII. European antisemitism culminated in the Holocaust, an extermination led by Nazi Germany, whose foundations were European. The Nazis could not have succeeded in their perfidy without the aid of collaborators. In the Middle East, there were attacks against Jews by many Axis sympathizers, yet the demand and institutional organization to take such action on a large scale was not ripe. This in no way detracted from the traditional Islamic denigration of the Jews. In 1939, the Muslim world as a whole had yet to arrive at the conclusion that Jews were their foremost enemy and their liquidation an Islamic obligation. The ingredients for such a conclusion existed in previous Islamist thinking. Although the variables of such an equation were extant, the sum of all its parts did not yet add up to taking exterminatory action. A "catalyst" was clearly absent. Simply put, Christians and Muslims theologically, as well as Arab nationalists on the secular cultural plane, saw Jewish self affirmative action to ensure equality as a national entity through Zionism as a violation of the "world order."

From the European perspective, German antisemitism was rooted in a secular, pseudo-scientific, racial theory condemning the Jews to extinction. It was not religious; a Jew could convert to Christianity, yet he remained a Jew. An assimilated Jew was most dangerous, spreading his "polluted" gene pool amongst superior and purer races, especially the "German Aryans." It is important to examine the transition from religious antisemitism to the more deadly secular type, which was justified through rationalist thinking and supposed scientific proof.

Paradoxically the Enlightenment nourished secular antisemitism, which was a challenge to the old feudal order. Trusting rationalism, human observation and scientific experiment, Enlightenment thinkers were sure they would arrive at truth and the best world possible. Questioning the diocentric world and jettisoning most of it, they believed in humankind's abilities of self-rule, equality and human rights. By the late seventeenth century, a process of questioning the old order was set in motion. 1789 brought the French Revolution; ten years later Napoleon was dictator over the secular French state. Catholic France allowed for Protestant and even Jewish equality, which was already acquired in 1790-91. There were equal rights for men only, equal obligations and expectations of full loyalty to the French nation state and its self-appointed ruler, Napoleon. Anthropocentrism triumphed, the people reigned, and religion was defeated. Religion was lucky to retain its status at all, and remained subordinate to the secular nation. Napoleon's forces spread anti-church, anti-aristocratic modernity by advancing on central Europe and later into Russia. By 1815, Napoleon suffered his final defeat, but rationalism continued to spread, changing the face of Europe and ushering in the Age of Nationalism.

The Middle East experienced a similar but much more limited process. Very much in parallel, the more secular leaning Mohammed Ali of Egypt swept through the Middle East, although he was far less successful at imparting similar secular ideals. Whereas the Enlightenment penetrated much of Europe prior to the French Revolution, the Middle East was fairly bereft of rationalist understandings. Religion and the Caliphate still held sway. Napoleon invaded Egypt in 1798; the French remained three years and were finally defeated with the help of Mohammed Ali. Originally sent by the Turkish sultan in a military capacity, Mohammed Ali gained power from the disintegrating traditionalist Mamluk regime in 1805. As the Pasha of Egypt, he consolidated his regime by 1811. The Ottoman sultan was less than thrilled with his meteoric rise, but still turned to Mohammed Ali to put down the Islamic fundamentalist Wahhabist rebels who captured Mecca and Medina in the Arabian Peninsula, invaded Iraq, sacked the Shiite holy city of Karbala and were on their way to Damascus.[96] Despite massive logistical problems, by 1818 Mohammed Ali and his sons successfully defeated the Wahhabist extremists. He then expanded his rule into Yemen, the Hejaz and Sudan.

Shoring up an economic and military power base, Mohammed Ali accelerated Egyptian domestic development using European personnel, technology, and educational models to build a modern state on the Nile. He established a central bureaucracy, wrenched control of Egypt's mainly rural economy from the previous Mamluk local rulers, expanded agriculture, confiscated *waqf* Islamic trusts held by specific religious authorities, revised the tax structure to benefit the state, modernized the army, constructed a navy, established industry where none existed, improved health care, rebuilt the port of Alexandria and poured the foundations for the modern Egyptian nation state. He particularly invested in modern education for his military. In many ways Mohammed Ali was the state, built as an extension of his power somewhat similar to Napoleon's France but not as advanced.[97]

In 1831, Mohammed Ali invaded the Levant and Syria on his way to Istanbul. European intervention halted the collapse of the Ottoman Empire and Mohammed Ali was forced into partial retreat by 1840, yet he began to set the Middle East free from the entrenched aristocracy and clerical ruling classes. Rationalism and secular modernity brought great improvements, first to Egypt and then to the Holy Land and Syria during Mohammed Ali's brief rule, especially where Christian *dhimmi* communities were concerned. The Jewish condition improved, but did not approach that of the Christian advance. Individual Muslim contempt for the Jews increased despite regime policies relaxing *dhimma* disabilities. The regime and the Capitulation Treaties specifically designed to protect European merchants also protected non-Muslim foreigners including quite a few Jews.[98] However, Muslim reaction against the Jews was particularly felt in the Safed pogrom of 1834.

The arch conservative Islamic religious backlash was not long in coming, leaving tens of thousands killed and injured, in particular Lebanese and Syrian Christians. By awarding equal rights to the *dhimmis,* those implementing secular Enlightenment-type legislation undermined the will of Allah, destroyed the Sharia and replaced the Divine with themselves as rulers. Both in France and in the Middle East, the elimination of institutionalized religious discrimination against the Jews in particular brought about an increased disdain for what was considered a non-deserving, theologically discarded minority.

In Europe the pro-Vatican Catholic, pro-aristocratic right wing challenged French secularism while Ottoman and Arab secular attempts were seen as a betrayal of Islam and the Caliphate. The secular state undermined Christianity and the European monarchies while in the Middle East the Caliphate was abolished by secular Turkish nationalism in the 1920s. Undeservedly, and in violation of theological dictate, the Jews, a despised minority, attained equal rights and were viewed as the beneficiaries of all these calamities befalling both the Christian and Muslim old orders. The baseline for secular revolutionary success was Napoleonic France and a century later Ataturk's post-WWI Turkey, now coinciding with massive British-French penetration into the Arab world.

Islam Incorporates Czarist/Nazi Indictments Against Jews

Devout Christians and Muslims were deeply disturbed by Jewish advances made in civil society, yet European rationalism and secularism would spawn the harshest forms of antisemitism, of the racist-genetic-type. Christian antisemitism mutated from relegating the "deicidal Jews" to what can be considered "the trash heap of history" with its accompanying degradations to justifying Jewish annihilation. Centuries earlier in the 1530s, Protestant founder and theologian Martin Luther declared, "Know Christian, that next to the devil thou hast no enemy more cruel, more venomous and violent that a true Jew."[99] Once Charles Darwin published *On the Origin of Species* many interpreted his findings as parallel to human anthropology. Antisemites compared Jews to evil bacteria in need of extermination. To quote the famous French-German antisemite Paul de Lagarde who accused the Jews of plotting to destroy the Christian-German faith:

> One would need a heart as hard as crocodile hide not to feel sorry for the poor exploited Germans and – which is identical – not to hate the Jews and despise those who – out of humanity! – defend these Jews or who are too cowardly to trample this usurious vermin to death. With trichinae and bacilli one does not negotiate, nor are trichinae and bacilli to be educated; they are exterminated as quickly and thoroughly as possible.[100]

Neither Eastern Christians nor Muslims countenanced a Jewish return to the Land of Israel. Such attempts could only be interpreted as a diabolical plot. The *Protocols* provided the answer of a Jewish Zionist plot to conquer the world through money, media, revolutions and a constantly expanding state. *The Hamas Covenant* echoes these accusations in Articles 22 and 32. With the success of the Jewish National Home and subsequent Israeli State, a "temporarily" defeated Islam embraced the character traits described by the Czarist antisemites. After all, Allah could never betray true Muslims; rather, the enemy was much more crafty, powerful and evil than imagined. Drawing back on Islamic texts and stereotypes was helpful, but not quite enough until full theological demonization provided the missing variable. Secular rational demonization was explained and justified using the *Protocols*.

The *Protocols* highlighted themes of Jewish perfidy and conspiracies against Europe and Christendom. The spirit of demonization is similar to those from antisemitic Islamic texts, in particular the Hadith accusation that the Kaybar Jewess Zaynab Bint al-Harith poisoned the Prophet Mohammed in 632 and that the Dajjal, or devil will be born of a Jewess.[101] Today's demonization includes accusation reversal whereby Israel or Jews are accused of being Nazis *(The Hamas Covenant* Articles 20, 31 and 32), which conflicts with the well-known fact of Islamists' own sympathies and cooperation with the Nazis during WWII (see Chapters III and IX "Jewish Nationalism" and "Czarist Nazi Integration" respectively). Emphasis on these themes increased after the war.

More significantly a "solution" was needed. Antisemitism changed its form evolving from the diocentric concept of theological evil with the possibility of repentance, to the anthropocentric human determination of the intrinsic inability to change one's genetic make-up and join "the forces of good." The Aryan race was positive, the Jews negative, and only one side would be victorious. Nazi Europe was said to have found the solution during WWII, but never completed the job. Theological Christian and Islamic antisemitism merged. Extremist Islam integrated theological antisemitism and the rational "scientific" Nazi explanation determining inherent Jewish evil and arriving at the solution for a full Jihad in the "struggle against the Jews," who were the ultimate enemy. Muslims were good, Jews were by definition "evil" with no chance of repentance and Islam was destined to achieve victory as determined by Allah. This ideology summarizes the world-view of the most influential Islamic ideologue, Sayyid Qutb (see Chapter II "Ideologues").

Europe "upgraded" its antisemitism to annihilating the Jews in WWII. Extremist Islam only fully picked up on the Final Solution after the Holocaust. For the Nazis, exterminating the Jews was a racial imperative. For the Jihadists, including the Muslim Brotherhood, Hamas and the like, exterminating the Jews mutated into a theological commandment exceeding the *dhimma* status and Charter of Omar.

We see this continuity when skipping forward for a moment to the twenty-first century. Popular Egyptian Brotherhood preacher and *Al Jazeera* Arab world TV star Sheikh Yusuf al-Qaradawi's book *Rulings on Palestine* (2004) demanded a continuing Jihad to destroy Israel. The book legitimized suicide-homicide bombings and warned Muslims that all Jews are their enemies. Echoing *The Hamas Covenant* Article 32, he succeeded at fully integrating theological and modern European antisemitism, solidifying the anti-Jewish Islamic hatred of today.[102] His later exhortations on *Al Jazeera TV* on January 9, 2009 demanded Zionist and Jewish destruction down to the "last one." Three weeks later he went further, condoning Jewish extermination at the hands of Hitler as Divine punishment to be continued at the hands of the "believers," today's Muslim Brotherhood.[103]

The best example of the totality in the demand for Israeli Jewish destruction is support for suicide-homicide bombings. These bombings were understood to be a successful tactic taken from the Shiites in Lebanon when battling Israel and the West. The Palestinian Authority Jerusalem Mufti Sheikh Ikrima Sabri and Chief Justice Sheikh Taysir al-Tamimi agreed fully with Qaradawi that suicide-homicide bombings against Israel were legitimate, and constitute martyrdom, not suicide. The 9/11 bombers and others were condemned, but attacks against Israeli civilians are still encouraged and condoned, even if the victims were children. Part of the logic behind the support is the false belief that all Israelis serve in the army and therefore the Palestinians have no choice but to kill with impunity until Israel accepts a *"hudna,"* an Islamic cease-fire of temporary duration, or surrenders on Islamic terms. Martyrdom is the "greatest Jihad," especially when directed against Jews and their state, which is deemed a military society deserving of eradication. Muslims believe a person will achieve life after death in the End Time if one dies as a martyr.[104] Only Jews deserve such a non-compromising war of destruction.

Even more frightening is the Islamic definition of "Jewish evil." Qaradawi's reasoning is not a racial theory, one that can be put to the test and disproved through scientific investigation, but it is an unshakable theological determination with no option to prove otherwise. Once described as an "extinct" or "fossil society,"[105] by the historian Arnold Toynbee, Jews no longer only face banishment to the historical garbage dump; they are now targets of elimination in the name of Allah. Islamic antisemitism exists since the advent of Islam but is now augmented with Czarist and Nazi accusations and hatreds. Jihadi Jew-hatred is most deadly. "Replacement Theology" became "Erasement Theology," except perhaps for a few "good" Jews willing to submit to Islamic superiority and the *dhimma*.

As we will see, Hamas focused attention on and embodied Jihadi antisemitic ideals in its theology while attracting media attention, world sympathies and recognition. Israel's destruction is Hamas' implacable demand, but then so is their oft-forgotten desire for the annihilation of world Jewry. Many fail to recall that the State of Israel is only one battlefront for Hamas, as made clear in their *Covenant*. Hamas and Jihadist Islamists have every intention to implement all they advocate yet much of the world appears oblivious to this fact.

Endnotes

1. Newby, Gordon Darnell, *A History of the Jews in Arabia, From Ancient Times to Their Eclipse Under Islam*, University of South Carolina Press, 1988, Chapters 4-5.
2. Ibid, pp. 78-82.
3. Ibid, pp. 83-86.
4. Ibid, pp. 89-96.
5. Koran 5:82.
6. Bat Ye'or, *Islam and Dhimmitude, Where Civilizations Collide*, translated from the French by Miriam Kochan and David Littman, Fairleigh Dickenson University Press, Teaneck NJ, USA, 2002, p. 346.
7. Newby, pp. 97-104.
    Wahhabist ideals form the theological basis for the Islamic State (ISIS/ISIS) today.
8. *The Koran*, translated by N.J. Dawood, The Penguin Classics, Penguin Books, New York, 1977.
    This is the specific English translation to be used throughout this work when quoting the Koran. The only exceptions will be quotes of the Koran within other texts when cited.
9. Koran 46:12 and 46:30-32.

10   Goitein, S.D., *Jews and Arabs, Their Contacts Through the Ages*, Schocken Books, New York, 1974, pp. 56 and 64.
     Haman was the advisor to the Persian king Ahasuerus according to the Book of Esther placed somewhere in the 6th to 5th century BCE. The Egyptian Pharaoh in the Exodus story is understood to be Ramses II who would have lived some eight centuries earlier. There is no chronological or geographical connection between the two. Another example is the Koranic confusion between Miriam the sister of Aaron and Moses in the Hebrew Scriptures and Mary the mother of Jesus in the Christian Gospels.
11   Koran 9:30.
12   Old Testament *(Tanakh)*, Exodus, Chapter 32.
13   Ibid, Numbers, Chapter 16.
14   Arlandson, James M., "Did Allah Transform Jews into Apes and Pigs?" *Answering Islam*, retrieved April 21, 2011, www.answering-islam.org/Authors/Arlandson/jew_apes.htm.
15   For an online abstract see Andrew Bostom's "Antisemitism in the Quran," April 11, 2008 at http://www.islam-watch.org/Bostom/Anti-semitism-in-the-Quran.htm. And selected articles from Palestine Media Watch, www.pmw.org.il.
16   Qutb, Sayyid, "Our Struggle With the Jews," in Nettler, Ronald, *Past Trials and Present Tribulations: A Muslim Fundamentalist's View of the Jews*, Vidal Sasson International Study of Antisemitism, Hebrew University, Jerusalem, Israel, 1987.
17   Bat Ye'or, *The Dhimmi, Jews and Christians Under Islam*, Fairleigh Dickenson University Press, 1985, Doc. 11, "The Jizya's Meaning," p. 189.
18   Vajda, Georges, "Jews and Muslims according to the Hadith," *The Legacy of Islamic Antisemitism*, Bostom, Andrew, ed., Prometheus Books, 2008, pp. 235-260.
19   "Excerpts from the Canonical Hadith Collections," *The Legacy of Islamic Antisemitism*, Bostom, Andrew, ed., 2008, pp. 235-260.
20   Vajda in Bostom.
21   Ibid.
22   Vajda and "Excerpts from the Canonical Hadith Collections," in Bostom.
23   Parkes, James, *Church and Synagogue*, A Temple Book, Atheneum, New York, 1969.
     "Jews in the Byzantine Empire," *Wikipedia*, retrieved May 17, 2011, en.wikipedia.org/wiki/Jews_of_the_Byzantine_Empire.
     "Corpus Juris Civilis," *Wikipedia*, en.wikipedia.org/wiki/Corpus_Juris_Civilis.

   Historically, restrictions against Jews and their relegation to a second class tolerated status were previously introduced in the Byzantine Empire, initially by Theodosius commencing in 404 and later reinvigorated by Justinian from 527-565. Taken together Jews were banned from all government posts, the civil service and the military. Jews could be the "decurion" or tax collector who would be forced to personally pay any deficits should he fail to collect the revenue demanded. In a slave driven economy Jews were first restricted in such rights and then banned completely from owning Christian slaves on penalty of death. All of the above were serious economic disabilities.
   Concerning religious practice and community organization synagogues could be taken for state usage especially in the sixth century. Under Justinian, Jews, pagans and heretics were dispossessed of rights throughout North Africa. In court Jews could not testify against Christians unless they were testifying in the name of the Byzantine state against an individual. The state claimed the right to interfere in Jewish religious courts as a matter of projecting universal power while banning the study of the Mishna and the reading of Torah in Hebrew and the "Shema" (seen as a nullification of the Trinity) in an effort to weaken loyalties and the specific characteristics of the community. Together such restrictions constituted the Servitus Judaeorum or "Servitude of the Jews." Lax enforcement often undermined the

desired effect and even reinforced social cohesiveness as the Jewish community battled for its religious and national heritage.

24 "Omar, Covenant of," *Encyclopedia Judaica*, Jerusalem, Keter, 1972, Vol XII, columns 1378-1379.
25 Chouraqui, Andre N., *Between East and West, A History of the Jews of North Africa*, Temple Books, Atheneum, New York, 1973, p. 46.
26 Lewis, Bernard, *The Arabs in History*, Harper and Row Publishers, New York, 1966, pp. 57, 77 and 78.
    Bat Ye'or, *The Dhimmi*, Doc. 5, "Jizya and Kharaj," pp. 175–180.
27 Bat Ye'or, *The Dhimmi*, p. 55.
28 Bat Ye'or, *Islam and Dhimmitude*, pp. 42-45.
29 Ibid, pp. 69-71.
30 Ibid.
    This is the overall major point made by Bat Ye'or. The jizya tax funded Jihad.
31 Ibid, pp. 50-51, 59 and 65.
32 Ibid, 89-90, 103-104 and 106-107.
33 Ibid, pp. 74-75 and 89.
34 Gilbert, Martin, *In Ishmael's House, A History of Jews in Muslim Lands*, Yale University Press, New Haven and London, 2010, pp. 66-68.
35 Ibid, pp. 67-68 and 85.
36 Bostom, *Antisemitism*, p. 33.
    Bostom quotes the 20th century Islamic scholar Mohammed Sayyid Tantawi.
37 Ibid, p. 64.
38 Bat Ye'or, *The Dhimmi*, Doc. 94, "Forced Conversions and Degradations," pp. 351-352.
39 Ibid, Docs. 84 and 85, "Forced Converts and Conditions of the Jews (1850)," and "Servitudes in Persia", pp. 331-336.
40 Bostom, pp. 141-147.
    "Persia," *Encyclopedia Judaica*, Vol XIII, Columns 301-319.
    Goitein, p. 81.
41 Bat Ye'or, *The Dhimmi*, Doc. 86, "Official Edicts of Protection (1875 and 1897)," p. 337.
42 Ibid, Doc. 85, "Servitudes in Persia," p. 336.
43 *London and Kirshenbaum*, Israel TV, Channel 10, June 6, 2010.
44 Ahroni, Reuben, *Yemenite Jewry: Origins, Culture and Literature*, Indiana University Press, 1986, pp. 114-117.
45 Ibid, p. 115.
    In reference to Judge Mohammed al-Sahuli's explanation of the "Latrines and Scrapers Edict."
46 Bat Ye'or, *The Dhimmi*, Doc. 88, "Expulsion of the Jews from San'a (1666)," pp. 339-340 and Doc. 99, "Trials and Sufferings in Yemen (1666)," pp. 361-364.
47 Aharoni, pp. 121-135.
48 Bat Ye'or, *The Dhimmi*, Doc. 90, "Edict Promulgated by the Imam Yahya of Yemen (1905)," pp. 340-341.
49 Ibid, Doc. 91, "Behavioral Distortions Resulting from Oppression (1910)," pp. 341-343, Doc. 107, "Return of the Exiles to Zion," pp. 376-382.
    Goitein, pp. 76-77.
50 Rejwan, Nissim, *The Jews of Iraq: 3000 Years of History and Culture*, Widenfield and Nicolson, London, England, 1985 pp. 155-160.
    "Iraq," *Encyclopedia Judaica*, Vol VIII, columns 1444-1461.
51 Goitein, p. 83.
52 Bostom, pp. 39-40.
53 Chouraqui, pp. 38-41.
54 Ibid, pp. 42-54 and 173.

55   Bat Ye'or, *Dhimmi*, Doc. 65, "Defenseless Dhimmis of Morocco (19th Century)," pp. 305-307 and Doc. 73, "Refusal to Emancipate the Dhimmis in Morocco," p. 317.
56   Bostom, pp. 48-49.
57   Bat Ye'or, *Dhimmi*, Doc. 58, "Sack of the Jewish Quarter of Fez (1820)," pp. 293-294.
58   Ibid, Doc. 59, "The Dhimma in Algeria and Morocco (early 19th century)," pp. 294-299.
59   Eisenhower, Dwight D., *Crusade in Europe*, Doubleday, Garden City New York, 1948, p. 129.
60   Ibid, p. 108.
61   Ibid, p. 128.
62   For a survey of the topic see Michel Abitbol, *The Jews of North Africa During the Second World War*, Wayne State University Press, Detroit, Michigan, 1989.
63   "Libya," *Encyclopedia Judaica*, Vol XI, column 202.
64   "Testimonies by North African Jews," displayed in the *Yad VaShem Holocaust Museum and Documentation Center*, Jerusalem, Israel.
65   For a brief yet compelling overview of this period see Gilbert 198-250. Several are noted here.
66   Libyan Jewish historian Maurice Roumani in Gilbert p. 286.
67   Ibid, pp. 282-310.
68   Bat Ye'or, *Dhimmi*, Doc. 112, "Jihad in Modern Times (1968)," p. 390-396.
69   Ibid.
70   Peters, Joan, *From Time Immemorial, the Origins of the Arab Jewish Conflict Over Palestine*, JKAP Publications, USA, 1984, pp. 145-147.
71   Safrai, Zeev, *The Galilee in the Time of the Mishna and the Talmud*, (Hebrew), Ministry of Education and Culture, Israel, 1985.
      See the entire book as a source. Any modern map of the Galilee clearly shows Hebrew roots from the Talmudic and Second Temple Periods in the names of today's Arab villages.
72   Peters, p. 85.
73   Kohen, Elli, *A History of the Turkish Jews and Sephardim: Memories of a Past Golden Age*, University Press of America, Maryland, 2007, p. 18.
      Bornstein-Makovetsky, Leah, "Suleiman I," *Jewish Virtual Library*, retrieved May 5, 2011, https://www.jewishvirtuallibrary.org/jsource/judaica ejud_0002_0019_0_19345.html.
74   Shor, Natan, *Toldot Tzfat*, (Hebrew), Dvir Company and Am Oved Publishers, Israel, 1983, pp. 182-198.
75   Bat Ye'or, *Dhimmi*, Doc. 100, "Description of the Status of Non-Muslims in Palestine (1700)," pp. 364-366.
76   Peters, "Dhimmi in the Holy Land," pp. 172-195.
      Bat Ye'or, Doc. 103, "Jews of Palestine before 1847," pp. 371-372, Doc. 31, "Visit to the Jews of Hebron (1836)," pp. 222-223, Doc. 33, "A Visit to Israel's Holy Places (1839), pp. 225 – 226, and Doc. 35, "Nineteenth Century Ottoman Palestine," pp. 228-241.
77   Quoted in Peters, p. 187.
78   Ibid, p. 188, italics in the original quoted source.
79   Goitein, p. 88.
80   Antonius, George, *The Arab Awakening*, Capricorn Books, New York, (1946) 1965, Chapters 1-3, pp. 13-61.
81   Bat Ye'or, *Dhimmi*, pp. 103-108.
82   For a review of the collapse of the Ottoman Empire and development of modern Turkey see Lewis, Bernard, *The Emergence of Modern Turkey*, Oxford University Press (Third Edition), New York, 2002.

Also see Erik J. Zurcher, *Turkey: A Modern History*, I.B. Tauris, London, 1997.
83   Ibid, pp. 93-97 and 103.
84   Bard, Mitchell G., *Myths and Facts Online*, "Arab/Muslim Attitudes Toward Israel," retrieved May 21, 2011, www.jewishvirtuallibrary.org/jsource/myths3/MFattitudes.html.
85   Bat Ye'or, *Islam and Dhimmitude*, pp. 113-117.
86   Ibid, pp. 134-141.
87   Influenced by French antisemitism in the late 19th and early 20th centuries were those Christian Arab intellectuals as represented by Naguib Azouri and George Antonius.
88   "Assyrian Genocide," *Wikipedia*, en.wikipedia.org/wiki/Assyrian_genocide. *Assyrian Genocide Research Center*, Rutgers University, www.ncas.rutgers.edu. Both retrieved July 8, 2015.
     Much more information can be found at the Assyrian Genocide Research Center.
89   Bat Ye'or, *Islam and Dhimmitude*, pp. 148-152.
90   This antisemitic document forged by the Czarist secret police is referred to as *The Protocols of the Learned Elders of Zion* or *The Protocols of the Elders of Zion*. Either designation is correct.
91   Ibid, pp. 134 and 152.
92   Rejwan, pp. 215 and 230-235.
     Sassoon Heskel was minister of finance in the 1920s, Dawood Samra served on the High Court of Appeals in the1930s-40s, while Menachem Daniel and his son Ezra Menachem Daniel represented the Jewish community in the Iraqi parliament during much of this period. Furthermore Jews were very active in the Iraqi communist party during the 1940s. Due to antisemitic outbreaks of violence some 90% of Iraqi Jews came to Israel in the early 1950s.
93   This is a continual theme underscored by Bat Ye'or when researching most Eastern Christian Churches, their clergy and policies.
94   Waqf lands in general are "endowed lands" or any territory previously captured by Islam. In this specific instance of the Temple Mount/Noble Sanctuary domain "waqf" refers to the institutionalized mosque holdings. In this case "waqf" properties are similar in definition to "church" properties in Christian societies.
95   Bat Ye'or, *Islam and Dhimmitude*, pp. 153-154.
96   The Wahhabist uprising of the early 19th century is reminiscent of the Islamic State (ISIS or ISIL) destruction reigned on Shiites and others in the early 21st century even if the Islamic State did not originate in northern Arabia.
97   Thompson, Jason, *A History of Egypt*, American University in Cairo Press, Cairo, Egypt, 2008, pp. 219- 234.
     Antonius, p. 22.
98   Antonius, pp. 21-34.
     "History," *Jews of Egypt Foundation,* retrieved October 14, 2011, www.jewsofegyptfoundation.com.
99   Dawidowicz, Lucy, *The War Against the Jews 1933 – 1945*, Bantam Books, New York, 1975, quoted on p. 29.
100 Ibid, p. 41, quoted in Dawidowicz.
101 Bostom, *Antisemitism,* pp. 62-63.
     Cook, David, "Anti-Semitic Themes in Muslim Apocalyptic and Jihadi Literature," *Jerusalem Center for Public Affairs*, 2007, retrieved February 29, 2016, http://jcpa.org/article/anti-semitic-themes-in-muslim-apocalyptic-and-jihadi-literature
     "Quran, Hadith and Scholars," *Kebenaran Islam* (Truth Islam), retrieved August 2, 2015, http://mengungkapkebohonganislam.blogspot.com/search?q=Hadith+and+Scholars
102 "Qaradawi Predicts a Muslim Apocalypse," *Jewish Chronicle Online*, May 2008,

retrieved November 10, 2011, http://www.thejc.com/news/uk-news/1770/qaradawi-predicts-a-muslim-apocalypse.
103 "Qaradawi, Yusuf," *Wikipedia*, retrieved November 10, 2011, en.wikipedia.org/wiki/Yusuf_al-Qaradawi.
104 Tamimi, Azzam, *Hamas A History from Within*, Northampton Press, Mass, USA, 2007, pp. 177-186.
105 "Toynbee, Arnold," *Wikipedia* retrieved November 11, 2011, en.wikipedia.org/wiki/Arnold_J._Toynbee.
 The eminent British historian saw Jews as irrelevant at best.

Map of the Caliphate as demanded by the Islamic State of Iraq and the Levant

This map only constitutes the borders stipulated by "Defensive Jihad," which is the first step toward world domination. Once achieved, "Offensive Jihad" captures the remaining parts of the globe and imposes Sharia law on all humanity.

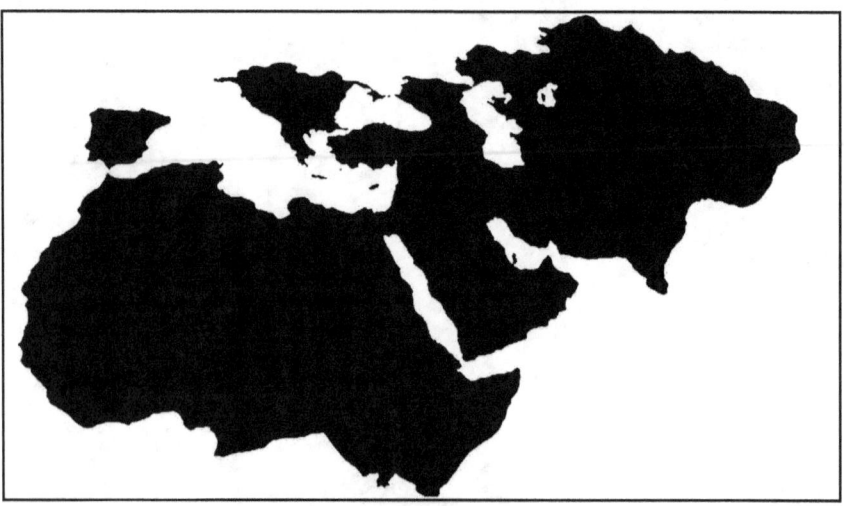

Credit: Modified from map by Débora Cabral, Wikimedia Commons

II

Arab Islamist Ideologues - Jihad Past & Present

Hasan al-Banna, Sayyid Qutb, and Abdullah Azzam

Overview

This chapter analyzes the theories and Jihadi activities of the three foremost Islamist ideologues, leaders who use Islam as a political and military tool: the Egyptians Hasan al-Banna (1906-49) and Sayyid Qutb (1906-66), and the Palestinian Abdullah Azzam (1941-89). All are at the heart of the Arab Middle Eastern Islamist movement. This troika established the intellectual and theological basis for Jihadist activities, which are determined to crush Western civilization and destroy Jewish existence. While there are other Jihadists of influence, their role is less central to the discussion at hand when addressing the birth, continuity and empowerment of such a movement in the Arab Muslim world. The Indian Sayyid Abul A'la Maududi and the Iranian Ruhollah Khomeini are part of the overall development of this Jihadist, anti-secular, anti-Western and antisemitic movement, yet in this work these two non-Arab Muslims are only mentioned in a peripheral manner. Their influence is taken into consideration, but they serve as lesser role models for Arab Islamists and have less direct involvement in the development of Hamas, as opposed to the other three thinkers. This does not detract from their Jihadist or antisemitic ideological impact. Other Arab Islamists include today's leading Egyptian Muslim Brotherhood theologian, Yusuf al-Qaradawi, the Grand Mufti of Jerusalem Haj Amin el-Husseini of yesteryear, and the modern Hamas leadership led by the late Ahmed Yasin alongside today's present political leader and often prime minister of Gaza, Ismail Haniyeh. These ideologues will be discussed elsewhere as their perspectives were either not foundational, or they were involved more in taking Jihadi action than in developing theological directives.

Part I
Hasan al-Banna
Founding the Muslim Brotherhood, Activating Jihad

Jihad and Arabization

Sheikh Hasan al-Banna (1906–1949) founded the Muslim Brotherhood in Egypt in 1928, and today's loyalists still hold him in great reverence. He was initially revered more for his opposition to secularism and Western influence in the Muslim world than for his short-lived, anti-Israel diatribes. Viciously anti-Israel and antisemitic, he lived to see the establishment of the Jewish State much to his dismay. During WWII, he supported Hitler's Nazi Germany and Mussolini's fascist Italy, advocating a German victory over the British in Egypt. The Brotherhood translated Hitler's *Mein Kampf* into Arabic and spread its message along with the Czarist *Protocols of the Elders of Zion*. Apparently, the Egyptian regime of King Farouk murdered al-Banna in February 1949 after the Brotherhood assassinated numerous officials. Their victims included Interior Minister Amin Othman Pasha, Chief of Police Selim Zaki Pasha, Chief Justice Ahmed al-Khazindar, and two prime ministers, Ahmad Maher in 1945 and Mahmoud an-Nukrashi Pasha in December 1948 after the latter attempted to repress Brotherhood activities.[1]

For al-Banna, Jihad was a religious command, even more than most sanctified obligations in Islam. Rabidly anti-Western and antisemitic, he urged believers to return to basic fundamentalist Islamic understandings. In his essay "On Jihad," he defined the sanctity of Holy War: "Jihad in its literal significance means to put forth one's maximal effort in word and deed; in the Sacred Law it is the slaying of the unbelievers, and related connotations such as beating them, plundering their wealth, destroying their shrines, and smashing their idols..." Responsibility lies with all believers: "It is initiated by us as a communal obligation, that is, it is obligatory on us to begin fighting with them after transmitting the invitation [to embrace Islam], even if they do not fight against us."[2]

Sheikh or Imam al-Banna quoted the above as an interpretation of Jihad with which he wholeheartedly agreed. Jihad, as he defined it for the Muslim Brotherhood, meant constant war, slaughter, abuse and plunder of the non-believers. Al-Banna influenced the Islamist theologian and intellectual Sayyid Qutb, who assumed leadership of the Egyptian Muslim Brotherhood in the 1950s and sharpened its anti-Western and specifically antisemitic message. Qutb, although not mentioned directly in *The Hamas Covenant*, profoundly influenced its antisemitic message and demands for Jewish elimination, as is shown below. Both men influenced Abdullah Azzam who had a direct impact in the formulation of *The Hamas Covenant.*

Evasion from active participation in Jihad constituted a major sin and guaranteed annihilation for those refusing to undertake the responsibility of a physical Holy War against one's enemies. Toward the end of "On Jihad," in a sub-heading entitled "Supplements to Jihad," he disabused everyone of the notion that Jihad can be less than the physical destruction of Islam's enemies. He mocked those who spoke of the "jihad of the heart," or the "jihad of the spirit," accusing them of cowardice. He described Jihad as a "struggle with the spirit so that it may be sincerely devoted to God in every one of its acts. So let it be known." For al-Banna, Jihad was the ultimate act, there could be no "supreme martyrdom and the reward of the strivers in jihad, unless he slays or is slain in the way of God." As if his words were insufficient, he concluded the Epilogue by declaring, "If you strive for an honorable death, you will win to perfect happiness. May God bestow upon us and upon you the honor of martyrdom in His way!"[3] Today, al-Banna's perspectives are among the most important in the Arab/Muslim world. His spirit permeates the Muslim Brotherhood, Salafists, al-Qaeda and the Islamic State (ISIS/ISIL). His vision is integral to Hamas and its *Covenant.*

Fully aware of Islamic military history, he recounted the victorious march of the Jihad of yesteryear in his tract "Between Yesterday and Today." Believers invoking "Koranic principles" brought about the defeat of "superstitious idolatry in the Arabian Peninsula and Persia," and "warred against guileful Judaism," which was a first step in greatly limiting its influence. A more difficult challenge was the battle against Christianity, weakening it in Asia and Africa and later confining it to Europe. Islam swept across North Africa, captured Spain and reached as far as France and Italy. By the mid fifteenth century, Constantinople was overrun. With Christianity restricted to central Europe, "Islamic fleets plowed the deeps of the Mediterranean and Red Seas, and both became Islamic lakes." Not only was there Islamization of conquered people, but, "They Arabized them, or succeeded in doing so to a degree, and were able to sway them and convert them over to their language and religion because of their splendor, beauty and vitality."[4]

For al-Banna, adopting Islam was not sufficient; one must be Arabized to achieve fulfillment and perfection. Transferring authority to non-Arabs, such as Persians, Daylamites, Mamlukes, Turks or anyone else, could never be as pure and successful because "they had never absorbed genuine Islam," and "had never been illuminated with the light of the Koran because of the difficulty they encountered in trying to grasp its concepts." For true Jihad to succeed, the Arabs needed to initiate and bring victory through Holy War, otherwise its message would be lost. By definition inferior teachers are unable to incorporate the vanquished with the Divinely revealed faith. Al-Banna then quoted the Koranic verse 3:118 warning the Arabs not to confide in "those who are not of you."[5] This is discussed in the Introduction to *The Hamas Covenant*. (See Chapter VI "*The Hamas Covenant* Analysis")

In the tract "Our Mission," under the heading "The Characteristics of Arabdom," al-Banna at first tried not to denigrate non-Arabs, but insisted Arab traits must be adopted "as a means of realizing the foremost task for which every people is responsible—the renaissance of humanity." Only an Arab/Muslim world was capable of reaching perfection. However if Arab/Muslim superiority is an eternal fact, why was the Arab world in such an inferior position in the first part of the twentieth century? He explained that science, rationalism and the resulting secularism defeated the Church in Europe. Through this export of corruptive materialism and permissiveness, European chauvinism as seen in World War I put the Muslim world on the defensive. The West continued to be victorious because the Muslim world was ignorant of its own religion. Of necessity, the first step in assuring victory was the re-education of Islamic people.

Hence, al-Banna set out on his Islamic mission to address several "General Aims." The "Islamic fatherland [must] be freed from all foreign domination" while "a free Islamic state" was to arise "acting according to the precepts of Islam" and "broadcasting its sage message to all mankind" in every region containing Muslims and Arabs. He called for the betterment and cleansing of society, especially in Egypt, where Europeans dominated and the people suffered from crime, hunger and disease while foreign companies monopolized the economy, most notably utilities. The solution lay in the expulsion of foreigners and the extension of Islamic law achieved through "Deep Faith," "Precise Organization" and "Uninterrupted Work." With great expectations of success, al-Banna proclaimed this battle against European influence to be "the loftiest of all missions," an Islamic system that offered "the Sacred Law of the Koran." The first step began with the internal revolt and cleansing within Islam. In the next stage, it was offered to other non-Islamic peoples in what was seen as the export of the revolution. Some could consider such moves against secularism in the Arab world today as a form of internal societal Jihad. The true universal Jihad then followed.[6]

Al-Banna spread the mission of Islam not just for the sake of humanity, but as an obligation "propagating it among men with argument and proof. But if they should persist in rash acts, outrages and rebellion, then with the sword and the spear!" Essentially there were three choices: "Islam, tribute, or combat." Adopting Islam made one a brother; tribute, or *jizya*, meant subjection to second-class *dhimmi* status, which Jews and Christians who would not convert willingly accepted. Combat referred to those who were neither Jews nor Christians and rejected Islam. Al-Banna bewailed the lack of the Jihadist credo in modern times, accusing his fellow Muslims of seeing Jihad as something from the past or making excuses concerning the ability of their societies to wage a Holy War. He accused others of giving lip service, but doing nothing. He called for an awakening of all Muslims to Jihad, whether implemented communally or individually.[7]

Jihad was a necessity and demanded implementation as a religious obligation; one who refused to follow Jihad committed a major sin of omission, carrying with it a guarantee of annihilation. Imposed warfare was unique to Islam. One battled for Allah in this world and thereby attained eternal life in the world after death.[8] As proof, al-Banna quoted two Koranic verses, 3:169-3:170 and 4:74 respectively. "Do not reckon those who have been slain in God's way as dead. Nay, rather they are alive with their Lord, granted sustenance, gladdened by the bounty God has given them, and rejoicing for those who have not yet joined them and remain behind, in that no fear shall be upon them, nor shall they grieve." And, "So let those fight in the way of God who sell the life of this world for the next, and he who fights in God's way and is slain or overcomes, We shall provide him with a mighty wage."

Al-Banna condemned the "slackers, cowards, truants, and opportunists" and urged them to change their course of action. Allah "encourages the fearful to the utmost degree to plunge into the uproar of battle and to face death unflinchingly and bravely, showing them that death must overtake them in any event, and that if they die waging jihad, they will receive the most magnificent recompense for their lives." The call to martyrdom in the name of Allah was clear. The martyr was assured a place in heaven with all its rewards, while anyone who did not participate in Jihad was flawed, a hypocrite, suffered the stigma of humiliation, and could only redeem himself by embracing this true religious understanding.[9]

In the final justification of his case, al-Banna reviewed the understanding of Jihad by legal specialists. The sanctified destructiveness of Jihad was so essential that he once again quoted the work of Hanafi jurisprudence mentioned above, "The Sacred Law it is the slaying of the unbelievers, and related connotations such as beating them, plundering their wealth, destroying their shrines, and smashing their idols." More significantly, a committed Jihadist went to battle after offering an invitation to accept Islam, even if there was no hostility on the part of those who rejected Islam as their way of life. It was considered a declaration of war against the Muslim world should a person, a group or a nation refuse to embrace Islamic beliefs. Neutrality did not exist.[10] This was and is the pinnacle of offensive Jihad.

For al-Banna, should Islamic lands, also called the *waqf*, ever fall under non-Muslim control, it became not only a communal obligation to declare Jihad, but an individual duty. The duty even applied to women, slaves, children and debtors, all of whom need not ask anyone's permission for responding to the mission to expel non-Muslim conquerors. A text from the Shafi'ite School proclaimed Jihad as a communal obligation, even when non-believers remain in their own territory. As explained above, such action became a personal necessity for everyone should outsiders invade Islamic territory. Although he quoted different commentators, al-Banna found the common denominator for sending troops to Jihad at least once, and possibly twice a year. To quote al-Banna's conclusion concerning defensive Jihad, he stated Muslims must:

> . . . agree unanimously that jihad is a communal obligation imposed upon the Islamic *umma* in order to broadcast the summons [to embrace Islam], and that it is an individual obligation to repulse the attack of unbelievers upon it. Today, Muslims, as you know, are compelled to humble themselves before non-Muslims, and are ruled by unbelievers. . . . Hence it has become an individual obligation, which there is no evading, on every Muslim to prepare his equipment, to make up his mind to engage in jihad, and to get ready for it until the opportunity is ripe and God decrees a matter, which is sure to be accomplished.[11]

The need for Jihad was explained in the sub-section "Why Do the Muslims Wage War?" It is "not as a tool of oppression," but "as a defense for the mission [of spreading Islam], a guarantee of peace, and a means of implementing the Supreme Message." When all live under Islam there will be peace since no reason will exist to wage a Holy War. The stipulations for declaring Jihad to regain *waqf* lands held by infidels brought in personal obligation and could be defined as a defensive Jihad; the capture of new lands is understood as offensive.[12] Al-Banna blended defensive and offensive Jihad and minimized the distinction between the two approaches.

In attempting to sound more humane, he approached certain moral issues of life and death in his limited sub-chapter entitled "Mercy in the Islamic Jihad." Al-Banna demonstrated supposedly gracious behavior in Islam by declaring it forbidden to kill women, children, old men, the wounded, or to disturb monks, hermits or any peaceful individual. Almost written as a disclaimer to the brutality of Jihad, this short explanation stands in contradiction to what was written both before and afterward. All these non-Muslims were to be forced to convert to Islam or accept a subservient status as *dhimmi* subjects. Death awaited them should they refuse.[13]

Jihad Today – The Battle Against the West

Today's Jihadists see their most strenuous battles to be against the West, meaning Jews and Christians, also called the "People of the Book." Al-Banna continues to serve as the Jihadists' inspiration. He used numerous quotes from the Koran and later Islamic sources all directed at the modern Muslim reader, yet one quote in particular continues to stand out. Al-Banna explained the "clear indication of the obligation to fight the People of the Book, and of the fact that God doubles the reward of those who fight them, jihad is not against polytheists alone, but against all who do not embrace Islam."[14] This subsection is titled "A Sampling of Prophetic Traditions on Jihad." It ranges over eight pages and is replete with quotes from the Koran and other sources. It ends with a reminder of the identity of the true enemies Islam faces in modern times: "there are precious Traditions on this subject and the like, as well as on campaigning by sea and its manifold superiority over campaigning by land, and on campaigning against the People of the Book."[15] What we find was a return to the Islamic glory of the Middle Ages and early modern period described above with special emphasis on sea power. Islam defeated Christianity before and will again. Muslims would be rewarded twice over in a double martyrdom for their Jihad against the West, leading the believer to understand the necessity of initiating hostilities in the name of Allah and Islam.

Today's Jihadists perceive a world divided in two, *Dar al-Islam* and *Dar al-Harb*, the abode of Islam and the abode of War. The only other legitimate mode of interaction is that of *hudna* or an Islamic cease-fire.[16] Under *Dar al-Islam* all live in submission to devout Islamic regimes loyal to Allah, such as a reestablished Caliphate or a theocratic Islamic state. *Dar al-Harb* is the institutionalized conflict against infidels, polytheists and idolaters including atheists, on the one hand and against any People of the Book ruling *waqf* lands, meaning Jews and Christians (as discussed in Chapter I "Negative Image of the Jew"). Islam captured many lands over the centuries, which are no longer in the Muslim domain today but rather held by Jews, the example being the State of Israel, or Christians who rule the Iberian Peninsula and most of the Balkans for example. These circumstances form a compelling reason for an immediate Holy War. This is the "defensive Jihad," as Abdullah Azzam explained. Once all infidels are defeated and *waqf* lands retrieved, the second stage of Jihad is implemented: the conquest of the remainder of the world, known as "offensive Jihad" as strongly advocated by Sayyid Qutb. In obedience to Divine dictate, Islam must spread, encompass all and dominate the world. Even those who declare themselves neutral are not immune if they fail to accept their place in the world of Islam, either as converts or in the *dhimmi* status. They too are the enemy. For the Muslim believer, Islam is the most perfect world order and must be spread by force if necessary, ensuring universal peace in the future and for all eternity.

The issue of *hudna* or an Islamic cease-fire is pivotal nowadays. Many consider it a cease-fire as in the Western understanding, which is an arranged truce to arrive at conflict resolution between two warring factions. Nothing can be further from Jihadist intentions. A *hudna* is arranged to gain breathing space for the Islamic side in order to recruit, retrain, reload, rearm and re-initiate hostilities at the time and place most fitting for a Jihadist victory. The *hudna* works almost exclusively to the advantage of the Jihadist side, since it is forbidden to accept a cessation of hostilities when they have the upper hand. Only if the Muslim side is weakened to the point of possible defeat may they accept a *hudna*. Islamists like al-Banna view Western Christian society as weak at its core, since secularism triumphed over the true belief in God. Islamists are convinced Westerners are not willing to fight and die for their beliefs, as Muslims are. They think Westerners do not believe in the Creator or the Almighty as the ultimate force of existence. To the Islamist, such a lack of commitment to Allah makes Westerners spiritually inferior and unable to persevere against Islam in the final conflict. Furthermore, even should they be religious, Islam is the only complete truth Allah gave and victory is its Divine destiny. *Hudna* can only be a temporary status and there is no intermediate area between war and peace, only a lull in the ever-continuing Islamic Holy War.

In the predestined final accounting, all people will live under Islam—the perfect society in its legal, economic, judicial and political expressions. As envisioned by Hasan al-Banna, Allah the Eternal will rule as the omniscient power over all humankind through his emissaries the Muslim Brotherhood Jihadists.

Part II
Sayyid Qutb
Theological Father of *The Hamas Covenant*

Jahiliyyah and Jihad

Although not mentioned in *The Hamas Covenant* by name, the Egyptian-born Muslim Brotherhood leader and Islamist Sayyid Qutb (1906-1966) heavily influenced the contents of the document, possibly more than any other thinker. After having studied in the US (1948-50), he became furiously anti-Western and returned to Egypt convinced of American decadence, in particular as related to sexuality. An ardent Islamist, he became the sworn enemy of the revolutionary secular Egyptian Arab nationalist regime of Gamal Abdul Nasser. He was eventually jailed and executed for his opposition. Hasan al-Banna and Abdullah Azzam made their contributions, but none left such a deep imprint on Jihadist commitment to complete Islamist worldwide victory as did Sayyid Qutb. No less significant is the simultaneous call to organize Muslims throughout the world for a universal war of destruction against the Jews. Such hatred, which fell somewhere between the persecutions of Czarist Russia and the full Nazi scheme for Jewish extermination, fueled the struggle against the Jews. Qutb was a prolific writer, particularly on internal Islamic issues. Here we will deal with his overall Islamist/Jihadi world-view and his two works most pertinent to *The Hamas Covenant*: *Milestones*, and his short yet incendiary antisemitic essay "Our Struggle With the Jews."

In his famous work *Milestones*, Qutb laid out the contours of the global conflict between the world of Allah's Islamic revelations as embodied in the Koran and the Divinely written Sharia law, which true Muslims follow. He contrasts the "true" Islamic world with other societies known as the "*jahili*," or "*jahaliyyah*," meaning "ignorant ones." Qutb presented the cosmic battle of the diocentric God-driven world against all others. The Messenger, the Prophet Mohammed, revealed the religion of Islam. Mohammed recited the Koran to enlighten humankind and bring all people under Allah's rule as Islam defines it. The Koran represented heavenly purity as a complete way of life, demanded obedience and called for action in the name of Allah.[17] The term *jahaliyyah* originally referred to the era before the advent of Islam in the early 600s CE. The dating of the Islamic calendar began with the *Hejira* or the first Muslims' emigration to Medina in 622 CE, forced to leave their hostile home-town of Mecca after the people there rejected Mohammed's message. The moment Mohammed spoke Allah's message, the world divided into two, which led to a universal clash between fundamentalist Islam, and the ignorant ones or *jahili*. The first thirteen years of the original Muslim community embodied Islamic purity and perfection[18] as they fought to protect their life from the pollution by the *jahiliyyah*.[19] This conflict continues to the present day.

Qutb's Islam was a religion of universal conquest including both people and land. One may either submit to Islam, or die at the hands of Muslims fighting a Holy War in the name of Allah's Divine will. According to Qutb, true Muslims were Allah's hand for implementing the Divine will. At the time of his writing in the 1950s, all nations and virtually the entire world were *jahili*, including those with a majority Muslim population. Qutb called for an Islamist war by the true believers and, despite the odds, he had no doubt they would emerge victorious. They were the God-fearing community demanding obedience to the invincible Allah who has all-encompassing power. Hence true believers must wage eternal Jihad until they reach final victory.

The Jahili in Other Cultures: East, West, Judaism and Christianity

Qutb saw all other cultures as inferior, beginning with the ancient worlds of Greece, Rome, India and China.[20] He believed no secular culture had the right to rule humankind, only Allah had that right. Man-made laws were completely powerless since they derived from a source other than Islamic beliefs,[21] which were superior to "valueless" man-made theories. Secular or anthropocentric human-centered societies had no right to exist and must be forced into the diocentrism of Islam. Anything other than the true worship of Allah and obedience to His Sharia law was "*shirk*,"[22] or the worship of others, to whom humans assigned godly attributes and perceived powers. Islamic loyalists were to be separate and dedicate themselves to Allah, never giving loyalty to any *jahili* system. Qutb believed the seventh century Christian Byzantine Empire was defeated because it was inferior to Islamic society. Muslims had warned the Zoroastrian Persians to convert; they refused and suffered destruction. The war was one of forced conversion and not intended for material well-being. As we know Iran, or Persia, is Muslim today. The superiority of Islam assures its final victory.[23] Worse yet were the idolaters in India, Japan, the Philippines and Africa who do not recognize Allah's existence at all.[24] Qutb inferred the command to conquer and educate these people in the ways of Islam.

Though the Crusaders of the Middle Ages suffered defeat, for Qutb the Crusading spirit remained in its continual efforts to influence today's Muslim world through Christian and most recently, Enlightenment values. In the modern era, the empires of Spain, Portugal, France and Britain all oppressed and exploited others through man-made laws, which allowed for inequality based on class and materialism.[25] The United States was the vilest of enemies, which Qutb first wrote in 1950 and intensified afterward, where *jahili* values such as materialism and elected officials dominated. In the US, religion was corrupt and Allah had no place. The Western Crusading spirit including Enlightenment values such as equality, still existed and invited defeat. Allah's victory would only actualize at some future date when all Muslims fought a Jihad under the victorious banner of Islam. Selfish animal instincts such as food, shelter and sex, drove the world of humans with no Divine authority to guide it into these base material and violent needs. Islam was the answer. Unbelievers who refused to accept the true revealed religion would go to Hell, drowning in filth, regardless of any good deeds done on earth.[26]

In the first page of his "Introduction," Qutb lambasted democracy as a failure, citing the capitalist West's need to borrow from the communist East's value system, which excelled in opposing human nature. Communism and capitalism were equally condemnable rebellions against Allah because their roots were in science where human beings assembled their theories. Man-made understandings would always fail, because such men worshiped materialism and were *jahiliyyah*, a revolt of ignorance against Divine rule on earth.[27] Islam was the only answer. Although having material possessions was legitimate, materialism and communism in particular stood in opposition to the non-animalistic morality of Islamic family values. Material values were *jahili* and backwards.[28] Atheist communism was among the most polluted of all systems; its belief in party principles resulted in the limiting of choice and human freedoms.[29] At times, East and West material systems might appear to have something in common with Islam, but commonalities were only coincidence and should not fool anyone. Islam demanded submission to the Creator and not the quenching of selfish desires. To quote Qutb, "The tree of Islam has been sown and nurtured by the wisdom of God, while the tree of *jahiliyyah* is the product of the soil of human desires."[30]

In Qutb's world there was no such thing as "Islamic Democracy" or "Islamic Socialism." Both West and East were contemptuous and to be rejected. Man needed to embrace "far reaching change—change from the ways of the created to the way of the Creator, from the systems of men to the system of the Lord of men, and from the commands of servants to the command of the Lord of servants." There could be no compromise, reform or gradualism when confronting the worthless, corrupting *jahili* system. One would only deceive himself in doing so. There were too many Muslims with defeated mentalities taking a defensive approach in confronting the East and West.

One did not need to seek "justification and apology" when "presenting Islam to people." Islam was permanent and would prevail over all.[31]

Qutb acknowledged that Jews and Christians, as People of the Book, or *dhimmis*, had a better starting point than atheists or materialists. They began as believers in God, but quickly descended into the *jahiliyyah* as other cultures influenced them. *Dhimmis* were not trustworthy, and Allah gave the Koran to the first generation of Muslims.[32] Qutb accused *dhimmis* of worshiping priests and rabbis, no longer being subservient to the Almighty, but rather to other men.[33] The Koran notes this in 9:31, "They worship their rabbis and their monks, and the Messiah the son of Mary, as gods besides Allah; though they were ordered to serve one God only. There is no god but Him. Exalted be He above those whom they deify beside Him!" As well, in Koran 9:34, which states, "believers, many are the rabbis and the monks who defraud men of their possessions and debar them from the path of Allah." Koran 9:29 stipulated that *dhimmis* were allowed to survive, provided they paid the *jizya* or head tax.[34] "Fight against such of those to whom the Scriptures were given as [they] believe neither in Allah nor the Last Day, who do not forbid what Allah and His apostle have forbidden, and do not embrace the true faith, until they pay tribute [jizya tax] out of hand and are utterly subdued."

True Muslims were ordered to battle against the People of the Book for the above reasons as quoted from the Koran. Furthermore, not only was it an affront to Allah that Christians worshipped Jesus as the Messiah, but Jews were accused of deifying Ezra the Scribe as the son of God in Koran 9:30. This ridiculous charge has absolutely no basis in Judaism. Qutb brought all of these thoughts together and came to a "logical" conclusion declaring all *dhimmis* must pay the *jizya* tax even if they did not live under an Islamic regime. Hence, "It may happen that the enemies of Islam may consider it expedient not to take any action against Islam, if Islam leaves them alone in their geographical boundaries to continue the lordship of some men over others and does not extend its message and its declaration of universal freedom within their domain. But Islam cannot agree to this unless they submit to its authority by paying Jizyah, which will be a guarantee that they have opened their doors for the preaching of Islam and will not put any obstacle in its way through the power of the state."[35]

Jews and Christians lived in *jahili* societies and must acquiesce to Islamic rule. Should they refuse, they were to be conquered, even if peaceful and not attacking Muslims or Islam. Muslims accused them of having other laws beside submission to the will of Allah. Both religions were accused of *shirk*, or "the association of other gods with God," and of "unbelief," in addition to allowing their clerics to rule them. Christians rebelled against the Oneness of Allah by believing in the Trinity while Jews revolted against Allah's "prescribed way of life."[36]

Most fascinating was Qutb's critique of American society in 1950. This quote, although extensive, is in full, as few can accurately paraphrase his anti-American revulsion as well as the original states it.

> During my stay in the United States, there were some people of this kind who used to argue with us - with us few who were considered to be on the side of Islam. Some of them took the position of defense and justification. I, on the other hand, took the position of attacking the Western Jahiliyyah, its shaky religious beliefs, its social and economic modes, and its immoralities: "Look at these concepts of the Trinity, Original Sin, Sacrifice and Redemption, which are agreeable neither to reason nor to conscience. Look at this capitalism with its monopolies, its usury and whatever else is unjust in it; at this individual freedom, devoid of human sympathy and responsibility for relatives except under the force of law; at this materialistic attitude which deadens the spirit; at this behavior, like animals, which you call 'Free mixing' of the sexes; at this vulgarity which you call 'emancipation of women,' at these unfair and cumbersome laws of marriage and divorce, which are contrary to the demands of practical life; and at Islam, with its logic, beauty, humanity and happiness, which reaches the horizons to which man strives but does not reach. It is a practical way of life and its solutions are based on the foundation of the wholesome nature of man.[37]

In summation these "were the realities of Western life," which made "the American people blush." American life was mired in "filth in which Jahiliyyah is steeped" as was "the evil and dirty materialism of the East."[38] Much of this was comparable to the *jahili* society at the dawn of Islam where war and oppression reigned. Complete moral degeneration, drinking, gambling and rampant fornication characterized these early Arabian societies most of all. Allah sent the morality and social justice of Islam to correct these wrongs. Muslims were the guardians of Allah's eternal political authority on earth.[39]

When speaking of other societies, America was the enemy, first and foremost. The US was an unrepentant empire like Greece, Rome and Persia of the ancient world, or Great Britain of modern times. Furthermore, it was Judeo-Christian in outlook, ruled by the People of the Book who knowingly rejected Allah's Islamic message because of their evil, lustful materialism. They sold out Allah for the corrupt rule of humankind, embracing the concept of equal rights, especially for women. They nullified any possibility of happiness through their denial of the need to submit to Allah and embrace Islam. America was *shirk* and *jahili*. No compromise could be made; the only solution was Islamic rule and the destruction of that present day, 1950s America.

The Jahili Infected Muslim World

According to Qutb, Muslim communities as Islamic entities were suspended once Allah's laws were no longer in effect, which meant that no true Islamic societies existed any more. Living with false laws and customs, the authentic Muslim communities vanished, leaving *jahili* Europe to lead the way.[40] The result was defeatist, defensive behavior in trying to explain away Jihad as a "temporary injunction related to transient conditions and that it is concerned only with the defense of the borders," and not a planned offensive in the struggle for Allah's eternal Lordship.[41] Such an apologist approach emanated from the line of defense taken against the Orientalists, Westerners, Eastern Christians or even Westernized Muslims, examining Islamic history from a non-Muslim, secular and/or Western perspective. Qutb was furious at what he considered "compromised" Muslims for taking a defensive approach. Islam, Sharia law and Jihad represented Divine justice and were to be imposed on the entire human race. No apologies were tolerated. Supposedly Muslim societies were *jahili* because they had given authority to secular institutions and human-designed legislation, standing in contrast to the true Islamic community.[42] Qutb aligned with those "who are standing firm on the issue that Islam is a universal declaration of freedom of man on earth from every authority except God's authority, and that the religion ought to be purified for God; and they keep writing concerning, the Islamic Jihaad."[43] There was no discretion, one must advocate for Jihad.

He accused defeatist scholars of using "religion" or "belief" to describe Islam, weakening commitment from the inside and being equally reprehensible.[44] Muslim secularism was another aberration, and was not based on submission to Allah. It was an un-Islamic, illegal society such as those of the polytheists, characteristic of *jahili* systems. All man-made theories and religions were outmoded and had nothing in common with Islam.[45] This included secular nationalism, or all other loyalties outside of the commitment to Allah. At the outset, family and tribal relationships had certain significance, but Qutb quickly pushed aside even these most intimate relationships when people rejected Allah. It was at the point of accepting or rejecting Islam at which the pride of lineage ended and true commitment began. In the Koran, the Prophet battled his relatives from Mecca when they refused to adhere to the word of Allah. By extension, a person's nationality, race, and physically embodied homeland deserved no form of loyalty. The only true homeland was where faith ruled, this being "*Dar al-Islam*" or the abode of Islam. Absence of faith ruled in "*Dar al-Harb*," or the domain of war. Everyone was freed from commitment to blood and earthly relationships. Loyalty to Allah defined the true Islamic society, not family, nation or tribe.[46] Qutb explained, "The noble conception of homeland, of nationality, and of relationship should become imprinted on the hearts of those who invite others toward God. They should remove all influences of *jahiliyyah*, which make this concept impure and which may have the slightest element of hidden *shirk*, such as *shirk* in relation to homeland, or in relation to race or nation, or in relation to lineage or material interests. All these have been mentioned by God Most High in one verse, in which He has placed them in one side of the balance and the belief and its responsibilities in the other side, and invites people to choose."[47]

According to Qutb, Islam was a political and social force promulgating a "system of law" and "the constitution of law and injunctions, rules and regulations." Through obedience to Sharia law with the unquestioning submission to and sovereignty of Allah over all Islamic communities is the only way absolute freedom was achieved. There could be no compromise in the form of separation between state and religion with loyalties to both,[48] nor could Allah's reign be restricted to the heavens as in the Deistic approach to belief.[49] Allah and Islam were one, ruling heaven and earth; there were no competitors or compromises.

Islam in Contrast to Secular Arab Nationalism and the PLO/Fatah

Although *Milestones* was written in the 1950s prior to the establishment of Fatah and the PLO, Gamal Abdul Nasser, ruler of Qutb's homeland Egypt, was the greatest advocate of secular Arab nationalism. Qutb spent over a decade in jail as a result of his tenacious intellectual opposition to Nasser's secular Arab nationalism, which he condemned as *jahiliyyah*. Qutb believed neither Nasser, the Free Officers, nor the parliament had the right to rule, as ruling was Allah's domain. Loyalties to nationalism and the homeland were *shirk*. Allah was the only true sovereign and guidance would only come through Sharia law. Extending these ideas to the Palestinian front, today's Fatah, PLO or any other non-Islamist Arab political party would be *jahili*.

The entire universe belonged to Allah. The "homeland" was wherever faith ruled. The homeland was any domain where the Muslim community behaved according to the Creed of the first community, seventh century believers led by the Prophet Mohammed and the initial caliphs. Qutb was speaking of insular Islamist communities residing in Europe, North and South America, the Far East and of course the Middle East, anywhere on the globe. He expanded the meaning beyond the usually acceptable *Dar al-Islam*, the domain or abode of Islam, to any place where there was effective physical Islamic rule. His definition now included Sharia-obedient Muslim communities worldwide. These communities resided inside *jahiliyyah* societies.

A true Muslim knew there was no God beside Allah the One and Only, with Mohammed as His Messenger. The Muslim community was pure and submitted fully to the will of Allah, the omniscient authority administering Divine law. There was no room for one's own consideration or judgment known as "*ijtihad*," Allah revealed what was best through the Sharia. Allah created the physical attributes of humans to bring bodily harmony, granting Sharia law as part of the rules of the universe to bring harmony to man's relations with the cosmos and each other. Should there be no adherence to Sharia law, discord would reign. The rule of men or anthropocentrism would only lead to violence, conflict and destruction. Diocentric Sharia law brought peace of mind.[50]

Human law for Qutb was *jahili* law. Anthropocentric legislation could only lead to conflict as opposed to the all-perfect Sharia, which embodied the ultimate truth. There can be no such thing as a "modern" Muslim society, as this too was *jahili*. The world was split into two: true Islamic civilized society governed by Sharia law embodying the one indivisible truth, versus backward *jahiliyyah* man-made corruption. The Sharia had its place on earth, ruling over men. There were types of *jahili* societies confining Allah's rule only to the heavens; this too was anti-Islamic and ignorant since the Almighty rules on earth through Sharia law. Koran 43:84 states, "It is He Who is Sovereign in the heavens and Sovereign in the earth."[51]

Qutb was not averse to material development; everything derived from a belief in God and as a result must be used to serve Allah's demands. Even if Muslim civilization was not dependent on material progress, poverty was punishment for those refusing to adopt Islam. The Islamic civilizing mission was not necessarily dependent on material progress, although that was not ruled out.⁵² Muslims needed to make no compromises with the *jahiliyyah*, no synthesis with socialism or democracy. Islam exuded truth, was a superior religion whose values and faith were above all and whose triumphant spirit would be victorious on every battlefield. With the perfect belief in one God, believers need not fear; their spiritual superiority would defeat ignorance as the forces of light always defeated those of darkness. The believer was eternally superior.⁵³

The believer was most superior in his understanding and concept of the nature of the world and for the belief in One God. These beliefs come to him from Islam, and were the most perfect form of understanding, the greatest truth. The picture of the world, which Islam presented, was far above the heap of all other concepts, beliefs and religions. No great philosopher, ancient or modern, nor idolaters or the followers of distorted scriptures, nor the base materialists, could ever come near to the perfect world of Islam.⁵⁴ The unbelievers may ridicule those of true faith, but through the tenacity of belief and subservience to Sharia law the true Muslim prevailed over false materialism and honor as embodied in the *jahili* world.⁵⁵

Projecting these principles into the late twentieth and early twenty-first century makes clear there can be no secular rule in any Muslim society. Khomeinist revolutionaries overthrew the Shah in Iran because he reigned illegally. Until the 2011 Islamic Awakening ("Arab Spring") revolutions, all Arab regimes were *jahili* and demanded liquidation, including the present PLO-led Palestinian Authority in the West Bank. The Hamas government in Gaza could be considered legitimate, since its law code is derived from Sharia law. As for the other Arab/Muslim regimes in transition, their legitimacy is yet to be tested. We can safely assume the short-lived hybrid Muslim Brotherhood regime that ruled Egypt in conjunction with the military from 2012-13 would be considered far too compromising and unacceptable for true Islam, although it was heading in the correct direction. Further Islamization was demanded to fulfill Qutb's requirements for a Sharia dominated society.

Women

For Qutb, the status and behavior of women formed a major pillar ensuring the stability of Islamic society. There was a clear division of labor between husband and wife. She was responsible for child rearing and imparting the morals and values to the next generation, which was critical for success. The woman was not to work outside the home, "thus spending her ability for material productivity rather than in the training of human beings." Women having a career was dishonorable, "backward" and "*jahili*." In particular, Qutb condemned what he saw as a degenerate licentiousness. He wanted to ensure a world without sexual sin, lest "free sexual relationships and illegitimate children become the basis of a society . . . [Preventing a world where] the relationship between man and woman is based on lust, passion and impulse and the divisions of work is not based on family responsibility and natural gifts." When material well-being was more important than family values, the society was backward regardless of advances made in industry and science. Extra-marital sex and homosexuality were condemned. The family unit was of the utmost importance, demanding sexual loyalty to one's husband even if a loving relationship was completely absent. Banishing animal desires, family peace and stability were paramount.[56] Qutb's revulsion at the behavior of Western women was another result of his American visit. The Islamic role of women was the traditional one of raising and educating children while administering the household. All fundamentalist Islamic communities were subjected to offensive *jahili* societal norms, but the true "back-breaking pressure" was directed against the Muslim woman.[57] It was imperative for her steadfastness to prevail. "Thus, only Islamic values and morals, Islamic teachings and safeguards, are worthy of mankind, and from this unchanging and true measure of human progress, Islam is the real civilization and Islamic society is truly civilized."[58]

What Qutb omitted is no less important. For him the Koran was Divinely revealed writ, no compromises, mistakes or misinterpretations allowed. He knew the Koran intimately. The Koran made clear the inferiority of women; wife beating was permitted if she did not adhere to her husband's wishes. The question was whether today's Islamists continued to allow such actions toward women. The Taliban is known for their brutal treatment of women today as seen in parts of Pakistan and Afghanistan, while the Islamic State is infamous for its vicious sexual conduct toward the female population in Syria, Iraq and Libya where rape is rampant. For Hamas, the Koran is its constitution. Here is what the Koran says about women in 4:34, "Men have authority over women because Allah has made the one superior to the others, and because they spend their wealth to maintain them. Good women are obedient. They guard their unseen parts because Allah has guarded them. As for those from whom you fear disobedience, admonish them and send them to beds apart and beat them. Then if they obey you, take no further action against them. Allah is high, supreme." And further in Koran 2:223 there is male sexual domination, "Women are your fields: go, then, into your fields as you please."

Islamists and true fundamentalist Muslim believers cannot ignore these very clear Koranic stipulations. Islamic commentators determined male superiority; wives must be obedient, serve men sexually and reproduce. Husbands are obligated to beat their wives if necessary. Qutb certainly had this in mind when condemning women's equality in Western societies.

Jihad in the Battle Against Jahiliyyah

Jihad in Sayyid Qutb's world-view was similar to that of Hasan al-Banna, both Muslim Brotherhood leaders and colleagues. The former clearly enunciated the reason for Jihad as, being "To establish God's authority in the earth; to arrange human affairs according to the true guidance provided by God; to abolish all the Satanic forces and Satanic systems of life; to end the lordship of one man over others since all men are creatures of God and no one has the authority to make them his servants or to make arbitrary laws for them."[59]

The Islamic Jihad was and will be implemented to extend freedom to all, to liberate men from living under the rule of other men and to guarantee all will serve only Allah. Qutb called for Jihad to spread Islam and its universal freedoms, not for nationalism and economic expansion. Offensive Jihad was and will be declared not due to any threat by others to Muslims, or because of aggression against *waqf* Islamic lands. Jihad remained eternal until total victory—the offensive taken in order to return all authority on earth to Allah. The idea of Jihad usage only for defensive purposes was modern apologetic reasoning and completely opposed to true Islamic consciousness. The homeland was wherever Allah's authority was established and not necessarily a national geographic region. Obviously, Muslims were expected to defend the home community of believers serving as the Islamic headquarters; however, Islam was "to be carried throughout the earth to the whole of mankind, as the object of this religion is all humanity and its sphere of action is the whole earth."[60]

Qutb's true Jihadist battled for Allah's values with no thought of personal gain, and brought universal freedom to all through these actions. There were no geographic limitations. The inevitable clash preceded against all *jahili* societies. This went so far as to include the People of the Book not yet living under the sovereignty of Allah's Islamic obedience. It was an Islamic obligation to annihilate all *jahili* systems and impose the Creator's will on everyone by violence if necessary, as all the world was His. "Islam is the way of life ordained by God for all mankind" and "Jihad in Islam is simply a name for striving to make this system of life dominant in the world."[61] The responsibility for Jihad never ended. None ceased in the battle against the surrounding *jahiliyyah* and the "Jihad continues until the Day of Resurrection."[62]

Non-Muslims only had themselves to blame according to Qutb. These infidels brought Jihad upon themselves by refusing to accept Islam. At the outset in the seventh century, Muslim adversaries were divided into three categories: those with whom they were at peace by treaty, those with whom there was a state of war, and the *dhimmis*, which in this case specifically meant the Jews. Shortly afterward, war was declared against all polytheists, including those with whom there were treaties. Upon expiration of those agreements or the *hudna* cease-fire, Muslims were to declare Jihad unless the polytheists accepted Islam. Two groups remained: those continuing the war and the People of the Book, who could decide to live in peace provided they paid the *jizya* tax and accept Islamic domination. Paying the *jizya* spared them Holy War. At the beginning of Islamic history, during the Meccan period, the Prophet Mohammed tried persuasion, but soon moved over to the Jihad offensive in the name of Allah. Islam was never meant to be on the defensive.[63]

Every individual was called upon to "freely" accept Islam, but if he rejected the offer "it is the duty of Islam to fight him until either he is killed or until he declares his submission." Muslims were now to liberate men from serving other men and impose the rule of God through Sharia law and abolish *jahili* societies. Both "preaching" and "movement" were melded to attain the Islamic objectives of obedience as demanded in the worship of Allah.[64] Persuasion through argument and coercion through force of arms led to political power. Without Allah's rule on earth, there could only be a "superficial peace" where Muslims were said to be secure. True peace was religion "purified for God, that the obedience of all people be for God alone and that some people should not be Lords over others."[65]

There were three basic categories of enemies to be crushed, those who attacked you, non-believers and the People of the Book. First, if attacked a Muslim was to fight back like in any society, but this was only a defensive first step. More importantly Muslims, who were subservient to Allah, were to take an offensive step against polytheists and the People of the Book until all submitted to the rule of Islam. The former were forced to convert, while the later paid the *jizya* and accepted the superiority of an Islamic regime. Such action led to the universal subjugation of one's will to Allah. Qutb's thinking was devoid of the "transient conditions" that were "concerned only with the defense of borders." Modern political secular boundaries had nothing to do with the eternal theological imperatives demanded for engaging in Holy War to impose Allah's universal lordship.[66] Qutb insisted that all people, including former non-believers, who engaged in Jihad were to be rewarded with eternal life, proof provided in Koran 3:74. "They ought to fight in the way of God who have sold the life of this world for the life of the Hereafter; and whoever fights in the way of God and is killed or becomes victorious, to him shall We give a great reward."[67]

In conclusion, the concept of a defensive war in fundamentalist Islam did not truly exist. Muslims must conquer the entire world for Allah. At first, Muslims accused the Orientalists of trying to define the early Medina period (622-624 CE) as defensive, but Qutb dismissed this idea as being done only for immediate tactical purposes. Such interpretations would only weaken the Jihadist resolve since the true reason for going to war was negated if a Muslim accepted the Orientalist defensive approach.[68] Temporary treaties or *hudna* Islamic cease-fires were exclusively tactical. Agreeing to a *hudna* allowed the Islamist side to rearm, retrain and reorganize in order to re-initiate hostilities and continue on the Jihadist march. Western cease-fires, representing anthropocentric values, were arranged to begin conflict resolution with mutual respect for the rights of all concerned. Upon the successful completion of negotiations, the parties would draw borders, give security guarantees, and establish diplomatic ties between the former warring parties. This mentality totally contradicted the final result expected from an Islamic cease-fire, or Allah's *hudna*.

Muslim believers were exhorted to make every effort to assure Allah's reign on earth, including the ultimate sacrifice of giving one's life for the cause. One became a martyr or "*shaheed*." There was no greater honor for a committed Muslim living by the Koran. "The honor of martyrdom is achieved only when one is fighting in the cause of God, and if one is killed for any other purpose this honor will not be attained."[69]

And finally, in the past and continuing today, Qutb's world was broken down into two spheres and a tactical temporary third element: *Dar al-Islam*, the homeland where Islam rules, is anywhere in the world where there are true believers, not necessarily within the framework of an Islamic state entity. For instance, *Dar al-Islam* can be a community of believers in Egypt, Syria, Paris, London, Brussels, New York or Israel. From their home communities, they are to continue the Jihad and confront their enemies in *Dar al-Harb*, the realm of war. Jihad was waged against all polytheists for conversion, and against the People of the Book who refused submission and payment of the *jizya* head tax wherever they may be. Jihad was not contingent on sharing a common border with a *jahili* people; the war of destruction was brought to them until they submitted to Islam. Consider 9/11 the perfect example.

When conditions require it, Islam may declare a *hudna* cease-fire. During a temporary tactical lull designed to rebuild the Islamist fighting force and obtain ultimate victory, adversaries facing the Jihadist are at times referred to as *Dar al-Hudna* as they hold territory.

Islam adheres to predetermination and believers are fully confident in the final victory over all *jahili* societies. Qutb concluded *Milestones* on an optimistic Islamist note by declaring solidarity under the banner of Islam propelled the defeat of the Crusades and Christendom of yesteryear.[70] With unity of purpose and steadfast belief, Muslims will gain dominance and crush the Crusading spirit of the modern Western world. He made clear Jihad was not a battle against "imperialism" but one to spread belief in all-powerful Allah and Islam.

Jihad is, "not a political or an economic or a racial struggle; had it been any of these, its settlement would have been easy, the solution of its difficulties would have been simple. But essentially it was a struggle between beliefs—either unbelief, or faith, either *jahiliyyah* or Islam."[71] The only answer is Holy War, until all believe or accept subjugation to Islam.

Qutb's Continuing Antisemitic Influence

"Our Struggle With the Jews" is a major source for twentieth century Islamic antisemitism. It combines elements of the infamous Czarist forgery *The Protocols of the Learned Elders of Zion,* together with Hitler's "final solution," but falls short of a racial edict calling for the extermination of all Jews. Instead the condemnation is determined on theological lines and the inherent nature of Jews, which is akin to racism, but not exactly genetic race theory itself. Just to differentiate, Qutb made it clear that converts to Islam are equal to Muslims, and will face judgment concerning their dedication to Allah and following His demands. For Qutb, an Arab Muslim was not superior to a non-Arab Muslim as advocated in *Milestones.* Still, Muslims could not trust Jews to embrace Islam, since a Jew's objective would be to become apostates and weaken Allah's final, perfectly revealed religion. The only reason Jews would adopt Islam would be to destroy it from within. Qutb drew on Koran 2:74-76 for evidence, using the following quote. "Do you really want them to believe you, when a group of them have already heard Allah's Word and falsified it knowingly, after having understood it? When they meet believers they say, 'we too believe'; but when they are alone with one another they say 'Do you tell them about what Allah has revealed to you, that they may argue with you about it before your Lord? Do you not understand?' Do they not know that Allah knows what they keep secret and what they proclaim?"[72]

This Koranic quote has little to do with Islam, but refers to Israelites who contradicted God's revelations during the Sinai Exodus. Qutb took it out of context to justify his accusations against the Jews. He explained that the Jews knew Islam to be the true way of life, but they undermined Allah's commands because subversion was inherent in Jewish nature. No doubt *The Protocols of the Elders of Zion* printed in Arabic in the 1930s-40s influenced Qutb. It likewise claimed that to absorb Jews into one's body politic or religion was to invite destruction from within. Nazism made the same accusation from a racial perspective. Nazi ideology emphasized the perfidy of the Jewish plot to genetically mix with the Aryan "super human race" to destroy the God given German physical and mental superiority ordained to rule the world.

Qutb's argument is simple. The Jews battled the Muslims and the Prophet Mohammed in Medina, Khaybar and throughout the Arabian Peninsula as recounted in the Koran and other Islamic religious texts, always rejecting Allah's message. The Koran accused the Jews of betraying their own Torah and killing the prophets Allah sent to them.[73] In the past, their Temple was destroyed twice and conquerors expelled them from the Promised Land. Now, explains Qutb, the Jews will be banished a third time for their continuing evils, this time when Islam destroys the State of Israel. These interpretations emanate from "The Night Journey" (Koran 17:2-8). The modern Jewish nation state and Zionism were one more Jewish ruse, conspiracy or form of weaponry to use against Muslims in the supposed eternal Jewish offensive against Islam.[74]

Ronald Nettler describes Qutb as a fundamentalist theologian uninterested in historical developments. Qutb believed the acrimonious relationship which existed between Jews and Muslims at the time of Mohammed was, always has been, and will continue into the future until the End of Days. There can be no change in the relationship; the Jews will continue to act the way they do. It is as a behavioral "given," with no possibility of reversal and thus the Jewish-Islamic conflict has no compromise solution. Islam is the only true religion (Koran 3:109). Islam eternally draws inspiration from its glorious past, in particular its defeat of Arabian Jews in the seventh century. Jews will forever be the enemy and will continually attempt to destroy Islam both spiritually and physically. The Jews began plotting against Mohammed and Islam in Medina after the *Hejira*; instead of joining Muslims in believing in the same God, they allied themselves with pagans. In Qutb's world, there was virtually a genetic programming of Jewish evil to destroy Islam.[75] It was just a matter of time before Islam defeated the Jews. Those remaining would have no choice but to agree to live under Islamic dominance. All others must perish.

For Qutb, in order to be victorious the entire Muslim world must unite under the banner of Jihad and Islam as defined by the Creed of Mohammed's first band of followers in the seventh century CE. Today's offensive is no different than its Koranic precursor, where Islam crushes Jewish conspirators and their polytheist idol worshiping allies. Muslims force polytheists to convert or face death, while Jews, as People of the Book, are given the *dhimmi* status of a heavily taxed second-class existence, but are allowed to survive because they believe in Allah. Yet throughout the centuries, Muslims eternally accuse Jews of consistent plotting to draw Muslims away from the true faith. In particular there were Jews who converted to Islam only to become apostates and draw the weaker-minded believers away from Islam as recounted in Koran 3:69. As co-believers in Allah, Jews are expected to side with Muslims, yet they are so taken in by Satan, preferring to destroy those of the true faith and rally polytheists against Islam.[76] Jews began the deception and evil plotting, which Christians later picked up and by inference made them no better in the long run.

The Jews faced punishment of isolation for being selfish, fanatical, hateful ingrates bent on fostering wars and destruction, while detesting the Messenger Mohammed for the good and truth he brought. Hence, despite their *dhimmi* status as People of the Book, such nefarious behavior earned the Jews their position in the Muslim world as the eternal unrepentant enemies of Islam, more so than polytheists. Secularization of Muslim societies throughout time was by definition the work of evil Jewish agents.[77] To quote, "Anyone who leads this Community [Muslims] away from its Religion and its Qur'an can only be a Jewish agent – whether he does this wittingly or unwittingly, willingly or unwillingly. The Jews will, then, be safe from this Community, so long as the Community is alienated from the one unique Truth from which it derives its existence, its power and its victory – the Truth of religious creed, the practice of belief and the Shari'ah... This is the Way and these are the landmarks."[78]

Due to Muslim mercies, the Jews survived despite breaking co-existence accords, such as the agreement at Medina. Qutb claimed that through plotting and scheming, a diabolical "Jewish genius" kept them alive over the centuries. The true Muslim community or Creed faithful of yesteryear defeated the Jews, but what happened since then and how the Jewish menace would be contained were outstanding questions. Believers must return to the Creed of the first Islamic community, and then Allah's victory over the Jews would not be long in coming.

True to form, Qutb saw specific Muslim weakness as emanating from seventh century Arabia when the great hypocrite of "spiritual vacillation," Abd Allah b. Ubayy, himself a Muslim, defended the Jews even after their betrayal of Mohammed after the Battle of the Ditch (or Trench). Ubayy was a Jewish agent from within. During the mid-twentieth century, Muslims betraying Islam were referred to as the "Brown British" as opposed to the true English, who are "White." But their objectives were the same concerning the destruction of Islam. These Brown British traitors were the power elite responsible for the debacle of the 1948 War against Israel.[79] The Jews worked with the hypocrites to attain their victory. As Nettler explains, "The most prominent modern form of ancient Jewish deception was found in the new, modern classes of deceivers: intelligentsia, communications people, professors, writers, Orientalists, which is a catch-phrase referring to those whose Islam was influenced by Western conceptions of the religion, politicians, and even sometimes those who were recognized as Islamic religious authorities and functionaries."[80]

Qutb blamed the Jewish saboteurs for historic Muslim divisions and inadequacies. Past and present were blended together, yesterday's hypocrites were today's modernists and the Jews were the same evil conspirators as always. The Jews engineered the Medina betrayal, the lies spread about the Prophet Mohammed, the assassination of the third Caliph Uthman and the resulting sectarianism, the challenges to the authenticity of the Koran and Hadith, the removal of Sharia law under the Ottoman Sultan Abdul Hamid II in the early twentieth century and the eventual abolition of Sharia law with the rise of the secular Turkish nation state. The West and apostate Muslims saw Ataturk, the father of the modern secular Turkish nation state, as a "hero."[81] This trend continued through "hero worship" of secular leadership in the Arab world in the 1950s (Nasser) and can be projected into the contemporary period as symbolized by Mubarak of Egypt, Syria's Assad regime, Jordan's King Abdullah II, the Palestinian Authority's Yasir Arafat of late, and so on. Islamists often accuse all of these leaders of being agents in the Jewish plot to collapse Islam from within. The Saudi editor who published "Our Struggle With the Jews" (1970) pointed out that such Jewish conniving was already believed through *The Protocols of the Elders of Zion*.

Qutb claimed the modern worldwide Jewish conspiracy was contained in the work of three Jews. "Behind the doctrine of atheistic materialism was a Jew; behind the doctrine of animalistic sexuality was a Jew; and behind the destruction of the family and the shattering of sacred relationships in society . . . was a Jew."[82] The Saudi editor added the following footnote upon marking an asterisk at the end of the above quote. "These three are, in order: Marx, Freud and Durkheim. And additionally, behind the literature of decadence and ruin, was a Jew—Jean Paul Sarte!!"[83] Comments by the Saudi editor became an integral part of the original work itself and through the influence of *The Hamas Covenant* reached a wide audience.

Qutb went on to claim, "The war which the Jews launched against Islam was longer, more extensive and of greater ferocity than the war which the polytheists and idol worshipers perpetrated – then and now." As a modern example, the battle against India presented, "the intensity of the struggle between the Hindu idol worshipers and Islam is vividly apparent, but it does not equal the viciousness of world Zionism which considers Marxism as a virtual branch of its (own activities)."[84] As was traditional among Islamists and their apologists, Zionism and Marxism were thrown together as two strategies in the same Jewish plot for world domination. Qutb lines himself up quite well with the Czarist antisemitic forgery *The Protocols of the Elders of Zion* and Hitler's *Mein Kampf.*

When writing in the mid-1950s, Qutb faced the uncomfortable existence of Labor Zionist rule in the State of Israel. The mass immigration of Diaspora Jews, or "in-gathering of Exile Jews," in the 1950s was comprised mostly of Jews from Muslim and Arab countries. They were free, no longer subjected to persecution and discrimination as stipulated by the detested *dhimmi* status Divinely ordained by Islam. They would no longer suffer persecution at the hands of Muslims for any other non-theological reason. Furthermore, Israel exuded a Western-style socialist equality and secularism, and even communism when considering the kibbutz movement. These political ideals were deemed threatening to fundamentalist Islamic societal values. Israeli egalitarianism included women, who obtained equal rights at the outset of the Jewish national movement. Not only had the *dhimmi* Jews asserted him and herself and defeated the Arab armies, they also brought secular, polytheistic, or "pluralistic," influences into the Middle East. The Labor Zionists were the adversaries on several fronts. Should the Arab Muslim world pick up on such equality oriented, secular "Jewish" traits, Islamists would face their greatest struggle yet, and it would be much worse than confronting the Israeli Army on the battlefield. Here, Qutb particularly took fright and made his most vehement, hysterical condemnations of the new Jewish offensive, this time in the guise of Israel. The existence of a Jewish State was an attack against Islam.

He explained it this way: "As for today, the struggle has indeed become more deeply entrenched, more intense and more explicit, ever since the Jews came from every place and announced that they were establishing the State of Israel."[85] Having been expelled twice for doing evil, the Jews will be ousted from the Holy Land once again, this time by the forces of Islam. "The Muslims then expelled them from the whole of the Arabian Peninsula." Over the centuries Jews continued doing evil and Allah sent others to punish them until "Allah brought Hitler to rule over them." For Qutb, Hitler was the hand of Allah punishing the evil Jews. He continued, "The Jews have returned to evil-doing, in the form of 'Israel' which made the Arabs, the owners of the Land, taste of sorrows and woe. So let Allah bring down upon the Jews people who will mete out to them the worst punishment, as a confirmation of His unequivocal promise in Koran 17:8: 'If you return, then we return.'"[86]

He saw the Islamic world taking the place of Hitler as Allah's hand in punishing the Jews. The Muslims were not frightened and would undoubtedly defeat the Jews. In referring to Jewish cowardice in the battles of the 1948 War, Qutb paraphrased from Koran 59:14, "They fight united only (in the safety) of protected towns or from behind walls. Their courage is great, among themselves. You think they are united but their hearts are scattered." This is an obvious reference to the kibbutzim, which are collective farms and the ultimate Labor Zionist symbol, in the Negev, which defended the newborn state of Israel against the Egyptian armored thrust toward Tel Aviv. The Egyptian offensive included troops from the Muslim Brotherhood.

Qutb added his own observations, "the Jews would fight the Muslims only from (the security of) fortified settlements in the Land of Palestine… Thus when the Jews lost their cover for one instant they turned their tails and ran away like rats. It was almost as though this verse had been revealed about them at that moment."[87] Qutb railed against the double Jewish conspiracy of communism and Zionism, symbolized by the kibbutz movement and the newly formed Israeli army. He did not explicitly mention the fact that many of the army's best commanders came out of the Palmach fighting force, a good amount of whom were members of the kibbutz movement, but this connection between Zionism and socialism was clear to him.

The comparison to rats came from Nazi propaganda, Goebbels in particular. What Qutb did not mention was the attrition of the Egyptian Army and the Muslim Brotherhood in the face of the kibbutz defenses, thereby halting the advance on Tel Aviv. Half a year later, the Israeli counterattack broke the back of the Egyptian Army and only through British intervention was Egypt able to secure the Gaza Strip in January 1949. Qutb convinced his future

Hamas warriors that the Jews were cowards and that victory on the battlefield and mass Jewish expulsion were just around the corner, ignoring the Egyptian and Muslim Brotherhood defeat.

Part III
Abdullah Azzam
Jihadi Scholar and Warrior

Afghani Jihad and World Conquest

Born in the British administered Palestine Mandate, Abdullah Azzam (1941-1989) rose to be the most active Jihadi of his time, both in word and deed. Later, his protégé Osama bin Laden surpassed him. Ideologically, it was Azzam more than anyone else who influenced bin Laden to take the Jihadist route. Azzam left the Jordanian held West Bank after the Israeli victory in the 1967 War, refusing to live under Israeli occupation. Joining the Jihad in Jordan, he fought for Israel's destruction in cross border raids, but was disappointed at the lack of religiosity on the part of his fellow warriors. In the 1970s he received a doctorate in Islamic jurisprudence from Cairo's Al Azhar University. Shortly afterward, he joined the Jihad in Afghanistan battling against the Soviet invasion, which began at the very end of 1979. The war lasted almost a decade and concluded with an Islamic victory. For Azzam, Jihad was a way of life about which he wrote and fought for extensively. Two of his most important works concerning Afghanistan, Palestine and world Jihad are his *fatwa*, or Islamic judgment, *Defence of the Muslim Lands,* and his short but very influential booklet, *Join the Caravan* published December 9, 1988, which was coincidentally the year anniversary of the outbreak of the "First" Intifada. Azzam was assassinated in Pakistan in November 1989 while organizing a continuation of the universal Jihad. There are two theories as to who killed him. Some believe that Western intelligence sources such as Israel or the US were responsible. Others claim it may have been an internal Jihadi settling of accounts, possibly over strategy or the next tactical move.[88]

Beginning in *Defence of the Muslim Lands,* Azzam explained that after "*iman*" or belief and faith, the most important obligation of any Muslim was to participate in Jihad. Holy War rests on two pillars, bravery and generosity. Their opposites were cowardice and miserliness, these being "the present condition of the Muslims, they have become as rubbish of the flood waters." The Jihad obligation is broken down into two distinct categories, offensive and defensive. Throughout the text Azzam quoted many Islamic scholars with whom he agreed. "*Fard kifaya*" was taking the offensive and a community obligation. The example he gave was in forcing payment of the *jizya* tax by the People of the Book. He then spoke of the community obligation to *fard kifaya*. Jihad could be deferred, but never abandoned. Most pressing for Azzam was the unconditional individual commitment, "*fard ayn,*" demanded in the case of the defensive Jihad to repel the invader.[89]

Four causes were given for the defensive Jihad: a "*kuffar*" or a non-Muslim infidel entered a Muslim land; a battle was enjoined; the imam called for a march into battle; or the *kuffar* captured Muslims. Due to the greater religious obligation to engage in such combat children did not need to ask approval from their parents to engage in Jihad. Neither did women need approval from their husbands, though they were forbidden to physically engage in battle nor did debtors need release from their creditors to join the Jihad. There was a personal Muslim commitment to Holy War beyond the immediate circle of those unable to repel the invader. Those physically closest were obligated to respond first, and the concentric circles expanded to the entire Islamic world if need be.[90] He reiterated this argument in *Join the Caravan,* in essence elevating personal involvement in Jihad to the level of an unbreakable dogma in the Islamic world. It was a communal sin of omission for everyone not participating in Jihad against non-Muslim infidels controlling *waqf* lands. The transgression was the same as if a person ate during the daytime Ramadan fast, or if a wealthy man refused to give charity. He quoted the scholar Qurtubi, "the Muslims are all a single hand against the enemy. This is the status of jihad until the inhabitants of the area have managed to repel the enemy that descended upon the land and occupied it, at which point the obligation is waived from the others." There should likewise be a general draft concerning a perceived threat.[91]

All Muslim lands were considered as one integral unit in the words of the theologian Ibn Taymia, "If the enemy enters a Muslim land, there is no doubt that it is obligatory for the closest and then the next closest to repel him, because the Muslim lands are like one land. It is obligatory to march to the territory even without the permission of parents or creditor, and the narrations reported by Ahmad are clear on this."[92] All such battles were compared to the early Muslim Battle of the Trench (or Ditch) in defense of Medina when the Prophet Mohammed exempted no one. This was a General March where all believers must fight and defeat non-believers. Azzam quotes Koran 8:39, "And fight them until there is no more Fitnah (disbelief and polytheism: i.e. worshiping others besides Allah) and the religion (worship) will all be for Allah alone (in the whole world)…"[93] The injunction for defensive Jihad was very strict, thereby allowing an attack against the enemy even should Muslim prisoners be killed in the process. The explanation given is "because the protection of the remaining Muslims from Fitnah (disbelief) and Shirk (polytheism), and the protection of the religion, 'Ard (earth) and wealth are more of a priority than a small number of Muslim captives in the hands of the Kuffar."[94]

A believer existed for Jihad. "Jihad was a way of life for the Pious Predecessors, and the Prophet (may Allah bless him and grant him peace) was a master of the *Mujahideen* (freedom fighters) and a model for fortunate, inexperienced people." The evidence is in the twenty-seven battles fought by the Prophet Mohammed. Jihad was purifying: "The virtuous Companions continued upon the path of the Noble Prophet (may Allah bless him and grant him peace) for the Glorious Qur'an had brought up this generation with an education of Jihad. They had been bathed in Jihad and cleansed of engrossment in worldly matters, just as a wound is bathed in water."[95] Alternatively, one could decide on self-destruction as seen in Koran 2:195: "So, the destruction lay in remaining with one's family and wealth and abandoning Jihad."

In *Join the Caravan,* Azzam not only dealt with the legal issues involved with Jihad, but commented on the reality of the struggle and built the political theory for its universal implementation. He began with a condemnation of his fellow Muslims. "Anybody who looks into the state of the Muslims today will find that their greatest misfortune is their abandonment of Jihad (due to love of this world and abhorrence of death). Because of that, the tyrants have gained dominance over the Muslims in every aspect in every land. The reason for this is that the Disbelievers only stand in awe of fighting."[96] Azzam listed sixteen reasons for Jihad against the tyrants among which were the fear of being damned to Hell, the need to establish "a solid foundation as a base for Islam," protecting Islamic dignity, the hope for martyrdom, the "achievement of the highest peak of Islam" through Jihad, and finally "Jihad is the most excellent form of worship, and by means of it the Muslim can reach the highest of ranks." A Muslim could never abandon Jihad, otherwise *fitnah* or disbelief would dominate. Despite this, it was a rarity for a Muslim to remain on the Jihadist path. A few Arab youth participated, but this was far from what was necessary in terms of men and money. The young Jihadists had motivation, but very little in the way of a Muslim education. Instead, Azzam bewailed the influence of the Orientalists who poisoned the minds of so many in the present generation.[97]

For Azzam, the essence of existence was Jihad. Jihad was compulsory; hence, those who refused to go but were fit to do so, deserved severe punishment. Even should a Muslim hesitate to engage in Jihad, Hell was his just reward. Many Muslims already lived Hell on earth through humiliation and persecution, much of it suffered at the hands of secular governments. To be successful, Jihad must be eternal and a way of life obeyed like a constitution. "Jihad and emigration to Jihad have a deep-rooted role which cannot be separated from the constitution of this religion. A religion, which does not have Jihad, cannot become established in any land, nor can it strengthen its frame. The steadfast Jihad, which is one of the innermost constituents of this religion and which has its weight in the scales of the Lord of the Worlds, is not a contingent phenomenon peculiar to the period in which the Qur'an was revealed; it is in fact a necessity accompanying the caravan which this religion guides."[98]

Sayyid Qutb was quoted saying Jihad was not a "transitory phenomenon;" otherwise why would it be emphasized to an extreme in the Koran? It was hypocritical not to engage in Jihad, which could be done in numerous ways such as: commencing with ideas, confronting falsehoods and as an armed force against evil.[99] Azzam knew many were looking for an exemption from the duties of Jihad. He therefore referred to the Koran passage 9:41 concerning repentance: "Go forth, light and heavy, and strive with your wealth and selves in the Path of Allah; that is better for you, if only you knew." He quoted different Islamic scholars, in an effort to arrive at the true meaning of this specific exhortation. The "light" were those who would and could participate in battle and the "heavy" were those who could or would not. If one was invalid and could not do battle, he was expected to help in education, supplies or in the rear lines. Only those who were too sickly or impoverished were exempt from battle. Even women could help in "education, nursing, and assisting refugees," although they were not allowed on the battlefield itself. Muslim men who refused to fight were hypocritical cowards more concerned about their personal material well-being and fear of death. The wealthy must spend their savings on Jihad expeditions, even should this mean supporting poor families willing to send their men on Jihad. Parents must send their sons on Jihad showing thanks to Allah for having given them children and as a guarantee of reward in the World to Come.[100]

While living in other countries, Muslims should not refrain from Jihad due to fear of the police, intelligence agencies, or security services. "Jihad is a certainty and fear of interrogation by the Intelligence is a matter of doubt." Nor should one show "fear of police authorities in the country whose passport he holds, even if he is sure that when he returns they will detain him and kill him or sever his limb, it is not an acceptable excuse before Allah because in this case he is obliged to forsake his country and live in the land of jihad."[101] He has only scorn and condemnation for those refusing Jihad,

"The adulterers, homosexuals, those who abandon jihad, the innovators and the alcoholics, as well as those who associate with them are a source of harm to the religion of Islam. They will not cooperate in matters of righteousness and piety. So whoever does not shun their company is, in fact, abandoning what he has been commanded to do and is committing a despicable deed."[102] He then condemned those who defined Jihad as something other than the physical war. The saying, "We have returned from the lesser jihad (battle) to the greater jihad (jihad of the soul) which people quote on the basis that it is a *hadith*, is in fact a false, fabricated *hadith* which has no basis. It is only a saying of Ibrahim Ibn Abi 'Abalah, one of the Successors, and it contradicts textual evidence and reality."[103]

In *Join the Caravan,* Azzam reemphasized the individual compulsory obligation of defensive Jihad to confront the invader without obtaining permission from anyone. A land based "people's Jihad" was necessary, but must be free of secular influences. Otherwise such an effort resulted in failure, similar to that of Egypt's Gamal Abdul Nasser whom he mocked. "Those who sit at the helm of leadership of people by the power of the first announcement of a military revolution accomplished behind the scenes in one of the offices of mediation can easily lose everything." But for those Jihadists who fight for their land, it will be far from easy for them to relinquish it. The true "*Ummah*" or people will always support Jihad, in particular when it involves protecting "the weak and oppressed in the land" and even more so when defending the honor of Muslim women.[104]

Azzam believed that one of the greatest motivational factors in Jihad was giving one's life for the cause and dying a "*shaheed*" or martyr. The Jihadist was "Hoping for martyrdom and the High Station in Heaven." He would receive his just reward in the form of dignity, forgiveness and seventy two "*houris,*" maidens or virgins.[105] Those who did not engage in Jihad feared death and loss of possessions. On the other hand, the fighters knew they reached the highest peak of Islam through Jihad, a form of worship. Azzam concluded "Jihad is the most excellent form of worship, and by means of it the Muslim can reach the highest of ranks" while proceeding to praise those who engage in face-to-face combat, quoting such action as "the most excellent of deeds."[106] Clearly there was a religious ecstasy by engaging in the righteous deeds of Jihad and slaughtering the enemy.

Afghanistan and Palestine

The Jihad in both Afghanistan and Palestine (battling the Jews in Israel) was considered defensive and demanded personal obligation. If those closest to the battle "fail to repel the *kuffar*" (infidel) then "Jihad spreads to those behind, and carries on spreading in this process, until the jihad is *Fard Ayn* (legal obligation) upon the whole earth from the East to the West." Azzam once again emphasized that the child, the woman, the debtor needed no one's permission to join the struggle.[107]

Afghanistan and Palestine were part of a greater universal problem, one of sin. "The sin upon this present generation, for not advancing toward Afghanistan, Palestine, the Philippines, Kashmir, Lebanon, Chad, Eritrea, etc, is greater than the sin inherited from the loss of the lands which have previously fallen into the possession of the Kuffar. We have to concentrate our efforts on Afghanistan and Palestine now, because they have become our foremost problems. Moreover, our occupying enemies are very deceptive and execute programs to extend their power in these regions. If we were to resolve this dilemma we would resolve a great deal of complications. Their protection is the protection for the whole area."[108] In *Join the Caravan,* Bukhara, Bulgaria, Sudan, Somalia, Burma, Caucasia, Uganda, Zanzibar, Indonesia and Nigeria were next on the list for Jihadi conquest. These problems were faced before. Muslims confronted Constantinople (Rome's Byzantines) Persia, Spain, Britain, the supposedly American puppet Nasser of Egypt of recent years, and the Russians in Afghanistan. In the end all of these countries and their ways were and will be defeated.[109]

Similar to al-Banna, Azzam ascribed greater responsibility and understanding to the Arabs as compared to other Muslims. For those who were able, the Arab Muslims were to begin their Jihad in Palestine - meaning all lands from the Jordan River to the Mediterranean Sea including the entire State of Israel, the remaining Muslims taking up arms in Afghanistan. "Palestine is the foremost Islamic problem" being in "the heart of the Islamic world," yet Afghanistan was more pressing at that moment and needed immediate attention. In Afghanistan, the *mujahideen* forces were engaged in a Holy War to establish an Islamic state, while "in Palestine the (Arab) leadership has been appropriated by a variety of people, of them sincere Muslims, communists, nationalists and modernist Muslims. Together they have hoisted the banner of a secular state."[110]

Furthermore Azzam preferred the *mujahideen* freedom fighters battling on their own without the help of any *kuffar,* or infidel power, as opposed to the

Palestinians who received aid from the Russians thereby aligning themselves with a big power player and inviting defeat. The military situation in Afghanistan was better, with open boundaries allowing Islamic forces freedom of movement. "The Palestinian borders are closed from all sides for anyone who attempts to infiltrate its borders to kill the Jews." He admitted the success of the Israeli army in its defensive abilities.[111] Overall, the Russians were much more of a threat in Afghanistan than the Jews were in Palestine. Efforts for victory in the former were the primary objective, defeating the Jews and completely clearing Palestine of their presence was next. The successful Jihad in Afghanistan led Azzam to believe that much could be learned from this singular experience. In the meantime the Palestinian situation was and continued to be simply too corrupt. The Arab world continually split itself between national entities, unwilling to come to each other's aid such as in Jordan and Syria.[112]

Worst of all was secularization, and in particular the Christianization of the struggle. "If only the Muslims had fought in Palestine, in spite of the corruption that was present in the early stages, and before the situation had become aggravated with the arrival of George Habash, Naif Hawatma, Father Capici (sic) [Capucci] and their likes, Palestine would not have been lost."[113] Habash led the Popular Front for the Liberation of Palestine (PFLP), and Hawatma led the Democratic Front for the Liberation of Palestine (DFLP). Both were Marxist organizations affiliated with the PLO and Yasir Arafat. Israel caught Father Capucci gun running for the PLO and jailed him in the 1970s. Azzam blamed Christians and Muslim secularists for the Palestinian debacle.

These Palestinian failures had a major personal impact on Azzam. According to the "Translator's Forward" in *Join the Caravan,* Azzam was unable to succeed in the Palestinian cause due to secularism and the lack of Jihadist commitment of its members. They played cards, listened to music and refused his overtures to teach them about Islam. When he asked what religion stood behind these *mujahideen,* they replied: "This revolution has no religion behind it." He left the secular movement, immersed himself in Islamic studies and years later joined the Afghan Jihad, exhibiting his disgust with the Palestinian façade of Jihad. He constantly said his "ultimate goal" was Jihad until victory in Palestine.[114]

Overall, "It is obligatory to fight (together) with any Muslim people as long as they are Muslims. It does not matter how bad or corrupted they are as long as they are fighting the Kuffar (infidel), People of the Book or Atheists." With the Palestinians the situation started becoming clear for Azzam and other Islamists in the late 1980s when Hamas began to challenge the PLO secularists.[115] This brought about the question as to whether a believing Muslim could accept help from "*mushrikun*," or disbelievers in the Oneness of Allah, polytheists, pagans, or idolaters, in fighting other *mushrikun*. Some Muslim scholars forbade accepting help outright, seeing it as corrupting the objectives of the Jihad. Others accepted such aid under very restrictive conditions. If Muslims were in great need, they could accept aid. Also, if the rule of Islam had the upper hand over any temporary *kuffar* ally, the non-believers needed a positive opinion of Islam and the Muslims had to feel safe concerning these non-believers. Even in all of this a Muslim was never to contradict Sharia law, even in war. The example he used was the directive expelling all Jews and Christians from the Arabian Peninsula.[116]

Azzam accused the Jews of the "blood libel"[117] against Muslim and Christian children. Tirades against the West, and the Jews as usual, were common in the 1980s when Azzam declared: "Today, humanity is being ruled by Jews and Christians. The Americans, the British and others. And behind them, the fingers of world Jewry, with their wealth, their women and their media. The Israelis have produced a coin on which it is written 'we shall never allow Islam to be established in the world.'"[118]

In the 1980s, Azzam learned a bitter lesson not only from the Russians, but also from the American *kuffars* as he recounted: "The Russians have taken five thousand two hundred Afghan Muslim children to rear them on the Communist ideology, and to sow heresy deep within them. The Americans have confirmed the opening of six hundred schools, and they are maintaining education and raising one hundred fifty thousand Afghan children inside and outside the country."[119] Azzam expanded the discussion to include making treaties with the *kuffar*. Although certain Islamic scholars rejected any treaty possibilities, one was forced to consider the stipulations of those who allowed such agreements. The Muslim side was forbidden to relinquish any land, hence this pre-condition necessitated a full unconditional Russian withdrawal from Afghanistan and the establishment of an Islamic state dictating the terms of a settlement to Moscow. This is exactly what he wanted to see happen with the Jews in Palestine; meaning the liquidation of the Jewish State was the only alternative acceptable to the Islamic warriors, in this case the Hamas Palestinian wing of the Muslim Brotherhood.[120]

Azzam used the pact of Hudaybia as an example of Mohammed's agreement to a ten year *hudna* with the Meccans. In cases where the Muslim side proved to be too weak to continue the battle, a temporary halt in hostilities was preferable. Upon regaining their strength, however, the Muslim forces should strike the enemy at will. Mecca was captured two years after the accord at Hudaybia. This was the paradigmatic *hudna* or the Islamic ceasefire designed to defeat the enemy, never seeking conflict resolution.[121] Using the *hudna* correctly was a fundamental tool for assuring Islamic victories throughout history. Even such a temporary agreement had harsh stipulations. Muslims cannot be given to outside corruptive influences of the enemy *kuffar*, nor allow Muslim lands to be ruled by them in the meantime. One was forbidden to abandon the *fard ayn*, the compulsory personal Jihad and there could be no abrogation of Sharia law. Azzam insisted that *hudna* was only a delay, but with strict limitations, in order to protect Muslims from the influence of the non-believers.[122]

The obligatory defensive Holy War was then to be re-engaged as emphasized before. All previously held Muslim lands would be re-conquered and Muslims would move over to the offensive Jihad directed by community obligation and engage in new conquests. Such was and will be the permanent order of the day as emphasized in both *Join the Caravan* and *Defence of the Muslim Lands*. Azzam's Afghan war was only the beginning. "The Jihad in Afghanistan will broaden until the entire world will be conquered because Allah has promised the victory to Islam."[123]

But the ultimate obligation of Jihad was in the internal essence of the definition of Islam, and the Koran itself was likened to a "constitution," as noted above. To re-emphasize, Azzam rejected any thought of compromise when pursuing the permanent universal Jihad: "The steadfast Jihad, which is one of the innermost constituents of this religion and which has its weight in the scales of the Lord of the Worlds, is not a contingent phenomenon peculiar to the period in which the Qur'an was revealed; it is in fact a necessity accompanying the caravan which this religion guides."[124] Jihad is eternal until final global victory and is not limited to any specific period of time, whether it be the seventh century when Islam was revealed or any other historical era.

Part IV
Al-Banna, Qutb, and Azzam

Comparisons and Impact on Hamas

When studying the *The Hamas Covenant*, (see Chapter VI, "*The Hamas Covenant* Analysis") many mistakenly believe it to be a Palestinian national document more than anything else. Hamas leaders Khaled Mashal and Ismail Haniyeh would like non-Muslims to believe the *Covenant* is a national document in order to lull the international community into complacency. Palestinian nationalism is only a veneer for the *Covenant*. The scope of the document relates to the need for world Islamic conquest as evidenced by Article 7, "The Universality of the Islamic Resistance Movement," and Chapter Two, "Causes and Goals." Hamas is at war not just against Israel, but identifies the Jews as the enemy and champions a harsh antisemitic policy demanding death to the Jews, which was made clear in the Introduction - Preamble and Article 7. Many believe Hamas to be "only posturing," as if a Jihadist movement would concede Allah's objectives. The development of Jihadist theory begins with Hasan al-Banna, continues through Sayyid Qutb and culminates with Abdullah Azzam, who had a direct hand in influencing *The Hamas Covenant*. All three writers constructed the pillars of faith. The three are compared below, and the course of their Islamist ideals, in particular that of Jihad, are tracked over time.

On the level of personal action, Muslim Brotherhood founder al-Banna actively worked against the British and the pro-Western Egyptian regime during WWII and in its aftermath through propaganda and assassinations. Qutb was much more of an intellectual and a writer. At first, he supported the overthrow of King Faruk by the Free Officers and Gamal Abdul Nasser, but he was quickly disappointed with the development of the Egyptian secular state. Qutb was jailed for over a decade as a result of his leading role in the Brotherhood and his furious opposition to the regime. Azzam, although a Palestinian, was the only one to leave that arena for a battlefield Jihad against the Russians in Afghanistan in the 1980s. All were Arab Muslims and each met a violent death. Al-Banna and Azzam were assassinated, and Qutb was executed.

The three were in complete agreement that Muslims were obligated to Jihad and that any evasion of Jihad was sinful. None accepted the notion of a Jihad of the spirit or heart. They condemned such an interpretation as cowardice, sinful and a willful lack of action deserving Allah's punishment. The most honorable death was in the name of Islam through Jihad. One became a martyr and Allah would justly reward him with the seventy-two heavenly virgins spoken of by Azzam. All agreed that Muslims must live under an Islamic regime governed by Sharia law.

Their greatest adversary was the People of the Book. All three writers emphasized battles against Christendom and the Crusaders of yesteryear. The modern struggle was against today's Christianity, Western imperialism and values now tainted with communism, socialism, capitalism or any sort of pluralistic or polytheistic ideal. In particular, Qutb was repulsed by these *jahili* influences peddled by Jews and Christians. He held a special hatred for the Jews as the Satanic betrayers of Allah, and against the United States for spreading a vile *jahili* anti-morality destined to destroy Allah's message of Islam. Azzam focused on the Russians in Afghanistan and then turned his attention to the next Jihad against the Jews in Palestine, without ever acknowledging the State of Israel. Al-Banna went so far as to quote a tradition promising a double martyrdom for Jihadists dying in the battle against the People of the Book. All three believed Jews and Christians were the greatest enemy. Nowadays, Qutb holds the most influence as an ideologue, with Jihadi Islamists violently opposed to the existence of both the "little Satan," Israel, and the "big Satan," the United States.

For all three, world order was bipolar and broke down into *Dar al-Islam*, the abode of Islam for Muslim believers, and *Dar al-Harb*, the abode of war against non-believers and polytheists. The People of the Book constitute a third category. Despite their enmity, Muslims could tolerate peace with the People of the Book on the condition that they agreed to live as *dhimmis* under Islamic dominance, subject to the *jizya* tax and the discriminatory stipulations in the Charter of Omar. Many believe al-Banna was clear on this point; however, in reviewing the Preamble to *The Hamas Covenant* there is the quote attributed to al-Banna declaring that "Israel" will continue to exist until destroyed by Islam. Some interpret the quote very narrowly to mean the State of Israel; others construe it more broadly to mean the People of Israel, or Jews everywhere which appears to be a more accurate understanding. Writing in "Our Struggle with the Jews," Qutb's anger boiled over at any leniency toward the reprehensible Jews. Knowing Allah was omniscient and controlled the universe one can sense his intense pain when confronted with the thought of continued Jewish survival in the wake of all their supposed anti-Islamic conspiracies. Qutb lived the contradiction of a resurrected Jewish State in the heart of the Middle East in his lifetime. For him, vindication from Allah came with universal Jewish annihilation. This was the Jihadi mission. Hence the calls for Jewish destruction in *The Hamas Covenant* are very much the voice of Qutb.

"Orientalism" is the analysis of non-Western societies; in this case it refers to Western scholars' analysis of the Muslim Middle East. Azzam and especially Qutb heavily attacked such analysis. They accused secular, defeatist Muslims of accepting the interpretations of such "scholars" to explain Islamic civilization and Jihad in historic anthropocentric terms. Qutb was furious at these Muslims for the excuses they made for Jihad, especially when they engaged in Western defensive apologetic terms such as "the right of self-defense." Instead, he wanted to convince the *jahili* societies of the beauty of Islam and the obligations Muslims have in ensuring victory. Orientalist thinking was weakening Muslim resolve from the inside, was a plot by the People of the Book, and harbored dire consequences for Jihad.[125]

Qutb emphasized women's domestic educational role in raising their children as faithful Muslims and Jihadists devoid of outside influences. Although not actual battlefield warriors, they could aid in the physical defensive Jihad, giving support as Azzam explained. They fully agreed that where possible, all women were personally obligated to defensive Jihad, not requiring their husbands' permission to participate in the Holy War; however, they were not to serve on the battlefield.

Nationalism, be it Arab, Palestinian or any other type, was secular and therefore not Islamic. Although pro-Axis during WWII, al-Banna did not see secular totalitarian Italian fascism or German Nazism as relevant models because those societies were militaristic, worshipped their leaders and revered the secular state.[126] Jihad was for Allah and to bring an Islamic peace. Al-Banna, Qutb and Azzam all condemned the "Crusading West" led by the People of the Book and reviled the atheist communist-socialist East. Qutb was the most violent in expressing his visceral revulsion especially for the United States, a nation where religion and state were separated and belief hopelessly corrupted.

On the local front, another major challenge was the rise of secular Palestinian nationalism as represented by Fatah and the Palestine Liberation Organization (PLO), both led by Yasir Arafat beginning in 1969. Al-Banna and Qutb objected to any sort of secular nationalism and were bitter enemies of the pro-British regime of King Faruk. Al-Banna battled both through assassinations resulting in his own death at the hands of Faruk's agents. In 1952, Nasser led the Free Officers in overthrowing the regime, a move initially supported by the Brotherhood. Not long afterward, Qutb and the Brotherhood found themselves jailed for their opposition to the secular "social" revolution of the Soviet Union and the East Bloc. Qutb was executed in 1966 for his anti-Nasserist efforts. Neither Jihadist could support a secular, non-Islamist Egyptian regime, both paid with their lives. Nor could either, had they lived, supported the PLO.

Likewise, Azzam reviled secular nationalism and ridiculed Nasser. He hated Palestinian Arab nationalism, but for political reasons he could not say so openly. Instead, Azzam's hand is obvious in compromises continually seen throughout *The Hamas Covenant*. As a Palestinian, he supported the PLO struggle against Israel, although he believed such efforts were doomed to failure due to their secular format. He believed secular nationalism led to a lack of commitment and more importantly caused the cessation of Jihad through the "End of Conflict" agreement of a two-state solution and Israel's continued survival. Secular leaders view conflict resolution as anthropocentric, between mortal humans, who make compromises and then live in peace. This humanistic ideal contradicts the word of Allah revealed through the Koran and Sharia law. The entire world must submit to Islam and diocentrism; no conflict resolution is permitted. Instead, Islam must dominate. *The Hamas Covenant* was written in 1988, but the theory has never changed and can be seen in Articles 2, 6, 7, 11, 13, 14, 15, 26 and 27. On one hand, the *Covenant* praises Palestinian secularists for their struggle in opposing the Zionists and Jews. On the other hand, the *Covenant* condemns Palestinians for secular nationalism, an ideal influenced by the East, West, Orientalists and Christians. All Palestinian issues must be solved within the Islamic vision, as just one more piece in the universal puzzle of Jihadi conquest. The PLO must adopt Islam as its way of life; if they do not they will forfeit Hamas support.

The Jihad in Palestine was considered "defensive" because the land is *waqf* land conquered by Muslim rulers in a previous era and transformed into Islamic lands for eternity. The demand for personal obligation in Jihad was extolled in this arena. Through Azzam's influence, *The Hamas Covenant* emphasized defensive Jihad and alluded heavily to the offensive next step without carefully detailing the ultimate goal the way Qutb did in *Milestones*. The offensive Jihad is purposely obscured so any non-Muslim reading the *Covenant* will doubt the ultimate objective of Hamas, and may even consider them as breaking with one of the main pillars of the Muslim Brotherhood. It must be recalled that the Jihadists in Afghanistan agreed to cooperate with the *mushrikun* non-believers to defeat the Soviets, but only temporarily. The *Covenant* consistently frames Hamas' objectives as part of the local Palestinian and Jewish/Israeli battle that hint at a conclusion with Israel's destruction. Delving deeper, it is obvious the frame is completely false. Azzam saw the battle for an Islamic Palestine as only one step in the overall world Jihad. This message may be played down, but it permeates the *Covenant* and Hamas. One must be cognizant of the full spirit of Islamist conquest linking the defensive and offensive Jihad into one as Azzam clearly expressed in both *Defence of the Muslim Lands* and *Join the Caravan*.

This brings us to the offensive Jihad, advocated by all three ideologues. The Jihadist Azzam fought a defensive Jihad, but longed for the offensive. A careful reading of *The Hamas Covenant*, in particular Article 8 makes it clear that this battle is not limited to destroying Israel and the Jews. The elimination of the Jews is one single military objective among many. Articles 3, 4 and the first sentence of Article 15 are general calls for Jihad, while Article 7 begins with the Islamic universalism of Hamas ideals. Article 11 declares the commitment to recapture all *waqf* lands across the globe. Hamas is an integral part of the Muslim Brotherhood demand for world conquest as a final goal. The following is the quoted slogan from Article 8:

> Allah is its goal
> The Prophet its model to be followed
> The Koran is its constitution
> Jihad its way
> And death for the sake of Allah its loftiest desire

There is nothing above mentioned about Arabs, Palestinians, Israel, Jews, or any other national group. It is a universal Islamic call for Jihad, whether defensive or offensive. Azzam and Hamas could pretend to build a localized Jihad in the guise of Palestinian rights, but neither of them betrayed the overall principle of world Islamic domination. This ultimate objective is obfuscated under the facade of Palestinian Arab nationalism. Defensive Jihad flows into offensive Islam, which was crystal clear in the aftermath of 9/11 when the Hamas leadership and Palestinian supporters expressed euphoria over the attack. Qutb would have been proud; offensive Jihad won the day. Article 36 concludes *The Hamas Covenant*: "The Islamic Resistance Movement [Hamas] depends on Islam as a way of life, its faith, and religion and supports whoever adopts Islam as a way of life." For Hamas, world Jihad is intrinsic to the Islamic way of life, and is the next step after the destruction of Israel and the defeat of the Jews.

Finally, there is the issue of *hudna*, or Islamic cease-fire, agreed to while battling the *jahili* societies. The three thinkers agreed that *hudna* was only a temporary arrangement permissible when the Islamic side was in danger of losing. The *hudna* allowed for rearming, retraining and the military reorganization of the Jihadi forces. With the *hudna* Jihad began anew, according to the Muslim military timetable. We were familiar with the ten-year *hudna* agreed upon at Hudaybia between the Prophet Mohammed and the Meccans. Muslim forces broke the *hudna* two years later, which resulted in the conquest of Mecca. Because Jihadi forces were battling in the name of Allah, any agreement reached between humans was subservient to the need to spread the Divine religion of Islam. Agreements may be broken if it serves the interests of the Muslim side. Any *hudna* was only a postponement of the eternal battle for universal subjugation to Islam.

In conclusion, these three thinkers are a continuum whereby Azzam becomes the hand of direct imprint on *The Hamas Covenant*, but he is far from the only one. It is important to realize that Azzam may emphasize the need for defensive Jihad, but he is no less an advocate of offensive Jihad, nor is the Palestinian Muslim Brotherhood, Hamas. The spirit of al-Banna and Qutb hover over the *Covenant* as well. Today, Hamas operatives are fully aware that offensive Jihad is an extension and completion of the "defensive" Holy War. Finally, there is the never-ending question as to why the Arab/Islamic world finds itself in such an inferior position to the West, and even other less developed nations. In the spirit of the above-mentioned ideologues, and *The Hamas Covenant*, the answer becomes clear. Muslims betrayed Islam, Sharia law and Allah and to redeem themselves must return to the righteous path of devotion and a true Islamic society. A non-wavering commitment to a physically conquering Jihad will cleanse the globe and lead the Arab/Muslim world back to the glorious days of Mohammed and the first believers. Only then will Islam succeed in universal conquest and regain its former glory.

Endnotes

1. Taheri, Amir, *Holy Terror, The Inside Story of Islamic Terrorism,* Sphere Books Limited, 1989, London, (Second Printing), pp. 42-44.
2. al-Banna, Hasan, *The Five Tracts of Hasan al-Banna*, translated by Wendell, Charles ed., University of California Press, Berkeley, 1978, Tract "On Jihad," pp 132-161.
3. Ibid.
4. Ibid, Tract "Between Yesterday and Today," pp. 13-39.
5. Ibid.
6. Ibid, Tract "Our Mission," pp. 40-68.
7. Ibid, Tract "To What Do We Summon Mankind," pp. 69-102.
8. Ibid, Tract "On Jihad," pp. 133-161.
9. Ibid.
10. Ibid.
11. Ibid, p. 150.
12. Ibid, pp. 151-153.
13. Ibid, pp. 153-155.
14. Ibid, pp. 138-146.
15. Ibid.
16. Generally defined as an "Islamic cease-fire" of no fixed time or no longer than ten years. In Jihadi Islam a "hudna" is only tactical or temporary, can be broken when necessary to defeat the infidel enemy and is never meant to achieve mutual recognition as part of conflict resolution. Other interpretations by more liberal, moderate Muslim commentators define "hudna" as similar to the Western interpretations. This author finds little basis for the "liberal, moderate" interpretation while the study at hand deals with Jihadi Islamic perspectives.
17. Qutb, Sayyid, *Milestones,* pp. 11-12, internet version from *Studies in Islam and the Middle East: SIME Journal,* retrieved January 5, 2010, majalla.org/books/2005/

qutb-nilestone.pdf.
18 Ibid, p. 32.
19 Ibid, pp. 12-14.
20 Ibid, pp. 10-11.
21 Ibid, p. 29.
22 Ibid, p. 39.
23 Ibid, pp. 45, 130-131.
24 Ibid, p. 75.
25 Ibid, p. 44.
26 Ibid, pp. 45 and 132.
27 Ibid, pp. 1-5.
28 Ibid, pp. 92-95.
29 Ibid, p. 75.
30 Ibid, pp. 117-118.
31 Ibid, pp. 119-124.
32 Ibid, p. 11.
33 Ibid, p. 54.
34 Ibid, p. 57.
35 Ibid, p. 66.
36 Ibid, pp. 76-78.
37 Ibid, p. 126.
38 Ibid.
39 Ibid, pp. 22-24.
40 Ibid, pp. 3-4.
41 Ibid, pp. 57-58.
42 Ibid, pp. 74-77.
43 Ibid, p. 62.
44 Ibid, p. 69.
45 Ibid, pp. 78-79.
46 Ibid, pp. 109-112.
47 Ibid, p. 113.
48 Ibid, p. 42.
49 Ibid, p. 89.
50 Ibid, pp. 80-87.
51 Ibid, pp. 89-90.
52 Ibid, pp. 96-97 and 101-102.
53 Ibid, pp. 127-129.
54 Ibid, p. 129.
55 Ibid, pp. 130-133.
56 Ibid, pp. 94-95.
57 Ibid. p. 126.
58 Ibid, p. 96.
59 Ibid, p. 63.
60 Ibid, pp. 55 and 63-65.
61 Ibid, pp. 66-69.
62 Ibid, p. 99.
63 Ibid, pp. 47-50.
64 Ibid, pp 51-53.
65 Ibid, p. 56.
66 Ibid, pp. 57-58.
67 Ibid, pp. 62-63.
68 Ibid, pp. 58-62.

Orientalists are Western and in general non-Muslim scholars who are accused by religious Muslims and the Arab world of studying Islam as an "object" while using Western and often secular analytical tools to do so. Jihadists accuse them of influencing how Muslims view themselves and Islam as Divinely ordained doctrine. This is discussed below and more intensely in later chapters, especially as pertains to the Christian Arab intellectual, Edward Said and his criticisms of Western studies of Middle Eastern society.

69 Ibid, p. 111.
70 Ibid, p. 147.
71 Ibid. p. 146.
72 Qutb, Sayyid, "Our Struggle With the Jews," in Nettler, Ronald, *Past Trials and Present Tribulations: A Muslim Fundamentalist's View of the Jews*, Vidal Sasson International Study of Antisemitism, Hebrew University, Jerusalem, Israel, 1987, p.72.
73 Ibid, p. 78.
74 Ibid, pp. 77, 83, 85 and 87.
75 Nettler, Ronald, *Past Trials and Present Tribulations: A Muslim Fundamentalist View of the Jews*, Vidal Sasson International Study of Antisemitism, Hebrew University, Jerusalem, Israel, 1987, Chapter 4 "The Jewish Goal of Islam's Destruction," pp. 32-39.
76 Ibid, pp. 73-74, "Our Struggle With the Jews."
77 Ibid, pp. 77-78.
78 Ibid, pp. 72-72.
79 Ibid, pp. 45-46, "Jewish Goal."
80 Ibid, pp. 46-47.
81 Ibid, p. 83, "Our Struggle With the Jews."
82 Ibid.
83 Ibid, footnote on page 83.
    Sarte was not Jewish.
84 Ibid, pp. 83-84.
85 Ibid, p. 85.
86 Ibid, pp. 86-87.
87 Ibid, p. 87.
88 "Abdullah Azzam," *Wikipedia*, retrieved January 15, 2010, en.wikipedia.org/wiki/Abdullah_Yusuf_Azzam.
89 Azzam, Abdullah, *Defence of Muslim Lands*, internet version found on *Religiscope*, Chapter 1, retrieved January 7, 2010, www.religioscope.com/info/doc/jihad/azzam_defence_3_chap1.htm.
90 Ibid.
91 Azzam, Abdullah, *Join the Caravan*, internet version found on *Religiscope*, Forward, retrieved January 10, 2010, www.religioscope.com/info/doc/jihad/azzam_caravan_1_foreword.htm.
92 Azzam, *Muslim Lands*, Chapter 1.
93 Ibid.
94 Ibid.
95 Azzam, *Caravan*, Part 1.
96 Ibid.
97 Ibid.
98 Ibid.
99 Ibid.
100 Ibid.
101 Ibid, Part 3.
102 Ibid.
103 Ibid, Conclusion.
104 Azzam constantly mocks the leadership of the secular Arab world. In particular, he saves his wrath for Egyptian President Gamal Abdul Nasser who was

considered one of the greatest heroes for Arab world liberation. In retrospect, Nasserism and the rise of the Baath particularly in Syria and Iraq could be considered the ideological leaders of the true "Arab Spring" of the 1950s and 1960s. For Azzam, Arab nationalism would certainly be deemed a failure.

105 Qazi, Farhana, "72 Virgins in Heaven: Fact or Fiction?" February 19, 2015, farhanaqazi.com/72-virgins-in-heaven-fact-or-fiction.
"Authenticity of 72 Virgins, *WikiIslam*, retrieved August 6, 2015, net/wiki/Authenticity_of_72_Virgins_Hadith.

Qazi believes the promise of 72 virgins for martyrdom is a myth. The promise of virgins can be traced back to Koran 55:56-78 in Chapter or Sura, "The Merciful." The virgins are described a "bashful," "fair as corals and rubies," and "dark-eyed." The number 72 originates from the Hadith of Jami' al-Tirmidhi who speaks of heavenly rewards. There are many Hadiths, interpretations and even a fatwa from 2005.

106 Azzam, *Caravan*, Part 1.
107 Azzam, *Muslim Lands*, Chapter 2.
108 Ibid.
109 Azzam, *Caravan*, Part 1.
110 Azzam, *Muslim Lands*, Chapter 2.
111 Ibid.
112 Ibid, Chapter 4.
113 Ibid.
114 Azzam, *Caravan*, Translator's Forward.
115 Azzam, *Muslim Lands*, Chapter 4.
116 Ibid.
117 The "blood libel" refers to the murderous false accusation that Jews kill Christians and in this case Muslim children as well, to use their blood to bake matza (unleavened bread) for the Passover festival.
118 Emerson, Steve, "Abdullah Assam: The Man Before Osama bin Laden," in *Osama bin Laden: The Past*, retrieved January 17, 2010, www.iacsp.com/itobli3.html.
119 Azzam, *Caravan*, Part 2.
120 Azzam, *Muslim Lands*, in Chapters 2, 3 and 4 all discussions comparing the Jews of Palestine (Israel) with the Soviets in Afghanistan speaks of destruction of the first and expulsion of the second. One can say the Russians and Communists are seen as "an" enemy while the Jews are "the" enemy.
121 Ibid, Chapter 4.
122 Ibid, Chapter 1.
123 Emerson, "Abdullah Assam."
124 Azzam, *Caravan*, Part 1.
125 See Edward Said's book *Orientalism* for a discussion in this light. Said was a Christian Arab who attacked Western scholars for focusing on Islam as the determining factor in understanding the Arab world. He believed such scholars misunderstood the Arab/Muslim world because of the narrowness of such a focus. Qutb's approach completely contradicted Said, considering any investigation of the Arab/Muslim world where Islam does not stand at the center to be a betrayal of Islam and Allah's message. For Qutb, Said would be an Orientalist of the worst type, a Christian *dhimmi* defending the Arab/Muslim world from the liberal secular standpoint while downplaying Islam in an effort to prove that Middle Eastern peoples have much in common with those in the West. Said advocated a "secular humanist" approach, certainly a perspective totally rejected by Islamists.
126 al-Banna, Tract "Toward the Light," pp. 109 and 113.

1947 Partition Plan
(replaces British Palestine Mandate)

Map for two-state solution.
Includes internationalization of Jerusalem and Bethlehem.

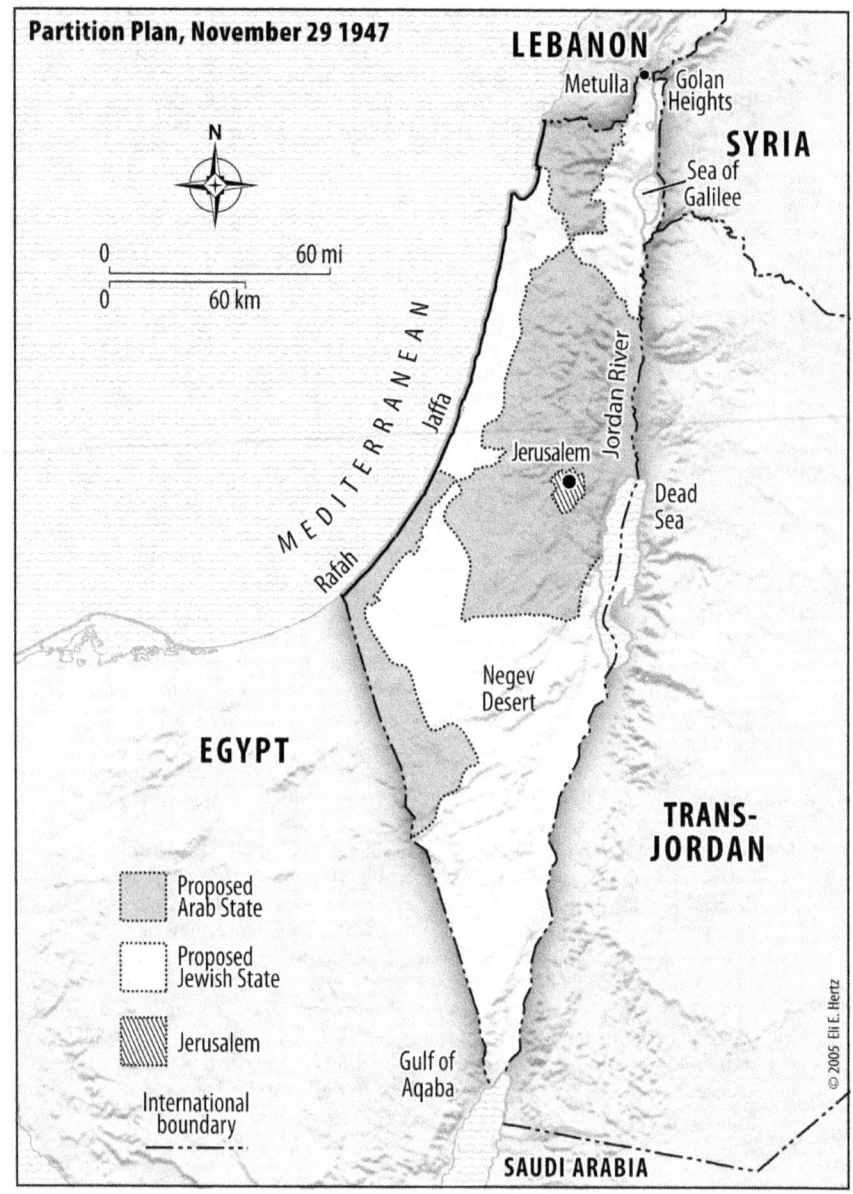

Credit: www.israelifrontline.com, "UN Resolution 181 - The Partition Plan"

III

Zionism

Jewish National Liberation Catalyzes Islamic Antisemitism to New Extremes

Overview

Jewish national liberation, better known as "Zionism," was and continues to serve as a lightning rod for Islamist antisemitism. Granting equal rights to Jews and other *dhimmi* minorities was painful enough from a religious perspective but the Jewish claim for sovereignty in the Land of Israel was considered the height of arrogance by a people rejected and discarded by Islam.

For the Ottomans, Zionism, similar to other national movements, was interpreted as an attempt at Jewish regional secession from the empire's territorial holdings. The Land of Israel, or Palestine, was a *waqf* land conquered by Islam, of which no other people or nation could lay claim, specifically not *dhimmi* Jews. On the secular Jewish and Arab scenes, the Balfour Declaration was a British pledge for a Jewish National Home while the MacMahon Correspondence committed London to the establishment of an Arab kingdom throughout the Middle East. These conflicting promises made during WWI led to competing territorial claims by these two national entities. By the early 1920s the Jewish world, the European powers and the Arab Middle East began to take Zionism seriously as a player on the global scene.

During the British Mandate period, Haj Amin el-Husseini, the Grand Mufti of Jerusalem, represented both the Muslim fundamentalist and Palestinian Arab national identity. He integrated a viciously antisemitic pillar of hatred into his leadership, eventually culminating in his full cooperation with the Nazis in WWII. He participated in the extermination of Balkan Jewry and was indicted after the war but managed to flee to Egypt where the Muslim Brotherhood and its leader Hasan al-Banna welcomed him as a hero. During Haj Amin's reign as Mufti of Jerusalem in the 1930s, Izz a-Din al-Qassam led the Islamic Black Hand terror group, attacking Jews, Christians, moderate Muslims and the British.

In 1948, the Arab State armies, Haj Amin's Palestinian forces, the Muslim Brotherhood, and other Arab irregulars, failed to destroy the State of Israel. They took their retribution out on the remnants of Jewish communities in the Arab Muslim world after most Jews in these other countries had already fled to the newborn Israeli State. Yesterday's *dhimmi* Jews now lived independently in the heart of the Muslim world despite their stereotypes as cowards, an unacceptable affront to the Arab Muslim world.

Muslims used theology to explain Jewish independence and reverted back to the seventh century idea that Jews satanically embodied ultimate evil. Secular Arab nationalism adopted Islamist ideas demonizing Jewish nationalism. Religious Islamists did not disguise antisemitism as anti-Zionism, but made it clear that the existence of Israel was only one manifestation of universal Jewish evil. Striding hand in hand with traditional Christian European antisemitism and Nazism, Islamists remained loyal to Haj Amin's vision and demanded Jewish elimination. Israel's 1967 battlefield success led to an intensified crescendo of hatred. A propaganda reversal by the Arabs heralded Israel as the new Nazis. Previously in the early 1950s the Israelis were accused of being communists and Bolsheviks, the ideological opposite of fascists and Nazis. Even during the Oslo Accords peace dialogue of the 1990s and continuing into the 2000s, the Hamas Islamists, and at times secular Fatah Palestinian Authority officials including Yasir Arafat, used antisemitic diatribes and declared Jihad against Israel and Jews in general.

Turkish Policy Toward Zionism 1882-1918

At the turn of the century, the Ottomans ruled in the Land of Israel or what was commonly known as Palestine. It is unlikely the Turks noticed any great change in immigration in 1882, though Jews mark that year as the advent of the First Aliya and commencement of modern political Zionism. At the time, those early Jewish agricultural pioneers did not know their descendents would refer to them as "Zionists" years later. Already from the end of the eighteenth century onward, the Turks experienced an increase in Jewish immigration to the Holy Land, in particular to the four holy cities of Hebron, Jerusalem, Tiberias and Safed and the coastal port of Jaffa. Ideas advocating a Jewish State already in the mid nineteenth century began with Jewish thinkers like Moses Hess and Protestants such as the Englishman Lord Shaftesbury and the American John Nelson Darby as well as certain English speaking consular officials in the Middle East.[1] Jews began to think in terms of modern nationalism while the European powers pursued policies further undermining the Ottoman Empire. There was a constant empowering of non-Muslim *dhimmi* communities who increasingly gained equal rights under European trade contracts and protection such as the Capitulation Treaties.

III Jewish National Liberation Catalyzes Islamic Antisemitism

Jewish emigrants began leaving Eastern Europe as antisemitism became more intense. They made their way to Central and Western Europe, the shores of North America, or almost anywhere else outside of the Land of Israel. Jews who came to the Land of Israel continued to endure the *dhimmi* troubles of living under Turkish rule. After noting the Jewish tenacity toward development in the region, Mehmed Sharif Rauf Pasha, governor of the Jerusalem district from 1877-1889, restricted Jewish immigration and the selling of land to Jews in his region. Bans spread to other parts of the country as well.[2] Despite having officially gained equal rights in 1856 during the Ottoman "Tanzimat" reforms the Jews continued suffering the *dhimmi* style abuse and humiliation, but at this point there was no real opposition to Jewish immigration as a national movement. Only after Theodor Herzl orchestrated the First Zionist Congress in 1897 did the Jewish nationalist movement make international headlines. Native Ottoman Jews and those already in Palestine were forbidden from purchasing state-owned lands. Herzl sought both political and financial support in the Jewish community while engaging in world diplomacy and hoping for international recognition and the opportunity to cut a deal with the Turkish Sultan Abdul Hamid II. Herzl planned to raise money to pay off the Ottoman Empire's debt in return for Jewish rights to the Land of Israel. As expected, the Sultan was averse to ceding any part of his empire; however, Jews could apply for Turkish citizenship. Although Herzl made little headway, even after his 1901 meeting with Abdul Hamid, the Sultan did address the Jews as a "nation" and apparently considered allowing land purchase in order to help fill Turkish coffers. But the overall policy of Ottoman obstructionism to Jewish immigration and development carried the day, and prevented many more Jews from arriving in the Holy Land.[3]

Herzl visited Palestine in 1898 and immediately saw the contrast between decay in Ottoman administered Jerusalem, and the enthusiastic pioneering spirit in the farming villages where Jews lived.[4] Jewish pilgrims were allowed to travel to the area provided they came on a visa and left a financial deposit guaranteeing their exit within thirty days, but the ban on new immigration remained.[5] Nonetheless, this opposition did not halt the flow of Jews into Palestine, which continued through the period of Turkish political instability and the overthrow by the "Young Turks" originally known as the Committee of Union and Progress (CUP) in 1908. That same year the Sultan himself became an opponent of Zionism at the urgings of Muslim officials who advocated a halt in land sales to foreign Jews. Anti-Zionist attitudes were expressed in the Jaffa Arab newspaper, *Falastin*. In particular, the Arabs perceived Jews as a weakling people attempting to buy land through stealth and lack of Arab awareness.

By 1914, there were two Arab notables from Jerusalem elected to the Ottoman parliament, one each from the el-Husseini and Nashashibi families, both insisted on ending land transactions with the Jews.[6] The Zionist Organization hoped for a change in policy with a secular regime in power, but to no avail. Neither pan-Ottomanism, which emphasized the equality of all citizens of the Empire, nor pan-Turkism, which concentrated on Turkish nationalism and extending unification with ethnic Turks throughout the world, found Jewish national objectives to be in their interest. The Young Turks behaved no different toward Jews and Zionism than the previous pan-Islamic leaning rulers.[7] According to the Ottomans, Zionism was seen as "secessionism" used by Western powers to detach parts of their empire, as already happened in the Balkans. Hence, imperial opposition rooted in the desire to hold the empire together influenced the Ottomans. Arab anti-Zionism often made common cause with the Turks through their joint Muslim identity.[8] Pan-Ottomanism gave the Jews supposed individual rights, while pan-Turkism and pan-Islamism entirely left the Jews out of their definition of citizens with rights. The latter contributed to re-strengthening the existing Turkish-Arab bonds through common religious ideals. Later, just prior to WWI, Turkish power faltered as a result of wars with Italy and the Balkan states. They needed Arab Muslim allies inside the empire, people who shared similar priorities and loyalties. Thus, it was important the Turks were not seen as making concessions to the Jews.

As for real estate, absentee *effendi*, or upper class Arab land-owners did sell private land to Jews. Within a short time the Jewish *dhimmi* image began to change as Jews worked the land, learned skilled trades and formed a national movement. The same *effendi* that sold land complained to the Turks of Jewish immigration and condemned the Jews for purchasing land. By 1909, the *effendi* charged the Jews with trying to displace the *fellahin*, or Arab peasantry, and lumped all Jews together as an adversary, whether Zionist or not. A Jew living in the Muslim world who supported Jewish nationalism, which was completely disconnected from Islam, was seen as a traitor and theoretically could face the death penalty. The Arab peasantry thought of Jews as *dhimmi*, even if the statute was nullified and no longer enforced. Even more significantly, if Jews could attain improved social standing, then why should they, the Muslim peasantry, be left behind? Certain "Ottomanist" Arab leaders and intellectuals protested as well, however they claimed to be only anti-Zionist and not antisemitic. The Arab peasants saw these protests as insincere and for the most part ignored them.⁹ They viewed Arabs who sold land to Jews as betrayers of the *dhimma* status. The Arab upper class *effendi* was worried by the latter conclusion even more so once the socialist Laborites took the helm of the Zionist movement by the 1930s. The Laborite atheism and socialism/communism were viewed as dual evils facing traditional Islam, and said to be comparable to the Jewish-pagan alliance mentioned in the Koran (see Chapter I "Negative Image of the Jew in the Muslim/Arab World").

World War I proved to be a major disaster for the Jewish community in Palestine, regardless of whether they supported the Zionist national idea or were ultra-orthodox communities awaiting the advent of the Messiah. Turkey joined the war in November 1914, but it took only a month for Djemal Pasha, the Turkish commander in Palestine, to issue expulsion orders for all those with Russian citizenship, condemning them as enemy aliens. Six to seven hundred Jews were shipped out before intervention by German Zionists halted the decree. As the Turks waited for another opportunity, 12,000 Jews departed on their own by early 1915 due to the threatening conditions. More decrees were on the way. Jews and Christians could no longer buy exemption from military service, a modern form of *dhimma* type taxation; they were forced into slave labor conditions, building roads and working in quarries, suffering mortal danger on a daily basis and often dying in epidemics or by starvation. Young Jewish men in particular were incarcerated in slave labor camps in Damascus, Bursa and Istanbul. Deha a-Din, the Turkish Secretary for Jewish affairs, banned all Zionist activities of any sort; he closed newspapers, schools and the Zionist Organization's Anglo-Palestine Bank. He confiscated Jewish crops and cattle, and increased taxes to impossible levels while the Turks encouraged Arabs to individually attack the remaining Jews. Within the first two years of WWI 8,000 Jews died, mainly from disease and starvation.[10]

Some Jews tried to prove their loyalty by joining the Turkish Army, but convinced no one. Despite these early efforts, the Zionist movement realized that Jewish national interests lay with the British.[11] Using newfound organizational skills and political strength, limited as it was, Jewish deportees organized the Zion Mule Corps in 1915, and later the 38th, 39th and 40th Royal Fusilier battalions, to fight against the Turks on the side of the British. The Zion Mule Corps, a transport unit, saw action in the Battle of Gallipoli, and some of the other battalions were involved in the offensive against the Turks in 1918 in Palestine.[12] After the first expulsions, Jews established the NILI[13] spy ring to supply information and to help plan the impending British offensive into Palestine from Egypt. Although fairly successful, the

Turks uncovered the NILI operation by the middle of 1917.[14] True to form the Turks expelled the remainder of the Jewish community from Jaffa and Tel Aviv already in March 1917,[15] further reinforcing Jewish pro-British sentiment. Following the Turkish massacres of one and a half million Armenians in 1915, the remaining Jews figured they would suffer the same fate whether during the war or in its aftermath. Hence a British victory was imperative. As a result of the overall suffering, the Jewish population dropped from 85,000 before the war to 58,000 in its aftermath.[16] Yet despite everything the "new" Jewish nationalists became warriors in complete contrast to their stereotype as understood by the average Muslim, be he Arab or Turk.

Palestinian Arab Muslims solidly supported the Turks. Jewish fear of the local Turks and Arabs ran high as those two groups worked together, forming one Ottoman Islamic front at the outset of the war. To quote the Palestinian historian Muhammed Muslih at length:

> Most members of the Palestinian nobility opted to identify with Ottomanism throughout the years of World War I. Thus, when the Ottoman Empire entered the war on the side of Germany against Russia, Britain, and France on November 5, 1914 many Palestinian notables remained loyal supporters of the sultan. No sooner had Sultan Muhammad Rashad proclaimed jihad (holy war) against the Allies at the start of hostilities than his call gathered momentum and support throughout Palestine.
>
> In Nablus, for instance, a big crowd gathered at the palace of the al-Nimr family with a huge camel which they slaughtered as a pledge of obedience to the sultan. In a resounding voice they all chanted: "God grant victory to the Prince of the Muslims our Sultan." Drawing on the support of local notables such as Sadiq Agha al-Nimr, the Ottoman state was able to recruit Nablusites to serve in the army. . . .
>
> In Jerusalem, the situation was not different. Here Shukri al-Husayni, Raghib al-Nashashibi, and other prominent urban notables also maintained their Ottoman patriotism and their support for the Ottoman regime.[17]

III Jewish National Liberation Catalyzes Islamic Antisemitism

Muslih further states that Arabs from Acre and Jaffa, along with the rest of Arab populations throughout the land, remained loyal to the Ottomans up to the end of the war. The famed Arab revolt led by Sharif Hussein in the Arabian Peninsula had little impact on either Palestinian or Syrian Arabs. Hussein himself fully broke away from the Turks and sided with the British after the Turks refused to make him governor of the Hejaz and recognize his hereditary rights to Mecca and the Hejaz.[18] "Ottomanism," or a renewed loyalty to the Ottoman Empire, could not satisfy the rising nationalist non-Muslim former *dhimmi* communities who often felt little allegiance to an oppressive regime now cloaked in broad-based secular nationalism where all were said to be equal. In practice Ottomanism played out in its most logical form, as a redirected hybrid state authority working for the benefit of all Muslims. It was a cross between the empire's attempt at secular nationalism and pan-Islam.[19] This fell very much in line with the pre-WWI thinking in Sharif Hussein's own family, that the possibility of a dual Turkish-Arab monarchy, modeled on the Austro-Hungarian example, could be the preferred solution as opposed to a clash with the authorities in Istanbul.[20] After all, Sharif Hussein was responsible for the Mecca and Medina holy sites.

The British MacMahon Correspondence (1915) and the Balfour Declaration (1917) made promises to both Arabs and Jews concerning post war arrangements. These conflicting obligations became the dominant points of controversy once the British and their allies won the war. No doubt the Arabs who supported Sharif Hussein felt betrayed when the British did not follow through with forming a unified Arab kingdom (Pan-Arabism). They expected a kingdom including all regions of Arab claims in the Middle East in return for their wartime efforts against the Turks. Instead they received a series of semi-independent entities either as British (Iraq, Palestine and the split off of Transjordan) or French mandates (Lebanon and Syria), which would receive independence at a later date, but these entities were not unified.[21] The Zionists obtained a "national home," or what the Jews hoped would be a state in the making, but with less commitment from Britain than expected.[22]

When the Ottoman Empire collapsed, the modern "secular" Turkish state formed under Mustafa Kamal (known as "Ataturk") solidified its Muslim-Turkish identity by eliminating its Christian population. In the northeast close to one and a half million Armenians were killed during WWI, a continuance of persecutions beginning in the 1890s. During the Greco-Turkish War (1919-22) an estimated one and a half million Orthodox Christians were expelled from the western part of the country and likewise, half a million Muslims were expelled or fled Greece for Turkey. Hundreds of thousands were killed and wounded on both sides.[23]

Islamic Fundamentalism and Palestinian Nationalism in the British Mandate

The collapse of Pan-Arabism and the establishment of the British-administered Palestine Mandate to advance the Jewish National Home catalyzed the Arab *effendi* into a new avenue of action, a localized Palestinian Arab nationalism eventually led by Haj Amin el-Husseini. Many try to present Palestinian Arab nationalism as only a secular nationalism, when in fact it was a mixture between regional Palestinian Arab identification and Islam. Haj Amin represented this integrated identity more than anyone else. He later became the Grand Mufti of Jerusalem, an Islamic clerical position he used as a springboard to achieve Arab and Islamic national interests.

Haj Amin was educated in Cairo, and received both an Arab and Islamic education serving as a Turkish officer during the war. Once the British were on the offensive he disappeared and made his way back to his native Jerusalem, re-emerging in its aftermath as an advocate of Pan-Arabism. In Jerusalem, he stirred up Arab mobs against the Jews and turned the Muslim Nebi Musa celebrations of 1920 into an anti-Jewish pogrom, killing several and wounding dozens. This violent move gained him hero status and notoriety among Arab Islamists. He was convicted by a British court for inciting the riots, yet amnestied. After promising future good behavior, he returned to Jerusalem, whereupon the British helped him become Grand Mufti, the leading Muslim cleric, although he was not even thirty years old. Politically savvy, he established the Supreme Muslim Council (SMC), developed official *waqf* lands (in this case lands held by the Muslim authorities in land registries), and built the largest political patronage system in the Palestine Mandate. He brilliantly outmaneuvered the older generation and its representative organization, the Arab Executive, which collapsed in 1934.[24] Overall sentiment was not just anti-Zionist but antisemitic as evidenced by the 1921 Haifa Congress of Palestinian Arabs which declared the inability of the Jew to live with others due to clannishness, demands for privileges, greed, wealth and overall plotting to take control of the country while driving its inhabitants into poverty.[25]

During the 1920s, Haj Amin renovated the mosques on the Temple Mount, which Muslims call the Haram al-Sharif, and constantly put out religious-political messages castigating Zionism, the British and the Mandate established in the name of the Jewish National Home. Still, the situation remained calm. Tensions arose over Jewish prayer rights at the Western Wall during Yom Kippur in October 1928 when the Jews placed a divider between men and women in violation of the status quo disallowing any physical changes in the prayer section. Finally the tension exploded with the summer riots and pogroms Haj Amin led against Palestinian Jewry in August 1929. A false rumor was spread accusing the Jews of planning to destroy the Al-Aksa Mosque and replace it with the Third Temple. This same Islamic battle cry has continued for close to a century. In a well-planned attack, Haj Amin and his followers slaughtered mostly non-Zionist Jews in Jerusalem, Hebron and Safed, as well as in other towns in outlying regions. No differentiation was made between ultra-orthodox Messianists, secular Zionists or any other type of Jew. The violence was purportedly directed against the Zionist movement, yet in the holy cities the Jewish Orthodox pietists who rejected Zionism paid the highest price.[26] Haj Amin and his followers no doubt knew the difference between these two groups, the former activist and the latter passive. However all Jews fell into the exact same domain, as *dhimmis* who raised their heads far too high and did not know their relegated place in Islamic society.

Zionism empowered the entire Jewish world, regardless of whether a Jew was religious or not. Jews living in the "national home" saw themselves as equal citizens not to be persecuted. The "new" Zionist Jew physically fought against persecution and dreamed one day of an independent state, which at the end of the 1920s appeared to be anything but assured. Nothing could have galled Islamic traditionalists more than Jews returning to theologically endowed *waqf* lands, captured by Islam and to be ruled in eternity by Muslims in the name of Allah. The Jews had League of Nations support for the Palestine Mandate "Jewish National Home" objective and were seen as the vanguard for a European-Western type society in the midst of the Arab/Muslim world. One either had to deny the Jews were a people in the national secular definition, as the PLO would do in the 1960s, or one had to declare them an evil nation to be punished, oppressed and destroyed if necessary, as advocated by the Koran, the Hadiths and continuing Islamic traditions. By no means a secularist, the Grand Mufti of Jerusalem Haj Amin and his followers chose the latter definition. In so doing they remained within Islamic traditions and with perfect logic would find their allies in the Axis Powers, most specifically Hitler's Nazi regime in the 1930s and even more so during WWII. Haj Amin and his supporters took a giant step beyond the oppressive medieval and czarist style antisemitism by allying themselves with Nazism and the "Final Solution" of Jewish annihilation in the 1940s.

Yet before Haj Amin found his way to the Nazis he was unquestionably a politician playing many sides of the same game. In the ensuing British investigation, Haj Amin pressed to have land sales to Jews and Jewish immigration halted while demanding an immediate Palestinian Arab State. The British initially agreed to the Passfield White Paper of 1931 basically accepting these demands, but reversed themselves and did not change their policies. An increase of Jewish immigration continued especially in the 1930s as a result of Hitler's rise to power. Retaining control, Haj Amin and his followers moved somewhat closer to the British, apparently as a tactical move, afraid of losing power and in the hope of one day gaining independence.[27]

Conflicting trends continued as tens of thousands of Jews arrived yearly to the Palestine Mandate in the mid-1930s as a result of Nazi persecutions and endemic antisemitism in Central and Eastern Europe. Land sales to Jews by the *effendi* upper classes continued behind the scenes while many of the same families publicly condemned Jewish purchases. Jewish immigration and investment brought economic boom to Palestine. It was one of the few successful economies during the Great Depression. In the 1930s, former Arab peasants in search of jobs moved to cities where Jewish and British Mandatory development prevailed. They came from all around the Middle East, some immigrating after their landowners sold the fields they worked. This simple lower class population was the most susceptible to Islamic messages, especially when those messages came from a Muslim cleric.[28]

Sheikh Izz a-Din al-Qassam was one of the most outspoken Muslim clerics who challenged the leadership of the Grand Mufti Haj Amin. Syrian born and Egyptian educated, al-Qassam considered Haj Amin to be overly cautious. Al-Qassam constantly preached Jihad, and inspired the more outwardly radical Islamists. He began a revolt against the Jewish National Home and the British without Haj Amin's support. Once pressures reached the point of no return, the Grand Mufti was forced to join. It is believed al-Qassam began planning the Islamic revolt in the period between the Yom Kippur incident of 1928 and the Mufti-initiated pogroms in the summer of 1929. Eventually, Izz a-Din al-Qassam became a symbol of the Islamic struggle for domination in Palestine and later against Israel.

By the mid 1930s the moderate Arab Executive opposition was vanquished (see above) but even Haj Amin and his Supreme Muslim Council (SMC) were seen as too accommodating to the British. Secular and religious ideological groups said to support the Mufti challenged him to take action against the British and the Jews. At this point, the radical, nationalist pan-Arab youth groups organized on the European fascist-Nazi models now found common cause alongside the extreme Islamists.

Leading the extremists, Al-Qassam stressed commitment to the purification of Islam, Muslim solidarity, xenophobia, and Jew-hatred. Proclaiming Jihad through the authority of the Koran, he organized a methodical armed struggle and developed religious justification for the 1929 massacres. He differed with the Mufti and SMC over their use of funds for the renovation of the Al-Aksa Mosque domain on the Temple Mount. He saw Haj Amin and the SMC as not confronting true issues, and focusing instead on renovating a mosque. For al-Qassam, Arab nationalism meant little, since Islam and Arabism were one. The Arab world was the birthplace of Islam, with the Arabs being the purest, most committed and faithful believers. For him the Palestine national and political struggle was a Muslim struggle, in which they would prove the superiority of Islam. He began by organizing the "Black Hand" terror group, mostly in the cities of Nablus, Jenin and Nazareth, to kill Jews in the northern districts of the mandate. He preached Jihad and self-sacrifice while training his followers in combat. He recruited lower class workers and youth from the Young Men's Muslim Association (YMMA) in the Haifa region where he lived. Al-Qassam's objectives were clear—abolish the Jewish National Home and British Mandate, force the present Palestinian leadership to resign, form alliances with Britain's enemies and accomplish it all through armed struggle.

Armed bands commenced terror actions in the spring of 1931 against Jewish farming villages and those traveling the roads. Tensions rose after Hitler's election in 1933, which resulted in increased German Jewish immigration. The advent of Nazism encouraged the Arabs that liberation was at hand and a global war would doom the Western European powers in the Middle East. In the wake of the 1929 massacres, the British caught a Jewish importer bringing 800 rifles and 400,000 rounds of ammunition through Jaffa port, in what came to be known as the "cement barrel incident." This situation proved to Muslims the Jews had no intentions of enduring victimization again. With less than one hundred followers, al-Qassam went on a major offensive in the autumn of 1935 after preaching Jihad in the Haifa region villages. He and his followers sold personal valuables and used the money to purchase guns, ammunition and explosives.

The British described al-Qassam as an extremely dangerous religious fanatic. Hiding out in northern Samaria, he led prayer services, preached Jihad, obtained local support and infused his loyalists with a "*mujahadin*" or a holy warrior identity taken from the days of the Prophet Mohammed and beyond. The group terrorized the land especially murdering Jewish farmers. He met his end after killing a policeman. The British now had a very good reason to pursue the Qassamists as criminals. They caught up with the group near the village of Ya'bad that November. Several people were killed, including the sheikh himself; others were arrested. The marauders made their point by fighting to the end and became martyr-heroes in the eyes of many Muslims. Al-Qassam's funeral procession was dramatic and impressive with shops closed and black flags draped along the route. His death became the basis of a new personality cult further exploited by Jamal al-Husseini and some in the Nashashibi clan to strengthen their young recruits. These groups were modeled on the Hitler Youth and the Italian fascist youth movements.

By early 1936 rebellion was in the air. Haj Amin urged caution, believing the time was not ripe for a full-scale revolt, and held to a temporary policy of cooperation with the British Mandate authorities. It is believed he approved of al-Qassam's activities, but could not be directly associated with them. It is noteworthy that Fatah's Yasir Arafat appears to have held a similar policy sixty years later in the 1990s when trying to politically smooth over Hamas' terror. Haj Amin had no disagreement on the need for an extreme anti-Jewish policy. Overall, the lower class hero al-Qassam undermined the Mufti who represented the *effendi*. Most likely al-Qassam saw Haj Amin as playing politics with Islam and not as the active Jihadist he should have been. Islamic radicalism shifted to the countryside while urban leaders, usually more moderate and representing wealth, began losing political power. By April, the Arab Higher Committee, representing virtually all Palestinian factions, with Haj Amin in its lead, called for a general strike. Within months the British established the Peel Commission, which became the first to recommend an Arab-Jewish two-state solution. The Islamists and Haj Amin rejected the compromise, although the moderates are said to have considered the idea, but in the end found it unacceptable. In the Jewish community, the Labor Zionist leadership was willing to discuss the issue but with many reservations, while the right wing Revisionists were not.

For the Qassamites the strike was not strong enough and any thought of compromise was even worse. Armed insurrection now broke out with the ghost of Sheikh Izz a-Din al-Qassam in the lead. The Black Hand terror campaign led by Sheikh Farhan al-Sa'di was renewed against Arab police officers working in the Mandatory Police and suspected collaborators. Often the terror was accompanied by *fatwas* from Islamic authorities in Damascus supporting the murders. Jews, Christian Arabs, British officials and many Muslim Arab politicians who were dubbed "traitors," such as Haifa's mayor Hasan Shukri and even the Mufti himself, received death threats. In 1938 moderates demanded protection after the Qassamites assassinated Lewis Andrews, the

III Jewish National Liberation Catalyzes Islamic Antisemitism

Acting District Commissioner for Galilee. When the British finally acted to break the back of the revolt they realized they had lost control over northern Samaria, western and lower Galilee. Britain banned the Arab Higher Committee and the SMC, exiled several Arab leaders, apprehended and executed al-Sa'di, when finally committing enough men to crush the armed bands. By 1939 they regained control.

During the revolt anarchy broke loose, the Mufti himself managed to flee to Lebanon in 1937, and the moderates led by the Nashashibi clan worked with the British to establish "Peace Bands" to defeat would-be assassins, many of whom originated with the el-Husseinis. Samaria and Galilee were inundated with the roaming armed units wreaking havoc, the most chilling attack taking place in Tiberias in 1938. Terrorists slaughtered nineteen Jews, eleven of them children, burnt down a synagogue and assassinated the Jewish mayor of Tiberias. By now many of the Arab upper class fled, especially the Christians. Law and order collapsed in the Mandate and the British were forced into an anti-insurgency counter offensive.[29]

In particular, Britain enlisted the aid of a policing force known as the Special Night Squads (SNS), made up of volunteers from the Jewish community under the command of the British officer Orde Wingate. Wingate believed in active defense, not just the restraint preached by many Jewish leaders to avoid clashes with the Arabs. The SNS guarded the oil pipeline originating in Iraq, cutting through the Lower Galilee and ending at the Haifa port. They kept Arab marauders on the defensive, brought increased security to areas in the north where anarchy ruled and, most importantly, continued a renewed tradition of Jews fighting back when attacked.[30] Jewish self-confidence returned and the image of the Jew as a coward was shattered.

For many this was a Palestinian Arab civil conflict as much as it was a rebellion against the Mandate. By the end, the el-Husseinis took up the radical position and battled the pro-British Nashashibis leading to internecine warfare between the two families, a conflict that continued for years, especially in the Jerusalem region. The massive British response killed and arrested enough of the Qassamites to put an end to the revolt by mid-1939. In the Judean hills they repressed the el-Husseini revolt, but extremism ruled the day, leading to the renewal of village clan feuds, financial extortions and attacks against Christians. In May, His Majesty's Government issued the White Paper severely curtailing Jewish immigration and land sales to Jews while recommending independence for a Palestinian State within five years. Despite the seeming loss in the field, the Arab revolt, or more historically correct the "Islamic" revolt, achieved its major objectives. The British had in essence cancelled the terms of the Palestine Mandate and the original objective of creating a Jewish National Home.[31]

The Qassamites were the first to integrate militant nationalism with extremist Islam. They became the role model for many groups due to their tenacious guerilla activities. These organizations included the secular PLO, established years later. However, it is more accurate to categorize Yasir Arafat and the PLO as behaving closer to the el-Husseini mode and doing their best not to make the ultimate sacrifice. Both Arafat and el-Husseini shifted sides, played politics and used extremist language while continuing to hold power in a balancing act, integrating an uncompromising Arab nationalism beholden to Islamic symbols and motifs. Sheikh Izz a-Din Qassam served as a model of personal sacrifice and behaved much more within the definition of an Islamic fanatic, intent on destroying all his enemies at whatever cost necessary. Eventually he became a folk legend. On the other hand, the wily Haj Amin mixed ardent nationalism with Islam and would live to fight another day, a lesson learned by Arafat in the 1970s. During WWII, Haj Amin el-Husseini integrated himself with the Nazi Axis from Iraq to Berlin and sought to help Hitler achieve his Final Solution in Eastern Europe. A brief survey of Haj Amin's WWII activities appears below. For a fuller review and analysis see Chapter VIII "The Czarist-Nazi Integration into the Palestinian Islamist Jihad."

Haj Amin el-Husseini's Alliance with the Nazis

Overall Nazi ideals were fairly well received in the Arab/Muslim world, and in particular in the Muslim Brotherhood. Likewise, Haj Amin el-Husseini felt a strong affinity toward the Axis powers and became their leading Arab/Muslim world ally. We find that Nazism also influenced the future Baathist leadership of Syria and the Egyptian Free Officers, led by Nasser in the 1950s, just like the rising Palestine national movement. By adopting Nazism, Haj Amin went well beyond the Charter of Omar stipulations. The Jews were not to be a second class *dhimmi* community allowed to exist upon declaring its loyalty to their Muslim overlords, but rather all Jews were to be annihilated. After the outbreak of the 1936 Arab Revolt in the Palestine Mandate, Haj Amin fled the following year to Lebanon and then onward to Baghdad. He joined the Iraqi pro-fascists led by Rashid Ali al Kilani in the overthrow of the overtly pro-British regime in April 1941.

The new Iraqi regime worked to attain an alliance with Hitler, but the British counter attacked too quickly forcing Rashid Ali and Haj Amin to seek

III Jewish National Liberation Catalyzes Islamic Antisemitism

refuge in Nazi Germany. As the regime collapsed in early June, incensed Iraqi mobs massacred 180 Jews in what became known as the "Farhud." Although forced out of the Middle East, Haj Amin was not to be deterred. The Mufti worked tirelessly to gain an audience with Hitler and secure an agreement whereby he and the Arab/Muslim world would be recognized as full Axis allies in the battle against the Jews, Britain and the Soviets. The opportunity arrived with their meeting on November 28, 1941 where a joint commitment was made to destroy the Jews and secure the Mufti as the Nazi representative in the Arab world. The Germans would engage in a Middle Eastern offensive at some unspecified future date. It appears Hitler was not particularly impressed with the Mufti's claim to leadership.

To prove his loyalties, Haj Amin helped organize the pro-Nazi Albanian and Bosnian Muslim units to aid Hitler in his war aims. The Mufti became intimate with the extermination camp commanders while his pro-Nazi Muslim forces were in training; however, these troops did not live up to expectations, ideologically or militarily. Haj Amin was successful as a public propagandist demanding Jewish annihilation. This is evident in numerous speeches and in particular in his viciously antisemitic pamphlet, *Islam and Judaism,* distributed to pro-Nazi Muslim troops. Most notably, his pamphlet accusses the Jews of paganism and attempting to kill the Prophet Mohammed. In particular, when using quotes from the Hadith he urges Muslims to slaughter the Jews.

Accused of war crimes, the Mufti was indicted after WWII but managed to flee to Egypt in 1945 where he was welcomed by the Muslim Brotherhood and would continue his activities in recruiting for the battle against Israel in 1948. He never recanted his demands for Jewish destruction.[32]

The Muslim Brotherhood and the Palestine Mandate 1945-48

As a result of the Arab Muslim revolt in the late 1930s, the Egyptian Muslim Brotherhood was forced to solidify its ideology in full support of the Qassamites. Jihad was an obligation in the battle against the Jews, all of whom were deemed Zionists, implacable enemies of Islam and lackeys of Western imperialism. The Brotherhood did its best to influence Egyptian policies and pushed for an official governmental condemnation of the 1939 White Paper believing the policy it advocated did not go far enough in putting an end to the Jewish National Home. After holding a conference in January 1939 declaring Islam as totally perfect and the only way of life, the Muslim Brotherhood began to grow in significant numbers beyond the 800 members counted three years earlier. Despite the growth of the organization, it was too late to join the struggle in Palestine, which the British had crushed.

In Egypt the leader of the Brotherhood, Hasan al-Banna and his deputy Ahmed Sukkari spent much of WWII in detention for their anti-British diatribes. The overwhelming British presence, coupled with a corrupt government not serving the people's interests, led to a massive increase in Brotherhood membership by the end of the war. Widely varying estimates claim between 100,000 to 500,000 Muslims joined the Brotherhood by 1945. At the time of the Israeli War of Independence it is believed the membership had at least doubled. The Brotherhood reached out to the Palestine Mandate, set up its first organizational branch in the fall of 1945 and two years later had a Palestinian membership of 12,000-25,000 and 25 branches, with Haj Amin el-Husseini as the nominal leader and al-Banna's hand-picked representative. This was somewhat of a false arrangement since the Mufti could not enter Palestine without being arrested because he was still wanted by the British for his wartime activities. Hence, he remained in Egypt. On the other hand, such a move reinforced the Mufti's standing as the leader of Palestinian Muslims and kept loyalties tied to the Cairo Center and al-Banna.

There was much discussion about Islam, morality, the Koran, social justice and the need for a war against the Jews in Palestine, but the physical effort demanded for engagement in the much acclaimed Jihad was lacking. Mahmud Labib, the Brotherhood's military commander, arrived in Palestine to unify the different youth movements of the Najjada and Futuwwa under his wing but before accomplishing his goal, the British deported him. The Arab Muslim militias lacked motivation, equipment, training, dedication and discipline during the 1948 conflict. These failed human capabilities became obvious during the early Palestinian Arab phase of the war. Their breakdown began with the rejection of the Partition Plan on November 29, 1947, and continued until January 1948, well before Israel declared independence and the Arab invasion began. These groups simply disintegrated although there were those who joined the Egyptian Brotherhood. Due to the Arab-Jewish irregular battles, much of the solid urban Palestinian middle class left the country during this period. Had Labib remained, it is possible he would have attained some form of unity.

There was no Palestinian Muslim Brotherhood unit. Cairo remained the center for political and military organization. In November 1945, groups began demonstrations against Jews, Zionism and the British. Ten to twenty thousand showed up for this first show of strength, which evolved into anti-Jewish, anti-Christian riots in Cairo and Alexandria. Muslims plundered

III Jewish National Liberation Catalyzes Islamic Antisemitism

Jewish and Christian properties, desecrated, and in some cases destroyed, synagogues and churches. Al-Banna condemned the violence, but scrupulously took no action to stop it. Instead, Muslims declared a boycott against Egyptian Jewry, since all were defined as "Zionists."

On the political level, the Brotherhood leadership made it clear to the Anglo-American Committee of 1945-46 seeking conflict resolution that no compromise was possible in Palestine. They saw themselves as a vanguard in the battle against the establishment of a Jewish State. Along with other Islamists they called for the establishment of an independent Muslim state in Palestine. The next Brotherhood-led Islamist demonstration involved 100,000 participants, took place shortly after the Partition Plan and called for the liberation of Palestine through blood. This was the largest demonstration in Cairo at the time and its effect reverberated throughout the region. Two months later al-Banna was bold enough to warn the UN Secretary General not to interfere in the Palestine conflict.[33]

The Brotherhood is said to have first penetrated into the Mandate with its Jerusalem office in 1943 under the name the "Makarem" society. During WWII, officials toured Lebanon, Syria and Palestine, immediately setting up several offices in Gaza after the war. They established their main Brotherhood headquarters in Jerusalem's Sheikh Jarrah neighborhood in May 1946. There were numerous Palestinian notables and delegates from outside of Palestine in attendance at a later convention in Haifa to discuss overall Middle Eastern issues. On the political level, the Jaffa National Committee urged cooperation with Arab nationalists, Christians and communists in opposing the Partition Plan of November 1947.[34]

In late October 1947, the Brotherhood began recruiting for the Jihad campaign and supposedly the first two days brought in 2,000 volunteers, with the number reaching 10,000 in early 1948. By early March, it was said 1,500 volunteers were inside Palestine. The numbers were wildly exaggerated with no more than fifty Jihadists training in Damascus, while a few others were involved in crossing the Egyptian Sinai border into the Negev. Apparently, only a few hundred Muslim Brothers enlisted in Colonel Abd al-Aziz's Volunteer Forces on the Southern Front in Palestine. Still, the political impact of the Brotherhood was much greater in forcing a reluctant Egyptian government to send an invasion force into the Negev to wipe out what was called "the Zionist terrorist gangs." The decision to invade the Negev occurred on May 11, four days before the Mandate expired. Previous to this decision, the Egyptian government already allowed Brotherhood military training under Cairo's supervision, wary that such abilities would backfire and be used against their regime. By joining the conflict, Egypt put the Brotherhood onto the battlefield within its own army, hoping to direct radical Islamic fire against the newborn Jewish State. The Brotherhood forced the Egyptian government to take the lead in its Jihad policy.[35]

In early April, before Israeli independence, the Muslim Brotherhood was involved in the attack on the religious kibbutz (collective farm) Kfar Darom in the Gaza region astride the road leading to Tel Aviv. Roving bands under the command of Colonel Al-Aziz infiltrated across the border from Sinai to gain information about Jewish defenses. There they met with local Arabs who assured them of a quick and easy victory and the Egyptians took the initiative. They charged the perimeter fence and broke through until reaching the defensive trenches. At that point, heavy fire held them back as they attempted a second breakthrough with armored cars, but that also failed. The withdrawal turned into a disaster when an Egyptian artillery shell fell short, killing their own men. The next day tanks appeared and one broke into the kibbutz, but a "Molotov cocktail" stopped it. Eventually, Jewish forces drove off the Brotherhood and their allies using explosive-filled *teffilin* (phylacteries) bags.[36] The attacks continued throughout the spring and failed. This being an untenable situation, Kfar Darom was abandoned by its Jewish defenders on July 8, at the outset of a major Egyptian offensive.[37]

When the Jews of Kfar Darom went to bury the bodies of Brotherhood members left behind they found matches, razor blades and a parchment the deceased wore around their necks, which declared the combatants as righteous Muslims fighting Jihad. Local Arabs later informed the Jews the matches were for burning the kibbutz to the ground and the razors were for castrating any prisoners taken. As brave and fanatical as the Muslim Brotherhood soldiers were, they did not have the ability to capture a fortified kibbutz. Al-Aziz was told to work on sabotaging water pipelines instead but he rejected the idea.[38]

Later Al-Aziz's brigade made its way to the southern outskirts of Jerusalem and took part in the battle of Kibbutz Ramat Rachel, almost capturing it in its entirety toward the end of May. The seesaw battle was witness to much looting and plundering by Arab villagers from Sur Baher and Al-Aziz's own troops, which led to a breakdown of discipline after the initial victory. Israeli forces counter attacked and retook the kibbutz in its entirety, only to lose most of it, except for the concrete dining hall, to the Egyptians shortly after. Israeli forces finally secured the kibbutz after Jordanian troops reinforced the Egyptians and ordered a withdrawal on May 25.[39] Besides doing battle in the Jerusalem region where they participated in limited activities in Silwan, Ramallah and the Battle of Kastel it is said the Brotherhood lent money to the basically defunct Arab Higher Committee for arms purchases. On the coast, they participated in the defense of Jaffa until its capture in May 1948. In addition the former Grand Mufti of Jerusalem Haj Amin el-Husseini, as an ally of the Muslim Brotherhood attempted to organize his Palestinian "Jihad Army" in the Jerusalem region under the command of his charismatic nephew Abdul Qader el-Husseini but when the latter was killed at the Battle of Kastel in April 1948[40] the military fortunes of the el-Husseini clan ended.

In the north Fawzi Qawuqji, another Arab Nazi collaborator who spent time in both Iraq and Hitler's Germany during WWII, invaded from Syria into the Galilee already in early January 1948, four months before the end of the British Mandate. Qawuqji led his several thousand man Arab Liberation Army (ALA) throughout the Galilee and reached as far south as the Nazareth region but was forced to withdraw in the face of counterattacks by Jewish forces. The ALA eventually retreated to south Lebanon as a result of the Israeli counterattack known as "Operation Hiram" culminating at the end of October 1948.[41]

Jihad and the Dhimma: 1948 and Beyond

As pointed out by many, the greatest defeat and resulting humiliation of Islam came as a result of Israeli independence in 1948. Israel survived despite five Arab countries—Egypt, Jordan, Iraq, Syria and Lebanon, along with contingents from other Arab lands—declaring war on and invading the newborn state. The two-state solution of "an Arab and Jewish State" to be carved out of Mandated Palestine failed when the Palestinian Arab State was not declared and remained stillborn. As a result of the war some regions allotted to the Palestinian Arabs were incorporated into Israel. Jordan annexed the West Bank and East Jerusalem while Egypt established a military administration in the Gaza Strip. Nineteen years later in 1967 when Israel's Arab neighbors once again threatened her with annihilation, the Six Day War ensued and Israel captured additional territories (see below). Nowadays, adherents of the Sunni Muslim Brotherhood alongside the Iranian Shiite Khomeinists[42] still agree on the need to eradicate Jewish independence. Jewish national assertion was deemed as nullifying Islamic superiority, and thus the continuing Jihad. Encapsulated within this epic are the former Jewish communities who lived under the *dhimma* in past generations, and immigrated to Israel en masse during this two-decade interim. Not only did the Jihad backfire, but these Jews integrated within the sovereign Jewish State on *waqf* lands. From the secular nationalist perspective, Jews were on

Palestinian Arab "national lands."

As shown above, Jewish independence was fortified and continues to thrive today because some 800,000 Middle Eastern and North African Jews immigrated or "made *aliya*" to Israel in the early days of statehood. Previously, Zionism was contained within a religious afterlife understanding represented in the Messianic arrival believed to promise Jewish national redemption and sovereignty.[43] Over the centuries these communities often lived under the harsh everyday *dhimma* afflictions and/or persecutions. In the post WWII era when harassed by Arab nationalism despite promises of equality these communities exhibited the highest percentage of emigration to Israel, often in the range of 90 percent. Most of Middle Eastern Jewry turned discrimination and their "out group" status into a socially cohesive force, shifting into a secular national redemption expressed through participation in the establishment and development of the modern Jewish nation state.[44] Many in Israel saw this as a true "ingathering of the exiles."

Secular Arab nationalism from the mid-1940s to the 1960s was no less insistent than the Jihadis in demanding Jewish national demise. Arab nationalism adopted the same absolutist beliefs, paralleling the Islamists beginning with their opposition to the Partition Plan of 1947, and their subsequent dispatch of troops to attack the newly declared Jewish State. The secular Arab beliefs were explained in the sixth point of the Arab League's eight-point "Declaration of Intervention" on May 15, 1948, as follows:

> Therefore, as security in Palestine is a sacred trust in the hands of the Arab States, and in order to put an end to this state of affairs and to prevent it from becoming aggravated or from turning into [a state of] chaos, the extent of which no one can foretell, in order to fill the gap brought about in the governmental machinery in Palestine as a result of the termination of the mandate and the non-establishment of a lawful successor authority, the governments of the Arab States have found themselves compelled to intervene in Palestine solely in order to help its inhabitants restore peace and security and the rule of justice and law to their country, and in order to prevent bloodshed.[45]

On the same day, the Secretary General of the Arab League, Azzam Pasha was much more direct. He saw the Jews as easy prey for slaughter. "This will be a war of extermination and a momentous massacre which will be spoken of like the Mongolian massacres and the massacres of the Crusades." To this he added the threat of annihilation to Middle Eastern Jewry stating explicitly, "There are over one million Jews in the Arab Lands. Their lives will be forfeit as well when we conquer the Jews."[46]

As regards the "Declaration of Intervention" the "sacred trust" or *waqf* of Islamic responsibility was and is transferred to secular Arab nationalism through the Arab States. The lack of a "lawful successor authority" compelled the Arab countries to intervene to guarantee the "rule of justice." Israel's acceptance of the Partition Plan meant there would be a twin Arab State in Palestine, yet no mention is made of its legitimacy in the "Declaration of Intervention." From the Arab Muslim perspective, the Jewish State by definition is an unlawful entity, but why the inferred negation of the Palestinian Arab State? The answer is found in UN Resolution 181, which called for linkage in the two-state solution, acceptance by one side of its own independence and lending support to the other's side's claim to sovereignty. In such a case, Palestinian Arab nationalism would be tarred with compromising the Islamic *waqf*, or Arab national lands, should it accept the two-state solution. This is an original sin of betrayal. It was best to reject both claims simultaneously and force a military showdown, whereby the Arab world would be victorious and any thought of recognizing Jewish sovereignty scuttled. Better yet, the invading Arab countries could divide Palestine for themselves in an effort to commence with the Pan-Arabist goal of one unified Arab nation instead of a plethora of small regional entities.

The Arab objective was not only the destruction of the Jewish political entity, but in the words of Azzam Pasha, the objective was "a war of extermination." The Mufti Haj Amin el-Husseini declared from Cairo, "I declare a holy war, my Muslim brothers! Murder the Jews! Murder them all!"[47]

Jordan in particular viewed all Palestine as its domain and did not accept the "right" of Palestinians to establish an independent entity. When the war ended, Jordan held the West Bank, East Jerusalem's walled Old City and most of the holy sites in what had been Mandated Palestine. Jordan refused to accept the Egyptian puppet-state dominated by Haj Amin el-Husseini, declared on October 1, in Gaza. Although formally recognized by other Arab countries, King Farouk and others would see to its early demise when Egypt extended its military administration to the territory.[48] Previously, in February 1948, the Arab League itself rejected a Palestinian government as advocated by Haj Amin and decided instead on the establishment of a committee to handle Palestine questions.[49] Seeking conflict resolution with the Jewish State, King Abdullah I of Jordan not only agreed to an armistice in April

1949, but initialed a peace agreement with Israel in late 1950. The information leaked and an Islamist assassinated Abdullah on the Temple Mount, or Al-Aksa the Noble Sanctuary domain, in July 1951.

The rise of Nasserism and the Syrian revolutionary Baath secular Arab nationalism solidified the anti-Israel front, yet neither wanted the Muslim Brotherhood at the helm of a Palestinian State. No peace agreement with Jordan would be forthcoming as the conservative monarchy, led by King Hussein beginning in 1953, was not in a position to concede sovereignty over a region deemed as belonging to the Arab nation. Both North African Nasserism and western Asian Baathism make claims to full Arab ownership of all lands throughout the Middle East. These two secular Arab nationalist doctrines would dominate the second half of the twentieth century.

Led by Egypt's President Nasser, Arab fury continued to boil in the interim period after 1948 and before the 1967 Six Day War. This was despite the cease-fire after the 1956 Sinai Campaign, Israel's complete withdrawal from Sinai and its demilitarization as part of the UN mediation to prevent hostilities. Below are declarations from the 1960s, demanding Israel's demise:

> Arab unity is taking shape toward the great goal – i.e. the triumphant return to Palestine with the banner of unity flying high in front of the holy Arab march. (Cairo Radio, 1963)[50]

> We swear to God that we shall not rest until we restore the Arab nation to Palestine and Palestine to the Arab nation. (Nasser, 1964)[51]

> Morale is very high among the members of our armed forces because this is the day for which they have been waiting – to make a holy war in order to return the plundered land to its owners. In many meetings with army personnel they asked when the holy war would begin – the time has come to give them their wish. (Egyptian General Abdul Mushin Murtagi, commander in Sinai, May, 1967)[52]

These are Jihadi sentiments from Egyptian sources, even if couched in more secular terms. There is to be "Arab unity" and a "holy Arab march" evincing the well known demand for Islamic unity and sacred entrance into a "holy war," or Jihad, as required "in order to return the plundered land to

III Jewish National Liberation Catalyzes Islamic Antisemitism

its owners." This is the concept of defensive Jihad in full (see Chapter II on "Ideologues") whereby Palestine is viewed as *waqf* lands, but in the Arab national sense is "restored to their rightful owners." The Arabs "shall not rest" until they achieve their objective. In this case, we are speaking of secular Pan-Arab nationalism led by Nasser, the same Nasser who executed Sayyid Qutb for his Muslim Brotherhood extremism. Secular Arab nationalism adopted Islamic absolutism and Jihad.

This is reinforced with the "total war," a continuation of the "defensive" Jihad ideal whereby an entire society is obligated to the military campaign, as we see below.

> We raise the slogan of the people's liberation war. We want total war with no limits, a war that will destroy the Zionist base. (Syrian Pres. Nuredime el-Atassi, May 1966)[53]

> The Arab people want to fight. We have been waiting for the right time when we will be completely ready. Recently we have felt that our strength has been sufficient and that if we make battle with Israel we shall be able, with the help of God, to conquer. Sharm e-Sheikh implies a confrontation with Israel. Taking this step makes it imperative that we be ready to undertake a total war with Israel. (Nasser, May 1967)[54]

> Our battle will be a general one and our basic objective will be to destroy Israel. (Nasser, May 1967)[55]

Syria's President el-Atassi defined the battle against Israel as a "people's liberation war" to "destroy Israel" in accordance with Nasser's declared objective shortly before the June 1967 War. Peace was never an option:

> We feel that the soil of Palestine is the soil of Egypt, and of the whole Arab world. Why do we all mobilize? Because we feel that the land of Palestine is part of our land, and are ready to sacrifice ourselves for it. (Nasser, 1962)[56]

> We say: We shall never call for, nor accept peace. We shall only accept war and the restoration of the usurped land. We have resolved to drench this land with our blood, to oust you, aggressors, and throw you into the sea for good. We must meet as soon

as possible and fight a single liberation war on the level of the whole area against Israel, imperialism and all the enemies of the people. (Syrian defense minister Hafiz el-Assad, 1966)⁵⁷

Nasser declared Palestine is part of Egypt while Hafiz el-Assad believed it to be inconsequential how much Arab blood is spilt. For Assad, all the land of Western Asia is Arab land, unified into a single whole known as "Greater Syria," as expressed in the Baath literature. Israel was to be destroyed "to restore the honor of the Arabs of Palestine," as noted below in tandem by Egypt.

> The Arab people is firmly resolved to wipe Israel off the map and to restore the honor of the Arabs of Palestine. (Cairo Radio, May 26, 1967)⁵⁸

The image of Israel and the Jew had changed. Suddenly Israel was more powerful than "empires" which previously fell to the Arab armies. But "full rights" in the Arab homeland would prevail and Israel was to be strangled as shown here.

> The noose around Israel's neck is tightening gradually . . . Israel is mightier than the empires which were vanquished in the Arab East and West . . . The Arab people will take possession of their full rights in their united homeland. (Egyptian newspaper Al-Gumhuriya, 1963)⁵⁹

Assessing the prevailing atmosphere in the Arab world in May 1967, Israel's Foreign Minister Abba Eban recorded the following impressions he had during an emergency cabinet meeting.

> There was no doubt that the howling mobs in Cairo, Damascus and Baghdad were seeing savage visions of murder and booty. Israel for its part, had learned from Jewish history that no outrage against its men, women and children, was inconceivable. Many things in Jewish history are too terrible to be believed, but nothing in that history is too terrible to have happened. Memories of the European slaughter were taking form and substance in countless Israeli hearts. They flowed into our room like turgid air

and sat heavy on all our minds. (Israeli Foreign Minister Abba Eban, May 1967)[60]

Originating from a Western background, Eban related the atmosphere to the Holocaust inspired exterminationist hatred of Nazi Germany and its accomplices. His perception of the Arab demand for Jewish annihilation was correct, but the catalyst was a mixture of the Arab Muslim fury over the breaking of the terms of the *dhimma* and the Arab clash with foreign Western powers who supported that same *dhimmi* group holding sovereignty in the Arab Middle East. The modern period sees the adaptation of Nazi solutions into the Arab world by way of the Muslim Brotherhood. This goes far beyond the murderous outbreaks of violence against the Jews of Iraq, Syria, Libya, North Africa and Yemen in the 1940s and continuing into the 1950s.

The Arab States mobilized and threatened Israel with annihilation by early June 1967. Israel struck first yet the Arab defeat was perceived as only a temporary setback to be rectified by Jihad. As a result of the Six Day War Israel captured Sinai and the Gaza Strip from Egypt, the West Bank and East Jerusalem from Jordan and the Golan Heights from Syria. As such, it was the exhortation of Muslim clerics hailing from 34 different countries and representing 750 million Muslims worldwide when they met in Cairo in October 1968 to decide upon Jihad and not conflict resolution. *"It is the religious duty of Moslems to liberate Jerusalem and to guard its holiness and its Arab character: No Moslem country could maintain relations with Israel. Collaboration with the enemy is a violation of the sacred doctrine of Islam."*[61]

Today the Khomeinist Iranian understanding of Jihad is sharp, clear and focused. Khomeini himself saw America as the "Great Satan" and Israel as the "Little Satan."[62] Most notably Jews are the enemy, as Khomeini himself explained: "From the very beginning, the historical movement of Islam has had to contend with the Jews, for it was they who first established anti-Islamic propaganda and engaged in various stratagems, and as you can see, this activity continues down to the present."[63]

Referring to Israel, Khomeini determined a "handful of wretched Jews" destroyed the Al-Aksa Mosque (a lie) and continued to occupy Muslim lands, this having "resulted from the failure of the Muslims to fulfill their duty of executing God's law and setting up a righteous and respectable government."[64] The government should obviously be Islamic, any secular type would be a betrayal. Worse yet, it is said, Jews rule these secular governments. To quote Amir Taheri in *Holy Terror*:

According to the theorists of Holy Terror it is possible to convert Christians to Islam, whereas Jews will never abandon their faith. Jews who pretend to have converted to Islam are "agents on secret missions." All those who tried to Westernize Muslim countries are said to have been Jews, starting with Khedive Mehmet Ali Pasha and including Kamal Ataturk, Reza Shah Pahavli, his son Muhammad Reza Shah and even Gamal Abdul-Nasser. Anwar Sadat was "a Jew who lived like a Jew and died like a Jew." Yasser Arafat, leader of the Palestine Liberation Organization, is also a Jew and a "secret agent of world Jewry." The principle is that it is impossible for anyone born into a genuine Muslim family to act against the fundamentalist interpretation of Islam; those who do not see Islam as the exponents of Holy Terror must by definition be Jews.[65]

Khomeini himself made the well-known accusation against the Shah,[66] while other condemnations of Jewish connections came by way of Sunni thinkers such as Qutb (see Chapter II "Ideologues") and spread throughout the Arab and Muslim world. Turning to Holy War, in his famous essay "Key to the Secrets" Khomeini explained what he meant by Jihad in what Taheri entitled "Islam is Not a Religion of Pacifists," published in 1942. The text was republished several times in the 1980s:

> Those who know nothing of Islam pretend that Islam counsels against war. Those [who say this] are witless. Islam says: Kill all the unbelievers just as they would kill you all! . . . Kill in the service of Allah those who may want to kill you! Does this mean that we should surrender [to the enemy]? Islam says: Whatever good there is exists thanks to the sword and in the shadow of the sword! People cannot be made obedient except with the sword! The sword is the key to Paradise, which can be opened only for Holy Warriors![67]

Iranian President Ahmadinejad, who reigned from 2005 to 2013, insisted on destroying Israel, as reported in the *New York Times* in 2005.[68] Ahmedinejad claimed he referenced Ayatollah Khomeini when he reiterated the Shiite regime's demand for the extermination of the Jewish State. The next day he addressed a crowd at an anti-Israel rally where he defended his remarks while the crowds shouted, "Death to Israel; death to the Zionists," and burned Israeli flags. Egypt and even the Palestinian Authority rejected his demand for Israel's destruction.[69] As of 2016, the Iranians continued to threaten Israel with destruction even under the "moderate" President Hassan Rouhani.

The PLO and PA Adopt the Hamas Jewish Stereotype

The Muslim Brotherhood was established in 1928 in Egypt and existed well before either the secular Fatah or Palestine Liberation Organization (PLO) came on the Palestinian scene in the mid-1960s. The PLO is a conglomerate of varying groups and represents an anti-Zionist secular Palestinian nationalism, while verbally distancing itself from Brotherhood antisemitic stereotypes. The PLO *Charter* and its supporters explain their respect for Judaism as a religion, but condemn Zionism as "imperialism" never forgetting to differentiate between Jews and the State of Israel. They maintain there is no geographic historical connection between Palestine/Land of Israel and Judaism, thereby denying one of the pillars of the Hebrew Scriptures or Old Testament (see Chapter VII *The Palestinian National Charter*, especially Article 20). Many third parties have bought into this line of reasoning over the years, especially those from the far left, and in part even some "human rights" organizations.

PLO Chairman Yasir Arafat (died 2004) was the most effective spokesman of this position. In addition to declarations from the *Charter*, his infamous November 13, 1974 speech at the United Nations was saturated with fierce anti-Zionism, while encouraging a supposedly friendly attitude toward Jews. He denounced "colonialism, imperialism, neo-colonialism and racism in each of its instances." Arafat denied Jewish nationalism, and said Zionism was a "scheme" for:

> . . . the conquest of Palestine by European immigrants, just as settlers colonized, and indeed raided, most of Africa. This is the period during which, pouring forth out of the West, colonialism spread into the further reaches of Africa, oppressing, plundering the people... This period persists into the present. Marked evidence of its totally reprehensible presence can be readily perceived in the racism practiced both in South Africa and in Palestine.[70]

In 1947 the UN recognized the legitimacy of Zionism, or Jewish nationalism, alongside Palestinian Arab national claims and decided on the Partition Plan, a compromise all Arabs rejected. Arafat testified to this rejection, claiming Palestinian Arab nationalism was legitimate while Jewish nationalism was not. "This General Assembly early in its history approved a recommendation to partition our Palestinian homeland. This took place in an atmosphere poisoned with questionable actions and strong pressure. The General Assembly partitioned what it had no right to divide – an indivisible homeland."[71]

PLO Chairman Arafat went on to describe Palestinian suffering as a result of the ensuing war and blamed the "colonialist settlers" because of their "dissatisfaction with the decision." The Arabs themselves were fully cognizant of their own decision to reject the Partition Plan and wage war. After losing the 1948 War, the Arab world narrative became one of Palestinian suffering and victimhood. After he described the 1967 Six Day War as the "result of Zionist aggression" and blamed the State of Israel for all ills in the Middle East, Arafat returned to focusing foremost on the Palestinian dispersion. Parroting *The Palestinian National Charter,* Arafat claimed Muslims, Christians and Jews would all live together peacefully in the state he envisioned, conveniently forgetting its overall definition as a Palestinian Arab State (Articles 1, 12 and 14 of the *PNC*). His statements meant Jews would need to accept a Palestinian Arab identity, or should they insist otherwise, equality in the state-to-be appears questionable.

Arafat said Palestinians would continue to "distinguish between Judaism and Zionism. While we maintain our opposition to the colonialist Zionist movement, we respect the Jewish faith," all the time denying Jewish nationhood. He closed by claiming, "Today I have come bearing an olive branch and a freedom fighter's gun. Do not let the olive branch fall from my hand." This is a similar statement as found in Article 31 in *The Hamas Covenant* where Jews can live a *dhimmi* existence; war and peace are contingent on whether Jews denounce Zionism, Jewish pride, collective identity and rights. This would be the life of the Jews—the same people Arafat defined as neither a nation nor a people.

A year later in November 1975, the UN General Assembly declared "Zionism is Racism" and despite the repeal of the decision in 1991, Arafat left the indelible worldwide impression of Jewish nationalism as equal with racial discrimination. Israel was tainted with the brush of colonialism, imperialism, exploitation and overall oppression. Arafat and the Palestine national movement legitimized additional diatribes and stereotypes of the Jew as the ultimate evil should Jews dare to be brash enough to not only seek independence, but stand up against Arab attacks. Similar to Hamas, Arafat blamed any aggression against Jews on Jews, even though they were the victims. Arafat gave major support to Hamas and borrowed from the Muslim Brotherhood through the delegitimization of the Jewish State, justifying it all through his supposedly anti-colonial, freedom loving, Third World liberationist perspective. Anti-Zionism and Israel bashing are easily translated into support for Hamas antisemitism. If Jews are responsible for Zionism as deemed evil by Arafat and the Palestinians, then it follows that Jews are intrinsically evil, since by Arafat's definition Jewish nationalism is responsible for colonialism, imperialism and universal oppression. It is of little significance that not every single Jew is included in this stereotype. (For more see Chapter VII, "A Comparative Analysis.")

Arafat ignored the detail that over half of the Jews in Israel were and are of Asian and African backgrounds, a group oppressed and despised by so many in the Arab Muslim world. Asian and African Jews became the backbone for building a new Jewish society no longer willing to have their identity dictated to them by those from their former states of residence. Even UN Resolution 242 in November 1967 attested to this fact concerning Middle Eastern "refugees." The refugees are a problem brought about as a result of the Arab rejection of the two-state solution and their 1948 invasion into Israel. Jihadism justified the invasion, as did Arafat. The resolution called for "achieving a just settlement of the refugee problem," devoid of any reference to whether the aforementioned were Arabs or Jews from Palestine or Jewish refugees from the Arab Muslim world.[72] The Arab invasion resulted in a Palestinian Arab refugee problem and the expulsion-flight of Jews from Arab countries.

Arafat, the PLO and their "anti-colonialist" allies use the veneer of modern secular analysis to condemn redemptive Jewish nationalism through the use of "Palestinianism." Arab Christian clergy originally advanced the idea of Palestinianism, as Bat Ye'or pointed out in her work *Islam and Dhimmitude*. The idea included "supersession" whereby Muslim and Christian Arabs adopted a secular Palestinian identity and replaced Israel while condemning the Jews as European invaders. The movement is understood to be theological, denying any Jewish claim to the Land of Israel. Instead, Christians and Muslims are to act together as a unified Palestinian Arab people, replacing the Jewish claim, or covenant. This newly constituted Palestinian people, never mentioned in antiquity, takes on secular sanctity as an indigenous group having dwelled eternally between the Jordan River and the Mediterranean Sea. Such a claim is anything but factual. On the subliminal Christian theological level, the Savior Jesus exchanges identity, replacing his Judaism with a secular Palestinian Arab identity—a complete fabrication. Next, all the Jews were identified as Judas Iscariot embodying the ultimate evil and condemned to damnation forever. This supposedly "secular" Palestinianism now takes on overwhelming theological significance, certainly for Christians, but for Muslims as well who see Jesus as a prophet.[73]

In the mid-twentieth century Arab Muslims began to redefine "Zionism," equating it to Nazism. They exclaimed the Jews were the Nazi oppressors and the Palestinians their victims. Muslims adopted this same originally antisemitic Palestinian Christian narrative as they moved from an Islamic identity to a victimized Palestinian identity. This is the height of "victim reversal" where Israelis and Jews are given Nazi traits while the Palestinians place themselves in the role of the Jewish victims of the German Third Reich. Let us not forget that the greatest Palestinian leader of all time, Haj Amin el-Husseini was a Nazi collaborator aiding in the transport of Balkan Jewry to the death camps. Hamas Islamists concurred with equating Jews to Nazis in Articles 20, 31 and 32 of their *Covenant*. Muslims identifying with the Palestine national movement reinforce this absolutism through extension of the *dhimma*. A secular form of absolutism nullifies the Jewish claim to independence when the national cultural homeland of Palestine takes on the holiness of *waqf* lands.[74] In their mind, no other national group can claim rights to the land, regardless of their attachment through history, culture and/or religion. Not only does secular Palestinian Arab nationalism rewrite history and theology, in particular the Hebrew Scriptures, or Old Testament, but also the Koran. It completely eradicates the memory of the Jewish connection to the Land of Israel as made clear in Article 20 of the *PNC*.

In 1932, the vehemently Jihadist pro-Nazi antisemite Indian Muslim Sayyid Abul A'la Maududi published his article "Suicide of Western Civilization." In his article, he explained Jewish suffering as Divine punishment for previous misdeeds and rebellion against Allah during the Exodus from Egypt despite their trek toward the Promised Land. Quoting Koran 7:137, which spoke of "the land We had blessed, and the most fair word of Allah was fulfilled upon the people of Israel," he justified the continued persecution of the Jews "by the tyrant rulers of Iraq, Greece and Rome." These leaders made the Jews homeless while they "wandered from country to country in utter humiliation. They were deprived of every authority. For the last 2,000 years, they are suffering so miserably with the divine curse that they find no place to live in with peace and honour all the world over."

Maududi's article shows a traditional antisemitic mentality by a non-Arab Muslim up until the advent of the State of Israel. Due to the Arab failure in

III Jewish National Liberation Catalyzes Islamic Antisemitism

the 1948 War the Muslims were and are humiliated through their defeat and establishment of Jewish sovereignty in the Land of Israel. Such a feeling of disgrace was intensified after 1967 when the remainder of the Land from the Jordan River to the Mediterranean Sea, including the Old City of Jerusalem, was captured by Israel. This leads to a devastating theological contradiction for Islam as the Jews were fulfilling the covenant and no longer must wander. In 1991, when Maududi's article was reprinted, a footnote was attached, which explained that "people doubt the authenticity of the Quranic prophecy" because the Israeli State continues to survive. The explanation blames America, England, Russia and France for Israel's existence, and Muslims are assured that when the Western powers are "incapable of supporting Israel" the "Arab countries around shall push this bundle of filth into the Mediterranean Sea" bringing about well-deserved calamity upon the Jews.[75] Maududi's hatred for Jews, admiration for Nazism and hope for an Islamic-style National Socialism impacted Arab Islamist movements and thinkers.[76] Islamists today view Israel's existence as a temporary phenomenon, one presenting a challenge and whose ultimate end is extermination. Islamists like Maududi understand the threat to Koranic prophesies and interpretations should Israel continue to survive. For them it is a "Zero Sum Game," Jihadi existence is predicated on Jewish destruction. This type of thinking is overtly similar to the WWII German racial understanding of the need to exterminate the Jews, but is set in an Islamic theological domain.

Historian Bernard Lewis believes anti-Zionism evolved into antisemitism at the outset of the twentieth century prior to WWI, when certain Arab intellectuals accused the Jews of using their enormous financial leverage and Western patronage to make gains at Arab expense. In addition, Russian Orthodox Christian influence and certain Western missionary activities impacted the Palestinian Eastern Christian churches so much so that these in turn catapulted the "crucifixion accusation" into Islamic thinking. Islam began to blame the Jews for Jesus' death and the crime of deicide despite the fact the Koran clearly denies the Jews crucified Jesus. Yet the deicide accusation made by certain Christian sects began its transfer into Islam despite the Koran's denial of Jewish responsibility for the death of Jesus. In fact, Koran 4:156-157 explains that the Jews falsely took credit for the death of Jesus. Hence began the deadly upgrade of Muslim anti-Jewish behavior into a condemnation of full "collective and hereditary Jewish guilt."[77]

During the Mandate period, the Arab leadership further attacked Zionism and Jews on the socio-economic level, denouncing them to the British as Bolsheviks. The Labor Zionist construction of a workers' movement, based on equality for all, greatly aggravated the Arab upper class *effendi* who feared such ideas would inspire their oppressed peasants and workers to rebel against the landowners and economic elite. They condemned the voluntary communal kibbutz framework as similar to the forced labor of the Russian *kolhozy*. The *effendi* saw Zionists and Jews in full alliance with the Soviets despite vicious anti-Zionism and antisemitism emanating from Moscow.[78] Certain aspects of European Christian demonization of the Jews made full inroads into the Arab Muslim world as attested to in the great Czarist forgery, *The Protocols of the Elders of Zion* where Jews were accused of wielding international influence by serving as communists and capitalists simultaneously. Arab Muslim leadership adopted this political stereotype of Jews as well.

The traditional Muslim stereotype of the Jew as the object of ridicule and embodying impoverished, humiliated, defeated cowardice in his very being, gave way in the face of twentieth century realities of Israeli military success, particularly in 1948 and 1967. Other characteristics, such as wickedness and slyness, remained a part of the Jewish stereotype. Traditionally, no one feared the hostile impotent Jew, but rather he was the object of laughter. Reconciling the old stereotype with the new reality, those of the traditional European Christian antisemitic determination of the Jew as a deadly "cosmic evil" became relevant, the emphasis now focused on Jewish craftiness, trickery and deceit.[79] In the Arab Muslim mind, the Jews transformed into an all powerful, Satanic, anti-Allah entity fulfilling the need for an all encompassing vilification to explain Arab defeats.[80]

It was against such a background, in the heydays of the Oslo Accords, that Arafat is thought to have jettisoned his openly "Palestinianist," anti-Zionist stance as described above, or so it seemed. In his letter to Israeli Prime Minister Yitzhak Rabin on September 9, 1993, he declared: "the PLO recognizes the right of the State of Israel to exist in peace and security." He then accepted UN Resolutions 242 and 338 calling for an end to the conflict and agreed to enter negotiations to arrive at conflict resolution.[81] Today it appears this was a well-planned tactical move to force Israel into concessions. Israel may be granted the "right to exist," but Arafat, by way of careful omission, never accepted the claim of Israel as the nation state of the Jewish People. Jews could live there, but the Jewish People had no legitimate claim to the Land of Israel, or in his perspective, "Palestine." Thus, there may be a two-state solution, yet any state other than Arab Palestine, as defined by *The Palestinian National Charter*, lacked legitimacy. By never changing the *PNC*, as the 1998 Wye Accords later demanded, agreements with Israel were contrary to its foundational principles and allowed for future cancellation of any such accords.

This echoes the "abrogation" clause in Koran 2:106 as a sacred understanding transferable to the international arena. (See Chapter IX "Conflict Resolution in the Shadow of Islamic Abrogation" for an explanation concerning the nullification of Koranic clauses).

Despite the lack of clarity in 1993 there was great hope for conflict resolution, yet the seven years of the Oslo process were not smooth. With numerous terror attacks, most carried out by Hamas, the Arafat-dominated Palestinian Authority continued its verbal assault against Israel, which set the stage for more conflict. In 1998, Deputy Minister of Supplies Abd Al-Hamid Al-Qudsi declared, "Israel did not change its strategy, which aims to kill and destroy our people." In 1999, Arafat's wife Suha accused Israel of using poisonous gas to induce cancer in women and children.[82]

Just prior to the failed July 2000 Camp David negotiations, mediated by President Bill Clinton between Yasir Arafat and Israeli Prime Minister Ehud Barak, the Palestinian Authority Minister of Supplies Abd El Aziz Shahian revealed the true objectives of the PLO. "The Palestinian people accepted the Oslo agreements as a first step and not as a permanent arrangement, based on the premise that the war and struggle on the ground [i.e., locally against Israeli territory] is more efficient than a struggle from a distant land… for the Palestinian people will continue the revolution until they achieve the goals of the '65 revolution…"[83]

Returning to the secular PLO/Fatah, and those who are said to be Israel's peace partners, the PA appointed Al-Aksa Mosque Sheikh Hian al Adrisi addressed 22,000 Friday afternoon worshipers with the following tirade igniting the Palestinian Low Intensity Conflict of 2000-2004, known as the Second Intifada, on September 29, 2000. The lie that anti-Zionism had no connection to antisemitism, as PLO Chairman Yasir Arafat constantly asserted, is suddenly revealed as beyond "inaccurate."

> It is not a mistake that the Koran warns us of the hatred of the Jews and put them at the top of the list of the enemies of Islam. Today the Jews recruit the world against the Muslims and use all kinds of weapons. They are plundering the dearest place to the Muslims, after Mecca and Medina and threaten the place the Muslims have faced at first when they prayed and the third holiest city after Mecca and Medina. They want to erect their temple on that place . . . The Muslims are ready to sacrifice their lives and blood to protect the Islamic nature of Jerusalem and Al-Aksa![84]

Two weeks later, the Palestinian Authority appointee Dr. Abu-Halabia of the "Fatwa Council" had this to say in Gaza: "The Jews are Jews, whether Labour or Likud, the Jews are Jews. They do not have any moderates or any advocates of peace. They are all liars. They are the ones who must be butchered and killed . . . Have no mercy on the Jews, no matter where they are, in any country. Fight them wherever you are. Whenever you meet them, kill them. Wherever you are, kill those Jews and those Americans who are like them . . ."[85]

Simultaneously on October 13, 2000, Palestinian Authority TV accused Jews of being outright murderers and terrorists as part of their essential being. "The Jews are the Jews. There never was among them a supporter of peace. They are all liars ... the true criminals, the Jewish terrorists, that slaughtered our children, that turned our wives into widows and our children into orphans, and desecrated our holy places. They are terrorists. Therefore it is necessary to slaughter them and murder them according to the words of Allah."[86] Notice the quote speaks of "Jews" and not "Zionists" or "Israelis," and religiously references Allah's commands.

The attempts through the Oslo Accords to achieve peace were exposed as a deception, as made perfectly clear in the words of Arafat confident Faisal el-Husseini in 2001. "We are ambushing the Israelis and cheating them . . . If we agree to declare our state over what is now 22 percent of Palestine, meaning the West Bank and Gaza, our ultimate goal is the liberation of all historic Palestine from the River to the Sea . . . We distinguish the strategic, long-term goals from the political phased goals, which we are compelled to temporarily accept due to international pressure."[87]

El-Husseini's statement eight years after the Oslo Accords was a throwback to PLO rejectionist resolutions taken in June 1974 in Cairo declaring what became known as the "Step by Step Approach" to Israel's destruction. To quote from three of the ten decisions (clauses 2, 3 and 4):

> The PLO will struggle by all possible means and foremost by means of armed struggle for the liberation of the Palestinian lands and the setting up of a patriotic, independent, fighting peoples' regime in every part of the Palestine territory which will be liberated.
>
> The PLO will struggle against any proposal to set up a Palestine entity at the price of recognition, peace and secure boundaries, giving up the historic right and depriving our people of its right to return and to self-determination on its national soil.

III Jewish National Liberation Catalyzes Islamic Antisemitism

> The PLO will consider any step toward liberation which is accomplished as a stage in the pursuit of its strategy for the establishment of a democratic Palestinian state, as laid down in the decisions of previous National Council meetings.[88]

The PLO would use any land from which Israel withdrew as a staging ground to continue the struggle for the "liberation" of all Palestine. Palestinians would not establish an entity at the price of conceding their historic full right of return. There would be nothing less than "self-determination on its national soil." The Palestinians "will consider any step toward liberation which is accomplished as a stage in the pursuit of its strategy," meaning any land obtained by the PLO in any manner, including negotiations, is an advance toward the realization of *The Palestinian National Charter*. The *PNC* to this day continues to call for Israel's destruction and a one state Palestinian Arab solution (See Chapter VII "A Comparative Analysis").

PLO Chairman Yasir Arafat himself was of this opinion as he continued negotiations for implementation of the Oslo Accords. In a Johannesburg mosque, six days after signing the Oslo I Accords in May 1994 commonly known as "Gaza and Jericho First," he declared, "The Jihad will continue, and Jerusalem is not [only] for the Palestinian people, it is for all the Muslim nation." He then compared the Oslo Accords to the agreement made between Mohammed and the Meccans in 628, whereby the Prophet agreed to a ten year truce, which the Muslims violated two years later when they captured the city and killed many of the inhabitants. This is the concept of *hudna*, or Islamic cease-fire, as discussed in Chapter I. Arafat continued his appeal. "We are in need of you as Muslims, as warriors of Jihad."[89] The Jihad went well beyond Jerusalem when Arafat addressed Arab diplomats in Stockholm in February 1996:

> The PLO will now concentrate on splitting Israel psychologically into two camps . . . We plan to eliminate the State of Israel and establish a Palestinian state. We will make life unbearable for Jews by psychological warfare and population explosion. Jews will not want to live among Arabs. I have no use for Jews. They are and remain Jews. We now need all the help we can get from you in our battle for a united Palestine under Arab rule.[90]

The Jews were to be destroyed as seen above, whether from the Palestinian Authority Fatah perspective, or from the Hamas point of view. The Palestinian Authority's Ibrahim Madhi had this to say in a sermon about the Jews in Gaza City in 2002 during the Low Intensity Conflict, or Second Intifada:

> The Jews await the false Jewish messiah, while we await, with Allah's help... the *Mahdi* and Jesus, peace be upon him. Jesus' pure hands will murder the false Jewish messiah. Where? In the city of Lod, in Palestine. Palestine will be, as it was in the past a graveyard for the Tatars and to the Crusader invaders, [land for the invaders] of the old and new colonialism . . .
>
> A reliable *Hadith* [tradition] says: 'The Jews will fight you, but you will be set to rule over them.' What could be more beautiful than this tradition? 'The Jews will fight you' – that is, the Jews have begun to fight us. 'You will be set to rule over them' – Who will set the Muslim to rule over the Jew? Allah . . . Until the Jew hides behind the rock and the tree.
>
> But the rock and tree will say: 'Oh Muslim, oh servant of Allah, a Jew hides behind me, come and kill him.' Except for the *Gharqad* tree, which is the tree of the Jews.
>
> We believe in this *Hadith*. We are convinced also that this *Hadith* heralds the spread of Islam and its rule over all the land...[91]

Sheikh Mahdi then continued, "Oh Allah, accept our martyrs in the highest heavens . . . Oh Allah, show the Jews a black day . . . Oh Allah, annihilate the Jews and their supporters . . . Oh Allah, raise the flag of Jihad across the land."[92]

In August 2002, the Palestinian Authority's Communications Minister Imud Falouji declared Yasir Arafat's claims, as well as claims made in the *Palestinian National Charter,* that the Jews were only a religious community but never a nation, as eternal falsehoods. The PLO completely contradicted itself and adopted the Islamist Jihadist perspective on the Jews, "The Jewish nation, it is known, from the dawn of history, from the time Allah created them, lives by scheme and deceit."[93] Just like Hamas, Falouji recognized the Jews as a nation, even if he detested them.

III Jewish National Liberation Catalyzes Islamic Antisemitism

In 2004, Sheikh Ibrahim Mudayris and Dr. Muhammad Ibrahim Maadi were among those who referenced the Koran (5:60, 7:166 and 2:65) when condemning the Jews as apes and pigs in speeches on the official Palestinian Authority TV. Mudayris went on to paraphrase *The Hamas Covenant* Article 7, once again with the demand to slaughter cowardly Jews who hide behind rocks and trees. By January 2005, he like many others, paraphrased Nazi vilifications of the Jews as a lethal disease by declaring, "The Jews are a cancer spreading in the body of the Arab nation and the Islamic nation, a cancer that has spread and reached the Arab institutions, the villages and the refugee camps." In May 2005 he said, " . . . the Jews are a virus like AIDS hitting humankind . . . Jews are responsible for all wars and conflicts . . ."[94] It is easy to see the similarity between accusations made in *The Hamas Covenant* Article 22 echoing Nazi ideals and the statements made through official Palestinian Authority media.

Falouji lined up with Muslim Brotherhood/Hamas beliefs and accusations concerning the Jews, while Sheikh Mudayris compared Jews to AIDS—a disease in need of extermination. Other peoples would be allowed to live a crippled *dhimmi* existence, but Jews would not have the opportunity to even be *dhimmi* because they would be dead. His statements fit perfectly with Hamas spokesman Ghazi Hamad's declaration to the BBC on November 8, 2006, that "Israel should be wiped from the face of the earth." He also compared the Jewish State to "a cancer that should be eradicated."[95]

The end game was and is provided by the modern spiritual leader of the Muslim Brotherhood, the Egyptian cleric Yusuf al-Qaradawi. At one point, al-Qaradawi was supposedly a liberal who urged interfaith dialogue, but he underwent an interesting transformation over the years. In 2006 he appeared as an anti-Zionist, but still attempted to prove himself not an antisemite by declaring: "Our war with the Jews is over land, brothers. We must understand this. If they had not plundered our land, there wouldn't be a war between us." And yet in the same clip from *Qatar Television* he continued explaining the conflict with Israel in religious terms. "They fight us with Judaism, so we should fight them with Islam. They fight us with the Torah, so we should fight them with the Qur'an. If they say 'the Temple,' we should say 'the Al-Aqsa Mosque.' If they say: 'We glorify the Sabbath,' we should say, 'We glorify the Friday.' . . . Religion must lead to war."[96]

Al-Qaradawi condoned and encouraged suicide-homicide bombings against all Israelis[97] regardless of age or sex. Condemning the Torah in 2009, he declared: "Everything in the Torah constitutes a call for war." He demanded Jewish destruction: "Oh Allah, take this oppressive, Jewish Zionist band of people. Oh Allah, do not spare a single one of them."[98] He showed his true hand with the demand he made on *AlJazeera TV* on January 28, 2009, for Muslims to continue in the footsteps of Adolf Hitler. "Throughout history, Allah has imposed upon the Jews people who would punish them for their corruption. The last punishment was carried out by Hitler. By means of all the things he did to them – even though they exaggerated this issue – he managed to put them in their place. This was divine punishment for them. Allah willing, the next time will be at the hand of the believers."[99]

Al-Qaradawi echoed Sayyid Qutb almost exactly when exhorting the "believers," his fellow Muslims, to continue Hitler's work. On January 30, 1939, the sixth anniversary of his rise to power Hitler made his famous speech to the Reichstag accusing world Jewry of instigating the coming conflict. "Today, I will once more be a prophet. If the international finance-Jewry inside and outside Europe should succeed in plunging the nations of the world into a world war yet again, then the result will not be the Bolzhevization of the earth, and thus the victory of Jewry, but the annihilation of the Jewish race in Europe."[100]

Nearly seventy years to the day after Hitler made clear his intentions to destroy world Jewry in an address to the German parliament, the foremost Muslim Brotherhood spiritual leader Yusuf al-Qaradawi reminded his followers of their duty to finish the Holocaust extermination process. Abdul Al'a Maududi, Hasan al-Banna, Sayyed Qutb, Haj Amin el-Husseini, the Ayatollah Khomeini and Abdullah Azzam would be proud.

Unfortunately for all of the above the rise of Zionism completely changed the game and the Jewish stereotype. Jews sought to build a nation state in the Land of Israel and were willing to physically defend themselves. On the other hand the Jewish national movement accelerated the process whereby traditional Islamic antisemitism absorbed Nazi stereotypes in what would become Jihadist antisemitism. Jewish national success forces much of the Koran into a theological contradiction and intensified loathing toward the Jews. In the specifically Palestinian Arab Muslim sense such attitudes are traced through Haj Amin el-Husseini, Izz a-Din al-Qassam and eventually Hamas of today. *The Palestinian National Charter* and Yasir Arafat attempted to distance themselves from such ideals but as seen above leaders of the Fatah dominated Palestinian Authority do not differentiate between Zionists and Jews, demanding death for both.

By the early 1960s secular Arab nationalists began popularizing Israel's extermination through the use of Jihadi terminology. Egypt's President Nasser reached a venomous pinnacle of hatred through such diatribes just prior to

III Jewish National Liberation Catalyzes Islamic Antisemitism

the 1967 Six Day War. Not playing their stereotyped role as commanded by the popular Arab Muslim script, the Jewish State defeated the Arabs in 1948, Egypt in 1956 and a coalition of Arab armies again in 1967. These victories humiliated the Arabs, since they suffered defeat by a people they believed deserved Allah's constant punishment. The Jewish State transformed the overall Jewish image into one of a tough opponent that could only be subdued by full Arab unity and preparedness. By the 1980s, Hamas would compare the Zionists to the Crusader and Mongol adversaries of yesteryear (*The Hamas Covenant* Articles 29, 34 and 35). The fact that Jewish sovereignty disrupted the Jihadi and *dhimma* initiative infuriated the Muslim Brotherhood and Hamas, making the Israeli State entity the front line enemy reviled more than any other. Muslim Arab nationalists secularized their hatred, but used the same terminology of extermination against the Jewish State. Justification for Jewish extermination was developed through "Palestinianism," inspired by both Christian and Muslim theological condemnations of Jews, yet defining the conflict as national secular.

The Oslo Accords of the 1990s appeared to legitimize a double nationalist claim of Jews and Arabs to what the Jews call the "Land of Israel" and the Arabs refer to as "Palestine." During the 2000-2004 Low Intensity Conflict (LIC) or Second Intifada (see Chapter V on "Hamas Ideological Victory") the secular Palestinian Authority under Yasir Arafat moved toward Jihadi Islamist expression. They came full circle, virtually adopting the Hamas stereotype of the Jews as an evil nation to be destroyed, one undermining the Prophet Mohammed and opposing Islam in perpetuity. In comparison, *The Palestinian National Charter* written in 1968 appeared moderate when it "only" denied Jewish nationalism, peoplehood or any connection to the Land of Israel (Palestine) and did not call for Jewish destruction. In the coming chapters we will take a more in depth approach when tracking the transformation of Fatah/PA into an increasingly Islamic organization.

Endnotes

1   Evangelical Protestant Zionism commences with the Protestant Reformation in the 16th century. Many of these ideas are drawn from interpretations of the New Testament Book of Romans, Chapter 11. By the 19th century and the impending of the Ottoman Empire this theological perspective took on political meanings as well, especially in Britain. The Second Coming of Jesus was expected as a result of the Jewish return to the Land of Israel or Palestine.

2   Peters, Joan, *From Time Immemorial, The Origins of the Arab-Jewish Conflict over*

Palestine, JKAP Publications, USA, 1984, pp. 202, p. 503 fn. 67 and pp. 204-205.

3   Laqueur, Walter Zeev, *A History of Zionism,* Schocken Books, New York, USA, 1972, pp. 42-47 and 97- 119.
    Peters, p. 209.
4   Ibid, p. 110.
5   Peters, p. 205.
6   Gilbert, Martin, *In Ishmael's House, A History of Jews in Muslim Lands,* Yale University Press, New Haven and London, 2010, pp. 140-142.
7   Landau, Jacob, *Jews, Arabs and Turks,* Magnum Press, Jerusalem, 1993, pp. 170-173.
8   Lewis, Bernard, *Semites and Anti-Semites,* W.W. Norton and Co., New York and London, 1999, pp. 167 and 170.
9   Peters, pp. 213-214.
10 Sachar, Howard M., *A History of Israel, From the Rise of Zionism to Our Time,* Alfred A. Knopf, New York, 2007, pp. 89-91.
    Laqueur, pp. 176-180.
11 Ibid, p. 234.
12 Sachar, pp. 112-115.
13 NILI is the acronym for "Netzach Yisrael Lo Yishaker" translated as "The Eternal One of Israel Will Not Lie."
14 Sachar, pp. 103-105.
15 Ibid, p. 113.
16 Ibid, p. 181 and Peters, p. 215
17 Muslih, Mohammed, *The Origins of Palestinian Nationalism,* Institute for Palestine Studies, Washington DC, 1989, pp. 89-90.
    Muslih further mentions numerous pro-Ottoman leaders who either fled with the retreating Turkish army or were deported by the British due to these loyalties.
18 Ibid, pp. 90-91.
19 Landau, p. 172.
    Pan-Islam is the vision of unifying all Muslims under one regime loyal to Sharia law.
20 Antonius, George, *The Arab Awakening,* Capricorn Books, New York, (1946) 1965, pp. 110-111 and 130-133.
21 Hourani, Albert, *A History of the Arab Peoples,* Faber and Faber, London, 1991, pp. 315-319.
22 Weizmann, Chaim, *Trial and Error,* Hamilton, LTD, London, 1949, pp. 252-262.
23 "Greco-Turkish War," *Wikipedia,* retrieved July 20, 2011, https://en.wikipedia.org/wiki/Greco-Turkish_War_(1919%E2%80%9322).
24 For an in depth review of Haj Amin's rise to power and the development of Palestinian Arab nationalism in its early phases see Yehoshua Porath's two volumes: *The Emergence of the Palestinian National Movement 1918-1929,* (1974) and *The Palestine Arab National Movement 1929-1939,* Frank Cass, London, 1977.
25 Gilbert, *Ishmael's House,* p. 149.
26 Porat, *1918-1929,* chapter "The Conflict over the Wailing Wall," pp. 258-273.
    Such ultra-orthodox adherents believe only in a Messianic redemption and reject all moves to build a state, secular or otherwise.
27 For a survey of the Mufti's tactical pro-British policies in the early 1930s see Yehuda Taggar, *The Mufti of Jerusalem and Palestine Arab Politics 1930-1937,* University of London, London, 1973.
28 General overview by Porat, *1929-1939.*
29 Lachman, Shai, "Arab Rebellion and Terrorism in Palestine 1929-1939: The Case of Sheikh Izz a-Din al-Qassam and His Movement," *Zionism and Arabism in Palestine and Israel,* Kadouri, Elie and Haim, Sylvia, eds., F. Cass, London, 1982.
    Morris, Benny, *Righteous Victims,* Vintage Books, New York, USA, 2001, chapter

III Jewish National Liberation Catalyzes Islamic Antisemitism

"The Arabs Rebel," pp. 121-154.
30 Morris, pp. 148-149 and Sachar, pp. 215-216.
31 Morris, pp. 151-160.
32 For further details and citations see Chapter VIII "The Czarist-Nazi Integration into the Palestinian Jihad," subsection "The Grand Mufti Haj Amin el-Husseini and the Nazis" in this work *Hamas Jihad*.
33 Mayer, Thomas, "The Military Force of Islam: The Society of the Muslim Brethren and the Palestine Question, 1945-48," *Zionism and Arabism in Palestine and Israel*, pp. 100-107.
34 Hroub, Khaled, *Hamas, Political Thought and Practice*, Institute for Palestine Studies, Washington DC, 2002, pp. 15-18.
35 Mayer, pp. 107-111.
36 Kurzman, Dan, *Genesis 1948: The First Arab Israeli War*, Sefel Ve Sefel Publishing, Jerusalem, (1970), 2005, pp. 257-259.
37 Morris, p. 238.
38 Kurzman, p. 259.
39 Herzog, Chaim, *The Arab-Israeli Wars*, Steimatzky, Israel, 1984, pp. 60-61, Morris, pp. 224-225, and Kurzman, pp. 334-339.
40 The Battle of Kastel in April 1948 marks the beginning of the collapse of the el-Husseini led Palestinian Arab militias. The Arabs won the battle and some fifty men remained to guard the village. However all left to attend Qader el-Husseini's funeral and Kastel was recaptured by Jewish forces a short time afterwards. This brings up many questions as to whether there was a loyalty to Palestinian Arab nationalism or if such commitments were only personal to one's commander.
41 Herzog, pp. 89-91.
42 Patterson, David, *A Genealogy of Evil, Anti-Semitism from Nazism to Islamic Jihad*, Cambridge University Press, New York, NY, USA, 2010. pp. 131-132

   Quote, "in 1937 Ruhullah Khomeini hooked up with the Brotherhood and studied the writings of Hasan al-Banna; thus the seed of the Brotherhood's evil was planted in the thinking of the future Ayatollah of Islamic Jihadist Iran."
43 Many of these ideas are best known to Jews from the writings of the famous rabbi Moshe ben Maimon (1135-1204) better known as the Rambam or Maimonides. In particular his 13 Articles of Faith showing belief in a Messianic End Time are often recited with morning prayers.
44 Although on a completely different topic, it took Sephardi/Mizrachi Jews from the Arab/Muslim lands some forty years to integrate socio-economically and gain political power in Israeli society. Today few would doubt their major impact and contribution to modern Israel although there continue to be claims that these communities are not fully represented in all avenues in Israeli society and in some cases suffer from discrimination.
45 Medzini, Meron, ed., *Israel's Foreign Relations, Selected Documents, 1947-1974*, Isratypeset, Jerusalem, 1976, p. 138.
46 "What if the Jews Lost Any War," quotes 5 and 7, *Peace for Our Time?*, retrieved July 22, 2011, www.peaceforourtime.org.uk/page146.html.

   There have been arguments over the years as to the timing of the "extermination" quote, whether it was expressed in the autumn of 1947 during the Partition Plan debate or on the day Israel declared independence. This author believes it makes no difference. The intention of the quote was clear and never retracted.
47 Sharan, Shlomo, and Bukay, David, *Crossovers Anti-Zionism and Anti-Semitism*, Transaction Publishers, New Brunswick, New Jersey, 2010, quoted on p. 31.
48 Gilbert, Martin, *Israel: A History*, Black Swan, London, 1999, p. 230.
49 Lebel, Jennie, *The Mufti of Jerusalem, Haj Amin el-Husseini and National Socialism*, Cigoja Stampa Publishers, English translation, Paul Munch, Belgrade, 2007.
50 Laqueur and Rubin, quoted on p. 167.

51 Sachar, quoted on p. 616.
52 Laqueur and Rubin, quoted on p. 172.
53 Ibid, quoted on p. 168.
54 Ibid, quoted on p. 173.
55 Gilbert, *Israel,* quoted on p. 373.
56 Laqueur and Rubin, quoted on p. 167.
57 Ibid, quoted on p. 168.
58 Ibid, quoted on p. 173.
59 Ibid, quoted on p. 167.
60 Gilbert, *Israel,* p. 369.
61 Lebel, italics in the original, p. 304.
62 Lewis, Bernard, *The Crisis of Islam, Holy War and Unholy Terror,* Phoenix, London 2004, p. 74.
63 Khomeini, Ruhollah, *Islam and Revolution I,* translated and annotated by Hamid Algar, Mizan Press, Berkeley, 1981, p. 27.
64 Ibid, pp. 46-47.
65 Taheri, Amir, *Holy Terror,* Sphere Books, Ltd, London, 1987, p. 192.
66 Ibid, footnote 10, p. 187.
67 Ibid, quoted from "Islam is not a Religion of Pacifists," p. 226.
68 Fathi, Nazila, "Wipe Israel Off the Map," *New York Times,* October 27, 2005, retrieved January 20, 2016, www.nytimes.com.2005/10/26/world/africa.26iht-iran.html.
69 Reported in the BBC on Oct. 28, 2005.
70 "Speech of Yasser Arafat Before the UN General Assemly, November 13, 1974," retrieved August 2, 2011, www.mideastweb.org/arafat_at_un.htm.
71 Ibid.
72 "UN Resolution 242," retrieved August 2, 2011, http://www.mfa.gov.il/mfa/\foreignpolicy/peace/guide/pages/un%20security%20council%20resolution%20242.aspx.
   At the very outset in late 1948 with the establishment of UNRWA, whose mission it was to deal with all refugees from the conflict, this included Jews as well. By the early 1950s Israel rehabilitated its own refugees, resettling them in regions within the sovereign state and not leaving them in camps.
73 This "Palestinianization" approach is best expressed by the Palestinian Christian organization "Sabeel." Aligned with the Mainstream Protestant churches Sabeel claims it is anti-Israel but not antisemitic. Often these Palestinian Arab Church leaders call for Israel's destruction. Sabeel is known for attempts at re-writing the New Testament by eliminating any reference to Jesus as a Jew. Overall Sabeel works to eradicate any connection between Jews and the Land of Israel.
74 Bat Ye'or, *Islam and Dhimmitude,* translated from the French by Miriam Kochan and David Littman, Fairleigh Dickenson University Press, Teaneck, New Jersey, 2002, pp. 317-321.
75 Maududi, Sayyid Abul A'la, *West versus Islam,* translated by S. Waqar Ahmad Gardezi and Abdul Waheed Khan, Markaz Maktaba Islami Publishers, New Delhi, 2005, pp. 66-68.
76 Patterson, pp. 58-64.
77 Lewis, *Semites and Anti-Semites,* pp. 172-173.
78 Ibid, pp. 177-183.
79 Such vilification of Jews in Europe particularly solidified in the Middle Ages when the Blood Libel accusation became common. During the Reformation institutionalized demonization of the Jews became central to the Lutheran Church with Martin Luther himself denouncing "Jewish evils." In Eastern Europe the Russian and Ukrainian Orthodox Churches purveyed a popular Jew hatred. See works by Christian scholars Father Edward H. Flannery *Anguish of the Jews,* Malcolm

III Jewish National Liberation Catalyzes Islamic Antisemitism 159

     Hay *Thy Brothers' Blood*, Reverend James Parkes *The Conflict of Church and Synagogue*, among others.
80   Lewis, *Semites and Anti-Semites*, pp. 128-130 and 204-205.
81   "Israel-PLO Recognition-Exchange of Letters between PM Rabin and Chairman Arafat-Sept 9- 1993," *Israel Ministry of Foreign Affairs*, retrieved May 4, 2010, http://mfa.gov.il/MFA/ForeignPolicy/Peace/Guide/Pages/Israel-PLO%20 Recognition%20-%20Exchange%20of%20Letters%20betwe.aspx
82   Bard, Mitchell G., "Myths and Facts Online, Arab/Muslim Attitudes Toward Israel," *Jewish Virtual Library*, subtitle "Fabrications of Abuses," quotes 2 and 3, retrieved May 17, 2010, www.jewishvirtuallibrary.org/jsource/myths/mf25.html.
83   "WHY DID YASSER ARAFAT SIGN THE OSLO ACCORD?" Quote 9, *Peace for Our Time?*, retrieved December 31, 2015, www.peaceforourtime.org.uk/page24.html.
84   Bard, "Myths and Facts Online," quote 2, retrieved May 17, 2010.
85   Ibid, quote 3.
86   Bostom, Andrew, *The Legacy of Islamic Antisemitism, From Sacred Texts to Solemn History*, Prometheus Books, Amherst, New York, 2008, quoted from Palestinian TV, p. 681.
87   Bogdanor, Paul, quote 74, "Understanding the Arab-Israel Conflict," from the *Jerusalem Report*, retrieved July 12, 2010, http://www.paulbogdanor.com/israel/quotes.html.
88   Laqueur and Rubin, quoted from *Resolutions* 2, 3 and 4, p. 222.
89   "Arafat's Johannesburg Speech," *Information Regarding Israel's Security*, retrieved July 12, 2010, www.iris.org.il/quotes/joburg.htm.
90   Bogdanor, quote 67, from the *Jerusalem Post*, retrieved July 12, 2010.
91   Alexander, Yona, *Palestinian Secular Terrorism*, Transnational Publishers, Ardsley, New York, 2003, taken from Memri Report #370, p. 58.
92   Ibid.
93   Bard, "Myths and Facts Online," quote 8, retrieved May 17, 2010.
94   Ibid, quotes 13 through 17 under subheading "Anti-Semitism," and including a similar accusation by Hezbollah, retrieved May 17, 2010.
95   Bogdanor, quote 84, from *PA TV* May 13, 2005, retrieved July12, 2010.
96   Appearance by Yusuf Qaradawi on Qatar TV entitled "Our War with the Jews Is in the Name of Islam" (originally from MEMRI clip #1052), quoted in Bostom, Andrew, ed., *The Legacy of Islamic Antisemitism*, p. 455.
97   al-Qaradawi, Yusuf, "[UNTITLED]," in Bostom, Andrew, *The Legacy of Jihad: Islamic Holy War and the Fate of Non-Muslims,* Prometheus Books, Amherst, New York, p. 249.
98   "Yusuf al-Qaradawi," *Wikipedia,* retrieved July 15, 2010, en.wikipedia.org/wiki/Yusuf_al-Qaradawi.
99   Ibid.
100   "Hitler Predicted Holocaust as early as January 30, 1939," in *Militant Islam Monitor*, retrieved July 15, 2010, www.militantislammonitor.org/article/id/3853.

1949-1967 Armistice Lines

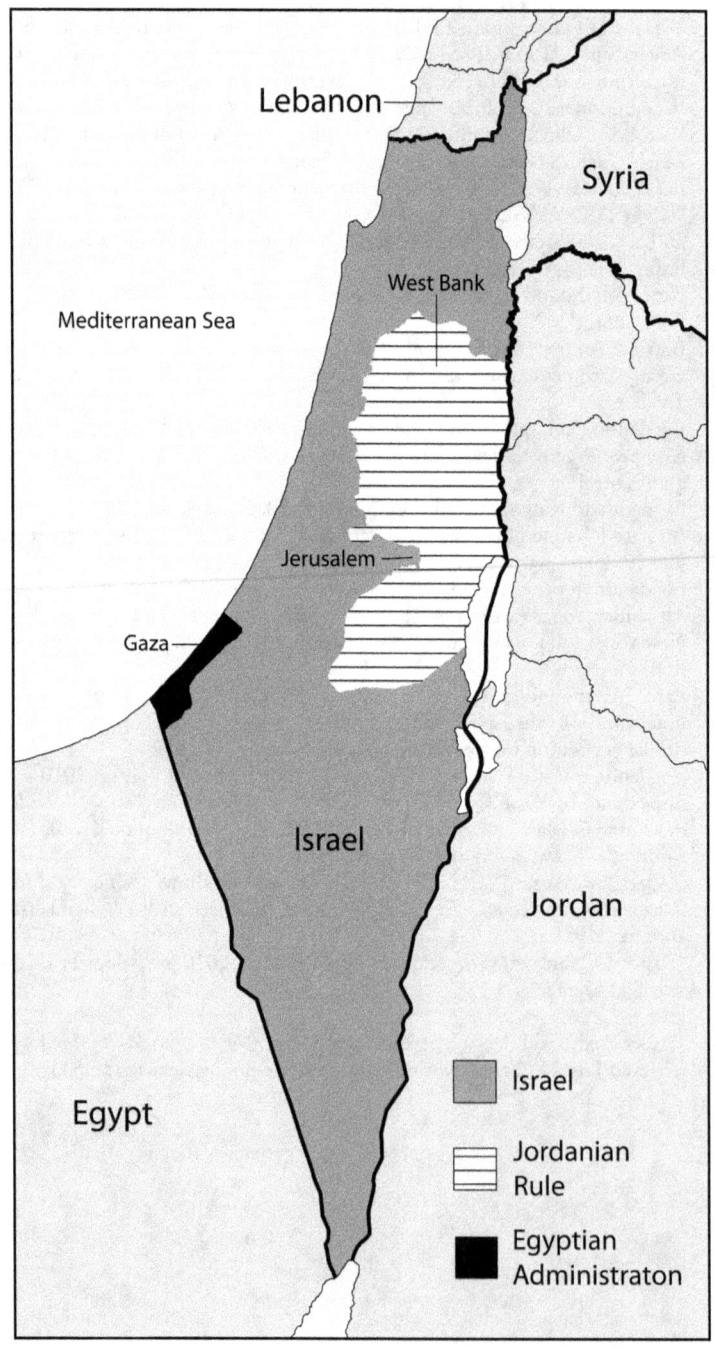

Credit: Modified from the United States National Imagery and Mapping Agency

IV

Development of the Palestinian Muslim Brotherhood/Hamas 1948-2000

Overview

The 1948 War represented a double failure for Palestinian Muslims. They neither destroyed the State of Israel, nor established their own state in the former Palestine Mandate. Jordan annexed the West Bank and Gaza remained under Egyptian administration. In Gaza the Muslim Brotherhood was harshly suppressed, first under King Farouk and later after Gamal Abdul Nasser took power. Secular Arab nationalism was triumphant, spawning Fatah and other secular Palestinian movements. In Jordan and the West Bank the Brotherhood was tolerated as a loyal opposition expected to stay within the limitations imposed on it by the authorities. Paradoxically, when Israel captured both Gaza and the West Bank in 1967, the Muslim Brotherhood activists were granted freedom of movement within the territories and were even allowed to cross into Israel.

Israel was sharply anti-Fatah/PLO and pursued Yasir Arafat as its most formidable enemy. On the other hand, the Islamists were allowed to build mosques, schools and a social welfare infrastructure in Gaza and the West Bank while organizing around religious study to strengthen their identity and opposition to the Jewish State. Freedom of religion in Israel allowed for expansion of Islamism in the Jewish State as well. The call to civil activity, or "*dawa,*" was seen as harmless and as a useful alternative to the more politically astute PLO operatives involved in physical attacks against Israeli security forces and civilians.

Two events shook the Palestinian Muslim Brotherhood in 1979; the Israeli-Egyptian peace treaty and the Khomeinist overthrow in Iran. The other side of the *dawa* theological commitment is Jihad, which until recently had remained dormant. By the mid 1980s the Gaza Muslim Brotherhood, led by Sheikh Ahmed Yasin, would move into the activist phase of military resistance against Israel. By December 1987, the Intifada broke out and the Palestinian Muslim Brotherhood now known as "Hamas" was born as an armed political entity.

By the early nineties, Israel gained the upper hand against the PLO and Hamas. A stark divergence of paths emerged between the two as the former engaged in peace talks while the latter saw any recognition of the Jewish State as a betrayal of sacred principles. Arafat and the PLO began the Oslo peace process while Hamas remained in vigilant opposition. The group initiated terror activities against Israel, but took little action against the PLO, which in time became the Palestinian Authority (PA) and Israel's "peace partner." From 1993-2000, Israel and the PA signed interim agreements granting the Authority more control over Palestinian daily life as Israel sought increased security. Many Palestinians accused the PA of corruption and betrayal for negotiating with Israel. As a result, countless Palestinians shifted their support to Hamas, although the general election boycott by Hamas during this period obscured their true support among the populace.

Arafat kept Hamas at bay until the collapse of the Oslo peace process at Camp David in 2000. Beginning in September, the Fatah-led PA and Hamas fought together against Israel in the Low Intensity Conflict (LIC), known as the "Second Intifada" even though the Islamists declared the corrupt Fatah leadership to be no less an enemy than the Zionists. Hamas began from a point of weakness, but realized it was a golden opportunity to continue gaining popular support at the expense of the despised PA. While the PLO/Fatah dominated PA negotiated but did not achieve a peace agreement with Israel, Hamas moved to replace them as the new Palestinian leadership and steadfast adversary of Israel.

The 1948 War to the 1987 Rise of Independent Hamas

As shown in the previous chapter, the 1930s saw an eruption of violent Islamist guerrilla and terror activities led by Izz a-Din al-Qassam against the Jewish National Home and the British-administered Palestine Mandate. The 1940s witnessed Muslim Brotherhood military activities with the Egyptian invasion of the newborn Israeli State. Neither was successful in the physical sense, yet both left an Islamic ideological impact on Palestinian Muslim society with al-Qassam as the ultimate hero. What secular Arabs attempted to define as a national conflict between Israel and the Arab world would retain serious elements of a religious Muslim-Jewish conflict evidenced by Brotherhood activity throughout the 1930s and 1940s. Upon the cessation of hostilities in early 1949, violent Islamist activism on the Palestinian front underwent a twenty year demise before reactivating itself after the 1967 War. Haj Amin el-Husseini blamed the Arab world for the Palestinian defeat while not uttering a word of accusation toward the Brotherhood.¹ In his "capacity" as Grand Mufti he continued to commend the Nazis, specifically Hitler, for the destruction of the Jews. He also made clear his demands for the destruction of the State of Israel, which he saw as a plot by "World Judaism" to expand and subjugate the Arab world.² Haj Amin remained the Palestinian symbol, one representing a murderous attitude toward the Jews and Israel, while lacking success both on the Arab national and Islamic political levels.

Two distinct pillars of Muslim Brotherhood ideology developed wherever the movement took root. The first included social welfare, education and the compassionate side of Islam representing everyday activities in alleviating the distress of the poor and downtrodden. The second invoked demands for Jihad and a resulting global Caliphate. It followed that Israel's existence was an outrage. The 1948 defeat was a catastrophe or "*nakba*" and this humiliation demanded a reversal—the elimination of the Jewish State. UN Resolution 194, Clause 11, calling for a refugee return in "peace" or for "compensation" was one way to rectify the situation, but this never came about. On the domestic front, poor Palestinian Muslims viewed the Brotherhood and later Hamas as radiating love, care and compassion for their plight. Secular Arab opponents, the West and Israel in particular were destined to become victims of vicious fanatical Islamic fundamentalism, to be killed for the glory of the Koran or to survive under an eternal Islamic rule. Suicide–homicide bombings were the Jihadi prelude to such subjugation. The Muslim Brotherhood always had a double mission of service to the community of believers and Jihad against infidels and *dhimmis*.

With no independent Palestinian State, Brotherhood activities took place in two very different locations: Egypt's Gaza Strip and Jordan's annexed West Bank. Muslims in Israel had no possibility to rebuild the organization. Activists in Gaza were absorbed into the Egyptian Brotherhood and those in the West Bank into the Jordanian branch. The Gazans were much more revolutionary, having already represented the Palestinian Islamist stronghold in 1948. In Egypt, Faruk's regime banned the Brotherhood the following year, forcing them to emphasize education and religion. Fortunes reversed when the Free Officers, led by Gamal Abdel Nasser, overthrew King Faruk's rule in 1952. The Brotherhood changed course into a more military direction and took the initiative, whereby they became the leading political force in the Strip. Conflict broke out with President Nasser's administration when violent protests were organized by the Brotherhood in alliance with the communists and the Baath secular Arab nationalists against a proposal to resettle Palestinian refugees in Sinai. Previously the Egyptian Brothers were outlawed after an assassination attempt against the Egyptian president, yet harsh repression in Gaza began only after the protests. Forced underground, this was a devastating blow halting the overall Jihadi objective for the liberation of Palestine.

The Brotherhood organized military cells, deemphasized ideology and concentrated on the future armed struggle. Such preparations served the activists well, especially in the aftermath of Israel's Sinai Campaign in late 1956 and the ensuing four month occupation of the Gaza Strip. The Brotherhood led the resistance and spawned the Palestine National Liberation Movement (PNLM) in the late 1950s, establishing the foundation for the secular Palestinian nationalist Fatah several years later. The Brothers saw this as a major mistake. The appeal needed to be to all Muslims, not just limited to the Arab States. Any break-off group emphasizing secular nationalism was seen as destined for failure. The central issue at hand was the original first "*qibla*" or "prayer direction," before Mecca, meaning the need to emphasize Jerusalem and, by extension, all of Palestine. The Brothers saw religious commitment as much more powerful than secular ideals. In 1960, the Palestine armed resistance and the Islamists split; the latter accusing the former of being impractical and insisting on building a new "liberation generation" and not engaging in immediate armed struggle. The PNLM, or what became Fatah, owes its origins to the Muslim Brotherhood, while the Islamists themselves were seen as a failure. The Brotherhood lost the initiative to their more secular counterparts and waited until the 1980s before becoming proactive once again.[3]

The PLO, led by its senior Fatah participant, became the foremost Palestinian resistance organization by the late 1960s. Beginning in the mid-1960s, there was hope of an Egyptian Nasserite victory over Israel, and at the time little expectation of an independent Palestinian State. Instead, Palestine would become a province of one of the surrounding Arab nations. Led by Yasir Arafat, Fatah and later the PLO embraced an abundance of ideological jargon reflecting Islam, Arab nationalism, Marxism-Leninism, and a plethora of developing world liberationist ideals. One basic objective was unshakable: the elimination of Israel and its Jewish population through military means. There were no true socio-economic ideals, as opposed to other left wing national revolutionary movements sweeping the Third World in the post WWII period. Violence was a value that existed for its own sake, and socio-economic issues were only to be confronted in the aftermath of victory over Israel.[4] The PLO was an empty barrel compared to the Muslim Brotherhood, and could not afford to focus on domestic concerns or *dawa* organizations serving the Palestinian population's needs. However, in aligning with Muslim Brotherhood thought, the PLO advocated and began implementing the armed struggle until final victory.

In parallel to the development of Fatah and the PLO, the height of Nasser's brutal anti-Brotherhood campaign culminated in 1965 after an attempted coup against his regime. A period of sharp repression followed, with sweeping arrests of suspects including Ahmed Yasin, who years later founded and led Hamas, and the execution of the leading Brotherhood member, Islamist theologian and ideologue Sayyid Qutb.

Brotherhood activities in the Jordanian-held West Bank were different. Whereas in Gaza, the Islamists were forced into a clandestine radical approach because of Nasserist opposition, their colleagues functioned under the Jordanian Hashemites as a loyal opposition throughout the 1950s, sharing a conservative, traditionalist platform opposed to the Egyptian revolutionary regime. After 1967, the two groups cooperated in the United Palestinian [Muslim] Brotherhood Organization. Ironically, Israel's "open bridges" policy with Jordan unlocked the borders into the Jewish State from both Gaza and the West Bank and allowed for expanding Brotherhood influence. Palestinian Islamists passed freely in and out of Jordan, Gaza and Israel. Their reach went beyond the Hashemite Kingdom, particularly among Israeli Arabs where Sheikh Yasin himself spent many Fridays preaching throughout the mosques in the Galilee and Negev.

The PLO and the Brotherhood worked together briefly after the 1967 defeat; however, controversy erupted between the two. Both established camps in the Jordan Rift valley from 1968-70 to facilitate raids across the Jordan River into the Israeli-held West Bank. The Muslim Brotherhood camps were situated in the north and administered independently even thought they flew a Fatah flag (See Chapter II "Ideologues" Abdullah Azzam). Recovering from the Nasserite period, the Gaza Muslim Brotherhood remained outside the guerrilla initiatives, and did not send volunteers to these camps. Furthermore, the Brothers were deemed reactionaries and clashed with those from the different PLO factions, most notably sworn atheists and leftists.

Fatah was officially established in 1965. Once they realized the impossibility of achieving victory on their own, they hoped to force the Arab world into major hostilities with Israel through constant border clashes. After Black September 1970 and the expulsion of the PLO from Jordan, the Brotherhood's military efforts were halted and they returned to education, proselytizing and organizing to rally the "*umma*," or people. Playing their cards wisely, the Brotherhood remained neutral during the clash, and in its aftermath returned to Islamic educational initiatives. The armed struggle in Palestine was not yet ripe.⁵ In particular, Sheikh Yasin introduced Sayyid Qutb's famous commentary *In the Shade of the Koran* for study without any objections from the Israeli authorities. Israel's non-intervention policies granting religious freedoms were a major improvement for Islamic activists as opposed to the continual Egyptian and Jordanian persecutions over the same issues.⁶ Israeli authorities were oblivious to the political implications of Islamic studies and so overly focused on Arafat and the PLO that they were blind to a much more powerful adversary organizing for a day of reckoning.

Nasser's anti-Islamist, pro-Arabist policies repressed the Gaza Muslim Brotherhood for fifteen years, forcing them to work clandestinely with their emphasis placed on educational preparation for the postponed Jihad. Fatah projected military strength in the face of the recent Arab defeat and continued to work openly, while the Brotherhood built mosques and infused the people with Islam from 1967-76. According to the Hamas narrative, this was the "hard core" solidification of the movement while under an oppressive Israeli regime. Gaza proved a fertile breeding ground for the Islamist message, since half the residents lived in refugee camps, the population density was among the highest in the world, and poverty was a common denominator for most. Yasin concentrated only on religious preaching and educational *dawa* organizations, believing the Israeli authorities, like the Egyptians before them, would not interfere. His religious center *al-Mujamma' al-Islami* was finally legalized in 1978 and quickly became the foundation for religious and educational Islamic institutions in Gaza. The *Mujamma* was composed of seven committees: preaching and guidance, welfare, education, charity, health, sports, conciliation, as well as establishing a women's association. By the end of the decade, the Islamic Center was the most influential unifying force throughout Gaza. They battled against social ills such as pornography, drug usage, alcohol and carousing between men and women. Mosque activities took on exceptional importance by hosting kindergartens, schools, medical facilities, vocational training, sporting clubs, social welfare and religious training.⁷

The Muslim Brotherhood realized it was on its own as far as gaining support from the pan-Arab movement and particularly Egypt. Nasser, for all of his bluster and mobilization of forces in 1967, would be remembered for admitting he had no plan for the liberation of Palestine and losing the Six Day War, while his successor, Anwar Sadat, was seen as betraying the Palestinian people when he signed the 1979 Peace Accords with Israel.⁸

Simultaneously, under both Labor and Likud governments, Israel allowed for the construction of new mosques and the expansion of religious activities while repressing the outlawed PLO and its nationalist initiatives. In 1967, there were 77 mosques in Gaza; twenty-two years later the number climbed to 200. Certain unifying factors also contributed to Islamist growth, whether it was the never-ending clash with the PLO, especially the Marxist PFLP, opposition to Israel, or support for the Iranian revolutionaries. Funding came internationally, mostly through the Saudis. The Palestinians were now exploiting the connections of the Brotherhood umbrella to attain international Islamic support for their cause.[9]

For the first decade after the June 1967 War, Israeli settlement activity in the West Bank—Judea and Samaria—was very limited, essentially concentrated just over the armistice lines in the Jerusalem region, or along the Jordan River and northwest Dead Sea shoreline. This was an area contained in what is known as the "Allon Plan." The average Palestinian did not experience a feeling of omnipresent Israeli settlement. The Likud victory and right/religious governments (1977-84) led by Menachem Begin and Yitzchak Shamir changed the equation. Jews began to populate the West Bank and Gaza in order to implement "the Greater Land of Israel" ideology based on the Biblical-Hebrew scriptural understanding that the Jewish People would develop all lands between the Jordan River and Mediterranean Sea, heralding in the End of Days. This meant large-scale settlement activity throughout the Palestinian areas. It is interesting to note that the Koran itself supports these ideas in 7:137 and 17:103-104. Led by Gush Emunim of the national religious movement, Jews often established settlements in the heart of heavily populated Palestinian regions such as the former Jewish Quarter of Hebron. This certainly was an accelerating factor in strengthening the Islamists appeal. In any case, it should be recalled that the viciously anti-Jewish Muslim Brotherhood existed decades before the territorial argument was enjoined over West Bank and Gaza sovereignty in the wake of the 1967 War.

Intensified conflicting claims over holy sites, in particular the Cave of the Machpela in Hebron and the Temple Mount in Jerusalem, further catalyzed what had been interpreted as a national clash into the religious domain. It did not help that the Israel Defense Forces (Israeli army or IDF) were unable to protect Palestinians against Jewish extremists.[10] Fears were further exacerbated when the Jewish right wing and religious nationalists advocated moving most of the Russian immigrant population into Palestinian areas. On the other hand, Israeli "deterrence" weakened with the murder of Israel's peace partner Anwar Sadat in 1981 by a Jihadi assassin and the corresponding success of the Iranian Revolution.[11] Soon afterward, the rise of the disenfranchised Lebanese Shiites brought about the formation of Hezbollah, whose Khomeinist

Jihadi commitments led them to demand the extermination of the Jewish State. By the 1990s, Hezbollah challenged Israeli forces in south Lebanon much more successfully than the PLO had in the 1970s and early 1980s.

While the Brotherhood built their social, educational and religious infrastructures, the Palestinian left and secular organizations failed in their attempts to rally the population to their agenda. Radical Islam catapulted forward with the success of Khomeini's Iranian Revolution in 1979 and the continuing popular resentment against Israel's occupation. The Muslim Brotherhood moved forward rapidly, but not fast enough for the right wing splinter "Islamic Jihad," a faction formerly part of the Brothers. Led by Fathi al-Shikaki and inspired by Khomeini, they demanded immediate military operations. For the first time, the Brotherhood faced a direct challenge from within. Should they continue with the one-track internal Palestinian social change, as had been the policy for years, or should they begin arming for a military showdown?

They devised a two-pronged approach. First, they organized militarily, collecting arms and ammunition, obtaining vital information and tracking down those working for Israel. Secondly, they launched popular protests against Israel's involvement in Lebanon, where in 1982 the PLO was routed and forced to evacuate. This led to a general strike in Gaza as a protest against overall conditions. In 1984, Israeli security forces became aware of military planning and arrested the head of the Political Bureau in the Gaza Strip, Sheikh Ahmed Yasin, and several others. A year later, Yasin was released in the "Jabril prisoner exchange," as a consequence of the 1982 War in Lebanon. It is interesting to note that three years prior authorities arrested Sheikh Abdullah Nimr Darwish, the Israeli Arab Muslim who led "The Jihad Family," for possession of illegal weapons. On the surface, even Israeli Arab Islamists were surpassing their Palestinian comrades in taking action, leading many to believe the Palestinian Muslim Brotherhood was betraying the cause and cooperating with Israel.

From 1984-87, Islamic Jihad became the activist role model and the orientation shifted to an armed struggle. Salah Shehadeh became the first military commander of the Palestinian Brotherhood. On the popular level protests continued as the most effective mobilization device, reaching levels of mass participation by 1985-86, especially among Islamized university students. The PLO was on the defensive, losing both funding and political control of the Islamic University to the Gaza Brotherhood. By the 1990s, the Gaza Islamic Center and Islamic University would not only unify, but began an internal Jihad to cleanse secular elements from Muslim society.[12]

IV Development of Hamas 1948 to 2000

While Fathi al-Shikaki left the social issues to the Brotherhood, he built a radical, Iranian Shiite-inspired Sunni activist organization eventually known as the Islamic Jihad whose sole purpose was to liberate Palestine. At the outset he paid a heavy price and was expelled from the Muslim Brotherhood in 1979 as a result of his actions. He proved himself in 1980, attacking Jews in the Hebron region, killing six and wounding seventeen, while forging an alliance with the great Jihadist Abdullah Azzam,[13] who later fought against the Soviets in Afghanistan and became Osama bin Laden's mentor.

A snapshot of Palestinian society just prior to the Intifada civil resistance or "shaking off" in 1987 provides a glimpse of an extremely frustrated society suffering from multiple contradictions. Israel had allowed the establishment of seven universities where previously none existed. Islamic associations flourished, as did social and professional institutions. The Muslim Brotherhood was encouraged as a counterweight to Palestinian nationalism as heralded by the PLO/Fatah and led by Yasir Arafat. Sheikh Ahmed Yasin, the charismatic Brotherhood leader who was also a quadriplegic confined to a wheelchair, was allowed to operate freely despite his rabid antisemitism and ability to spread Islamist ideology. Yasin challenged Arafat and the PLO, demanding Islamic behavior. He was no less anti-Israel than they were, but added a vicious, overt antisemitism.[14]

In contradiction to all logic, the Islamists continued to receive at least tacit support from Israel's right wing leadership, apparently because of the government's focused hatred toward Yasir Arafat and his vanquished PLO, the ultimate foes in the eyes of Prime Minister Menachem Begin and the Likud. Israel may have believed in its own ability to assure security, but Hamas was winning the hearts and minds of Palestinians through education, religious activities, social welfare programs and its universal Islamic ideals. To Muslims, this meant the religious world was their natural ally, and not the secular Arab States who conceded Israel's "illegal" existence. Hamas was instituting an unofficial regime, encroaching on PLO/Fatah influence and the Israeli civil administration. Most Palestinians were loyal to the PLO/Fatah for years, but remained under the Israeli radar. Hamas now sought to replace Palestinian secular loyalties with Islamic ones and eventually gain full dominion over all of Palestine through Israel's destruction.

By the early 1980s Palestinian incomes skyrocketed from where they had been prior to 1967. In Gaza, the per capita income went from $80 to $1,700. In the first ten years of the Israeli occupation, the Gaza GNP grew by an annual 12.1 percent and in the West Bank by 12.9 percent. By contrast, the Israeli average increased 5.5 percent. Roads, electricity and health care all improved. All this slowed significantly as a result of the world economic slump of the early 1980s, bringing hidden frustrations to the surface. By 1987, some 120,000 Palestinians or 40 percent of the work force sold their skills to Israel. There was no national Palestinian economic development to absorb their labor. Added to this were those who had relocated to the Persian Gulf and particularly Iraq years before and had sent home remittances, but now were forced to leave as a result of the Iran-Iraq War (1980-88).[15]

The average Palestinian worker was building Israeli infrastructure and the Jewish nation state as opposed to his own, whether by choice or due to lack of choice. Soon, internal frustrations reached a boiling point. Contradiction over loyalties, such as the economic necessity to work in Israel versus national commitment to build a Palestinian State, opened the door for extremist, violent ideologies. Physical and ideological breakout was imminent in order to regain collective self-esteem. Fatah and the PLO were going nowhere, especially in the wake of the crushing defeat in Lebanon in 1982. Expectations grew as did the gap between education and personal economic well-being and the receding possibility of obtaining independent national and religious recognition. All of this brought Palestinian society to a breaking point in what is known as "relative deprivation."[16] The Islamist mix of world Jihad coupled with an uncompromising antisemitism and anti-Israel rhetoric acted as a catalyst for a sweeping rebellion, not only against Israel but against the Fatah/PLO.

The die was cast—social change and the armed struggle took place simultaneously. Delay could no longer be countenanced; the dual priorities of Hamas would come to fruition at the opportune moment. Palestinian historian and political analyst Khalid Hroub listed three major reasons for Palestinian national despair: the defeat of Yasir Arafat and the PLO in the 1982 Lebanon War, the loss of Arab world interest and support for the Palestinian cause, and the internal socio-political pressure built up over twenty years of the Israeli presence in the West Bank and Gaza. Furthermore, a new generation had grown up since 1967. They saw Israel as an arrogant adversary and were not afraid of a confrontation. They chose an Islamist religious fundamentalism as their true identity, replacing the secular Arab nationalism and socialist ideologies adopted by the PLO and much of the Arab world—ideas they saw as outside of the realm of Islam and tainted with failure.[17]

Beginning in 1983, the Muslim Brotherhood began preparing for the confrontation with Israel and commenced by building organizational infrastructure. Catapulted by the Iranian Revolution and the Jihadi successes in Afghanistan against the Soviets, operations began in the spring of 1987 with the active Jihad designated for November. However, the movement sputtered and was desperately in need of a conflict to prove itself. The infamous December 8th traffic accident in Gaza, where several Palestinians were killed by an Israeli driver, and the resulting violent protests, came as a godsend to the Islamists and allowed not only for attacks against Israel, but demands for her destruction. The PLO, on the other hand, was pressured by the US to recognize Israel's existence in the framework of a two-state solution. Arafat inferred recognition of Israel a year later and lost credibility with Palestinians in exchange for gaining Western support.[18] Battle-born Hamas entered the world with a birth certificate completely differentiating it from the secular PLO/Fatah.

By late 1987, signed communiques in the name of the Islamic Resistance Movement (IRM) accompanied actions strengthening a spirit of unity and resistance. Over a two month period there was a shootout with escaped Palestinian prisoners that killed four, a clash on the campus of the University of Islam in Gaza where dozens of students were wounded, the stabbing death of an Israeli in Gaza City, and a traffic accident between Israeli and Palestinian vehicles in December resulting in quite a few Palestinian casualties. Public outrage reached new heights and exploded the day after the accident.[19] Mass demonstrations and rioting broke out and the IRM used the acronym of those three letters in Arabic to form the term "Hamas," meaning "zeal," in its overtures to the public. Interestingly, the word means "violent theft" in Biblical Hebrew.[20]

The continuing protests were referred to as an *"intifada"* or in popular terms referred to as an "uprising." The first official meeting and communique issued came from Sheikh Yasin, Abdul Aziz al-Rantisi, Salah Shehadeh, Muhammad Sham'ah, Isa al-Sashshar, Abdel Fattah Dukhan and Ibrahim al-Yazuri.[21] Mosab Hassan Yousef claims the seven participants to be Ahmed Yasin, Muhammad Jamal al-Natsheh, Jamal Mansour, Hassan Yousef, Mahmud Muslih, Jamal Hamami and Ayman Abu Taha.[22] One list or the other or a combination of the two made up the founding members of Hamas. The first communique was posted a few days later in Gaza and in the West Bank within a week. The loosely coordinated Palestinian Muslim Brotherhood solidified into the activist Hamas group with preexisting associated Islamic organizations joining the centralized body. Political action and armed resistance against Israel, and especially against "Jews," manifested. Hamas planned to destroy the Jews with Allah's help, as indicated in three excerpts from that first communique on December 14, 1987.

Our steadfast Muslim masses:

Today, you have a date with God's powerful decree against the Jews and their helpers. Nay, you are an integral part of this decree that, God willing, ultimately shall uproot them.

Let the Jews understand that despite the chains, prisons, and detention centers, despite the suffering of our people under their criminal occupation, despite the blood and tears, our people's perseverance and steadfastness shall overcome their oppression and arrogance. Let them know that their policy of violence shall beget naught but a more powerful counter policy by our sons and youths who love the eternal life in heaven more than our enemies love this life.

It [the intifada] comes to awaken the consciences of those among us who are gasping after a sick peace, after empty international conferences, after treasonous partial settlements like Camp David [1978-79]. The intifada is here to convince them that Islam is the solution and the alternative. Let the world know that the Jews are committing Nazi crimes against our people and that they will drink from the same cup.[23]

From these excerpts one learns that the Jews are the enemy of Allah and that homicide-suicide attacks are on the way by "youths who love the eternal life in heaven more than our enemies love this life." The Camp David Accords between Israel and Egypt in 1978-79 were rejected in their entirety. There could be no peace between Israel and another Arab country such as Egypt, nor could there be a blueprint for an initiative to grant autonomy to the Palestinians as agreed upon in the "Frameworks for Peace" section as a first step prior to "the final status of the West Bank and Gaza and other outstanding issues by the end of the transitional period."[24] To quote the first communique "Islam is the solution and alternative," there is no other. The Jews were equated with the Nazis and Hamas would accept no compromises to resolve the conflict.

Hamas and the Intifada 1987-93

In retrospect, Hamas was officially established as a result of its first communique at the end of 1987. Although in truth the organization existed for decades as the Palestinian Muslim Brotherhood. Under intense pressure due to its conflict with both Israel and the PLO, the foundational *Hamas Covenant* was presented as Islamic holy writ in August 1988. Only after the *Covenant* was published did the Israeli leadership become fully cognizant of Hamas' unbounded hatred for both the State of Israel and the Jewish People. It was now understood that Hamas words were to be equated with actions.

Finally, following the lead of the Islamic Jihad militant terrorist splinter group, Hamas became the activist force behind the Intifada. Hamas evolved into a composite of the Fatah PLO and Islamic Jihad on the military front by mixing the previously rejected Palestinian Arab nationalist element with Jihadi extremism. Both found expression in *The Hamas Covenant (HC)* and quite possibly Israeli security concerns were only fully awakened once Hamas announced its embrace of Palestinian nationalism (see *HC* Introduction-Preamble, Articles 6, 7, 11, 13, 14 and 15). The integration of Palestinian nationalism was conditional on the Islamist interpretation of what Palestinian nationalism meant, explicitly its subjugation to the Islamic world view, values and way of life (see *HC* Articles 25, 26 and 27). Hamas made its bid to lead the resistance and to either replace the PLO and Islamic Jihad, or to absorb both of them in the long run.

In juxtaposition stood the *dawa* organizations dealing in religion, education and social welfare. Many saw redeeming value to Islamic charities, including those of Hamas. Even Israeli governments noticed the good works as a result of such activity. In particular the right wing and religious were favorably inclined, as opposed to their confrontational policies targeting Arafat and the PLO. In reality, *dawa* activities never nullified the Islamic armed struggle and instead often led to a greater commitment by the masses, enhancing violent actions against Israel. As pointed out by Matthew Levitt, donations made for *dawa* social welfare organizations are inextricably linked to funding terrorism. There is no contradiction in the Islamist mindset to use social welfare monies either for poverty relief, for the armed struggle against Israel or any other enemy. Charities for the *dawa* also funded armed resistance, operated by the same Hamas leadership playing a double role.[25] To quote Levitt on the blurring of lines between social welfare and terrorism:

Inside the Palestinian territories, the battery of mosques, schools, orphanages, summer camps, and sports leagues sponsored by Hamas are integral parts of an overarching apparatus of terror. These Hamas entities engage in incitement and radicalize society, and undertake recruitment efforts to socialize even the youngest children to aspire to die as martyrs. They provide logistical and operational support for weapons smuggling, reconnaissance and suicide bombings. They provide day jobs for field commanders and shelter fugitive operatives.[26]

For instance bombers of the past two decades often began as moderates serving the community, which confused and blurred the lines between moderates and fanatics.[27] Both types can be construed as the *dawa* or "calling of Islam." They are not mutually exclusive and can be understood as two sides of the same Muslim commitment of total belief, action and the spreading of the Koranic, Sharia message.[28] The integrated objective of Hamas was and is to achieve full control over civil society while battling and defeating Israel. There is no pandering to a "Jihad of the soul" as Jihad clearly advocates homicide-suicide bombings and terror operations.[29] One may add that the Islamist flock views such leadership as having Divine authority given by Allah; there is no possible contradiction between the two functions, but rather they complement each other.

Israeli policy was anti-PLO and anti-Palestinian nationalist and thereby sought to foster Islamist elements in their battle for the Palestinian street until the summer of 1988, half a year after the outbreak of the Intifada. Ludicrous as it sounds, at the outset Israeli policy makers expected the Islamic fundamentalists to balance or even defeat the supposedly more dangerous secular nationalists, believing all of this was in Israel's interests. Yasin's antipathy toward secularism and the PLO was no secret, he condemned them as "pork eaters and wine drinkers" while taking them to task for allowing women the right or ability to speak at all, since "a woman's voice" was considered "indecent."[30] In early 1988, people began to fear curtailment of religious freedom should Israel take overt action against Hamas. Their fear was confirmed after Israel verified information concerning arms and explosives caches from captured activists who admitted association with Hamas. After that, Israel initiated a full-scale sweep against the Islamists in July.[31] Publication of the

Covenant confirmed suspicions that Hamas was potentially more dangerous than the PLO.

Overall, the Intifada was seen as a turnaround of Palestinian behavior since 1948. Palestinians were taking responsibility for their own future, beginning with those territories originally captured by Egypt in Gaza, and by Jordan in the West Bank. At the outset, everyone participated despite the lack of a centrally organized command. Fatah, the left wing PFLF and DFLP all worked alongside the Islamic Jihad and Hamas. Two brothers from Ramallah, Muhammad and Majid Labadi, established a grass roots national secular command known as the United National Leadership of the Uprising (UNLU). Accompanied by defensive barricades barring Israelis from Palestinian villages and East Jerusalem neighborhoods, stones, gasoline bombs and posters declaring the "revolution" were their major offensive weapons. Eventually Israeli security forces retook these areas.[32]

At first, the Intifada took the PLO/Fatah by surprise but with time Arafat gained control over events. Somewhere between $120-$300 million would be funneled annually to activists.[33] Hamas and the UNLU organized strikes together, with the former adding extra strike days of their own. Taking a tougher line did not work against the Islamists, but instead had the opposite effect leaving Hamas with a reputation of being more resilient than Fatah and the secular nationalists. Hamas operatives were seen as less likely to break during interrogations and outstripped Fatah in recruiting adherents from the prison populations. The secularists promised material well-being for a successful revolution while Hamas "only" promised spiritual rewards in the next world. Simultaneously, Hamas continued its social activism, working for the reopening of schools, running charitable operations and enforcing civil law after Palestinian police officers working under Israeli auspices resigned.[34] Arafat understood the competition he faced was over who set the ground rules for the continuing uprising. He appealed to the Muslim Brotherhood throughout the Arab world to recognize him and the PLO as the sole representative and leader of the Palestinian people and the Intifada.[35]

Hamas and the secular PLO nationalist adherents of Fatah, the leftist PFLP and DFLP were all battling a common enemy in Israel, but early on they were in competition with each other. The PLO called business strikes on the first day of each month, and Hamas called business strikes on the ninth day. The PLO was corrupt and power hungry while Hamas raised the banner of religion and Jihad. On both sides, personal vendettas were carried out against individuals and families under the guise of resistance and the average

Palestinian often suffered from the ensuing chaos.³⁶ To quote Mosab Yousef, son of Hamas leader Hassan Yousef:

> In the initial years of the first Intifada, ideological differences kept Hamas and the PLO on very separate paths. Hamas was largely animated by religious fervor and the theology of jihad, while the PLO was driven by nationalism and the ideology of power. If Hamas called a strike and threatened to burn the stores of anyone who stayed open, PLO leaders across the street threatened to burn the stores of anyone who closed.³⁷

Both sides fought for recognition as the leader of the uprising, hoping to gain Muslim Brotherhood support throughout the Arab world. Eventually, the largest and most powerful branch in Egypt answered Arafat's call for recognition of himself and the PLO as the sole Palestinian leadership. Hamas would have to work much harder, proving their abilities, as they did within the next two decades. Arafat's move, and recognition by the Muslim Brotherhood, exposed the PLO as not being a strictly secular organization, as it claimed. In retrospect one can surmise that the upstart Hamas was not taken seriously, a major blunder to be rectified by the Egyptian Brotherhood only in later years.

Although Hamas made inroads in the West Bank, their power still emanated from the Gaza Strip. In the 1988 summer crackdown throughout the Palestinian areas, Israel apprehended 120 out of 200 members of the Hamas military wing. The following May, the Hamas commander Salah Shehadeh and spiritual leader Sheikh Ahmed Yasin were arrested, yet Hamas continued to hold a much harder line than Fatah and the more moderate UNLU.³⁸ The UNLU was said to be the operative wing of PLO-Tunis, carrying out Arafat's orders from his headquarters in his Tunisian exile. Overall, UNLU leaflets usually called for strike days, demonstrations, non-payment of taxes or explained the usefulness of Molotov cocktails and rock throwing. On the moderate side, some leaflets expressed hope of working together with pro-peace organizations in Israel. Not everything the UNLU did was to the liking of the PLO, and Hamas in particular would hear nothing of such moderation.³⁹ Hamas carried out terror attacks usually with knives, metal rods, rocks and gasoline bombs but not live ammunition. Suspected Palestinian collaborators with Israel paid a price either through punishment or execution, and Israel's security services lost many of their informants in the Palestinian areas.⁴⁰

IV Development of Hamas 1948 to 2000

By the summer of 1990 the Intifada was slowing down but was revitalized when the Palestinians' greatest patron, and Yasir Arafat's closest ally, Iraq's Saddam Hussein invaded the Western-backed Kuwaiti sheikhdom and shortly afterward announced his intention to obliterate Israel with chemical weapons and capture Jerusalem. Palestinians cheered as Saddam fired 39 rockets into Israel and threatened to exterminate the Jewish state.[41] Saddam was a constant ally illustrating a commonality of interests between Palestinians and Iraqis as concerned confronting the West, whether it was Israel and/or the US. Arafat and Saddam were the closest of allies, both seen as the flag bearers of overall Arab nationalism. The price was steep, Iraq was in ruins from US and Allied bombings in the 1991 Gulf War while over a thousand Palestinians lay dead from clashes with Israel. Israeli casualties were disproportionately lower with dozens killed. Thousands more, mostly on the Palestinian side, were injured. Both sides were tired, with exhaustion setting in on the Palestinian side.

Almost as if on key, Israel's ultra-right religious "Temple Mount Faithful" announced their intent to start building the Third Temple during Succot, or Feast of the Tabernacles, in October 1990. Police prevented Jewish activists from entering the Al-Aksa domain on the Temple Mount in Jerusalem, but tensions were not relieved. A massive clash ensued between Israeli police and Arabs on the Temple Mount, leaving 19 Palestinians dead and dozens injured on both sides. Jews were forced to leave the Western Wall plaza as the Palestinians hurled stones at them from above. In the end, the police regained control on the Mount. Although the narratives differ, one point became clear: the Intifada was re-galvanized as attacks against Israelis increased.[42] Most believe the Intifada ended with the Madrid Conference a full year later, and in the aftermath of Saddam Hussein's defeat by Allied forces in the 1991 Gulf War.

While Israelis argued amongst themselves over whether to engage in peace talks or not with the Palestinians, the popular uprising made clear to the world that the present situation could not continue. Although there were Palestinians who favored a two-state solution, many others sought Israel's annihilation. Unfortunately for the Palestinians, the seeds of internal conflict and possible self destruction were planted in the ensuing clash between Hamas and the PLO. In 1991 more Palestinians died by the hand of other Palestinians—approximately 150 deaths—than by Israelis—approximately 100 deaths. This was not only a matter of killing collaborators and criminals, but it was a result of violence between Fatah and Hamas. The Palestinian economy, so dependent on Israel, was in ruins. With Palestinian workers absent on strike days and during curfews, Israeli employers eventually brought in replacement foreign workers driving up Palestinian unemployment. To make matters worse, 300,000 Palestinians were banished from Kuwait after Saddam Hussein's invasion resulting in the loss of $400 million in remittances.[43] Another crippling blow came with the demise of the Soviet Union by the end of 1991 leaving Arafat and the PLO without financial and diplomatic superpower support to stand up to their adversaries.

From March 1990 to July 1992, after the collapse of the second Likud-Labor national unity government (NUG), Likud Prime Minister Yitzchak Shamir led the most hard line right wing-religious government in Israel's history. A major policy cornerstone was settlement activity. By the time the opposition Labor moderates won the 1992 elections and Yitzhak Rabin became premier, there were well over 100,000 Jews living in Palestinian areas. Rabin's strategic outlook was based on the "Allon Plan," meaning a territorial compromise with the Arabs to ensure security, democracy and a Jewish majority within Israel's final borders.[44] Originally the plan called for an agreement with Amman, but when Jordan relinquished all responsibility for the West Bank in the summer of 1988 and the future Oslo Accords, the Palestinians, and in particular the PLO, would replace the Hashemite Kingdom as the future peace partner.

Already in 1990 Fatah was faced with a choice between an alliance with Hamas, who demanded 40 percent representation on the Palestine National Council, or to move toward conflict resolution as the Americans demanded. Fatah opted for the latter—which was a betrayal of principles as far as Hamas was concerned.[45] In 1991, pro-Fatah Palestinians went to the Madrid peace talks as part of the Jordanian delegation, but besides symbolic "Palestinian" participation little was accomplished. Two years later, Israel led by Prime Minister Rabin and the PLO headed by Chairman Arafat engaged in the Oslo process ostensibly to bring about a two-state solution by the end of the decade. Much of the Israeli right objected and Hamas remained in steadfast opposition to any compromise recognizing Israel's existence.

Despite serious losses through death, injury and incarceration, Hamas was not idle during these two years. From the start, Hamas condemned Madrid as a sellout and clashed with the PLO, often resulting in injuries and even deaths. By December 1992, Israelis felt the intensity of Hamas' Jihadi terrorist pressures. After several attacks, and in particular the murder of border policeman Nissim Toledano, Rabin's moderate Labor cabinet endorsed a roundup of 1,600 fundamentalists and the banishment of 413 Hamas and Islamic Jihad activists, sending them across the border to south Lebanon. This action spiked terror activities against Israel into early 1993. The exiled activists spent their time networking and building a more cohesive Islamic resistance, while establishing connections through world media and cultivating international sympathy. Their activities resulted in a UN condemnation of Israel embodied in Resolution 799 demanding the immediate return of all those expelled. The Hamas activists also forged ties with Lebanon's Hezbollah. A year later, the exiled Islamists returned home as heroes.[46]

The Intifada impacted Palestinian society on a social and economic level. Although impoverished by strike days and curfews, a new social order and solidarity were developing. By November 1988, many Palestinians followed Arafat and Fatah in support of peaceful engagement with Israel through renunciation of terrorism and the acceptance, at least in theory, of a two-state solution. Overall, Israelis were skeptical of PLO intentions due to mixed messages; the PLO's declarations and actions were both conciliatory and hostile at the same time.[47] Neither the Israeli government nor the average citizen was convinced of Arafat's sincerity, yet it was a shift in course.

Still, to many Palestinians, Islamic Jihad seemed to be the answer. Coupled with *dawa* social welfare activism was the Jihadist and viciously antisemitic ideology of the Muslim Brotherhood. Sheikh Yasin was always forthright in his discussions with Israeli security officials, clearly stating that Islam will be victorious and the Jews returned to their *dhimma* status.[48] Previously in the 1950s, Yasin took a stand against secular Palestinian nationalism and any attempt at Arab national resistance, as he believed victory would only come through Allah and Islam. Still, he would not condemn such national secular efforts outright. Yasin was a major influence in constructing the carefully worded *Hamas Covenant*, and fully emphasized the above-mentioned ideals in its final draft in August 1988.[49]

Hamas understood that the traditional Brotherhood demand of Islamic solidarity was too amorphous; nationalization, or specifically a "Palestinianization," of the conflict was the answer to rally the masses. The question asked was whether Palestinianization was a limiting strategy when the Hamas objective was to capture the Palestine national movement, Islamicize it, declare victory and engage in a form of endless "*hudna*" with Israel. Was Islam just a strategy to be used in the overall objective of defeating the PLO and its secular allies? This being the case, Hamas would be forced to settle for an Islamic Palestine alongside Israel, "Islamic," being only an adjective. To do so was a theological "sell-out."

Their other option was using Jihad as the strategy for the defeat of the PLO and Israel. In this scenario, the battles for Palestine were limited to one piece in an Islamic world puzzle, but the local Jihad could only be won if packaged in Palestinian national terms. Palestinian nationalism was a tool for victory to be discarded in the aftermath, while the Palestinian-Israeli conflict was reduced to one front in the Islamic reach for global hegemony. The State and Islam would be synonymous, a breeding ground for global Jihad where Muslims were trained and garrisoned worldwide to bring about total victory. The eventual merging of the Palestinian locale with the universal Islamic state was the final objective. Here "Palestinian" was only an adjective used to indicate from what region of the world the Jihadi warrior originated.

If Islamic Palestine was the objective, then *The Hamas Covenant* would be reduced to a tool to obtain a pragmatic objective - the nation state. But if the *Covenant* is taken at face value, then capturing any part or all of Palestine (Land of Israel) is only a preliminary step leading to eventual world conquest. Would Hamas become pragmatic? If so, then was Islam betrayed in the name of secular Palestinian Arab nationalism? Or rather, was Hamas using secular Palestinian nationalism as a veneer for an Islamic victory and an eventual Islamic state or regional entity? Hamas solidified its ranks, held up well under fire despite the arrest of most of its activists, retained its turf—in particular Gaza despite PLO pressure, and expanded its influence. In retrospect, Hamas claimed victory in the Intifada.

Many believe Hamas was forced to face reality and became a pragmatic Islamic movement, abandoning its *Covenant* and enemies list. In this scenario, even the Jews were no longer to be considered hostile, only the Zionists who were defined as part of a world imperialist scheme would remain the implacable enemy. Hamas' belligerency was to be directed at those who attacked it first.[50] At differing times Hamas could appear quite flexible. In 1988, during the Intifada, one Hamas leader Mahmoud al-Zahar met with then Defense Minister Yitzhak Rabin, Foreign Minister Shimon Peres and IDF Chief of Staff Dan Shomron. He made it plain that hostilities would end provided Israel withdrew from the West Bank, Gaza and East Jerusalem.[51] Such an interpretation meant Hamas was evolving into an Islamic PLO confining its struggle to the Palestinian front against Israel, but allowing no hostilities beyond. On the other side stood Arafat and the PLO appearing even more "moderate," yet the question remained as to whether Hamas had truly given up the armed struggle or if such secular nationalism was only a facade and tactical move to hide a more Islamist perspective of never-ending struggle until victory—couched in secular terms. Israel took such PLO declarations at face value testing their intentions regarding the possibility of peace and the two-state solution. Concerning Hamas, Israel understood a halt in hostilities was not a prerequisite for a peace agreement, but for a *hudna,* or a temporary Islamic cease-fire until the Islamists could strike back and win.

Oslo Accords and the PA Mini-State 1993-2000

The militant Hamas line was countered by a much more conciliatory PLO policy shift. Back door negotiations between the PLO and Israel through Norwegian intermediaries during the Labor party's first year in office resulted in the Oslo Accords, as embodied by the Declaration of Principles. To facilitate matters Israel recognized the PLO as the legitimate representative of the Palestinian People. The Israeli government and the PLO signed the Oslo Declaration of Principles (DOP) in September 1993 on the White House lawn in essence with the PLO representing a state-in-the-making until elections. Hamas reviled the dealings between Israel and the PLO, and responded by striking Israel with Islamic terror just prior to and immediately after the signing of the Declaration. Still exiled to south Lebanon, the Hamas leadership took the Declaration as an affront, as did Sheikh Yasin who watched the ceremonies on TV from an Israeli prison. Everyone realized Oslo was also meant to destroy the Hamas movement. Fatah prisoners were freed from Israeli jails, while Hamas inmates remained incarcerated. Due to their uncompromising demands for Israel's destruction the Lebanese Shiite Hezbollah replaced the PLO as a more natural Jihadi ally for Hamas.[52] During this time, Hamas avoided any clash with the PLO. Hamas "moderates" like Sheikh Hassan Yousef who opposed Oslo and did not trust Israel or the PLO, realized the time was not ripe to challenge Arafat, so all remained calm.

The Declaration of Principles contained several main points, mostly taken from the previous "Framework for Peace in the Middle East" negotiated as an integral part of the Camp David Accords peace agreement between Israel and Egypt in 1978-79 after the 1973 Yom Kippur War. The main points used from the Camp David Accords were as follows:

- Acceptance of UN Resolution 242 and 338 calling for peace, security and recognition of all states in the region, most obviously including Israel.
- There was to be a five-year transitional period of Palestinian self-rule or autonomy in the West Bank and Gaza.
- A self-governing authority would be elected.

- There were to be Israeli withdrawals to "security locations" as the Palestinian self-governing authority exercised its control.
- Jordan and Israel were to arrive at a peace agreement.
- Israel, Jordan, Egypt and the elected local representatives were to negotiate a final status agreement for the West Bank and Gaza to ensure the "legitimate rights of the Palestinian people."
- The elected representatives of the West Bank and Gaza were to approve the final status agreement to be implemented after the five-year transition period.

The major disagreement between Egyptian President Sadat and Israeli Prime Minister Begin in the late 1970s was over the status of Jerusalem. Handing in two separate letters to President Carter, they refused to compromise. Begin declared Jerusalem "One city indivisible, the capital of the State of Israel." Sadat insisted, "Arab [East] Jerusalem is an integral part of the West Bank."

The DOP of 1993 was negotiated directly with the PLO and Arafat, who fourteen years prior rejected the Camp David Accords. The outstanding differences were Israeli agreement to a Palestinian police force, thereby removing Egypt and Jordan from the scene, and Israeli agreement to negotiate directly with the Palestinians over "Jerusalem" while confronting the other issues of "refugees, settlements, security arrangements, borders, relations and cooperation with other neighbors, and other issues of common interest." All of these matters were to be negotiated as part of the permanent status accord. In 1993, Arafat and the PLO accepted the Camp David Accords after a few significant updates.[53]

On the other hand, Hamas fully rejected the American brokered 1978-79 Camp David Accords as explicitly stated in Article 32 of its *Covenant*. For the Islamists, the DOP represented a Palestinian national capitulation, and a larger Muslim capitulation to what was seen as an American-Israeli dictate. For Israelis the DOP involved a major concession whereby Israel agreed to negotiate the status of East Jerusalem which was annexed along with the Old City and its holy sites in 1967 immediately after the war. Jerusalem as Israel's "united capital" was now in doubt. Though Israelis saw the DOP as a major concession to the Palestinians it held no significance for Hamas which continued to demand Israel's destruction.

Oslo I took place in early summer 1994. It was the first implementation of an Israeli withdrawal from the vast majority of Gaza and the Jericho municipal region. Arafat entered Gaza from Sinai by vehicle on July 1 and established the Palestinian Authority (PA), bringing with him thousands serving as his police and security forces. Palestinian prisoners were released from Israeli prisons, but Hamas and Islamic Jihad prisoners remained incarcerated. Israelis hoped the terror would end at this point, while many saw the change as favorable toward peace between the two adversaries. Israel and the PLO made a deal that resulted in what came to be known journalistically as "Land for Peace." The Palestinians obtained land and Israel obtained peace. It was a step-by-step process where each side tested the other during a five-year period of interim agreements. This was far from the real story, as Israel sought full security, while the PLO demanded sovereignty in its quest for an internationally recognized state. Each expected the other to help it achieve its objectives because they were deemed as mutually beneficial to both sides. Questions were continually raised as to how much sovereignty and security could be attained, and at what price?

Prior to the May 1994 signing there were terror attacks against Israelis and a major "retaliation" of personal revenge in February by Barukh Goldstein, a resident of the Jewish Quarter in Hebron, who killed 29 Palestinian worshipers at the Ibrahimi Mosque, also known as the Cave of the Machpela, a site holy to both Jews and Muslims. In April, the Islamists stepped up their attacks with bus bombings in Afula and Hadera, inside Israel proper. While Israel worked to repress its own right wing religious extremists, the PLO did not always try to contain Hamas. The PA only responded when directly threatened, as happened in November 1994 when Arafat gave the order to open fire on the Islamists during a Hamas show of strength in Gaza. The result was fourteen deaths and dozens of injuries. Thus, the PA further consolidated power.[54] It was unclear whether the PA was willing to clash with Hamas as part of guaranteeing Israeli security. Despite all their disagreements, both fought side-by-side during the Intifada and Fatah did not want to be seen as repressing patriotic acts of resistance against the Israeli occupier. On the other hand, the PA was obligated to cooperate on security with Israel as agreed upon by the Oslo principles.

Settlement issues were postponed until a permanent status agreement could be reached. Although Rabin refused to build in most of the Palestinian areas, there was continued construction in communities just past the 1967 Green Line, most notably in the Jerusalem area and in East Jerusalem itself. Palestinians saw this as a permanent encroachment on their eventual state.

For the Islamists, construction was less of a problem since the existence of pre-1967 Israel was just as much of an affront as the settlements. Any Israeli actions seen as contradictions stirred the masses, worked to the Hamas advantage and against the PLO "appeasers."

Although viewed by many as corrupt and oppressive, Arafat and the PLO still embodied Palestinian nationalism and, despite promises to the contrary, *The Palestinian National Charter* remained unchanged—still calling for Israel's demise. This deterred neither Arafat nor the Israelis from signing Oslo II in September 1995, constituting an implementation of the DOP and continuation of Oslo I. By early 1996, Israel was to withdraw from the heavily populated West Bank Arab cities, relinquishing full civilian and military control to the PA. This region was designated as "Area A." Outlying Palestinian villages, where Israel still retained security forces but the PA took over civil authority from the military administration, became "Area B." All other remaining territories where full Israeli control remained intact were deemed "Area C." As of this writing, the Oslo II arrangement is still the major determining factor in the everyday lives of Palestinians in the West Bank, as some 96-98 percent live in Areas A and B.

Hamas, fearing the possibility of conflict resolution and a two-state solution arrangement, went on the offensive with a concerted wave of attacks and suicide bombings against Israel. The best known are the bus bombings: Tel Aviv in October 1994, Beit Lid in January 1995, Ramat Gan and Ramat Eshkol in Jerusalem in the summer of 1995, the dual suicide-homicide attacks against the Jerusalem #18 bus during consecutive weeks in February-March 1996, and the Purim Dizengoff Center explosion immediately afterward. In addition, there were kidnappings and executions of Israeli soldiers and numerous shootings and attacks on civilians in urban areas and on the roads. Prime Minister Shimon Peres, who took the reins of government after Rabin's assassination by the fanatical religious Jewish assailant Yigal Amir in November 1995, carried out all the withdrawals despite harsh right wing religious opposition. Sensing Arafat was not halting terrorism in the winter of 1996, he refused to withdraw from the Arab neighborhoods of Hebron. A few months later Peres lost the May elections to the right wing Likud. Benyamin Netanyahu became prime minister and implemented the Hebron withdrawal in January 1997 with certain security upgrades for Israel.[55]

By 1995 the die was cast—Hamas became the most implacable of all enemies. According to Mosab Yousef, "The transition of Hamas into a full-blown terrorist organization was complete. Many of its members had climbed the ladder of Islam and reached the top. Moderate political leaders like my father

[Hassan Yousef] would not tell the militants that what they were doing was wrong. They could not; on what basis could they declare it was wrong? The militants had the full force of the Qur'an to back them up."⁵⁶ Furthermore Israel appeared clueless. "But they [Israel] never made an effort to find out who or what Hamas really was. And it would be many painful years before they would begin to understand that Hamas was not an organization as most people understood organizations, with rules and a hierarchy. It was a ghost. An idea. You can't destroy an idea; you can only stimulate it. Hamas was like a flatworm. Cut off its head, and it just grew another."⁵⁷

The world saw the PLO-Israel conflict as political, one with possibilities for compromise. To the contrary, Hamas Islamicized the clash. It was absolutist, not just in theory but in practice. Allah was invoked in everyday actions, all the land belonged to Allah and Israel's existence demanded termination. Those "racist leaders of Hamas" took it a step further insisting "Allah had given us the responsibility of eradicating the Jews." Supposed moderates like Hassan Yousef accepted such a policy, even if they did not take action toward its realization.⁵⁸ In other words, everyone in Hamas was either active or complicit in an attempted policy of annihilating the Jews, not just the State of Israel.

Hamas strengthened its power and prestige while its leaders were in Israeli prisons. Not only were they allowed to organize prayers and Islamic study sessions, they ruled much of the prison population with an iron fist, including torturing suspected Palestinian "collaborators." False accusations were made of sexual misconduct involving multiple women and bestiality. Confessions were extracted under excruciating pain and the "convicted," many of them former Hamas supporters, served as an example of an absolute power wielded by the incarcerated Islamic leadership.⁵⁹

In January 1996 following the DOP stipulations, elections were held and Arafat was chosen president by an overwhelming majority of Palestinians with 88.2 percent of the vote, while Fatah took 55 of 88 seats in the legislative assembly. Only four declared Islamists were elected. Hamas boycotted the elections and total PLO domination ensued.⁶⁰ Feeling pressure from the PA, unsure of electoral support and working grass roots as an alternative to the Fatah secular regime, Hamas strove to rebuild itself on the social-political front while keeping up the armed struggle. This set them apart from the PA, who were seen as oppressive, corrupt American-Israeli lackeys betraying Islam and the Palestinian people. They were even compared to the pro-British Palestinian Arabs of the late 1930s, who were condemned as "peace gangs" because they sought to end the uprising of 1936-39.

Israeli Prime Minister Netanyahu did his best not to engage the PLO in further negotiations, preventing Israel from having to cede anything. When Israel opened an exit gate from the Western Wall tunnel to the Muslim Quarter in September 1996, the PA responded with "popular" and regime-coerced violence against Israel forcing Netanyahu's government to the negotiating table. Previously in a secret arrangement the Palestinians received an enormous underground extra prayer room in the Al-Aksa domain in return for the opening of the gate, yet the PA controlled press whipped up an atmosphere of confrontation. Once again Arafat used violence effectively.[61] As a matter of ideological principle, Netanyahu's right-religious administration promised not to relinquish any land west of the Jordan River—the "Greater Land of Israel." By January 1997, the Israeli prime minister implemented the Labor-negotiated, and Likud-renegotiated, withdrawal from most of Hebron but leaving Jewish areas under Israeli control. More significantly, by October 1998 Netanyahu signed the Wye Accords, ceding another 13 percent of the West Bank (the Biblical Judea and Samaria), to the Palestinian Authority.[62] Ideologically and practically the Israeli secular right adopted the Laborite territorial compromise and autonomy solutions, leading toward a two-state solution, even should their territorial concessions be more limited.

On the Palestinian side, the PA gained credibility by obtaining concessions from Israel and the eclipse of Hamas appeared in the making. Working to arrest the peace process, the Islamists executed two massive Jerusalem terror attacks in the summer of 1997, but failed to halt progress. Despite the attacks, Netanyahu's government was credited with ensuring more security for the average Israeli than the previous Labor coalition.[63] The PA did not implement the Wye Accords, and Israel only partially so.[64] Arafat promised to amend all clauses in the *PNC* calling for Israel's destruction, and even convened the PLO National Council in the presence of American President Bill Clinton in December 1998, where the vote was in favor of deleting all anti-Israel references and euphemisms. Yet *The Palestinian National Charter* was never amended.[65] Hamas terror attacks meant to cut short the official peace process appeared ineffective; however, the Islamist influence remained strong. Inside the Palestinian territories Fatah physically had the upper hand, but ideologically it was all Hamas. As opposed to Hamas, for whom the *Covenant* is also known as the "Charter of Allah" and cannot be amended, the PLO can revise their *Charter* by a two-thirds vote (see *PNC* Article 33), indicating it was deliberate PLO policy not to make the necessary changes.

In the 1990s Hamas already understood the need to topple the Fatah-led PA. There were clashes on several levels beyond ideological; in particular the

IV Development of Hamas 1948 to 2000

focus was the class conflict between the wealthier Fatah power elite and the more proletarian Hamas. Within Fatah there was rising resentment against the PLO Tunis leadership, which began with the commencement of the Oslo process. The PA viewed Hamas terror activities against Israel as undermining the Oslo Accords and on the international level, as calling their own legitimacy into question as the actual representative of the Palestinian People. By 1996-97 the PA incarcerated much of the Hamas leadership, accusing them of plotting the assassination of PA officials, including Arafat himself. It appeared such plans even had Sheikh Yasin's approval. According to Gaza Preventive Security Chief Mohammed Dahlan, now that the Hamas military wing was neutralized, the question arose as to whether to cajole loyalty among their activists and absorb Hamas into the PA, or to destroy them completely. The PA and Israel could not agree on either policy and in the end nothing was decided. There is speculation that Arafat did not know who to trust less, Hamas or Netanyahu, and hence paralysis set in. The survival of Hamas spelled big losses for both Israel and the PA in the future.[66]

In the meantime, Hamas activist Ibrahim Makadme established a doctrine that was gaining ground. He advised not to attack the Fatah-dominated PA to prevent Hamas from destruction. Rather, all attacks should be directed at Israel, who in turn would demand action by the PA to ensure security. When the PA would fail, or for whatever reason not fulfill its mission, Israel would retaliate against the Arafat regime, forcing the ultimate demise of Fatah, the PA and the Oslo Accords. Then Hamas would move to capture political power.[67] As we know, the doctrine succeeded in Gaza in 2007 and was on its way to victory in the West Bank. Israeli and US intervention kept Fatah and the PA in power in the West Bank (see Chapter V "Hamas Ideological Victory").

The late 1990s saw a surge in diplomatic activity concerning Hamas. When two Israeli Mossad agents failed to assassinate Hamas leader Khalid Mashal in Jordan in 1997 and were themselves apprehended, the resulting prisoner exchange with Israel resulted in the release of Sheikh Yasin once again. He quickly traveled throughout the Arab world and Iran, preaching the dual message of an anti-PLO domestic policy and the struggle against Israel. Once Kuwait was no longer the center of Hamas activities, which occurred in 1990 as a result of the Gulf War, Jordan became the external hub. Hamas was forced from Amman in 1999 and relocated to Damascus due to increasing PA, Western and Israeli pressures to expel the Islamists. Hamas had rejected the Wye Accords, condemned Jordan's peace agreement with Israel in 1994, and moved closer to Khomeinist Iran, all in contradiction to Jordanian foreign policy objectives. Domestically, Palestinian refugees identified much more with the half-blind, half-deaf, quadriplegic Sheikh Ahmed Yasin who lived on $600 a month in a Gaza low-income neighborhood, than they did with the bloated bureaucracy and lavish lifestyle of many Fatah operatives, or with PA Chairman Arafat, whom they suspected might not enforce their demand for Palestinian refugee return to Israel.[68]

A Labor government led by Prime Minister Ehud Barak succeeded Netanyahu in July 1999 and attempted not only to implement the Wye Accords, but to make peace with Syria by withdrawing from the Golan Heights, though they failed on both fronts. In May 2000, Israeli forces hastily withdrew from the south Lebanon "security zone." The Iranian sponsored Hezbollah Shiite militia moved up to Israel's northern border demarcation as Israel's predominantly Christian South Lebanese Army (SLA) allies collapsed, a humiliating flight in the face of Islamic pressures. The Palestinians were emboldened.

On the internal Palestinian scene, Arafat refused to hold elections as long as the Palestinians were still "at war" with Israel. Hamas continued as an underground organization without official representation. Frustration and unrest were rampant in the Palestinian areas, especially on May 15—what is known as "Nakba Day" marking the catastrophic Palestinian failure in the 1948 war. Intense violence dominated the Palestinian front for eight days with Arafat refusing to calm the situation; rather he preferred confronting Israel with an angry Palestinian populace. Instability continued into the summer and through the crucial Camp David July 2000 summit, where Barak met Arafat in an attempt to attain a final status agreement. Although Barak offered to return well over 90 percent of the West Bank, withdraw completely from Gaza, divide Jerusalem and accept a limited, but symbolic return of several thousand Palestinian refugees coupled with participation in refugee compensation, Arafat rejected all these proposals outright without making counter suggestions. In essence, Arafat remained entrenched in his original positions demanding full refugee return, including all descendants, totaling into the millions, Israeli withdrawal to the 1967 borders including Jerusalem, with a possible land swap, and full Israeli responsibility for the conflict. Although this was the veneer of a two-state solution, it spelled the end of Israel as a Jewish national entity and left it devoid of defensible borders. President Clinton, who brokered the talks, blamed Arafat for the impasse. Barak and Clinton were looking for an "End of Conflict" scenario and mutual recognition of two national entities, Jewish and Palestinian Arab. Arafat moved closer to the Hamas position without explicitly demanding the dismantling of the Jewish state.

There was no movement for the remainder of the summer while tensions soared in the Palestinian areas. Barak's government collapsed just prior to the talks, leaving him vulnerable in his role as prime minister leading a minority government. Declaring he would never relinquish Judaism's most sacred site, the Temple Mount, Likud opposition leader and former general Ariel Sharon ascended the mount with the permission of Israeli and Palestinian security forces in late September.[69] This was the flash point Arafat awaited and now the Second Intifada, or what will be referred to in this analysis as the Palestinian "Low Intensity Conflict" (LIC), ensued. Beginning the next day on September 29, 2000, it took four years before Israel contained the "Al-Aksa" or Second Intifada,[70] which was really a Palestinian LIC and terror offensive.

Arafat knew Israel could never accept a full refugee return, since such a move constituted the destruction of the Jewish State. A two-state solution would begin alleviating Palestinian suffering and give hope, but "Yasser Arafat had grown extraordinarily wealthy as the international symbol of victimhood." In playing up what can be called his Che Guevara guerrilla leader image to the hilt "he wasn't about to surrender that status and take on the responsibility of actually building a functioning society."[71] Catastrophe ensued as Arafat and his allies became a Robin Hood in reverse, making increasing Western media gains through the sacrifice of Palestinian blood. At the time, many thought Ariel Sharon's jaunt to the Temple Mount triggered a "spontaneous eruption of Palestinian rage," but, as in many media events, a later revelation of the facts proved first impressions incorrect. The day before, Sheikh Hassan Yousef had attended a meeting with Fatah Secretary General Marwan Barghouti to discuss a joint uprising in response to Sharon's expected visit. Palestinian Security Chief Jibril Rajoub authorized the visit, fully complicit in the plot to paint Sharon as the catalyst for the planned violence. Yousef agreed to a show of unity and even watched from a distance with his son Mosab. Sharon did not enter any mosques although there was a heavy police presence. Demonstrations in response to Sharon's visit were limited and Mosab went to Galilee on vacation. A day later, after Friday prayers, everything exploded and Arafat used Hamas as his scapegoat.

Although formally the PA was still holding back Hamas terror activity against Israel, the two now aligned themselves closer together than at any time since the 1980s. Arafat and the PLO had been far more effective in crushing the Hamas military wing than Israel previously was, especially through mass incarceration. With their forces jointly working against the common Israeli enemy wide-ranging violence ensued, planned and led by the Palestinian Authority leadership who now shifted sides, to an alliance with Hamas.[72]

In the past, Arafat and the PLO tried to be everything to everyone: moderate secular nationalists to the Israelis, Americans and the West who favored a two-state solution, and non-compromising nationalists and committed Muslims to their Palestinian electorate and Arab Muslim world allies. During the Intifada in 1987-91, many West Bank activists advocated acceptance of a negotiated settlement with Israel. At the time, those activists and their ideas were suppressed and the PLO even floated the possibility of a mini-state confederation with Jordan. On the other hand, the hope was to re-ignite the conflict and bring about Israel's destruction in the name of Islam. As Arafat spoke of peace, terror activities continued. A prime example was the 1990 failed Tel Aviv beachfront attack and Arafat's continued support for Saddam Hussein.[73] Sensing Arafat's weakness and fears of Hamas ascendancy, Israel chose to negotiate with him over the Oslo Accords, fully expecting to find an anthropocentric partner bent on mutual recognition, peace and cooperation. At that point, Arafat supposedly faced a dilemma concerning war or peace. In retrospect, Arafat as leader of the PLO reached the fork in the road where he returned to his Islamist understanding of the conflict, with any compromise deemed unacceptable.

Under the watchful eye of Arafat's Palestinian Authority, during his sermon on that fateful Friday, Sheikh Hian Al-Adrisi addressed worshipers at the Al-Aksa Mosque on Jerusalem's Temple Mount declaring the Jews to be the "enemies of Islam." Shortly afterward, the PA-appointed Dr. Abu-Halabia of the Fatwa Council urged Muslims to murder Jews wherever they were found (see Chapter III "Jewish Nationalism"). The West Bank, Gaza and East Jerusalem exploded.

Although there were those who advocated a more radical approach, a review of Hamas policy reveals a fairly moderate attitude toward the PLO Fatah-administered PA during the seven-year Oslo period of engagement with Israel from 1993-2000, despite the Islamists' vehement opposition. Hamas even went out of its way to house the first PA security forces when they arrived in Gaza. At the time, the Hamas decision not to physically challenge Fatah for control was seen as proof of the organization's practical responsibility, despite the November 1994 killings and later assassinations.[74] Hamas continued to strengthen its network of educational, religious and social services while leaving the political front to the PLO/PA. The only real attempt to challenge the PLO/Fatah paradoxically took place amongst Palestinians held in Israeli prisons where there was competition for loyalties. When forced to face the question of a referendum or even elections, the Islamists vowed to respect the will of the people whether they won or lost. There was much debate surrounding the issue, but in the end they decided not to participate in the voting, because to do so would legitimize the Oslo process.[75] Theoretically, elections could advance the Islamic agenda through democracy regardless of the outcome. Should Hamas lose they could try again, perpetually playing the democracy card of "the people's will." But should Hamas win, would free and open elections take place four years into the future, or would the absolute reign of Islam bar any seemingly democratic (pagan infidel) electoral process?

Both Sheikh Yasin and his deputy Abdul Aziz Rantisi only favored temporary cease-fires or a *hudna* with Israel. Rantisi was explicit in comparing these to the Hudaybia truce of extremely short duration between Mohammed and his enemies, while Yasin spoke of a time period lasting no longer than fifteen years. Others claimed any peace treaty was similar to a *hudna* and could be signed with Israel since such agreements only reflected the balance of power at the time of signing, and would be subsequently annulled when the opportunity for victory arose. An example of such one-sided "pragmatic behavior" is the Oslo II period in the second half of 1995, when Israel agreed to fully relinquish Area A and civilian control in Area B to the Palestinian Authority with a withdrawal timetable going into late March 1996.[76] Whether Hamas truly restrained itself without coercion from the PA at any time is not certain, but after the Goldstein attack in Hebron in early 1994, and following Israel's targeted removal of explosives expert Yahya Ayyash in January 1996, there are those who claim Hamas only then decided to break the *hudna* and engage in a full-scale homicide-suicide bombing campaign against Israeli civilians. This is said to be particularly true concerning the February-March 1996 terror campaign. Previously, Hamas claimed it only targeted the Israeli military and settlers.[77]

When Israeli and PA interests came together after the early 1996 bombings, both knew Hamas had to be contained or the Oslo Accords were doomed. The PA cracked down, arresting 900 Hamas activists, which included a raid on the al-Najah University campus. Shifting gears, by 1997 Fatah attempted to co-opt Hamas into joining the PA regime, but failed. Realizing the grass roots nature of the movement, Arafat pressured mosque organizations and charities, scrutinizing their every move. The Islamists refused the bait and took no military action against Fatah, to prevent giving the PA a reason to annihilate them.[78] After a sharp internal debate, Hamas refused to take part in the 1996 presidential and legislative elections. There was the question of how much support the Islamists had for their rejection of the Oslo Accords versus the need to represent their constituents. Hamas took the middle way and tested public support by challenging Fatah in student and professional organizations throughout the West Bank and Gaza. Here, Hamas gained some 40-50 percent of the vote and claimed a similar level of support from the Palestinian population at large. The official Palestinian pollster Khalil Shikaki estimated the Hamas faithful at 18 percent, while political analyst Khalid Hroub argued for 30 percent.[79] While it is impossible to know the general level of support for Hamas in the late 1990s, the next elections held in 2006 for the parliament awarded Hamas 44.45 percent of the popular vote and an electoral landslide. Hamas took 74 seats out of 132 in the Palestinian Legislature. Fatah and the smaller secular parties had a slight popular majority, but this did not come to fruition in the tally for representation because of too many split votes among the secular and nationalist candidates.[80]

On the world Islamic front, Hamas continued its ideological purity supporting Islamist parties in the Arab world and the Islamist uprisings in Algeria, Tunisia, Afghanistan, Chechnya, Bosnia and Kashmir. In sub-Saharan Africa, Hamas advocated victory for Omar al-Bashir's repressive and genocidal Sudanese Muslim Brotherhood regime over the Black African Christians and animists in the south of that country[81] while citing the need for world Islamic actions against supposed Zionist threats to Sudan, Ethiopia and the Bab el-Mandab region in Somalia. Khomeinist Iran rewarded Hamas with full support for its total rejectionist stance against Israel, beginning what later can be seen as a strategic relationship.[82] This is in line with the expressed ideology of supporting Islamic movements worldwide *(HC* Articles 2, 3, 5, 7, 8 and 23).

Decisive Crossroads: Summer 2000

Towards the year 2000 and what is commonly called the "Second Intifada," but in essence was a Low Intensity Conflict (LIC), Hamas policies were of a tactical pragmatism developed in the name of the overall Islamic objective of victory. Despite declarations to the contrary, Hamas would not take on the PA in a military confrontation for fear of defeat, nor did the organization participate in the first elections in 1996. They would not participate due to possible ideological constraints, as well as the need to shore up their own support and not only enter, but exit any election with a victory. Hamas did not want a coalition with Fatah or any other arrangement where they would be responsible for self-rule, construed as a compromise with Israel and an ideological sell out. Holding to an Islamic theology after winning an election meant confrontation with Israel, an eventuality demanding more preparation. Mundane matters would include sanitation, road repairs, street lighting and taking responsibility for jobs and the economy; these last two were at least in part dependent on Israel.

In contrast, Hamas allowed for the PA to rule and fail. Corruption, poor economic management and the complete lack of *dawa* organizations, or a socio-economic blueprint for development, undermined the PLO from the outset. Arafat never made the transition from "freedom fighter" to statesman and took no action to improve the everyday lives of his constituents. He continually emphasized "resistance" as an option while threatening war and martyrdom should Israel not heed his demands. Moderates criticizing him for not making peace were silenced while the hardliners urging military action were free to express themselves. Peace dividends were not to be had and massive frustration mounted.[83] Despite appearances to the contrary, by the end of the seven year Oslo negotiation period Arafat's PA merged closer to Hamas than ever before. It might be more correct to say that in one way the PA resembled the Islamic Jihad—an organization committed to victory, but devoid of welfare policies. Besides religious devotion, the one great difference between the Islamic Jihad and the PLO directed PA was that the former was purist while the PA was overwhelmingly corrupt.[84] For Arafat, taking an uncompromising position toward Israel while derailing Oslo could only lead to a reunification of joint efforts in a PLO-Hamas alliance, embracing Arafat as the leader of a joint command. Here we have a combination of ego and a return to certain basic Muslim Brotherhood understandings.

This confused many in the West, in particular concerning ambiguities between peace-making and continuing the never-ending "armed struggle." At the time, an "End of Conflict" two-state solution with vast financial aid was on the table under the auspices of President Clinton, who did his utmost to advance as favorable a permanent status agreement as possible for the Palestinians. Arafat engaged in "*taqiyya*" and "*kitman,*" which is lack of truthfulness in the service of the Islamic cause, a mode of behavior he used in the national secular struggle. *Taqiyya* is "lying" while *kitman* is an "omission," thereby altering the truth.[85] Arafat's behavior included both during negotiations and the subsequent four year LIC as he secularized these concepts; however, it is safe to assume he retained the same Islamic mindset. Moving toward Hamas may have only been a tactic at the outset, but what Arafat may not have counted on was the absorption and/or later defeat of the PLO by Hamas as opposed to a unifying PLO hegemony under the PA. This internal Palestinian Hamas success will be discussed and analyzed further in the next chapter alongside Arafat's behavior and less-than-forthright negotiating tactics.

For good reasons, Hamas delayed its entrance into the official political arena. Although battered and exhausted by both Israel and the PLO, Hamas sensed the upcoming failure of the Camp David 2000 talks, as most did, and would take its chances maneuvering under the PA administrative umbrella. Should Arafat come to a permanent status agreement with Israel ensuring a two-state solution, allowing virtually no refugee return and anything less than full Islamic control over the Temple Mount/Noble Sanctuary in Jerusalem, Hamas could count on overwhelming support to challenge the PA on the grounds of betrayal, not to mention increasing corruption and oppression. Hamas was moving toward an upgrade either at the PA's expense, or Israel's. By September 2000 it appeared to be a win-win situation should the Palestinian Islamists consolidate their support in the street while battling Israel and either working with or confronting the PA when necessary. And to clinch it, one only needed patience and correct timing to implement the Makadme doctrine whereby Hamas attacked Israel to evoke a punishing Israeli retaliation against the PA for not curtailing terrorism. A weakened PA would then be overthrown by Hamas.

Endnotes

1. Hroub, Khalid, *Hamas Political Thought and Practice*, Institute for Palestine Studies, Washington DC, 2002, p. 14.
2. Lebel, Jennie, *The Mufti of Jerusalem, Haj Amin el-Husseini and National Socialism*, Cigoja stampa publishers, English translation, Paul Munch, Belgrade, 2007, pp. 299-300.
3. Mishal, Shaul and Sela, Avraham, *The Palestinian Hamas, Vision, Violence and Coexistence*, Columbia University Press, New York, USA, 2000, p. 18. Hroub, pp. 19-24.
4. Rubin, Barry and Rubin, Judith Colp, *Yasir Arafat, A Political Biography*, Oxford University Press, New York, 2005, pp. 26-29.
5. Mishal and Sela, pp. 17-18 and Hroub, pp. 29-31.
6. Tamimi, Azzam, *Hamas, A History from Within*, Olive Branch Press, North Hampton, MA, USA, 2007, pp. 36-37.
7. Mishal and Sela, pp. 18-20.
8. Tamimi, pp. 12-19.
9. Mishal and Sela, pp. 21-23.
10. Morris, Benny, *Righteous Victims*, Vintage Books, New York, USA, 2001, p. 572. Mishal and Sela, p. 25.
11. Morris, pp. 566-569.
12. Mishal and Sela, pp. 21-24. Hroub, pp. 29-34.
13. Tamimi, pp. 43-44.
14. Morris, pp. 561-564.

IV Development of Hamas 1948 to 2000

15 Ibid, pp. 564-566.
16 Ted Gurr explains this theory in *Why Men Rebel*. The concept of "relative deprivation" is the "perceived discrepancy between value expectations and value capabilities" of individuals or groups. More simply put this is the gap between one's everyday reality and one's expectations. When the gap grows too wide, a level of frustration is reached whereby violence ensues. This analytical tool will be used more fully in the conclusion, discussed in Chapter X.
17 Hroub, pp. 35-36.
18 Tamimi pp. 45-53 and 61.
19 Hroub p. 39.
 Morris p. 573.
20 "Word of the Day/Hamas the Terror Movement that didn't do its Hebrew Homework," retrieved February 2, 2015, www.haaretz.com/news/features/word-of-the-day/1.608751.
 "Hamas," Even Shoshan, Abraham, *The New Hebrew Dictionary* (Hebrew), Kiryat Sefer, 1992, Jerusalem, Israel, Vol. I, p. 405.
21 Hroub, pp. 36-39.
 Mishal and Sela, pp. 20-23.
22 Yousef, Mosab, *Son of Hamas,* Tyndale House Publishers, USA, 2010, pp. 19-20.
23 Hroub, "Hamas First Communique," pp. 265-266.
24 "Camp David Accords," *Israel Ministry Foreign Affairs*, retrieved August 12, 2011, http://www.mfa.gov.il/mfa/foreignpolicy/peace/guide/pages/camp%20david%20accords.aspx
25 Levitt, Matthew, *Hamas: Politics, Charity and Terrorism*, Yale University Press, New Haven and London, 2006, pp. 2-5.
26 Ibid, p. 5.
27 Yousef, pp. 9-12.
28 "Dawah," *Wikipedia*, retrieved July 8, 2011, en.wikipedia.org/wiki/Dawah.
29 Levitt, p. 9.
30 Schiff, Ze'ev, and Ya'ari Ehud, *Intifada*, translated by Ina Friedman, Simon and Schuster, New York, 1990, pp. 224-225.
31 Ibid, p. 238.
32 Morris. p. 575.
33 Ibid, p. 576.
34 Up until 1987-88 there were Palestinians working under Israeli auspices in the police in Gaza and the West Bank. They were paid by Israel. These police were mostly a continuation of the Jordanian paid force which existed prior to the 1967 Six Day War.
35 Schiff and Ya'ari, pp. 234-236 and 239.
36 Yousef, pp. 32-34.
37 Ibid, p. 33.
38 Schiff and Ya'ari, pp. 238-239.
39 Morris, p. 579.
40 Ibid, pp. 583-584.
41 Yousef, pp. 45-46.
42 Morris, pp. 584-585.
43 Sachar, Howard, *A History of Israel From the Rise of Zionism to Our Time,* Alfred A. Knopf, New York, 2007, pp. 977-986.
 Morris pp. 612-615.
44 This is best outlined in Yigal Allon's article "The Case for Defensible Borders" in *Foreign Affairs Quarterly,* Fall, 1976, www.foreignaffairs.com/articles/.../israel-case-defensible-borders. Allon was commander of the Palmach in 1948, deputy prime minister and minister of education in Golda Meir's government (1969-73) and foreign minister in Yitzhak Rabin's first government (1974-77).

According to the plan Israel for the most part was to retain the region in and around Jerusalem, the Jordan Rift Valley along the Jordan River and the northwest coast of the Dead Sea. The remaining 60% or so of the West Bank was to be demilitarized and handed over to Jordanian civilian administration. A road through Israel's Lachish region would connect to the Gaza Strip allowing the Jordanian-Palestinian entity development of port facilities on the Mediterranean coast. The Jordan River was Israel's security border while the Jordanian-Palestinians were to enjoy economic development. Although unofficially discussed with the Jordanians in the early 1970s the plan was never awarded Israeli government approval and was not implemented.

45 Tamimi, pp. 189-190.
46 Hroub, pp. 159 and 193.
Morris, p. 618.
Yousef, pp. 49-52.
47 Laqueur, Zev and Rubin, Barry, eds., *The Israel-Arab Reader*, Penguin Books, New York, NY, 1995, "The Palestine National Council: Political Resolution and Declaration of Independence," pp. 537-546.
Morris, p. 608.
48 Eldar, Shlomi, *Getting to Know Hamas*, Keter Publishers (Hebrew), Israel, 2012, p. 25.
49 Ibid, pp. 66-68.
50 Hroub, pp. 44-52.
51 Eldar, pp. 26-27.
52 Yousef, p. 52 and Eldar, pp. 70-71.
53 "Declaration of Principles," including the Annex and accompanying letters, *Israel Ministry of Foreign Affairs*, retrieved August 12, 2011, http://www.mfa.gov.il/mfa/foreignpolicy/peace/guide/pages/declaration%20of%20principles.aspx.
54 Hroub, p. 55.
55 Ross, Dennis, *The Missing Peace*, Farrar, Straus and Giroux, New York, USA, 2005, Chapter 4, "From Oslo to the Palestinian Authority," pp. 122-136, and Chapter 7, "The Interim Agreement," pp. 188-208.
Sachar, pp. 989-1015.
Morris pp. 626-627.
56 Yousef, p. 57.
57 Ibid, pp. 57-58.
58 Ibid, pp. 58-59 and 63.
59 Ibid, pp. 97-99.
60 "Palestinian General Election 1996," *Wikipedia*, retrieved August 14, 2011, en.wikipedia.org/wiki/Palestinian_general_election,_1996.
61 Rubin and Rubin, p. 177.
Ross, pp. 263-268.
The underground prayer space was known by the popular term "Solomon's Stables" and today is called the Marwani Mosque.
62 For a review see Ross, Chapters 15-17, "The 13% Solution," "Prelude to Wye," and "The Wye Summit," pp. 349-459.
63 Ross, pp. 353-356.
64 Rubin and Rubin, pp. 181-183.
Israel released 250 terrorists and began its first withdrawal in northern Samaria, but later reversed the decision.
65 Ibid, pp. 167-168 and p. 183.
In 1996 the vote to change the Charter was 504-54, but it was never amended.
66 Eldar, pp. 73-84.
67 Ibid, p. 73.
68 Tamimi, pp. 99-133.

IV Development of Hamas 1948 to 2000

69   For most Muslims the Al-Aksa Mosque and Dome of the Rock represent the third holiest site and city (Jerusalem). For Jews the mount is the site of the First and Second Temples of yesteryear and where the future Third Temple will exist in the Messianic End Time.
70   Sachar, Chapter 36 "Ehud Barak's Two Years," pp. 1024-1045.
Ross, Chapters 23-24, "From Stalemate to Camp David," and "The Camp David Summit," pp. 591-711.
Morris p. 659.
Rubin and Rubin, pp. 185-205.
71   Yousef, p. 126.
72   Ibid, pp. 125-134.
73   Rubin and Rubin, pp. 109-123.
74   Hroub, p. 108.
75   Ibid, pp. 211 and 220-227.
76   Ibid, pp. 69-84.
77   Ibid, pp. 242-247.
78   Ibid, pp. 106-108.
79   Ibid, pp. 228-233.
80   "Palestine Legislative Elections 2006," *Wikipedia*, retrieved, August 14, 2011, en.wikipedia.org/wiki/Palestinian_legislative_election,_2006.
81   An estimated 2.5 million black African Christians and animists were murdered by the Sudanese Muslim Brotherhood Jihadist regime from 1983-2006.
82   Hroub, pp. 166-180.
83   Rubin and Rubin, pp. 149-162.
It was pointed out that by the mid-1990s the Palestinian GDP was a dismal $2.9 billion, Arab donor nations only contributed 8.6% of outside assistance (out of $2.5 billion). Palestinians were given work permits on a more limited basis due to continued violence against Israel.
84   For a deeper rendering of Arafat's corruption in particular see Lew, Uzrad, *Inside Arafat's Pocket*, 2005, in Hebrew. Lew worked closely with Arafat and in particular with Mohammed Rashid, Arafat's financial advisor from 1997-2001 in the hope of investing Palestinian monies wisely in order to bring about a stable, economically viable Palestinian entity which would live in peace with Israel. Lew claims at least $300 million was stolen by Arafat and in particular found its way into Swiss banks.
85   "Lying (Taqiyya and Kitman)," *The Politically Incorrect Truth About Islam*, retrieved August 15, 2011, www.thereligionofpeace.com/Quran/011-taqiyya.htm.
"Taqiyya And Kitman: Are Muslims permitted to lie?" *Nairaland Forum*, November 23, 2011, retrieved January 5, 2016,
www.nairaland.com/809331/taqiyya-kitman-muslims-permitted-lie.

Israel 2016

After the full return of Sinai to Egypt in 1982.

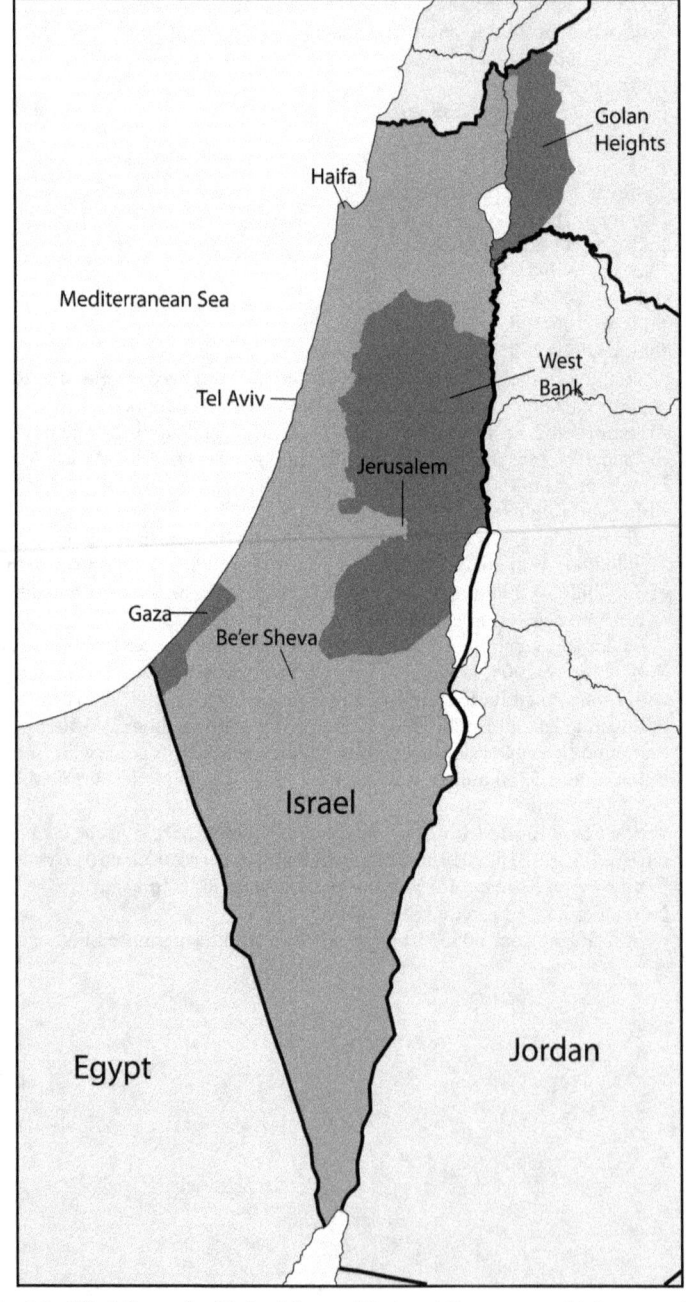

Credit: Modified from the United States National Imagery and Mapping Agency

V

Hamas Ideological Victory, Resistance, and Pragmatism 2000 to 2016

Overview

In the year 2000 US President Bill Clinton invited Israeli Prime Minister Ehud Barak and the PA/PLO Chairman Yasir Arafat to the Camp David Summit to secure what many thought would be a permanent status agreement invoking a two-state solution between Israel and the Palestinians. When Arafat arrived, he had little intent to compromise. Rather, he used the negotiations as a tactical move for consolidation and radicalization of the overall Palestine national movement. Orchestrating the failure of the negotiations, he positioned Hamas as his "point man" in the planned uprising against Israel.

Arafat lined up ideologically with the Islamists; both *The Palestinian National Charter (PNC)* and *The Hamas Covenant (HC)* call for Israel's demise. The PLO/Fatah strove to harness Hamas' commitment and activism in their joint struggle against Israel, and in doing so unwittingly reinforced Hamas' power in the Palestinian street. Hamas turned the tables and proved dominant during the ensuing four year Second Intifada, or Low Intensity Conflict (LIC). Using the plan advocated by the Hamas strategist Ibrahim Makadme, Hamas attacked Israel knowing full well the PA would be held responsible for all security breaches emanating from Gaza and the West Bank. As a result of these attacks, Israel destroyed PA security forces, viewing them as complicit with the perpetration of Hamas violence. Once the PA was weakened, Hamas moved to solidify its leadership among the Palestinian population.

From the outset, Hamas declared a two-front war against both Israel and the corrupt secular PA regime. Israel attained military success, but its deterrence was weakened. The PA was discredited while Hamas achieved respectability and gained legitimacy as a positive leadership alternative. In the 2006 Palestinian legislative elections Hamas claimed victory, yet PA Fatah President Mahmoud Abbas repressed its influence. Abbas replaced Arafat after his death in late 2004. Discussions and attempts at national unity were accompanied by street battles between Fatah and Hamas. By June 2007 the Hamas military captured the Gaza Strip, banishing Fatah. With US and Israeli help, the PA regime reconstituted itself and regained full control of the West Bank.

In June 2006 Hamas flexed its muscles once again and abducted Israeli soldier Gilad Shalit in a daring cross-border raid from Gaza. The move pitted the more pragmatic political leadership of Hamas Prime Minister Ismail Haniyeh against its own Jihadi military faction "Izz a-Din al-Qassam," led by Ahmed Jaabari. Jaabari's success was a result of defiance against "political Hamas." Over the next few years, Hamas strengthened its hold on Gaza, using both the *dawa* calling for educational and social services and the Jihadi demand for conflict. In particular, conflict included rocket attacks against Israel. Eventually tensions exploded into the "Cast Lead" operation in 2008-09. Hamas suffered massive military failure, but won the propaganda war against Israel, portraying the Palestinians as victims and leaving its own bravado behind. Two factors came into play: Jaabari initiated much of the conflict and outmaneuvered Haniyeh, and when the dust settled the Hamas political echelon shifted to become more Jihadi and less *hudna*-oriented or "pragmatic." In short, all of Hamas became even more activist militant after the 2007 takeover.

Israel imposed a partial blockade in the wake of the Hamas electoral victory, but such a move was circumvented by the ever-expanding tunnel operations under the border between Gaza and Sinai. Legally enforced Islamization continued and Hamas sought both Iranian and Turkish support. Islamists led by the IHH Turkish NGO and closely aligned with Hamas organized the "Marmara Flotilla" in 2010 to openly break the Israeli naval blockade of the Strip, which intensified after Cast Lead. Once again Israel succeeded on the ground, but lost in the media. By the time of the 2011 Islamic Awakening, journalistically known as the "Arab Spring," Hamas imposed its will on Gaza, facing opposition only from the more extreme Salafist/al-Qaeda groups. Overall Hamas enjoyed a hero's status having been among the first Islamists to overthrow a secular regime in the Arab world. By the year's end Haniyeh and the civilian leadership shifted further to the right, declaring their ultimate objective as world conquest through Jihad. As 2012 drew to a close, Hamas served as a role model for much of the Arab world and would rule in Gaza for years to come.

Periodic intensive rocket attacks into Israel resulted in limited responses with operations "Pillar of Strength" in November 2012, and "Protective Edge" in July 2014. Such actions only damaged Hamas temporarily. More troubling was the loss of Egyptian support when General a-Sisi overthrew the Muslim Brotherhood regime in July 2013, and Hamas faced a growing challenge by smaller, more fanatical organizations such as the Islamic Jihad, the Resistance Committees, al-Qaeda and the Islamic State. By the summer of 2014 Hamas had a formidable arsenal, including medium range rockets which they began firing at Israeli urban centers, especially Beersheva and metropolitan Tel Aviv. On the other hand, Israel hesitated to destroy Hamas fearing the rise of even more radical leadership from the fanatical Jihadi opposition referred to above. Although somewhat weakened Hamas remained firmly in power in Gaza through 2016.

The Low Intensity Conflict (LIC) of 2000 to 2004, The Second or "Al-Aksa" Intifada

In most Western circles it is commonly understood that the US and Israel were interested and even believed in the possibility of peace in the summer of 2000; the Fatah dominated PA was perceived as firmly in control throughout the Palestinian areas. Conversely, Yasir Arafat and the Palestinian Authority saw the Camp David meeting as much more of a tactical move. He refused to make concessions and understood that Israel could never accept his demands, in particular regarding refugee return, a demand including all refugee descendants, which would result in millions of Palestinians flooding into Israel. As pointed out previously, the Palestinians are virtually the only refugees who can pass down their status from one generation to the next for eternity, taking advantage of the United Nations Relief Works Agency (UNRWA) to advance their goal.[1]

By entering the Oslo process, many saw Arafat as accepting the two-state solution, although his words and actions rejected the State of Israel and Jewish national legitimacy. His rejection was abundantly clear, as seen in Article 20 of the *PNC*. On the tactical level Arafat accepted the need for negotiations, and recognized Israel as a state—at least for the present. He technically negotiated, but still held the idea that Jewish nationalism was not legitimate. The only peaceful prospect was Israel negotiating itself into oblivion, which was a non-option. Consequently, Palestinians believed the Jewish nation-state had to be destroyed, whether by war or peace. Destruction would be inevitable if Israel agreed to demands for a withdrawal to the indefensible 1967 borders, and full refugee return. The "red-line moment" or point beyond which one cannot make any more concessions arrived in 2000 when Israelis understood they were being asked to abdicate national existence. At that time Arafat publicly re-embraced the Jihad concept, which he likely never abandoned, under the semantics of the "armed popular revolution to liberate Palestine" as urged in *PNC* Article 9. This is in full concurrence with *The Hamas Covenant* Article 11, which demands re-conquest of the Islamic *waqf* lands, including the State of Israel. In essence, the West Bank, Gaza, and East Jerusalem were all non-issues since Jewish sovereignty, even in Israel's pre-1967 borders, was by definition considered illegal by both the PLO and Hamas.

The Palestinian LIC took most of Israel's political leadership by surprise. The PA leadership and Arafat referred to the same conflict as the "Al-Aksa Intifada," citing its point of initiation, giving it religious overtones and historically linking it to the disturbances led by Haj Amin el-Husseini in 1920 and 1929 at the Western Wall and the Al-Aksa mosque domain. However, Israel's security forces were prepared, since they foresaw the failure of the Camp David 2000 Summit from the outset. On September 29, following Friday afternoon prayers rioting broke out throughout the Palestinian territories and quickly spread to the Israeli Arab population a day later. Initially most violence was focused in East Jerusalem. Likud leader Ariel Sharon's visit to the Temple Mount the day before was played up as the trigger for what Arafat hoped would be a Palestinian popular uprising demanding independence. As long as there was violence Arafat could not agree to any compromises, hence he stoked the flames. Publicly he blamed Hamas and other extremists for the assaults against Israel, yet simultaneously his security forces were involved in attacking the Israeli military and police. Initially he played a brilliant tactical game, placing himself and the PA in the middle between the Israeli "aggressor" and Islamist "extremists."

In 2001, PA Minister of Communications Imad Falouji spoke of the PA preparations for the uprising while on a visit to the Ein Hilweh refugee camp in south Lebanon. He stated, "It had been planned since Chairman Arafat's return from Camp David, when he turned the tables in the face of the former US President [Clinton] and rejected the American conditions." Falouji, a former Hamas activist now cooperating with Arafat continued, "The PLO is going back to the 60s, 70s and 80s. The Fatah Hawks, the Kasam Brigades, the Red Eagles and all the military action groups are returning to work." In December 2000, Al-Falouji explained the PA prepared for mass violence in the immediate aftermath of the Camp David failure.[2]

Israeli Prime Minister Ehud Barak and what remained of his government scrambled to contain the hostilities in the Palestinian territories and in Israel proper. By mid-October the collision with Israel's own Arab population calmed down after ten days of riots, where thirteen Arabs and one Jew were killed. The government established the Orr Commission to investigate the violence. The Egyptian-initiated Sharm el-Sheikh conference convened in October to end the violence met with little success.

In December, Clinton presented his "Parameters." A month later in January 2001, Israelis and Palestinians met in Taba, Egypt in an attempt to end the clash and return to the negotiating table. Attempts at peace failed. The Fatah/PLO-controlled PA took an uncompromising line and the conflict intensified. From Camp David in July 2000 to Taba in January 2001, the four basic issues were land, refugees, security and Jerusalem. The land issue was based on the 1967 borders. Israel proposed keeping 6 percent of the West Bank and the Palestinian negotiators, without Arafat's consent, agreed to let them keep 3.1 percent. In addition there were to be minor land swaps by the two sides. They also discussed Israel leasing 2 percent of the land. In a "non-paper," or unofficial document, Israel suggested taking back 25,000 refugees over a three-year period, but that the remainder would be rehabilitated in their host country, or reabsorbed elsewhere should they not want to live in the future Palestinian State. The Palestinians insisted all refugees and their descendants had the right to return to their homes in accordance with the recommendations in UN Resolution 194, Clause 11, from December 1948, which states:

> . . . the refugees wishing to return to their homes and live at peace with their neighbors should be permitted to do so at the earliest practicable date, and that compensation should be paid for the property of those choosing not to return and for loss of or damage to property which, under principles of international law or in equity, should be made good by the Governments or authorities responsible . .

From Israel's perspective, expecting Palestinian refugees to "live at peace" was completely unrealistic after over fifty years of suffering, publicly blamed on Israel as a matter of policy. Palestinian refugee descendants were not part of the resolution, and needed to be treated equal to refugee descendants from all other conflicts as defined by the United Nations Higher Commission for Refugees (UNHCR); meaning they have no refugee status. Yet the United Nations Relief Works Agency (UNRWA) awarded favorable discriminatory extra rights to Palestinian Arabs over the years. Since the rights came from the UN, it was a measure that would ultimately assure Israel's demise. Compensation to be given "by the Governments or authorities responsible" could certainly refer to the Palestinian Arabs themselves and the surrounding Arab nations that invaded Israel in 1948. They rejected Resolution 181 and initiated the conflict. It is true that claims can be made against Israel on the local battlefield level, but here adjudication would be necessary to determine whether individuals were forced from their homes for no reason. Realistically, the US and Western world offered to help resettle and rehabilitate refugees in their countries of residence, or to facilitate their move to the newly declared Palestinian State to-be. The costs would be in the tens of billions of dollars. In Clinton's "Parameters" he suggested the Palestinians retain 94 to 96 percent of the West Bank, but this made no difference since Arafat concentrated on full refugee return.[3]

As for security arrangements, Israel insisted on keeping troops along the Jordan River, although the Jordan Rift Valley was to be given over to Palestinian control. The new State of Palestine was to be non-militarized and allow Israeli early warning stations and international forces to strengthen security. Jerusalem was to be a split city between Israel and Palestine, serving as a dual capital. Issues arose over sovereignty at holy sites, in particular over the Kidron Valley and the Mount of Olives, known as the "Holy Basin." The Israeli concept was to work within an international forum, an idea unacceptable to the Palestinians. Israel sought control over the Jewish Quarter and Western Wall, while the Palestinians were to be given jurisdiction over the Christian and Muslim Quarters of the Old City. The Temple Mount/Noble Sanctuary proved to be the most problematic holy site as the Palestinians, Arafat in particular, insisted Jews had no historic or religious connection to the site. Even Clinton took offense at the falsehood, claiming Christians also believed that both Temples stood on Mt. Moriah. It should be pointed out that the Koran supports this perspective as well in "The Night Journey" 17:2-8. Citing numerous *fatwas,* Arafat insisted on full Islamic control of the site. One can understand that Arafat felt he had no discretion in the matter, and if he took any action contrary to the Islamic legal opinions he would be branded a traitor. Even a suggestion to divide the Mount/Sanctuary, with the upper public area going to the Palestinians and Israel retaining the lower archaeological site underneath was rejected.[4] Arafat's argument was Muslim *waqf* oriented and to him Islam superseded Judaism and Christianity.

The Taba Talks ended without agreement in late January 2001 immediately after President George W. Bush took office. The failure to compromise was of little significance since neither Arafat nor Barak attended the talks, and Arafat himself had already set the tone—no deal was possible. Some say the Taba discussions were too close to the Israeli elections to achieve a successful outcome. Others believed the continued violence overcame attempts to make peace, and that the asymmetry in casualties—279 Palestinians and 41 Israelis killed alongside thousands wounded, mostly on the Palestinian side—prevented Arafat from disengaging from the conflict. The Israel Defense Forces (IDF) was accused of not following demands by the Barak government to ease up security restrictions on the Palestinians, particularly roadblocks, which resulted in more conflict. Though Arafat faced the most dovish Israeli government and negotiating team ever presented, it was clear even his advisors knew he did not want to finalize an accord.

To cap it off, at the Davos World Economic Forum Arafat publicly insulted Israeli Minister for Regional Cooperation Shimon Peres[5] and condemned Israel by declaring "the present Israeli government has been waging a savage, barbarous war and fascist military aggression against our Palestinian people . . . using weapons forbidden by international conventions . . . [including] depleted uranium." Arafat made these remarks at the conference despite the fact that Peres was in attendance—a man whom Arafat shared the Nobel Peace Prize with in 1994. Israelis were livid; any idea of peace with Arafat evaporated. Ten days later, the Likud leader Ariel Sharon won the prime ministerial elections against Ehud Barak by a landslide.

It is common knowledge Clinton considered Arafat responsible for the failure of the Camp David 2000 initiative and the follow-up talks ending in Taba.[6] The question remains as to why Arafat "abandoned" the peace process. Simply put, he could never tell the Palestinian refugee descendants that there would be no refugee return, since refugee return represented the winning card in Israel's destruction. Arafat believed the final Palestinian victory would be brought about with the support of the Arab world. As far as the Temple Mount/Noble Sanctuary in Jerusalem was concerned, he made absolutely no concessions to the Jews whom he saw as having no claim to the site.

Arafat played the media card to its fullest; the Palestinians were victims and the Israelis the oppressors. He worked his way toward international hero status as if fighting for the meek against the powerful. As radicalization set in, the PA took a much harder line despite declarations of trying to rein in Hamas. The "secular" PLO adopted a more Islamic absolutist perspective of the conflict, leaving little or no weight to Israeli Jewish narratives concerning the Land of Israel and especially the Temple Mount. On the secular level, the shift was back to the full and literal understanding of the *PNC*. Palestinians retained expectations of at least an independent state in the pre-1967 territories of the West Bank and Gaza, and a return of the refugees to Israel while envisioning full control over East Jerusalem and the Temple Mount/Noble Sanctuary. A victorious future was foreseen. Israel would collapse as a bi-national state with an Arab majority and be re-united with the West Bank and Gaza as an Arab State. Palestinian leaders, especially Arafat, were not making public speeches to the contrary, or preparing the people for compromise. The

PA directed rising expectations and popular anger at Israel despite the Oslo peace process.

The PLO/PA thought in terms of political gain, yet Hamas was to be the greatest beneficiary. The PA moved toward a mixed *PNC* /Islamist understanding embodying uncompromising conflict until victory. Fatah invoked the *PNC* in full, most specifically, the armed struggle mentioned in Articles 7-18, and those delegitimizing Israel and condemning all compromises as found in Articles 19-23. The Fatah secular language of struggle was no less ardent than calls for Jihad.

The left leaning Labor Prime Minister Barak called for a prime ministerial election, the Knesset was unaffected, and in February the Likud's Ariel Sharon defeated him by a landslide of 62.4 percent to 37.6 percent. The election results reflected Israeli frustrations and the vote put in power a security-minded hard-liner known for his pro-settlement activities. In early 2001 after Barak resigned as party leader, Shimon Peres led Labor to join the national unity government, with the Likud as the senior partner. Now the battle with the Palestinians was even more fully enjoined.

Arafat sought to absorb his erstwhile Hamas adversaries in the hope of riding the Islamic tiger to victory over Israel. The PA chairman thought he was using a beaten, exhausted Hamas for his own purposes, but as it turned out the Islamists got the better of the PLO/Fatah despite their designated inferior role. The Palestinian LIC rejuvenated a moribund Hamas. The Islamists had a solid worldview in Islam and Sharia law, never deviated from full rejection of Israel's right to exist, and were unswerving in their demand for Jihad. The *dawa* organizations consistently provided social services and Hamas retained the people's support. Regardless of abuses suffered for supporting the Islamists, Hamas commanded much greater loyalty than the PLO/PA, despite the fact that Arafat regained and increased his popularity, especially after calling for war against Israel. Copying Hamas strengthened Arafat, no matter how corrupt or abusive his regime.

Arafat blamed everyone else, never took responsibility and insisted to the international community that he was working to halt the violence. He continued to deceive the West and directed attacks against Israel, but quickly lost control as clashes with the IDF led to mounting casualties and Palestinian rage. Hamas led the attacks, but was soon in competition with the Al-Aksa Martyr Brigades, a radical activist Fatah subsidiary group comprised of Arafat's personal guards and unwittingly funded by the US and the international community. Hamas representative Sheikh Hassan Yousef met with the PA's Marwan Barghouti and Arafat on a weekly basis, although by now the

Islamists were in the lead. Working to show a civilian side to the uprising, the Palestinian leaders led mob charges at Israeli positions, abandoning their flock just at the moment they feared the IDF would open fire. Fatah's Al-Aksa Brigades, Hamas, the Islamic Jihad and other political militia groups were involved, but quite often individuals initiated and carried out attacks. Every organization wanted to claim martyrs. Iraq's Saddam Hussein paid out some $35 million to families for their sacrifice, stating, "Ten thousand dollars to the family of anyone killed fighting Israel and twenty-five thousand to the family of every suicide bomber."[7]

Fatah/PLO was now following the Hamas activist lead, integrating together in a ferocious battle against Israel. Two of the most powerful emotions were now blended together: the impulse to love and to kill. From the Islamic perspective infinite love for Allah brought young people to murder in Allah's name. They murdered not only enemies but themselves. People were willing to die multiple deaths to achieve a full fusion with the beloved god of Jihad, Allah, as seen in *The Hamas Covenant* Articles 8 and 15. In the secular national sense, suicide and homicide bombers gave their lives for the revolutionary cause and their ultimate leader, Yasir Arafat—an almost deified figure. Such actions went well beyond the demand for the armed struggle found in *PNC* Articles 9, 10 and 15. With the adaptation by Fatah of the Hamas modus operandi, there was an Islamization of the conflict, one previously viewed much more in the nationalist Third-World Liberation context. And, lest one forget, Arafat and the PLO/PA leadership were constantly involved in "*taqiyya*" (lying) and "*kitman*" (omissions) when explaining events to the West. As far as Arafat was concerned, any claim or accusation was considered acceptable when working to defeat the Israeli Jewish adversary. Western observers and diplomats were constantly willing to excuse Arafat due to his military and political weakness, so much so they were complicit in accepting his deceitfulness and often demanded concessions from Israel. For many Western statesmen, peace was not only an objective, but had become a value in its own right. Arafat's weakness became his strength, "wrapping him in Teflon" so to speak. No matter what he did Arafat had an "acceptable" excuse to justify his actions. Arafat was not a Third World revolutionary. He was not interested in state building and behaved more in line with Islamist Jihadi thinking than any secular national liberationist understanding.[8] Hamas understood his game, knew he was untruthful to the West and waited for the day of reckoning when the PLO would either be crushed or absorbed by their Islamist ways, as seen in *The Hamas Covenant* Article 27.

The four year Palestinian-Israeli clash known as the Second Intifada but which will be defined here as a Low Intensity Conflict (LIC), can be divided into four stages. In general, the initial LIC was directed more against the IDF/police and Jews living in the West Bank during 2000 and 2001. In the spring of 2001, there was a shift to a terror offensive perpetrated against Israeli civilians, which continued into 2002. The third period of "planned anarchy" came once Arafat was surrounded by Israeli forces in his Mukata'a Ramallah headquarters from 2002 to 2003. By 2004 Arafat spoke of the victory of demographic factors, invoking the "womb of the Palestinian woman" as his ultimate weapon. This last shift from the armed to the demographic struggle exposed his overall objective of a one-state solution, forcing Israel into a non-viable bi-national state.[9] This dovetailed completely with the Hamas objectives. There was no more pretense of recognizing Israel's existence or its right to exist. Palestinians abandoned the expected moderate stance of accepting Israel's right to exist as part of a bi-lateral understanding, and turned to Hamas absolutism. The Palestinian military offensive against Israel failed, but demographic entanglement could succeed. The Israeli settler movement demanded the expansion and development of new settlements throughout the West Bank and Gaza in response to Palestinian terror attacks, thereby further burying any possibility for a two-state solution. Hence, Palestinian Arab and Jewish populations became more "entangled" and played into the Palestinian demographic offensive. In essence, Arafat ran a united command according to Islamist guidelines, while engaging in *taqiyya* and *kitman* to handle Western complaints.

Commencing on September 29, 2000, attacks were directed against Israeli security forces and Jews living in the West Bank (Judea and Samaria) and the Gaza Strip. Stirred up by Palestinian religious and secular authorities, rock-throwing youth attacked Jewish worshipers in the Western Wall Plaza and a less-than-alert Israeli police force stationed not far from the Mount. The violence quickly escalated when Israeli police broke into the Al-Aksa compound. On the first day, four Arabs were killed and 160 wounded; 14 Israeli police were injured.[10] Marwan Barghouti and Arafat organized well-planned simultaneous riots, which spread throughout the West Bank and Gaza, while PA security personnel attacked Israeli forces when they tried to intervene.[11] The Palestinian territories quickly descended into chaos, Israeli resolve was strengthened as evidenced by Sharon's election in early 2001, and the PA/Hamas alliance shifted its tactics to a much more damaging and painful initiative against Israel.

Although many saw Arafat as a hero-liberator, Mosab Yousef, the son of Hamas leader Hassan Yousef, had this to say: "Yasser Arafat made it clear that he wanted to be a hero who was written about in the history books. But as I watched him, I often thought, *"Yes, let him be remembered in our history books, not as a hero, but as a traitor who sold out his people for a ride on their shoulders. As a reverse Robin Hood, who plundered the poor and made himself rich. As a cheap ham, who bought his place in the limelight with Palestinian blood."*[12]

When considering Sharon's Temple Mount excursion, Yousef said, "Conventional wisdom among the world's governments and news organizations tells us that the bloody uprising known as the Second Intifada was a spontaneous eruption of Palestinian rage triggered by General Ariel Sharon's visit to what Israel calls the Temple Mount complex. As usual, the conventional wisdom is wrong."[13] Yousef concluded, "Yasser Arafat and the other PA leaders had been determined to spark another Intifada. They had been planning it for a month, even as Arafat and Barak had been meeting with President Clinton at Camp David. They had simply been waiting for a suitable triggering pretext. Sharon's visit provided just such an excuse. So after a couple of false starts, the Al-Aqsa Intifada began in earnest and the tinderbox of passions in the West Bank and Gaza were inflamed once again."[14]

By the spring of 2001, the Palestinians initiated a "terror offensive" against all Israeli Jews. For the first five months there were few attacks within Israel's 1967 borders directed solely at civilians, but there were bombings in Jerusalem and Hadera. Beginning in March 2001, Fatah/Hamas and allies targeted civilians inside Israel and achieved infamy through unforgettable massacres perpetrated by suicide bombers. Bombings, stabbings and shootings occurred daily. The list below is a survey of the worst attacks.

On June 1, the bombing at the Dolphinarium in Tel Aviv killed 21, and injured 120. On August 9, the Sbarro Pizzeria bombing in Jerusalem killed 15 and injured130. On December 1, an explosion on Jerusalem's Ben Yehuda Street killed 11 and injured 180. On the next day, the Haifa #16 bus bombing killed 15 and injured 40. Ten days later on December 12, a public bus serving the non-Zionist ultra-orthodox community of Emmanuel in the West Bank was attacked, killing 11 and wounding 30.

The terror attacks continued into 2002, making it the worst year ever; however it led to the Israeli operational responses known as "Defensive Shield" and "Determined Way." Until now, Israel had only partial success sending in forces to specific Palestinian towns and villages to root out terror. It took several other attacks before Israel decided on sweeping counter terror operations. On January 17, 2002 there was an attack on a Hadera bar mitzvah celebration where 6 were killed and 35 wounded. On March 2, in another attack on the ultra-orthodox Beit Yisrael neighborhood, where mothers and children were waiting outside a synagogue in Jerusalem, resulted in killing 11 and injuring 50. On March 9, the Moment Café was attacked in Jerusalem's Rechavia neighborhood, killing 11 and injuring 54. On March 20, a suicide-homicide bomber blew up a bus on Highway 65 in Wadi Ara, killing 8 and wounding 30.

On March 27, terrorism reached a crescendo when a suicide-homicide bomber blew himself up at what came to be known as the Netanya Seder Passover Massacre, killing 30 and wounding 140. It was Hamas' most deadly attack. On March 31, Hamas continued the offensive with the Matza restaurant bombing in Haifa, killing 15 and injuring 40. Hamas took credit for most terror operations, although the Islamic Jihad, Al-Aksa Brigades and Tanzim also took part, as sometimes more than one organization demanded accolades for the carnage.

After the Passover Massacre the Sharon government took definitive action and initiated "Defensive Shield," a broad sweeping operation expected to continue for eight weeks. The operation enabled Israel to scour Palestinian West Bank cities and towns, arresting suspects, uncovering explosives laboratories and seeking out terrorist leadership. As the internationally recognized chairman of the Palestinian Authority Arafat could not be arrested and used his invulnerable status to provide a safe abode for dozens of suspected terrorists, hiding them in his Ramallah Mukata'a headquarters. He was allowed to travel from the Mukata'a but should he do so, Israeli forces were ready to pounce on the terrorists he was protecting. In April Israel did major battle in the Jenin refugee camp against the well organized, dug-in positions located in civilian zones. Not wanting to use air power, the Israeli army suffered 23 dead, while 52 Palestinians were killed, of which 7 were civilians. Initial Palestinian reports claimed five hundred killed, but the UN investigation proved otherwise. Israel's anti-insurgency, anti-terror campaign only began to bear fruit in 2003 when successful terror attacks dropped significantly. Terrorism continued during the Israeli offensive begun in 2002. There was another bus bombing on April 10, killing 8 and injuring 22. Two days later in an attack on Jerusalem's Jaffa Road, 6 were killed and 104 wounded. On May 7, in Rishon LeTzion, 15 were killed and 55 wounded. On June 5, a bus from Tel Aviv to Tiberias was attacked, killing 17 and wounding 38. On June 18 a Gilo bus exploded, 19 were killed and 74 wounded, and the next day at a Jerusalem bus stop in French Hill there were 7 killed and 50 wounded.[15] On July 31, the infamous Hebrew University cafeteria bombing killed 9 and wounded 85.[16]

Over a two-year period Israel and the Palestinians went from "almost" peace to downright war, even if it was the low intensity, terror vs. anti-insurgency/anti-terror type of conflict. In Palestinian society, the disabled PA and the secular Palestinian organizations led by Fatah followed the Hamas lead. In Israel the population lined up behind Prime Minister Ariel Sharon and the Likud, who initiated national unity governments with Labor as a junior partner. The world did not stand still. In the meantime Israel received condemnation for using "excessive force." Even immediately after 9/11, an event which would eventually change world attitudes toward Israeli actions, Sharon found himself in conflict with the Bush Administration over the Israeli military reply and American urgings of restraint. All of this took place before the Defensive Shield anti-terror IDF response. Israel's capture of the *Karine A'* weapons shipment originating in Iran brought about a significant shift in the American and Western position. George W. Bush was convinced Arafat was a liar when he denied any connection to the shipment.[17] Much more important was the PA and Fatah's developing Iranian alliance following in the footsteps of the Hamas-Hezbollah working relationship, which began in 1992-93. As the PLO cooperated with Hamas, both received support from Hezbollah and Iran.

From April 2 to May 10, 2002 there was the standoff at the Church of the Nativity in Bethlehem when Hamas, the Al-Aksa Brigades and Fatah operatives occupied the Church to avoid arrest by the IDF. Some 220 men were holed up inside, not all associated with the operation. In the end 26 militants were sent to Gaza and 13 exiled to Europe.[18] Over the first half of 2002, the Americans sharply shifted gears, placing blame on Arafat and publicly suggesting his removal.[19] That same summer the Israeli government finally decided on building a security or "separation" fence mainly along Israel's 1949-67 armistice line or "green line" with the West Bank, but at times spilling over to the east to include Jewish settlements not far from the demarcation. The bombings had their effect, ideology or not—Israel's political right was under massive pressure from the populace to separate mainstream Israeli society from the Palestinians. What Arafat had not done at the negotiating table Hamas succeeded in doing by leading the way through armed resistance and terrorizing Israeli civilians.

Remaining in his Ramallah Mukata'a headquarters and surrounded by the Israeli army, Arafat no longer controlled his forces from the top down. Not only had Israel taken most of the West Bank, but it was destroying the power of the Palestinian security forces, and exposing the Chairman's communications. Israel tapped his land and cell phone lines and observed anyone coming to visit. This time frame became a period of what can be termed "planned anarchy." All Palestinian fighters and terrorists were on their own, and Arafat could not give orders from the top.

Such pressure helped solidify PA/Hamas cooperation, the best example being when the PA released engineer Saleh Talahme from jail—a man who was involved in the Hebrew University bombing. Arafat's release of Hamas terrorists was in violation of previous agreements with Israel. Once freed, Talahme rebuilt the Al-Qassam Brigades, assembled explosives and worked with Bilal Barghouti recruiting homicide-suicide bombers. Israeli forces eventually killed Talahme in December 2003.[20] Hamas operated in a much more centralized discreet manner. Israeli security rounded-up suspected terrorists, but new Hamas operatives replaced them. Political power and the operational center were in the hands of Khalid Mashal. He ran the Hamas office in Damascus, Syria. It took time, but the IDF and Israeli intelligence broke the back of Hamas through firefights and arrests. In 2006, Hamas organizer Ibrahim Hamed was apprehended and the other Hamas operational leaders in the West Bank were exposed. These men had advanced university degrees, and led quiet, professional, and mostly secular lifestyles.[21] Because Hamas was working much more innocuously underground than Arafat, it took considerably more effort to break the Hamas operational network than any comparable Fatah initiative.

The clash with Israel entered its most critical stage from mid-2003 until early 2004. Arafat and the PA were expected to line up with the Bush "Road Map" approach calling for a two state solution in three steps: 1) A Palestinian halt to terror and an Israeli halt to settlement building; 2) An interim Palestinian state with temporary borders; and 3) A permanent status agreement.[22] Realizing Arafat was not a partner, Bush and Israel pressured for the appointment of Mahmoud Abbas, also known as Abu Mazen, as prime minister in March 2003. Abbas and Arafat clashed over policy. The new premier was seen as far too moderate in agreeing to end the violence and in being receptive to the idea of demilitarization of a future Palestinian State. Abu Mazen fell from power six months later. By early 2004, Arafat still spoke of the armed struggle, but now openly called for Palestinian victory through "the womb of the Palestinian woman," an admission that where military resistance had not succeeded, demographics would eventually overwhelm Israel's Jewish population. This was especially true should Israel not withdraw from the West Bank and de facto be forced into a bi-national state. Arafat remained under siege in Ramallah, preferring the image of a fighter and not that of a peace partner since many Palestinians equated peace to betrayal.[23]

During this period there was much discussion about a *hudna,* or an Islamic cease-fire. Israel accepted a cease-fire in wake of the "Road Map" and agreed to lift the closure of the Palestinian territories. The US put considerable pressure on Israel to agree. It was another possible point of reconciliation between Fatah and Hamas whereby all would be allowed to recover from the Israeli offensive, yet minor attacks against the Jewish State, as agreed to by Abbas, would continue. During mediation attempts between Israelis and Palestinians, arch terrorist Abdullah Kawasmeh, who was responsible for the deaths of 52 Israelis and the wounding of hundreds, was killed by Israeli forces. From Israel's perspective, if Hamas and Fatah still advocated certain operations, then so did Israel. Everyone understood the nature of a *hudna*, and Israel continued its anti-terror sweep into Hebron, arresting some 120 suspects. Abbas attempted to revive the peace process with Israel, but was unable or unwilling to contain Hamas. Mapping out future actions against Israel included the production and transport of rockets and plans for suicide-homicide bombings. The Jerusalem bus bombing by Hamas in the middle of August, which killed 20 civilians and wounded dozens more put an end to the *hudna* and the planned Israeli withdrawal from four West Bank cities, which was scheduled to take place the next day. This terror reorganization offensive by Fatah and Hamas was only partially successful due to Israeli preemptive actions.[24] The last surge in attacks against Israel began, but was contained by the spring of 2004.

Hamas was on the rise, absorbing Arafat and Fatah more fully into their way of thinking and operating. Fatah-style secular nationalism was seen as weak and greatly discredited in the face of Israeli demands, but even more so because of corruption and exploitation of the Palestinian masses. By now the Israeli leadership concluded it was necessary to eliminate the spiritual and operational command of Hamas. Hence, in the early spring of 2004, Sheikh Yasin and Abdul Aziz al-Rantisi were killed in targeted removals. The pivotal Fatah commander Marwan Barghouti was arrested two years earlier while others were liquidated. Arafat was suffering from a terminal illness, lost control of the PA/Fatah, and died in November 2004.[25] Militarily, Israel clearly gained the upper hand.

From the Israeli perspective, the security forces responded too late in the effort to regain control. The best example was the targeted removal of Salah Shehadeh, the Hamas mastermind of terror attacks against Israeli civilians. Responsible for dozens of terror operations and over 100 dead and wounded, the initiative to kill him was postponed eight times for fear of civilian casualties. Finally, in July 2002, the Israeli air force killed Shehadeh. Unfortunately,

15 other people died in the collateral damage when the explosives dropped on his Gaza residence proved too powerful. Shehadeh, and those like him, planned and operated terror attacks from civilian areas using non-combatants as human shields, hoping the fear of causing civilian casualties would halt any Israeli actions against them. For a while it worked, until Israeli civilian casualties mounted from continuing terror attacks and the government was left with no choice. Jihadi groups such as Hamas had no qualms about using their own populations as hostages while planning to kill Israeli civilians. In particular, their wives acted as unarmed bodyguards. Terrorists hoped that the presence of wives alongside the husband-bomber-masterminds would prevent their elimination. These women, and other unarmed adults, in essence were fully responsible when acting as human shields for their husbands. Children were exploited as unwilling or unsuspecting hostages. Israel tried its best to only kill terrorists, but eventually reconciled that collateral damage incurred was the moral responsibility of Hamas, Fatah or whoever used innocent people as human shields. In the end, killing terrorists who planned attacks against Israelis saved both Israeli and Palestinian lives, though collateral damage was an unfortunate corollary to certain operations.[26] As far as Hamas was concerned, everyone killed became *shaheeds,* or "martyrs," whether it was the perpetrators themselves, their bodyguards, wives, children or other innocents. It is clear in *The Hamas Covenant,* Article 18, that a Jihadist role is demanded of wives and children, who are instructed in Jihadist education. Anyone participating and dying in the Jihad would be a *shaheed,* whose death for Allah was most glorious, as described by the Hamas motto in Article 8. Such battles and deaths are to be repeated innumerable times until final victory is achieved as exhorted in Article 15 of *The Hamas Covenant.*[27]

When counting casualties, general estimates for the Second Intifada/LIC included over 1,000 Israelis and more than 3,000 Palestinians killed in direct conflict with each other, and three times as many wounded. Psychological damage on both sides is hard to access, but lasting scars remained. Furthermore, Palestinians killed hundreds of their own people during internal clashes, especially between Fatah and Hamas. Officially, the conflict is said to have lasted from September 28, 2000, to February 8, 2005 (when Mahmoud Abbas took office), or almost four and a half years—approximately 1,600 days.[28] Despite massive societal disruption, especially for the Palestinians, we are speaking of fairly low casualties—averaging less than four deaths per day.

In the aftermath Sharon's Likud-led national unity government decided in conjunction with the Bush Administration to disengage from the Gaza Strip and the northern West Bank. Many, especially in Hamas, interpreted this initiative as Israeli weakness in the face of Palestinian resistance. In the Bush-Sharon letters of April 2004, the US and Israel synchronized policies. Israel would leave the above-mentioned territories while the US made clear four major policy decisions: 1) The 1949-1967 armistice lines were not sacrosanct and could be altered; 2) The construction of the security fence was legitimate and could be moved; 3) Israel had the right to defend itself from terrorism wherever need be; and, 4) There was to be no US support for Palestinian refugee return to Israel proper. Many saw Israel's removal of some 8,000 Jewish residents from Gaza as a step to avoid demographic entanglement with the 1.3 million, mainly Hamas-supporting Palestinians who dominated the region in 2005. Israel contemplated, but never implemented, a Defensive Shield-style military operation in Gaza, as was done in the West Bank. Since 1994 and Oslo I, Israel ceded control of 85 percent of the Gaza Strip to Arafat and the PA. Israel fully withdrew from Gaza at the end of the summer 2005. In Palestinian eyes, Hamas gained credibility while Israel was weakened and the PA discredited.[29]

End of the LIC
Hamas Gains Power Despite Fatah, Israeli and Western Opposition

Abbas succeeded Arafat, and in January 2005 he won the presidential elections, running virtually unopposed. Yet Hamas overwhelmed Fatah in the parallel municipal elections held throughout the next few months. The Oslo peace process was declared dead and Fatah lost legitimacy. Hamas demanded continued resistance toward Israel and an end to PA corruption, while countering American and Israeli overtures for calm. Israel and the US vacillated over free, open elections for the 132 member Palestinian Legislature, and finally agreed to such. In January 2006, Hamas won a landslide taking 74 seats to Fatah's 45 even though the popular vote was almost even. Hamas leader Ismail Haniyeh was appointed prime minister. Credited with purist ideals, Hamas popularity grew slow and steady. They were viewed as honestly and selflessly administering social services while living amongst the people and not above them. With the peace process and secular nationalism judged as failures, the alternative was to return to basics—Islam and its revival. Belligerence toward Israel played a major role. It was believed the Jewish State would still retain control over Palestinian lives through any international peace agreement. Hamas' rejection of the peace process removed the stigma of a weak, dominated refugee population, especially concerning Gazans, who now saw themselves as vindicated in the battle for victory.[30]

According to Azzam Tamimi, Hamas and Islam were triumphant and everyone else was defeated, in particular Israel. The US set conditions for recognizing Hamas, including the renunciation of terror, the recognition of Israel as a state, and the acceptance of previous Palestinian-Israeli agreements including Oslo I, Oslo II and the Wye Accords. The Hamas victory proved the Palestinian people democratically rejected these agreements, at least for the present. Hamas stipulations to initiate a short-term process of non-violence, or the temporary *hudna* Islamic cease-fire, demanded Israel withdraw to the 1967 lines, remove all settlements and abandon East Jerusalem, including the Old City and Jewish Quarter. Hamas made it clear that recognition of Israel's right to exist would *never* be forthcoming, even with agreement to a long term truce.[31] Hamas stipulated there must be a multi-religious state in Palestine based on Islamic Sharia law as would be legislated by the end of 2008 in Gaza. The vision meant Jews and Christians would live under the *dhimmi* strictures stated in *The Hamas Covenant,* Article 31. While Israel and the West worked on sanctions in the hopes of crippling the Hamas victory, the Arab world and even Russia began to embrace the Islamist officialdom.

Hamas offered a national unity government (NUG) to Abbas, but he turned it down at first. He attempted to bolster his own security forces, especially the Presidential Guard, and sought to rally Fatah business interests. The ensuing clash over power, wealth and influence led to initiatives by both sides to force a NUG on their own terms. Paralysis and violence between the two set in for over a year. Fatah controlled the official security forces and Hamas rebuilt their militia. Abbas repeated the demand for the acceptance of Oslo coupled with the Saudi Arab peace initiative of 2002 calling for pan-Arab recognition of Israel's right to exist as part of a peace package, provided Israel made wide-ranging concessions including withdrawal to the 1967 borders.[32] Hamas continued to insist on its prerequisites for a *hudna*. Hamas leader Ismail Haniyeh toured Arab countries, including Egypt and Iran at the end of 2006, raising millions for the cause. Meanwhile, the West moved for sanctions to pressure the Islamists. Abbas demanded new elections in 2007, although the legislature had a four-year term scheduled to expire in 2010. Violence ensued between Fatah and Hamas, bringing Palestinian society to the brink of civil war by February and March 2007.[33]

Recent Palestinian narratives show the people fully supporting Hamas, especially in Gaza due to its firm stance against all adversaries. Hamas was vindicated with the Mecca Agreement negotiated by Saudi Arabia in February 2007. The agreement advocated a national unity government consisting of 24 cabinet posts, whereby Hamas took eleven ministers, Fatah six and independent groups seven. Haniyeh led the government as prime minister under President Abbas. At the time it was thought the accord would save the Palestinians from a civil war. President Abbas, the US and Israel lost virtually all influence. Hamas respected only international agreements not seen as damaging to the Palestinian people's interests and swore no recognition of Israel. The Islamist interpretation was clear, the Oslo Accords and the two-state solution were abandoned. The Saudis apparently held out hope that their 2002 peace plan recognizing Israel could be implemented after a full Israeli withdrawal to the 1967 lines.[34] They most likely expected Hamas to become part of such an initiative.

Israeli journalist Shlomi Eldar, who has covered Hamas activities extensively over the past twenty years, believes there is always a pragmatic side to the organization even if at times it is obscure. By 1994, after returning from his Lebanese exile, the future prime minister, Ismail Haniyeh—a middle level activist at the time—explained his willingness for a twenty-year *hudna* in return for freeing all prisoners in Israeli jails, a full withdrawal to the 1967 borders, and a fair solution to the refugee problem. Should Israel agree, somehow this could be construed as acceptance of Israel's existence.[35] *The Hamas Covenant* insists on Sharia law and *waqf* claims of eternal Islamic dominance over all lands contained within the borders of the State of Israel. Such theological dogma cannot be reconciled with a compromising political diplomatic progress or "pragmatism." No concession was made by the Islamists, the *hudna* cease-fire was only temporary, and the issue being addressed was Israel's short term existence and not Israel's fundamental *right* to exist. Such a Hamas shift was tactical, leaving the option open for renewed hostilities sometime in the future.

The 2006 Hamas electoral platform advocated more individual freedoms for Palestinians while allowing for differing opinions. Hamas made no demands for a state from the Jordan River to the Mediterranean Sea, the implementation of Sharia law, or for Jihad to destroy the State of Israel. In an apparent overture to the West, Mahmoud al-Zahar even hinted that *The Hamas Covenant* could be changed, as if it was only a political document. Such Hamas overtures were and are misleading because of the inherent contradictions between Islamism and pragmatism, when possible policy changes are explained away as being done in the interests of the Palestinian people. Any seeming policy moves in contradiction to Islamic objectives are subject to a *hudna* and cannot be construed as part of a permanent status accord to ensure peace. Israeli Prime Minister Ehud Olmert understood this in the aftermath of the Hamas election victory when he demanded an end to terrorism and a cancellation of *The Hamas Covenant*, knowing full well the power and ramifications of such a document of Divine intent overrode all other considerations.[36] One must recall that the *Covenant* is also known as the "Charter of Allah." The issue was simple: nothing was done in the interest of true conflict resolution. To do so meant to take into account Jewish national legitimacy in the eyes of the Islamist regime and not just to act in the interests of Palestinian Muslims at any given moment.

Hamas supporters like Tamimi believe the Saudis moved to force a NUG to forestall Iranian influence and intervention on behalf of Hamas. He dismisses Saudi concerns, stating categorically there is no evidence Iran or their Hezbollah allies ever had any intention of influencing Hamas. Secondly, as he worked out of Damascus, Mashal stated he would accept a Palestinian State in the West Bank and Gaza alongside Israel provided there be full Palestinian refugee return to Israel. Tamimi and Hamas advocates believe this proves a complete departure from *The Hamas Covenant*,[37] when in essence these are major steps toward Israel's destruction. Such a denial of Hamas intentions can only be described as ludicrous if one is hopelessly naive or deviously deceptive - invoking both *taqiyya* and *kitman*. A complete Israeli withdrawal and full refugee return are two major steps for the destruction of the Jewish State, the same as advocated by Arafat, the only difference being that the PLO/PA leader was willing to conclude peace under such conditions while Hamas would only agree to a *hudna* for a limited time. Such differences are of little importance when the final objective is the demise of the State of Israel.

Western researchers and analysts such as Michael Jensen viewed the *hudna* in terms of a Western cease-fire whereby the sides agree to a mutual recognition of legitimacy, despite clear evidence presented demonstrating the opposite. Jensen provided interviews with Hamas leaders Sheikh Yasin and Abu Shanab where Israeli withdrawal was demanded from Gaza, the West Bank and East Jerusalem and freedom of movement was guaranteed between the three regions, meaning open travel across Israel proper. Furthermore, all Palestinian prisoners jailed as terrorists by Israel were to be released, and all Israeli settlements removed. In return, Israel would receive a *hudna* expressly limited in time, and others would later decide whether to reignite the conflict, or continue the truce. Mahmoud al-Zahar and Abdul Aziz Rantisi were emphatic about the need for a *hudna,* but never spoke of an "End of Conflict" scenario. A commentary by Daniel Nepp made the point that all cease-fires begin this way, and that Israel was rejectionist for considering the *hudna* to be temporary.[38] Such a claim is deceitful as Hamas constantly made and makes clear it is fully cognizant of Israel's existence as an enemy to be destroyed.

Hamas totally rejects Israel's legitimacy as a state entity, even within the 1967 borders, as a matter of Divine dogma.

Hamas never nullified its belief in the Islamist manifest destiny of world conquest. As for Tamimi's first claim that there is no evidence of Iranian intent to influence Hamas and the PA, we have Arafat's weapons order smuggled in the *Karine A'* arms shipment in 2002 from Iran and a later Iranian arms smuggling attempt in 2011, when the *Victoria* weapons shipment destined directly for Gaza and Hamas was intercepted by the Israeli navy.[39] Until Hosni Mubarak's overthrow in February 2011, Egypt itself battled Hezbollah and Iranian operatives in Sinai who were trying to smuggle weapons into Gaza.[40] The second claim is even more absurd, since full refugee return to Israel would mean the end of the state. The halt of hostilities was to be only during a limited *hudna* period, which upon completion allowed for an overall Jihadist military offensive. Mashal's remarks simply break down Israel's destruction into steps culminating in an Islamic state and Jewish disabilities under the *dhimma* restrictions—basically, a twenty-first century Charter of Omar.

Backtracking to the 1990s, Hamas already behaved as a semi-state underground entity prior to their overthrow of the Palestinian Authority in Gaza by military means in 2007. This situation is known as "dual sovereignty" when the loyalties of the populace are not given to the official leadership but to an opposition group who may or may not seek to oust those in power. As clearly stated in its *Covenant,* Hamas never made secret its desire to Islamicize Palestinians, though in practice the policy was applied in particular to refugees. The Israeli deportation of Hamas activists in December 1992, which lasted for a year, reinforced the policy objective of replacing the PLO at the helm of the Palestine national movement as the exiled leadership. The objective was now heavily influenced by their Hezbollah hosts who themselves were conquering the Lebanese State step by step.[41] Acting as a shadow semi-state to Arafat's PA, Hamas had representatives in Jordan until 1999, when relations were broken due to Western pressures. The diplomatic relationship with Syria and Iran was on a quasi-state level.[42]

The Palestinian relationship with Iranian Shiite extremism began in the 1970s. Many exiled Iranians returned home to the Khomeini Revolution in 1979, including Abu Al Hassan Bani Sadr, Mustapha Mohammed Najjar and Muhsen Rafiq Doust, who later became the president, defense minister and Revolutionary Guards' leader, respectively. All grew up in south Lebanon alongside the Palestinian Sunni refugee populations, and their families were often guests of Fatah. Fatah and the Shiite Amal organization were allies.

When Khomeini took power in February 1979, Arafat arrived in Tehran to discuss the Iranian Revolution and the liberation of Jerusalem.[43]

The alliance did not last long. Arafat's support for Saddam Hussein in the Iran-Iraq War from 1980 to 1988, and his pre-Oslo peace gestures toward Israel in the late eighties brought Hamas in as the militant Islamic substitute. Holding firm to an anti-peace policy, Hamas found a natural partner in the Iranian Shiite regime. Tehran responded with financial and military support while Syria became a major ally despite the Assad regime's repression of Syrian Muslim Brotherhood forces in the Hama massacre of 1982 when over 20,000 were killed. By the mid-1990s there was a four-way alliance of Hamas, Hezbollah, Iran and Syria. Iran donated some $3 million to Hamas martyr families and prisoners while Palestinian Islamists continued to reap the benefits from Arafat's pro-Saddam policies.[44] Hamas was already a powerful non-state actor with three major allies when the organization wrestled Gaza from the PA.

Despite attempts at a NUG in the early spring, the Palestinian civil conflict was not long in coming and exploded for eight days in June 2007. Hamas built its military strength from within the ministry of interior policing apparatus and through outside funding, while the Fatah/PA faced increasing economic pressure from Israel, which withheld tax revenue transfers as a result of the Hamas legislative victory.[45] In Gaza, 6,000 dedicated fighters representing the people's will crushed Fatah's 22,000 man armed force with between two to three hundred killed, a fairly low figure considering the levels of animosity. Hamas had better pay, better training and better morale than the PA forces who suffered from low salaries, deficient training and corrupt leadership. Hamas secured Gaza while Fatah, after calling in Israeli and American support, took control of the West Bank and rebuilt its security forces under US guidance.

The origins of the conflict began with the Hamas electoral victory in January 2006, tensions having risen a full year earlier when Mahmoud Abbas replaced the deceased Arafat. When previously functioning as prime minister in 2003, Abbas advocated limiting and even halting the conflict with Israel, while Arafat became an Islamist ally. The fact that Abbas became PA president did not serve Hamas' interests, nor did the three conditions he presented: renunciation of violence against Israel, the acceptance of Israel's right to exist, and the honoring of all accords signed with her. The US, the EU and the moderate Arab States suspended foreign aid and imposed sanctions as a result of the Hamas victory. In particular, Israel boycotted Hamas and imposed economic sanctions on the Gaza Strip in the wake of its overthrow of the PA and

V Hamas Ideological Victory 2000 to 2016   221

the continuing demand for Israel's demise. Hamas did offer Israel a one-year *hudna*, but insisted on rejecting all other demands.

The US and Israel attempted to strengthen Abbas and the PLO/Fatah by undermining Hamas, but failed miserably. In December 2006, battles between Hamas and Fatah began in Gaza, but were halted in early 2007 when the two sides attempted a NUG negotiated by the Saudis in Mecca. The conflict exploded again in May, killing dozens. The Hamas-Fatah Mecca Accord called for dialogue to replace armed conflict, honoring all agreements including those signed with Israel, continued reforms in the PA and cooperation, pluralism and respect for all Palestinian political factions.[46] In June, Hamas won a full Gaza military victory, but lost to Fatah and Israeli security forces in the West Bank. Abbas dismissed Prime Minister Haniyeh, thereby breaking up any national unity arrangement and declared emergency law, sparking legal disputes between the two. Force on the ground determined the outcome. Hamas ruled in Gaza and Fatah in the West Bank, consolidating its control with Israeli security help and American military training directed by General Keith Dayton. Economist Dr. Salam Fayyad was appointed PA prime minister in an attempt to gain Western favor, rehabilitate its administration and to battle corruption. Two Palestinian entities evolved: "Hamastan" in Gaza and "Fatahland" in the West Bank. Fatah was accused of betraying the cause because the democratically elected, Hamas-dominated Palestinian legislature no longer convened, and Abbas ruled by executive fiat. Hamas leader Mahmoud al-Zahar threatened Fatah with retaliation should further action be taken against its operatives in the West Bank, but to no avail. Hamas waited to even the score until November 12, when 200,000 pro-Fatah demonstrators took to Gaza's streets in remembrance of Yasir Arafat. Hamas fighters opened fire on the participants, killing 6 and wounding over 80. Both sides consolidated control in their respective territories, repressed activities by their adversaries and inflicted casualties on the other side.[47] Years later in the spring of 2011, Fatah and Hamas reached a reconciliation agreement for a NUG, but implementation never followed.

The Hamas move in 2007 was emboldened by its Lebanese Hezbollah allies who fought Israel in a month-long summer war the previous year and suffered major damage, yet managed to launch 4,000 rockets into Galilee and were heralded as victorious throughout the Arab world and in much of the West. Likewise, Hezbollah was encouraged to take on Israel after Hamas was credited with forcing Israel from the Gaza Strip in the August 2005 Disengagement. Taking its cue from both, in June 2006 the Hamas Izz a-Din al Qassam military leader Ahmed Jaabari and the Islamists' affiliate known as the Popular Resistance Committees killed three Israeli soldiers and captured Corporal Gilad Shalit during a cross border raid, while the PA still ruled in Gaza.[48] Shalit was held for five years and eventually released in October 2011 in exchange for one thousand Palestinian security prisoners held by Israel, mostly Hamas members. Such a successful operation encouraged Hezbollah to open hostilities on Israel's northern border in July 2006, by killing several Israeli soldiers and abducting two others who died of their wounds. Five weeks of battle ensued between Israel and Hezbollah in what became known as the "Second War in Lebanon."[49]

Jaabari's Izz a-Din al-Qassam military wing, in alliance with the splinter Popular Resistance Committees and Army of Islam, outmaneuvered the Hamas political leadership. The politicians attempted to calm tensions and were in direct contact with Israeli Prime Minister Olmert when Gilad Shalit was abducted. When Israel responded with the massive "Summer Rain" retaliation bombings, Jaabari and the Hamas military became the de facto decision makers in Gaza. Shalit was released years later only when Jaabari gave the okay. In essence, it was a double coup. A year later, Jaabari's forces defeated Fatah, making it clear to the politicians that the military wing was the true ruling body in Gaza. Haniyeh's prudence lost out to Jaabari's Jihad.[50]

By 2008 the PLO/Fatah-Hamas split looked to be permanent and what had been a major crisis in the Palestinian territories stabilized into two semistates. There were increased rocket and terror attacks from Gaza, and Israeli retaliations helped Hamas solidify power. Fatah demonstrations were repressed with gunfire, and, despite denials, Gaza became an Islamic entity by implementing Sharia law in late December 2008, just days before Israel initiated the "Cast Lead Operation" to halt Hamas attacks.[51]

None of this was a surprise. Just prior to the 2007 Gaza civil conflict, Palestinian journalist and political analyst Zaki Chehab wrote that Hamas was continuing to deny Israel's right to exist while pushing a tough Islamist line in its domestic battles with the PLO/PA. They had the support of the people. To quote Chehab: "The Hamas electorate is unlikely to tolerate any diversion from the political and religious principles which Hamas has consistently advocated. By maintaining this rigid position, Hamas must realize that it risks losing a significant range of support unless it comes out and says categorically that it will accept UN resolutions and other agreements signed by the PA and the State of Israel." Furthermore, Hamas failed at government in partnership with Fatah. Chaos and corruption reigned in both Gaza and the West Bank, militias taking the place of a well-organized security force.[52] Israel also received messages from its own security forces on the ground at the still-functioning Gaza crossings manned by IDF Druze soldiers prior to the 2006 elections a year and half prior. Arabic being their first language, Palestinians spoke to the Druze soldiers freely and made it clear their hatred for Fatah and their intentions to support Hamas.[53] There was no reason to believe that the populace would stop short of sweeping Fatah out of Gaza entirely.

On the peace-making front, in November 2007, Israel and Fatah engaged in the American-sponsored Annapolis discussions in an attempt to arrive at a two-state solution. The West and Arab moderates supported the initiative with the moderate centrist Kadima-Labor coalition led by Prime Minister Olmert from 2006-09. Both sides re-committed themselves to the three-step Bush "Road Map" and agreed to a timetable resolving the core topics of borders, security, Jerusalem and refugees by the end of 2008. Donor nations were to give $7.7 billion as part of the Palestinian Reform and Development Plan 2008-10. Serious international supervision was necessary to avoid the massive corruption experienced during the Arafat years.[54] Economic progress advanced in the West Bank, but not for the forty percent of the Palestinian population residing in the Gaza Strip. Western and Israeli sanctions continued as a result of the Hamas military overthrow of the PA regime in Gaza and the Hamas refusal to recognize previous agreements, renounce violence and negotiate with Israel.

On the other hand economic growth in the West Bank was estimated at six percent in 2009, and was seen to be about eight percent in 2010, but much of this was donor-generated according to the World Bank. Under the Olmert government, an extensive amount of progress was made between Israel and the PA concerning borders, security and issues of sovereignty.[55] Negotiations set in for over a year, but no final agreement was reached when Benjamin Netanyahu and the Likud took power at the end of March 2009.

Hamas rule in Gaza was economically devastating, with 30 percent unemployment in 2007, rising to 40 percent the next year. The West Bank saw 18 percent unemployment in 2007 rise to 19 percent in 2008.[56] To circumvent the Israeli demand that all goods be sent through the overland Kerem Shalom/Rafiah Crossing to avoid contraband entering the Strip, Hamas and other operatives dug hundreds of tunnels under the Egyptian-Gaza border in a coordinated system to smuggle in weapons, ammunition and banned commercial goods. Israel eased the blockade significantly in June 2010 after the *Mavi Marmara* "flotilla" incident the previous month. Gaza did not keep pace with West Bank development. Israel played a major part, not only in aiding with security and the removal of roadblocks, but with tourism development in Bethlehem, bringing an estimated 1.5 million visitors in 2009, and Jericho in tandem with such joint Israeli-Palestinian business ventures as "Olives for Peace." Israeli President Shimon Peres, who was elected in 2007, remained directly involved in such peace ventures in the hope of achieving conflict resolution, at least with the Abbas Fatah regime. Ramallah served and continues as the West Bank economic hub. Investor confidence was high enough to enable work to begin on the new urban development of Rawabi north of the city.[57] In August 2009, Prime Minister Fayyad announced his plan for an independent Palestinian State *to exist alongside Israel.*[58]

For Hamas, the acceptance of a two-state solution by the officially recognized Palestinian Authority represented a near death blow to their ideology, forcing the demand for an immediate Jihad against those allowing the relinquishment of any part of the Divinely endowed *waqf* lands. Ideologically at least, Hamas was and is committed to a civil war to destroy the PA/Fatah and then to continue in its victory over Israel. The first half of 2008 proved critical. Hamas did its best to undercut the Annapolis peace initiative through increased rocket attacks against Israel. Already in January it undermined the blockade by destroying the iron border wall originally built by Israel between Egyptian and Gazan Rafiah. There was increased construction of tunnels from the Egyptian side to avoid the official border crossing and inspections. Hamas and its Muslim Brotherhood allies gained stronger footholds in the Sinai Peninsula and made their case on the Qatari-based *Al Jazeera* TV station, claiming they represented "the people," as opposed to the corrupt semi-military secular regimes ruling throughout the Arab world. Egypt was accused of being anti-Islamist, although President Mubarak refused to fully confront Hamas or the Muslim Brotherhood and their Bedouin allies in Sinai. To do so could backfire and stir the Islamists in Cairo. It was better to allow a smoldering border war with Israel as a safety valve to alleviate pressures.[59]

In early 2008, there were increased rocket attacks, including medium range Grads, on the northwest Negev region of Sderot, Ashkelon and the surrounding agricultural villages. This led to Israeli retaliations, especially in early March when 120 Palestinians were killed, though 80 percent of those killed were directly involved in the battle. Hamas spokesmen begged for a "*tahadiya*" or "calm" while others pleaded for a *hudna* to last five to ten years, but never retracted the demand for Israel's destruction. Hamas TV broadcasts requested that civilians enter homes of Hamas activists to act as human shields to deter Israel from considering targeted removals of those responsible for rocket attacks and terrorism against the Jewish State. Israeli Defense Minister Ehud Barak even consulted judicial experts to determine how legal it was to eliminate terrorists should they surround themselves with supposedly innocent civilians.

Civilian participation as human shields became official Hamas policy. Israel learned from its previous dilemma during the 2000-04 LIC and decided that Hamas operatives needed to be eliminated despite the fact that civilians, including women with children in tow, were threatening to sacrifice themselves alongside the Hamas military and political leadership. Israel deemed human shields responsible for their own fate as well as those of their children.

In March 2008, Israel was again planning for a full-scale assault on Gaza with the intent of destroying the Hamas government. While some Israelis expected a cease-fire could be arranged, others, like Israeli Member of Knesset Yuval Steinetz, believed Hamas was walking in the footsteps of Hezbollah by becoming "a forward Iranian position for raining down rockets on Israel." For many Israelis and in particular the leadership, Iranian activism against Israel remained a constant. Israel continued to be in conflict not only with Hamas but also Iran.[60]

Although Hamas gained a reputation for aiding its civilian population through the *dawa* social organizations, over the years it proved itself much more Jihadi and significantly less socially sensitive, especially should economic relief come through cooperation with Israel. A case in point was the terror attack at the Dor Alon fuel depot at Nahal Oz that April, where two Israeli workers were murdered while making shipments into Gaza. The plan was to get Israel to halt fuel supplies to Gaza and then complain to the international bodies and world media of boycotts and strangulation. This was a continuation of the Hamas activist policy, often aided by Fatah, of forced unemployment of its Gaza population. The continuous terror attacks against the Erez Crossing and Industrial Zone in 2004 brought its eventual closing for security reasons. This proved to be the most successful example of such a policy when overall 19,000 Palestinians lost their income, 15,000 no longer entered Israel to work, and another 4,000 lost their jobs in the industrial zone itself, which was known for Palestinian-Israeli joint ventures and cooperation. In January 2005, there was another attack—this time on the Karni border crossing in an attempt to strangle this last major economic lifeline of supplies after the demise of the Erez enterprise.[61]

Despite its adversarial relationships with Israel, the West, much of the Arab world and the PLO/PA, Hamas persevered and retained the support of the people especially in Gaza, right up until the "Cast Lead" operation at the end of December 2008. Furthermore, Sharia law became the law of the Gaza mini-state[62] and although possibly annoying some, it did express the will of the people. Many would ask and theorize as to the secret of Hamas' success—an organization seen as enjoying vast support while remaining uncompromising in its demands for war against Israel and for a universal Jihad. The answer lies partially in the *dawa* actions, inextricably linked with politics and the armed struggle or Jihad. *Dawa* social welfare activities are not only the foundations of much support for Hamas, but are used as indirect bribery to gain political power. For example in the Bethlehem area during the local elections in May 2005, 35 percent of the electorate claimed "poverty and unemployment" to be the major issues they faced. Hamas food distribution proved vital for their victory in the city council, where they took five of seven seats reserved for Muslims while the Islamic Jihad took a sixth.[63] Although outsiders could claim the Palestinian population was unaware, the Hamas system was "integrative" and "mutational" regarding the fusing of *dawa* social services and Jihad. It is much more difficult to accuse traditional Muslims of being naive. Islamist ideals were fully re-integrated into what had become a seemingly more secular Palestinian society now forced to mutate into a fanatically violent Islamist entity playing its part in the universal Jihadist struggle—its sector being Palestine/Israel. Full radicalization came side-by-side with welfare activities while frustration amongst youth soared, especially when Israel worked to repress the 2000-2004 LIC with arrests, incursions, firefights, roadblocks and curfews. Palestinian casualties and economic distress mounted and fueled greater motivation for homicide-suicide bombings. *Dawa* activities obligated many young people to Hamas and its terror. The bombers and their families received full support financially, spiritually and socially, while they became media heroes in a sub-culture of terror reverence.[64]

The positive aspects of social services were the first step in attracting and radicalizing the population. Next was martyr worship and death for the sake of Allah. The sacrificial martyr ideal resonated not only with young men, but with mothers and children as well. Rim Salih al-Rayashi, the first Hamas female suicide-homicide bomber and mother of two recalled wanting to be a "martyr" from the time she was in the second grade. She had this to say before her explosive death, "I have always told myself: Be filled with every possible grudge for the Jews, the enemies of your religion, and make your blood a road leading to paradise. I began to try and do my utmost since the second preparatory grade."[65]

Hamas often used children to transfer weapons and explosives. Summer camps were built for elementary school pupils where a network of future martyrs, often drafted from the poor, were indoctrinated and trained. Recruitment continued through mosque organizations and higher education, in particular at al-Najah University in Nablus. Potential martyr *"shaheed"* recruits craved death. Hamas ran its own economy, sponsoring jobs, commercial outlets, education, medical and social services. Of those citizens involved, many showed complete loyalty to these fanatical Hamas values, even more than they showed loyalty to their own families. They viewed their economic, physical and spiritual wellbeing as one indivisible whole, all attributable to the Palestinian Muslim Brotherhood—Hamas—of *dawa* and human sacrifice in the name of Allah.[66] *Dawa* charities were established worldwide to acquire monies for social welfare, but these funds frequently ended up supporting terrorism.[67]

Hamas achieved two major objectives: unbounded commitment and worship from their adherents, coupled with the establishment of themselves as a second and more dominant sovereign than the corrupt, ideologically-devoid Palestinian Authority. From "dual sovereignty," Hamas would make the move toward grasping full control of Palestinian society through political and military means. Charity for *dawa* good deeds was a tool used as a tactic to gain overall funding for Jihad. Hamas made no differentiation between its humanitarian and military wings—two sides of the same coin. This point was reiterated by Prime Minister Haniyeh once again in his December 2011 speech commemorating the founding of the organization. Integrated social welfare and educational activities brought about an unswerving loyalty and love for Hamas, Allah and Jihad.

With Hamas rule in Gaza, this dual system of social welfare activities and Jihad remained intact right through Israel's Cast Lead Operation. Still, a question remained. Could a true Muslim Brotherhood regime such as Hamas administer the Gaza Strip mini-state while simultaneously engaging in the *dawa* social programs with the accompanying Jihad, and still retain the loyalty of their citizens? The Gaza clash and its ramifications would be a major test for activist Islamist Jihadi values and implementation. It was clear Hamas was at a serious disadvantage when confronting Israel, but when viewed as part of the global Muslim Brotherhood, victory would be theirs even if only at some distant future date.

The "Cast Lead" Gaza War and Repercussions

From June 19, 2008, Egypt arranged a half-year cease-fire between Israel and Hamas, including the supposed halt of Hamas arms smuggling through the 500 tunnels connecting under the border between Egyptian and Gazan Rafiah. The agreement broke down in early November when Israel discovered a Hamas tunneling operation apparently designed to cross into the Negev and abduct soldiers, as had been the case with Gilad Shalit two years earlier. For the next six weeks, there were low-level rocket attacks and Israeli retaliations. On December 19, Hamas refused to renew the cease-fire under its previous terms and declared its official demise. Over the next few days, Hamas fired rockets at Israel intermittently. When Hamas spokesmen openly suggested possibilities for renewing the *hudna,* but made no commitments, Israel authorized pin-point air strikes. Escalation ensued and over sixty rockets landed in Israel just two days before the Israeli air offensive on December 27, inaugurating the Cast Lead operation. This was coupled with a ground assault from January 3-18, 2009. What is referred to at times as the "Gaza War" proved to be a serious test and even more so as a turning point for Hamas.[68]

Israel caught Hamas by surprise with its initial air strikes, killing some 140 of its members and, within the week severely damaged Gaza's infrastructure. By the time ground operations commenced about 400 Palestinians were killed, an estimated one quarter of them civilians. Hamas commanders were targeted and the organization made a systematic effort to defend their leadership by calling on civilians to act as human shields for the military and political echelon. Much of the Hamas leadership took refuge in the basement of Gaza's Shifa Hospital using patients and medical staff as a guarantee that Israel would not attack or capture them. They proved correct. Israel dropped leaflets and made phone calls urging Palestinians to move from areas slated for attack. Gaza, however, had little in the way of bomb shelters, so there were few options of where to flee. Bombings and shelling from the air, ground and sea attempted as much precision as possible, but some 50 percent of all Palestinian casualties were civilians despite efforts to avoid non-combatant involvement. This was especially true once the battles were enjoined in heavily populated areas such as Beit Lahiya, Beit Hanoun and Gaza City. Hamas responded with intensified Qassam and Grad rocket fire into southern Israel. They reached not only Sderot, Ashkelon and Kiryat Gat, but further to the Negev capital Beersheva and the main port of Ashdod, servicing Tel Aviv and the center of the country. South central Israel faced paralysis and Hamas shelling became a major threat to the country as a million people were within rocket range. Overall, damage was not heavy, but the constant alerts brought much of the Israeli economy in the region to a halt as schools were closed, rockets landed throughout civilian areas and workers stayed home. Facing the IDF Hamas was outmaneuvered and outgunned. Many Hamas fighters melted into Gaza's civilian zones. According to Israeli estimates based on statements by Hamas Interior Minister Fathi Hamad, 709 Hamas and allied militants were killed out of a total of approximately 1,400 Palestinian dead. Other estimates halve the amount of Hamas combatants killed and do not include in their figures the approximately 250 Hamas police officers who died in the fighting. Israel considers Hamas police armed combatants, while Palestinian and human rights organizations do not. Israel suffered 13 dead, 10 of whom were soldiers.[69]

There is little doubt concerning Israel's military success. Hamas could not match the Israeli military. Hamas lost around 600 men, or over 10 percent of its armed forces, with the usually accepted ratio of two to three times that amount wounded. It spelled paralysis for an organization when 35 percent of its fighting force is put out of action. Hamas miscalculated as badly as Arafat did in September 2000, and as Hezbollah's Nasrallah had in the summer of 2006.

The larger question remained as to what objectives did Hamas have and were any accomplished? Hamas did not expect a major Israeli retaliation for what they considered were only moderate provocations. Similar to Hezbollah, Hamas aggressive condemnations and actions against Israel were broadcast to its own population, thus serving as a unifying factor, one immersed in theological invective. Jews and the Zionist entity were to be constantly harassed and when the time was right, destroyed. It should be noted that upon acceptance of the cease-fire, in its pragmatic moment, Hamas was under attack ideologically and politically by even more extreme Islamist groups for having temporarily jettisoned the Jihad too early.

It must be noted that Israel was accused of "indiscriminate" attacks against civilians. Non-combatants were caught in the crossfire at times, but it was virtually unavoidable when Hamas consciously mixed combatants with civilians, using the latter as human shields. Half of the casualties on the Gaza side were Hamas men under arms. A quick calculation is necessary to determine what is indiscriminate. The Gaza Strip population at the time was about 1,400,000, of which an estimated 6,000 were armed Hamas members, this yielding a percentage of 0.43 percent, or less than one half of a percent of the overall population as Hamas fighters. According to different sources there were approximately 1,400 Palestinian deaths during the conflict. If all of the deaths were the result of indiscriminate or at random killing, taking into consideration Hamas membership made up less than half of a percent of the total Gazan population, that would mean a total of six Hamas combatants (0.43 percent) would die alongside 1,394 civilians, (99.57 percent). Yet by Hamas' own admittance their security forces, including police, made up half of those killed, thereby proving Israel to be ***over 118 times*** more accurate than indiscriminate when taking aim at the Hamas militia terrorists. Even if the total number of Hamas fighters killed was in the range of 350 deaths, according to the pro-Palestinian Israeli B'Tselem human rights group, we still notice a rate of accuracy 58 times greater than random. If we add in B'Tselem's Hamas police casualty count, the accuracy rate is overall 100 times greater than random. The Palestinian Centre for Human Rights claims almost 500 were killed if we include the police, who made up a bit more than half the deaths of those under arms, this still leaves us with a low percentage of civilian casualties. But here too, the number of casualties is over 80 times greater in accuracy than random. It is interesting to note that the Hamas military wing never published a casualty count as relates to their own combat and police loses.[70] This raises many questions and more than a few suspicions.

Yet on the media and humanitarian front, Israel was touted as the big loser. Borrowing from the late Yasir Arafat, Hamas played the victim card, as opposed to Nasrallah's Hezbollah style bravado. The Hamas spokesmen highlighted casualties and the major material damage done to infrastructure and thousands of buildings, including numerous homes. Israel did not allow reporters to embed with its troops until the last few days of the conflict when the Supreme Court upheld journalists' suit against the IDF. Because of the lack of reporting, Israel lost valuable first-hand evidence by third parties on the ground, which would endorse the Israeli army evidence of massive Hamas booby-trapping of civilian structures and its mixing of combatants in civilian environments. Despite overwhelming military superiority, there was no final push for total conquest of the Gaza Strip to include the arrest or elimination of all Hamas activists including its political elite led by Ismail Haniyeh. The Gaza population solidified its support around Hamas and Israel lost the information war, an arena considered much less important than the physical battlefield.[71]

The UN-sponsored Goldstone Report was issued in the wake of the conflict. The report claimed Israel used excessive and unnecessary force. Any claim of use of wanton, or indiscriminate fire can be dismissed as shown above. The report released on September 15, 2009 was a tremendous diplomatic and media victory for Hamas. Most of the accusations concerning human rights violations were against Israel. In particular, the title of the report "Human Rights in Palestine and Other Occupied Territories" already indicated the full delegitimization of Israel's right to exist, since the "other occupied territories" can only mean Israel proper— areas under Israeli control within the 1949-67 armistice lines prior to the 1967 war. From the outset, the article was creeping *dhimmization* of Israel by broadly hinting that the Jewish State was an illegal entity, calling any territory under its control "occupied." By logic, if there is doubt as to Israel's legality, then any action Israel takes to defend itself is also doubted. An illegal entity has no right to defend its existence because by definition it should not exist.

Beyond the report's title, Hamas continued to succeed on the diplomatic front. The investigation did not deal with the deliberate Hamas shelling of Israeli civilian areas prior to the mid-2008 cease-fire. Former UN Human Rights Commissioner Mary Robinson refused to head the committee of investigation when asked to do so, declaring the Human Rights Council mandate to be one-sided since it was initially not to include any review of Hamas activities. In the end, South African Judge Richard Goldstone chaired the committee, and did get a broadened mandate to investigate Hamas violations, but the UN Human Rights Council never officially endorsed the extended jurisdiction. Under such conditions, Israel refused to cooperate with Goldstone while Hamas agreed to talk. Investigators interviewed Palestinian witnesses from Gaza and quoted them as "credible sources," despite the fact that we do not know how credibility was determined. Hamas was cleared of any accusations of using civilian areas for launching rocket attacks against Israel or of allocating civilian structures for military activity even though films of such attacks were extant.[72] Israel's withdrawal from Gaza in 2005 was not mentioned at all. It was not noted that Israel's "blockade" was only imposed after the Hamas military overthrow of the Palestinian Authority and the massive rocket fire into Israel intensified, including 8,000 rockets since 2001. Nor was it mentioned that Palestinian tunneling essentially circumvented any "blockade" rendering it ineffective.

Israel was likewise accused of targeting civilians, a charge later recanted by Goldstone himself in an Op-Ed in the *Washington Post* in April 2011. He said Israel did not target civilians and such damages were due to commander errors. Furthermore he admitted that the IDF was investigating charges of illegal behavior by its soldiers toward civilians. He admitted Hamas investigated nothing. The retraction did Israel little good as the other three members of the committee undermined Goldstone and stood by the original report.[73] As far as the Goldstone committee was concerned the conclusions were completely logical as evidenced by the title of the report.

Despite the Hamas military failure during Cast Lead, their victory cannot be overstated in political and diplomatic terms. Media constantly highlighted Israel as the criminal and barely mentioned Hamas. The UN, including quite a few European countries, endorsed the Goldstone Report although the Americans refused to do so. In the immediate aftermath of the conflict, Israel enjoyed a momentary solidarity visit by EU leadership who clearly understood the threat of Islamic extremism, but there was no real implementation of security concepts.

As seen by Israel the UN Human Rights Commission was and is a deeply flawed organization itself, having housed members from dictatorships such as Qaddafi's Libya. As well, they completely ignored Hamas' vicious antisemitic and genocidal tones against Jews and Israel. The antisemitic diatribes are not only contained in *The Hamas Covenant,* but constantly repeated in the Hamas media—an active policy designed to rally the masses for a future conflict.

In the aftermath of Cast Lead, Hamas suffered a Jihadi setback, but strengthened its relationship with Khomeinist Iran, its Lebanese ally Hezbollah, and the Sunni Jihadist Sudanese regime, all of whom worked together to ensure military supplies via the tunnels originating in Egypt. In the continuing bid to retain world sympathy the Islamists highlighted their own victimization at the hands of Israeli Jews, portraying them as criminals of the worst type. Anti-Israel and antisemitic demonstrations led mostly by Muslims, but including the far left and the anarchist fringe, swept much of Europe and even parts of North America and Australia. People gathered on the streets in an uproar throughout the Arab world, angry against Israel, Jews, the West and their own governments. No Arab country except for Sudan came to Hamas and Ismail Haniyeh's aid. These same Sudanese killed between two and three million Black Africans, both Christians and Muslims, over the past thirty years. Hamas and the Muslim Brotherhood continued their efforts at undermining the Arab secular dictatorships.[74] Militarily Hamas needed a *hudna* - time to recover.

In the meantime the Americans elected a new government. The Bush Republicans were out and the Obama Democrats took office, two days after the Cast Lead cease-fire. Obama, whose father was from Kenya, began his foreign policy initiatives with overtures supporting the Muslim world. Early on he toured the Middle East, beginning in Turkey and proceeding to Cairo, where he addressed issues involving Islam, democracy and peace in the region—most importantly peace between Israelis and Palestinians. He discussed the PA's need to govern through state institutions and admitted Hamas had partial support from the population. He asserted, "Hamas must put an end to violence, recognize past agreements and recognize Israel's right to exist." He further addressed issues of freedom of religion and women's rights.[75] Obama called for a Palestinian State alongside Israel. The speech brought a response from Israeli Prime Minister Netanyahu who made his own proposal at Bar Ilan University, where he spoke of peace with the Arab world and outlined a two-state solution involving a demilitarized Palestinian State. In this vein he addressed the PA/Fatah leadership, not Hamas.[76]

By the summer of 2009, although not openly declared, it was apparent the Obama administration was interested in a two-state solution based on the 1967 lines, expected Jerusalem to be split as the dual capital of Israel and Palestine, was not keen on Israel stationing troops along the Jordan River for security and expected General Dayton's PA police force to handle all potential threats. It was not clear whether Israel would be allowed "hot pursuit" when tracking terrorists.[77] Obama would eventually confirm these policies, in particular his commitment to the 1967 lines, in his speech to the State Department in May 2011. Hamas took no solace from these clear American foreign policy stances, insisting that no compromise with Israel was in the offering and that the only solution was Israel's destruction. For Hamas, the two-state solution discussed between the PA/Fatah and Israel was completely out of the question. Prime Minister Haniyeh made this simple point in December 2011 on the twenty-fourth anniversary of the official founding of Hamas, when he stated, "Today we say it clearly. Armed resistance and armed struggle are the strategic way to liberate the Palestinian land from the sea to the river." He continued to say Hamas would never accept a Palestinian State if it only included the West Bank, Gaza and East Jerusalem, but could consider a temporary *hudna* should Israel make these concessions. His remarks were directed as much at the PA and President Abbas as they were at Israel.[78]

On the internal front, Islamization of Gaza continued and could be expected to intensify in light of Islamist successes throughout the Arab world during the 2011 Islamic Awakening uprisings. Although commencing in June 2007, Islamization in Gaza deepened by the end of 2008 onward. Women were forced to wear the hijab traditional Islamic dress, were not allowed to ride on motor scooters or dance and could be arrested for immodest dress or even laughing in public, as dictated in the last sentence of Article 19 in *The Hamas Covenant*. Extremists groups such as "The Swords of Truth" forcibly closed down hip-hop dancing and mixed bathing water parks. The Hamas government denied any involvement and condemned the activities, but they never captured the assailants. Palestinian scholar Dr. Khaled Hroub and Israeli Arab journalist Khaled Abu Toameh viewed Hamas rule as becoming similar to that of the Taliban in Afghanistan. Politically PA President Abbas agreed, insisting Hamas was working to establish an Islamic emirate, an accusation dismissed by Hamas officials.[79]

From the end of the Cast Lead operation and continuing into 2012, Hamas was curtailed in its abilities to attack Israel, although at times there were unprovoked rocket attacks. Israel retaliated and Hamas usually blamed other factions for the escalation while condemning the Israeli response. Attacks originating in Gaza were also less effective as Israel's newly developed Iron Dome defense apparatus began intercepting some of the longer-range missiles during the winter of 2011-12. Holes remained in the system, as not enough batteries were deployed and it could not intercept short-range projectiles such as Qassam rockets.

As Israel increased its defensive capabilities, Hamas upgraded its international standing by working through the freely elected Sunni Islamist Turkish regime led by three-time Prime Minister Recep Tayyip Erdogan while beginning to apparently break with the increasingly isolated Shiite Iranian regime. Hamas effectively played the suffering hero role during the flotilla episode on May 31, 2010, when Israeli commandos boarded a ship sponsored by Turkish Islamists as it attempted to crack the naval blockade on Gaza. The point of the Israeli blockade was to halt further contraband, in particular medium-range rockets, from entering the Strip. Israel offered to allow all non-contraband supplies into Gaza provided the ships were first searched. All supplies would be brought to the Israeli port of Ashdod and what was permissible shipped overland into Gaza. Leading the flotilla was the largest vessel, the Turkish registered *Mavi Marmara*. Many of the hundreds aboard the ship were IHH activists from the Turkish Muslim Brotherhood. The ensuing high seas clash left nine Turkish civilians dead and dozens injured on both sides as IHH activists put up a well-documented resistance, nearly beating to death several of the commandos. Ankara adopted the Hamas ruled Gaza Strip as a "client-state," acting as the patron for Hamas in the international arena. Hamas received much-needed support from a country with excellent Western credentials, meaning NATO membership. The Turkish-Hamas bond was cemented and made world headlines for the next month.

The UN-sponsored "Panel of Inquiry of the 31 May Flotilla Incident" known as the Palmer Report, declared Israel's blockade legal, but was critical of how the navy carried out the boarding operation and the resulting casualties. The report also questioned the humanitarian intentions of the IHH actions. In the aftermath, there was an attempt at Israeli-Turkish reconciliation with no positive results. Diplomatic relations between Jerusalem and Ankara remained severely damaged while Hamas came out the winner with a new ally in Turkey, the most powerful Sunni Muslim nation in the Middle East if not the world.[80] Flexing Turkish muscles, Erdogan's government leveraged the continuing belligerency against Israel as a springboard to enhance its power in the region. He threatened Greece, Cyprus, Syria, Kurdish insurrectionists and any combination of the aforementioned parties.[81]

Sharia Law, Hudnas, and the 2011 Islamic Awakening

With the Arab Islamic Awakening in full bloom during 2011 Hamas tightened control over Gaza despite the difficult economic situation. Gaza was and is considered impoverished, with an estimated 70 percent of the population living below the poverty line and 40 percent unemployment. Massive tunnel-smuggling operations under the border with Egypt brought some relief. A wide variety of supplies arrived, including weapons, rockets, fuel, animals, food and even vehicles. A serious class gap was developing amidst a black market economy, a recipe for rampant corruption and more instability.[82]

Still, there was no immediate threat to Hamas control. No real alternative existed, nor did any group have popular support or military prowess anywhere near that of Hamas. There was little opposition to speak of and Islamism was the rising force in the Arab world. Only a Salafist or al-Qaeda type of group was capable of challenging Hamas, not secular or liberal democratic initiatives.

On the internal Palestinian scene there was much discussion over a reconciliation and reunification of forces between Fatah and Hamas in the spring of 2011. Islamist Gaza was failing on the military and economic fronts while the West Bank PA regime was enjoying security and economic benefits, especially from the US and EU for cooperating with Israel and Jordan. The Arab Islamic Awakening suddenly endangered Hamas foreign policy when the organization was caught in a contradiction with its foreign headquarters in Damascus as a show of support for Syria's Assad regime. Assad's secular state was engaged in a civil war against the Muslim Brotherhood where tens of thousands died within the first year. Hamas could no longer be seen as a Syrian ally supporting Assad and his Iranian Shiite allies against Sunni Islam. Hamas relocated to Qatar.

Fatah, on the other hand, promised to declare its state in the UN by September 2011, following the two-state plan devised by Prime Minister Salam Fayyad in August 2009, entitled "Ending the Occupation, Establishing the State." The Palestinian State was to be delineated by the 1967 lines alongside Israel with East Jerusalem serving as its capital.[83] The unilateral Fatah initiative did not succeed due to American opposition in the Security Council. The

Western powers urged direct negotiations between Israel and the PA to arrive at a permanent settlement of all issues.

Hamas was greatly relieved when the PA move at the UN failed; the last decision they needed was an internationally sanctioned two-state solution with Israel. But they did not need to worry, as it appeared that politically Abbas could not cede any ground on the issue of refugee return. Over the years even parts of the Fatah leadership had become more militant in demanding Israel's total demise, dismissing the two-state solution as irrelevant.[84] Such rejection bolstered support for Israel's own right wing, which was fond of making the argument that there was no reliable partner interested in two-states for two peoples. However in 2010, 71 percent of Israelis and 57 percent of Palestinians still favored a two-state solution,[85] although many believe these numbers to be declining in recent years. In any event, settlements were never a major concern in Hamas thinking, but rather Israel's existence was the issue. Fatah/PLO hardliners were in agreement with the Hamas view. For those supporting the two-state solution, any strengthening of either Hamas or the Israeli right wing worked to their disadvantage.

Hamas scored its own major victory in 2011 when it forced the lopsided Gilad Shalit prisoner swap in October. One thousand Palestinian security prisoners, many of whom were involved in murderous bombings, were released for a lone Israeli soldier. Most of the released prisoners were Hamas operatives, but others were from Fatah and other non-Islamist organizations. Against the backdrop of the Islamist surge throughout the Arab world, Hamas stock rose enormously with celebrations not only in Gaza, but in the Fatah-controlled West Bank. The PA was undermined and Israel humiliated by the deal. Both began to reclaim international standing after peace talks in Jordan and discussions at the yearly economic conference in Davos where Israeli President Peres met Prime Minister Salam Fayyad in January 2012. Hamas, however, remained defiant and Islamist popularity soared throughout the Middle East.

There is much agreement that Israel misplayed her cards, but it is accompanied by a fair amount of controversy as to what should have been done. According to Shlomi Eldar, political Hamas, led by Haniyeh, was surprised when they were notified of the abduction of the Israeli soldier, to be used to barter for Hamas prisoners. The rebellious Izz a-Din al Qassam commander Ahmed Jaabari led the operation to kidnap the soldier in conjunction with Islamist splinter groups and forced the political wing to offer Israel fairly moderate conditions whereby Shalit was to be returned.

Khalid Mashal put together a "non-paper" or unofficial proposal, for "Hamas-Israel—Peaceful Coexistence" on September 8, 2006. Previously on May 25, just before the Shalit incident, Hamas and Israel were in contact through a third party to discuss a twenty-five year coexistence arrangement. Hamas first envisioned non-violence and mutual obligations to negotiate a two-state solution based on the 1967 borders and certain agreements as regards taxes and commerce. Issues such as refugees and the final status of Jerusalem would be postponed for a future date.

Eldar writes that the September "non-paper" set out two tracks: one dealt with the narrow issue of prisoner exchange, and the other embraced a broader strategic arrangement. In the first track, Hamas demanded one thousand prisoners in return for Shalit. In the second track the exchange would include less than two hundred prisoners, but included the longer-term coexistence accord. Mashal wanted to tie together the Shalit/prisoner issue with a general halt in hostilities. While Israeli Prime Minister Olmert was willing to work on both fronts, he refused to link them together. Discussions began in May and continued into the autumn of 2006, much of it during the five weeks of the Second Lebanon War against Hezbollah. In the end, Olmert chose neither option and Israel continued military action against Gaza leaving some five hundred Palestinians killed and a thousand injured while Hamas rockets landed uninterrupted in the western Negev. Olmert understood that Izz a-Din al-Qassam and Jaabari were the real decision-makers while Mashal, Haniyeh and political Hamas were developing strategies and making promises they could not fulfill since none lined up with the Hamas reality. Olmert has been criticized for not consulting with other cabinet members, especially Defense Minister Amir Peretz, or even the IDF Chief of Staff Dan Halutz, concerning the proposals. As for Hamas, in the end Jaabari got the upper hand, forced the release of a thousand prisoners in exchange for Shalit and scuttled any possibility of long-term non-violence.[86]

The Hamas proposals were said to be even more moderate than the PA Fatah platform, yet Israel chose to continue its relationship with the latter, putting no trust in the political Islamists' wide-ranging proposals.[87] Several points must be understood. Most importantly, Hamas never showed intentions of arriving at a permanent status agreement, nor could they. It is clear, as Eldar himself admitted, that Khalid Mashal put himself in direct confrontation with Sharia law concerning the need for a defensive Jihad to re-conquer *waqf* lands as explicitly stated in Article 11 of *The Hamas Covenant*. Should Israel have agreed to such wide-ranging military, political and religious concessions there is, in essence, more of a guarantee of non-compliance than continued acceptance by Hamas during or after the twenty-five year *hudna*. Theologically, Hamas is obligated to renew hostilities when sensing victory is on the horizon. Sheikh Yasin was forthright in stating the Jews would have to concede their state and return to their *dhimmi* status. There are those who suggest Hamas could change its *Covenant*. To do so would turn it into a secular national document and bring the wrath of Islamist jurists and the Muslim Brotherhood leadership down on Hamas heads. A weakened Israel without conflict resolution constitutes a major step when implementing the Jihadi final objective to destroy the Jewish State. On the ground, Ahmed Jaabari's military wing included the rulers and enforcers who objected to any moderation. Inviting a tactical overture was in full contradiction to the essence of the Hamas identity, theological understandings and Islamist world-view. When looking beyond the immediate future, and even if Mashal was sincere, Israel could never take the deal.

The Hamas regime was determined to achieve the objectives they made so obvious in their *Covenant*, as evidenced by their condemnation of and demand for Israeli and Jewish obliteration (*HC* Introduction-Preamble, Articles 7, 17, 20, 22, 28, 30 and 32). Their next objective was the internal struggle to destroy any possibilities of conflict resolution by moderates as represented by certain Fatah factions willing to accept a two-state solution (*HC* Articles 11-15 and 32). In the process, they would ensure the demise of secularism in the Arab Muslim world, in particular amongst Palestinians (*HC* Articles 25-27). Over the years, Israel was either dismissive or oblivious of these Hamas objectives. Analysts noted the rise of a deadly antisemitism, but many considered it a temporary phenomenon somehow to be negotiated away, and not a long-term Hamas strategy to achieve its primary objective of Jewish annihilation. By the end of 2011 there was an increasing awareness that Hamas represented mainstream Muslim Brotherhood thinking demanding Israeli and Jewish destruction, and not a permanent peace agreement with mutual recognition on both sides.

Hamas continued to influence the Fatah-dominated PA and not the reverse. Hamas viewed itself as a full member in the Muslim Brotherhood (*HC* Article 2) within worldwide Jihad, as evidenced by Prime Minister Haniyeh's November 2010 speech in Gaza. He never said otherwise.[88] With the so-called Arab Spring making its stamp as the Islamic Awakening, Hamas rode a wave of popular support within the universal Islamist domain. With the dawn of 2012, Hamas influence was on the rise, eclipsing Fatah and reflecting trends in the Arab world. A Muslim Brotherhood candidate, Mohammed Morsi was freely elected as Egypt's president in early summer 2012. Hamas ideals were in tandem with those prevalent throughout the Middle East. Now there was an ally at the helm in Egypt, not an adversary.

By October 2012, Hamas was working several initiatives on the overall Middle Eastern front and waiting to exploit religio-political opportunities. Egypt fell to the Muslim Brotherhood and the more extremist Salafists through elections less than a year after President Hosni Mubarak's overthrow. On the eastern front, with under-reported increasing destabilization in Jordan, Islamist opportunities for regime change looked possible as their influence was manifest. King Abdullah II was pressured to institute reforms and eliminate widespread corruption. The issue was not exclusively socio-economic but must be viewed in the prism of the secular Arab nationalist and Islamist clash. The Jordanian monarchy had always rested on its Bedouin foundations, but such support was waning. The king initiated reforms, but they were neither fast enough, nor deep enough; even the loyal Bedouin tribal alliances, known as pillars of support, were fraying.[89] Supposedly, they too were seeking Islamic answers. Palestinians constituted over 60 percent of Jordan's population and were always considered the most potentially destabilizing part of society. For years there was a process of "Jordanianization" as a counter to "Palestinianization" in the nationalist sense of the term. Furthermore Hamas influenced Palestinians living in Jordan and should one add in the continuing Islamist inroads among the Bedouin sector, the future of Jordan as an Arab nationalist monarchy might be in doubt. Reading the trend, Abdullah met with Hamas leaders at the beginning of the year in Amman. The meeting was a major rapprochement since the expulsion of Hamas leaders from Jordan to Damascus in 1999. Should Jordan fall to the Muslim Brotherhood or Hamas, it would only be a matter of time before the West Bank followed. Such a scenario greatly increased Israeli security concerns. The Islamization of Jordan, a Hamas policy goal, was within reach.

The turning point began in late October when munitions smuggling through Africa by way of Sinai into Gaza was at an all-time high. The Israeli air force was credited with destroying a major arms factory in Khartoum, Sudan, known for supplying the Hamas arsenal. Hamas responded with rocket fire and Israel replied in kind. Despite attempted ceasefires, the exchanges continued for three weeks. The heightened clash culminated when over 100 rockets landed in Israel, forcing Jerusalem into the "Pillar of Defense" operation on November 14. Mild in comparison to Cast Lead, the eight-day Israeli initiative once again took Hamas by surprise, striking over 1,500 targets in the Gaza Strip. The Gaza regime reported 133 Palestinians killed and 840 injured, less than a tenth the amount killed in Cast Lead. Hamas fired 1,456 rockets into Israel, but this time they fired not only Qassams and Grads, but Iranian Fajr-5s with a range of up to 75 kilometers or 45 miles. Central Israel and Tel Aviv became targets. Israel's Iron Dome anti-missile system was partially successful and took out 421 Hamas rockets. The rockets were becoming a strategic threat to paralyze the country, yet no ground offensive was launched into Gaza. Hamas was crippled by Israeli air strikes, yet claimed victory when Israel decided to cancel a ground assault.[90] Realizing Hamas had an ally in Egyptian President Morsi, it is quite possible Israel decided to limit the conflict. A ground assault might have been a litmus test for Morsi's Islamist credentials, possibly forcing a military response even if against the wishes of the Egyptian command. Cairo negotiated a cease-fire between Israel and Hamas, but little changed.

Hamas soon suffered two more setbacks. By late November 2012 the Palestinian Authority received non-member state observer status at the United Nations by a vote of 138 in favor, 9 against and 41 abstentions; clearly an overwhelming majority.[91] The Palestinian State was and is within the context of the two-state solution with Israel. To receive full membership the newly "declared" State of Palestine was further in need of Security Council approval, which could only happen with US approval. Washington demanded a negotiated settlement to the conflict between Palestinians and Israelis. The two sides were to recognize each other and solve the major outstanding issues of security, borders, Jerusalem and refugees before the US would support an independent Palestinian State. Recognition of Israel would mean a defeat for Hamas.

The second setback for Hamas came in early July 2013. The Egyptian military overthrew Hamas' newfound sponsor Morsi after a year in office. Egyptian relations with Hamas quickly soured. General Abdul Fattah a-Sisi and the Egyptian army claimed they were dealing with Islamist terrorist infiltrations sponsored by Salafist and al-Qaeda affiliates in Sinai. There were suspicions of an Islamist-inspired Bedouin rebellion, and reports of other radicals bringing supplies into Gaza. It is not clear whether Hamas was cooperating with or simply not purging these fanatics. Hamas influence flowing out of Gaza into Egypt by way of Sinai was a fear turned into reality. From 2011 to 2013, Israel suffered increased attacks on its Negev border from terrorist cells in Sinai, the origins of many being Gaza. The Egyptian military sought to reassert state authority in Sinai and believed the Muslim Brotherhood administration was responsible for the deterioration. Morsi was arrested and accused of aligning himself with Hamas and Islamic terror.

In the summer of 2013, the Egyptian military began work to halt operations in the 1,200 tunnels connecting Sinai to Gaza, apparently taking action to close them down. Not only did this affect freedom of movement for terrorists, but the Gazan economy took a heavy hit. Consumer goods no longer arrived. According to Hamas, the Gaza economy lost $450 million toward growth, the improved 32 percent unemployment rate was expected to rise to 38 percent, and construction would seriously slow down. In early 2013, Gaza was still recovering from the Pillar of Defense operation damages. Officially, trade had to be conducted literally above ground, and overseen by Israel and the Palestinian Authority, as stipulated in the Oslo Accords. Fuel was in short supply and the Gaza power plant functioned intermittently or not at all, leaving sewage to flow freely in the streets as pumps idled.[92] The official explanation was the refusal to pay the PA taxes, which the people saw as excessive. One might also consider the ideological angle. Why would Hamas agree to work with Israel—their sworn Jewish Zionist enemy? As for the PA, why would Hamas help them make money, when they were the compromised secular nationalist regime responsible for concessions to Israel and the West? Hence there was no reason to cooperate with either Israel or the PA, unless the situation became dire and then any arrangement would only be short-term.

Trying to remedy the situation, Hamas officials continued turning to many countries for aid, especially seeking reconciliation with Iran. Relations cooled when Hamas closed its Damascus office and implied support for the Sunni Jihadist rebels battling the Tehran-sponsored Assad regime. Hamas regularly broke up demonstrations thwarting any attempt by opposition forces to organize. The siege mentality brought social cohesion and Hamas continued to enjoy popular support.

On the PA/Fatah side, negotiations with Israel's Netanyahu government resumed. The US originally planned for a permanent status agreement by the following May, but the possibility of a far-reaching interim agreement was thought more attainable. There was a growing gap between the PA West Bank leadership and its constituency. Many saw the PA as detached from reality in trying to reach an agreement with Israel, especially concerning the demand for full refugee return. During his reign Yasir Arafat refused compromise on the refugee point, knowing Israel could never allow such a condition. Hamas demanded full refugee return, while the PA/Fatah was accused of betrayal by implying far-reaching compromise on this issue. Life was and is difficult in Gaza, but despite economic gains made by the Fatah/PA in the West Bank, the Abbas administration continued to be perceived as selling out its people for material gain. Support for Hamas in the West Bank was substantial and held solid in the refugee camps across the river in Jordan.[93] Hamas was on the defensive economically and militarily, but not out of the game. Ideologically, the Palestinian public supported the Islamists and blamed all others—in particular Israel, Egypt, the PA/Fatah and the US—for their suffering.

Extending beyond the immediate region, Hamas foreign policy played Khomeinist Shiite Iran against Erdogan's said-to-be moderate, but increasingly Sunni-Islamist Turkish regime. Hamas sought financial and diplomatic support from both, as evidenced by Prime Minister Haniyeh's February 2012 visit to Tehran. His visit came at a time when many believed a full realignment was in the works with the Sunni regimes led by Ankara. Turkey and Iran were increasingly at loggerheads over Middle East influence regarding the Sunni-Shiite clash in Iraq, the northeast Saudi Arabian oil producing region, the Arab Persian Gulf states, Yemen and Hezbollah influence in Lebanon. The Syrian civil war pit the Iranian supported pro-Shiite Alawites and other minorities against the majority Arab Sunni community. The Syrian Sunni rebel forces contained Islamist activists supported by Erdogan's Turkey and the Muslim Brotherhood in Egypt, Tunisia, Libya and elsewhere. Hamas attempted to retain both as allies. Haniyeh expressed general support for Iran during his visit while other Hamas officials lined up with their Brotherhood allies. Hamas sought an alliance with both regional powers, but in the end may be forced to choose one side. They can be expected to support their Sunni brethren. However in its conflict with Israel, Hamas simultaneously anticipates support from the Iranian Shiites and Arab Sunni fundamentalists.

Among Gaza's population it is difficult to gauge satisfaction or rejection of Hamas leadership. Free speech does not exist in the Sharia-dominated society. Certain Western analysts, such as Jeroen Gunning, believe Hamas commitment to Islam may be interpreted as a commitment to democracy, since the people want Sharia law. In this sense, "democracy" is in the form of a double contract between the Hamas leadership and its dedication to Sharia law, and the need for the political elite to ensure constant popular support. Free will is to be respected, but obviously is not defined in the Western liberal sense. Sheikh Yasin himself understood all freedoms to emanate from submission to Allah, meaning no secular political representation can be allowed as far as Hamas is concerned, hence secular democracy cannot exist.[94] Gunning concluded Hamas is not anti-democratic, nor anti-West or anti-modern, but rather draws on Western concepts originating with John Locke and integrates them within Islamic law and the people's will. He acknowledged that from a religious standpoint, Hamas cannot compromise, yet pragmatism does exist—the proof being the acceptance of *hudna* with Israel. He admits the contradiction, but takes up a secular, as opposed to Islamic, interpretation of Hamas actions and future policies. He expects socio-economic factors to dominate and lead to democracy.[95] This partially explains support for Hamas if one reviews the need for popular support, yet democracy is nowhere in sight. There is no platform for opposition views or elections. Rather we are seeing tyranny by the Islamist majority.

In addition to the Islamic-secular Arab nationalist split there were large gaps between the rich and poor. Until late 2014 Gaza was not under siege due to the Israeli naval blockade, as many believe. The underground tunnels between Egyptian and Gazan Rafiah served as a passageway for imports deemed necessary to at least maintain material satisfaction among the population at large. Financing for imports was and is obtainable from outside sources, whether from the wealthy Persian Gulf Arab States, Iran and/or Turkey. On the military front, Hamas freely imported weapons and particularly medium-range rockets through the tunnels rendering the Israeli blockade far from effective. Building supplies meant for civil projects were redirected for tunneling activities into Israel's western Negev border region, where soldiers and civilians would be abducted and/or killed in future Hamas operations. The best example is the thousands of tons of cement earmarked for civilian construction projects, which were appropriated for building offensive tunneling instead. Periodically the IDF discovered the tunnels once they came close to penetrating under the Israeli border.

Hamas rules Gaza since 2007. Even in the best of times, from mid-2011 to mid-2013, democratization did not take place. Before being overthrown, even the Islamist Morsi began increasing tunnel blockages on the Sinai side of the border. President a-Sisi initiated massive tunnel destruction by late 2014, and, to ensure full security, the Egyptian army began clearing a 500-meter wide swath of land on the Rafiah border. The objective was to disconnect Hamas from its Muslim Brotherhood allies battling Egyptian government forces in Sinai. By winter 2015 even PA President Abbas supported the tunnel crackdown.[96] In the earlier 2013-14 stage, Hamas felt serious economic and political pressure forcing them into a more prudent position in relation to Fatah. This was exacerbated due to the large economic gap between Gaza and the West Bank.

There were constant talks and attempts at an internal Palestinian rapprochement between Fatah and Hamas since the last parliament took office in 2006. An agreement was signed for the third time in April 2014 calling for a national unity government and elections, which never took place. Abbas encountered problems almost immediately because peace talks with Israel implied recognition of Jewish national existence. Within days of the Palestinian reconciliation there were calls rejecting Fatah's approach. Should Hamas accept Israel's right to exist, they would be absorbed into the secular nationalist Fatah, but with a more Islamist veneer, rendering them hypocritical. The Hamas alternative must include a rejection of Israel, although a *hudna* can be considered for short-term tactical necessities. Hamas cannot agree to a permanent status accord with an entity considered illegal on theological grounds.

Once again the game changed quickly when the Gaza border heated up in June and July 2014, and Israel responded with the Protective Edge operation. The causes were numerous, beginning with Hamas military and economic weakness as a result of the continuing Rafiah tunnel closings, the demand for elections, and the strengthening of the West Bank PA through US, EU and moderate Arab State support. Israel and the PA policing forces continued security cooperation, this translating into repression of Hamas activities on the West Bank. The Israeli naval blockade and Egyptian tunnel closings severely restricted military contraband entering Gaza. With less civilian goods arriving, Hamas suffered from decreased tax revenues as well. A violent Hamas response was calculated to attain increased aid, especially from the Arab and Muslim world.

In June 2014, two Hamas operatives abducted and murdered three Israeli teenagers in the West Bank. The terrorists were eventually killed in a firefight in Hebron three months later. In retaliation Jewish extremists abducted and brutally murdered an East Jerusalem Palestinian teenager. Police apprehended the perpetrators within a month and the three accused murderers stood trial, were found guilty and are awaiting sentencing as of late 2015. The event spurred riots in Jerusalem's Arab neighborhoods, particularly in the north of the city where the teen lived. Tensions were extremely high on both sides.

Hamas rocket fire into the Negev renewed in June, apparently with some connection to the West Bank and Jerusalem events, although certainly in support of Palestinian Arabs living in those regions. At the beginning of July it appeared Israel's Kerem Shalom region adjacent to the southern Gaza Strip was in immediate danger. Rocket attacks intensified, yet the Likud led Netanyahu-Lapid government (2013-15) held back, waiting for the right moment to unleash large-scale aerial bombardments. The air campaign began on July 8, 2014 and a week later Israeli ground forces entered Gaza, first to confront Hamas along the border and then to destroy the tunnels.

During the fifty-day conflict, between 2,100 and 2,200 Palestinians died, and another 10,000 to 11,000 were wounded. As in previous operations Israel was accused of targeting civilians and indiscriminate bombings. According to Palestinian sources there were approximately 1,600 civilian deaths, and only 500 Hamas combatants and other Islamists killed. Israeli army statistics claim 1,170 civilians and 957 Hamas fighters died, including affiliates. The discrepancy comes when counting young men of combat age who were not immediately identifiable as armed militants. According to Palestinian sources, slightly less than 24 percent of those killed were armed Islamists. Using Israeli statistics, almost 45 percent killed were from the Islamist groups, and 55 percent were civilians. Palestinian sources claim the Gaza population to be around 1.8 million. An estimate of Hamas strength and support is said to be about 20,000 or 1.1 percent of the population. Some claim the number of Hamas members to be as high as 40,000, but that number appears greatly exaggerated. Using Palestinian statistics in the former and Israeli stats in the latter, one realizes Hamas and associates took 22 to 41 times greater casualties than they would have had the bombing been at random. Israel pursued pinpoint bombings, with over 5,200 airstrikes, yet casualties remained low due to evacuation warnings given by the military when attacking embedded Hamas positions located in civilian neighborhoods. Warnings greatly limited casualties, resulting in less than half-a person killed and two wounded per air strike.

Israel accused Hamas of again using its own people as human shields and of almost exclusively targeting Israeli civilians, a charge supported by PA President Abbas. Abbas further castigated Hamas for murdering 120 Palestinian youths who were said to have violated curfews, and condemned the Hamas execution of another thirty Palestinians as "collaborators" supposedly aiding Israel.

Hamas fired some 4,500 rockets out of an estimated arsenal of 10,000. Others were destroyed in ammunition depots and an estimated 30 percent remained in the aftermath. These were directed almost exclusively at Israeli civilian targets, but many fell in open areas or were taken out by the Iron Dome missile defense system. During combat Israel discovered that Hamas dug at least 32 cross border tunnels with the intent of abducting Israeli soldiers and civilians and bringing mayhem to the northwestern Negev. Border battles and Israel's incursion into Gaza to destroy these tunnels resulted in 72 killed, over 90 percent of which were IDF soldiers. Both sides suffered population displacement, hundreds of thousands of Gazans left their homes, and thousands of Israelis headed north out of rocket range.[97]

Both sides claimed victory from the 50-day battle, but neither won outright. Israel caused Hamas vast material damage, but politically Hamas enjoyed increased support from the Palestinian public, recovering from those pressured weary days leading up to the conflict. Had elections for president been held at the beginning of October 2014, the Palestinian Center for Policy and Research reported that Hamas leader Haniyeh would have defeated the Fatah PA incumbent President Abbas by 55 to 38 percent in an overall ballot. In Gaza it was 50 to 47 percent, and in the West Bank 57 to 33 percent. A month previous Haniyeh polled an overall 61 percent, indicating there was serious slippage. Hamas retained solid support from the population and could be expected to take legislative elections by similar margins. In a repeat performance paralleling previous Israeli strikes, Hamas lost military and economic strength but gained politically. Mahmoud al-Zahar reiterated the Hamas demand for Israel's destruction by declaring the need to build an Islamic State in "all of Palestine."[98]

In the spring of 2014 Hamas was on the defensive, exercising prudence to assure survival. This did not imply an ideological reversal, but rather a delay in implementation. Before and after Operation Protective Edge, Hamas retention of power in Gaza itself was a matter of how much opposition they faced, how well their opponents were armed, and most importantly, the level of force and the speed in which the Hamas regime would impose its heavy hand on whatever challenges arose. As of mid 2015, Hamas continued enjoying majority support leaving little room for the opposition. Hamas applies the Iranian Ayatollahs' regime template to retain power. The use of firepower against demonstrators commenced in 2007, continues into the present, and is expected to persist in the future. As long as a sizable part of the population is Islamist and the security forces support the regime, Hamas will rule in Gaza. Egypt is hostile but has little impact on Hamas Islamist ideological influence in the Palestinian arena and beyond. The greatest threat to the Hamas regime comes from the fanatics on the Islamist al-Qaeda influenced right, including Islamic State (ISIS/ISIL) types. Knowing Fatah stands no chance of taking power in Gaza, Israel almost paradoxically supports a continuation of a militarily-weakened Hamas rule.

Overall, Hamas held up well despite the Israeli and Egyptian economic squeeze, the internal Palestinian political challenge led by PA President Abbas, and the fanatical military threat of al-Qaeda and Islamic State-type organizations in Gaza. The *dawa*, calling both for social works and Jihad, retains its allure and Hamas survives despite massive difficulties. As of 2016, Hamas continued to advocate the Islamization of the PA, the destruction of Israel and Jews, and supported world Islamic domination. They have not succeeded with any of these goals, but they remain intact within future policy objectives. Hamas declared a *hudna* after Israel's 2014 Protective Edge operation, but none should expect it to last more than a few years. Instead of working on economic development through an alliance with the more moderate Arab nations, we can expect a Hamas re-ignition of the border conflict with Israel in the future.

Endnotes

1. Lapidot, Ruth, "Legal Aspects of the Palestinian Refugee Question," *Jerusalem Center for Public Affairs*, September 1, 2002, retrieved January 10, 2011, www.jcpa.org/jl/vp485.htm.

    It is said that the Sahrawi refugees fleeing the Western Sahara War between Morocco and the Polisario guerrillas (1975-91) enjoy the same status. The number of refugees is disputed and stands between 50,000 – 200,000.

2. "Palestinian Authority Admits: Warfare Was Planned," *Associated Press*, March 4, 2001, retrieved March 17, 2011, arabterrorism.tripod.com/admission.html.

    Reported by the semi-official PA newspaper *Al-Ayyam*, December 6, 2000.

3. Ross, Dennis, *The Missing Peace*, Farrar, Straus and Giroux, New York, 2004, Chapter 25, "The Denouement – From Camp David to the Intifada to the Clinton Ideas," pp. 712-758 and Appendix "Clinton Parameters," December 23, 2000, pp. 809-813.

4. For a concise review of these events see Enderlin, Charles, *Shattered Dreams*, Other Press, New York, 2002, Chapters "Checkmate" and "Chain Reactions," pp. 263-361.

    Rubin, Barry and Rubin, Judith Colp, Yasir Arafat, Oxford University Press, 2005, Chapter "The Moment of Truth 2000," pp. 185-217.

    Also, the overall story of Middle East peace-making attempts by Dennis Ross in *The Missing Peace*. Enderlin apportions blame between the two sides although leans pro-Palestinian. The Rubins and Ross hold Arafat responsible for the breakdown in the peace process.

5. Peres served previously as prime minister from Sept. 1984 until October 1986 and again from November 1995 to June 1996 after Yitzhak Rabin was assassinated. He also served in numerous cabinet positions beforehand and afterwards, including at defense and the foreign ministry.

6. Rubin and Rubin, pp. 185-217.

    Ross, pp. 753-758

    Both review Arafat's refusal to make peace. At Taba nothing changed. Neither Arafat nor Barak were present.

7. Yousef, Mosab Hassan, *Son of Hamas*, Tyndale House Publishers, Inc., 2010, pp. 135-145.

8. Rubin and Rubin, pp. 218-221.

9. Ne'eman, Yisrael, *Mideast on Target*, "Arafat and Netanyahu Face the Sharon Evacuation Plan," February 3, 2004, http://me-ontarget.org/pws/page!5978. Website is the author's blog, hence there is no retrieval date.

10. Enderlin, p. 290.

11 Morris, Benny, *Righteous Victims*, Vintage Books, New York, 2001, p. 666.
12 Yousef, p. 127.
13 Ibid.
14 Ibid, p. 132.
15 Hanus, George D., *The Compendium*, Gravitas Media, Chicago, USA, 2002. Statistics from terror attacks taken from pp. 14-35. Usage permission thru telecom granted Nov. 25, 2015.
16 "Terrorist Bombing at Hebrew University Cafeteria," *Israel Ministry of Foreign Affairs*, retrieved March 18, 2011, http://mfa.gov.il/MFA/MFA-Archive/2002/Pages/Terrorist%20bombing%20at%20Hebrew%20University%20cafeteria%20-.aspx.
17 "Karine A Affair," *Wikipedia*, retrieved March 18, 2011, en.wikipedia.org/wiki/Karine_A_Affair.
    "Karine A," *The Washington Institute for Near East Policy*, retrieved March 18, 2011, www.washingtoninstitute.org/templateC05.php?CID=1471.
18 "Church of the Nativity," *Palestine Facts*, retrieved March 18, 2011, www.palestinefacts.org/pf_1991to_now_church_nativity_2002.
19 Bush, George W., "Rose Garden Speech on Israel Palestine Two-State Solution," June 24, 2002, retrieved February 23, 2012, http://www.americanrhetoric.com/speeches/gwbushtwostatesolution.htm.
20 Yousef, pp. 211-213.
21 Ibid, pp. 217-221.
22 "President Discusses Roadmap for Peace in the Middle East," March 14, 2003, retrieved December 28, 2015, http://georgewbush-whitehouse.archives.gov/news/releases/2003/03/20030314-4.html.
23 Tamimi, Azzam, *Hamas: A History from Within*, Olive Branch Press, Northampton, MA, USA pp. 202-203.
    Ne'eman, Yisrael, *Mideast on Target*, "Israel Forces a Hudna," June 24, 2003, www.me-ontarget.org/pws/page!13819.
    Ibid, "Dismantling the Terror Infrastructure – the Only Issue," July 28, 2003, www.me-ontarget.org/pws/page!13827.
    Ibid, "Mideast Cycle of Violence is Linear," August 22, 2003, www.me-ontarget.org/pws/page!13830.
24 Ibid, "Hudna Cease-Fire or Disarmament?" May 11, 2003, www.me-ontarget.org/pws/page!13805.
    Ibid, "Hudna Lunacy," June 23, 2003, www.me-ontarget.org/pws/page!13818.
    Ibid, "Israel Forces a "Hudna," June 24, 2003, www.me-ontarget.org/pws/page!13819.
    Ibid, "The Islamic Movement Considers a Hudna," July 1, 2003, www.me-ontarget.org/pws/page!13822.
    Ibid, "Dismantling the Terror Infrastructure - The Only Issue," July 28, 2003, www.me-ontarget.org/pws/page!13827.
    Ibid, "Hudna Collapse," August 9, 2003, www.me-ontarget.org/pws/page!13829, and "Mideast Cycle of Violence is Linear," August 22, 2003, www.me-ontarget.org/pws/page!13830.
25 Tamimi, pp. 206-207.
    Ne'eman, "Removing Sheikh Yasin," March 22, 2004, http://me-ontarget.org/pws/page!6021.
26 Ibid, "The Schadeh Elimination," July 25, 2002, www.me-ontarget.org/pws/page!13683.
    Ibid, "After Schadeh: The Morality Debate," July 26, 2002, www.me-ontarget.org/pws/page!13684.
    In these blogs Shehadeh's name was spelled as "Schadeh." The author is referring to Shehadeh, the Hamas operative.

V Hamas Ideological Victory 2000 to 2016

27 al-Qaradawi, Yusuf, "The Prophet Mohammed as a *Jihad* Model" and "Those Who Oppose Martydom Operations and claim that they are Suicide are Making a Great Mistake," in Bostom, Andrew, ed., *The Legacy of Jihad*, Prometheus Books, Amherst, New York, USA, 2005, pp. 248 and 249 respectively.

28 The numbers are well known and numerous sources - Palestinian, Israeli or international, agree on the figures. This author takes issue with the date given determining the end of the LIC and believes the Palestinians faced total exhaustion by mid-2004, essentially ending this stage of the conflict. Too many emphasize the relevancy of Arafat's death in November or the election of Mahmoud Abbas to the Palestinian presidency in early 2005 as indicating this LIC was finished.

29 Tamimi, 204-207.
"Exchange of letters between PM Sharon and President Bush," *Israel Ministry of Foreign Affairs,* April 14, 2004, retrieved February 23, 2012, http://mfa.gov.il/MFA/ForeignPolicy/Peace/MFADocuments/Pages/Exchange%20of%20letters%20Sharon-Bush%2014-Apr-2004.aspx.
Ne'eman, "Likud Referendum," May 2, 2004, http://me-ontarget.org/pws/page!6010.

30 Tamimi, pp. 209-222.
"Palestinian Legislative Elections 2006," *Wikipedia*, retrieved March 18, 2011, en.wikipedia.org/wiki/Palestinian_legislative_election,_2006.
Ne'eman, "Hamas Victory: Following the Trend," January 28, 2006, http://me-ontarget.org/pws/page!5095.

31 Tamimi, pp. 225-226.

32 "Document: The Arab Peace Initiative 2002 Al-Bab," retrieved July 20, 2011, www.al-bab.com/arab/docs/league/peace02.htm.

33 Tamimi, pp. 229-234 and 252-254.

34 Ibid, pp. 257-261.

35 Eldar, Shlomi, *Getting to Know Hamas*, Keter Publishers (Hebrew), 2012, pp. 49-50.

36 Ibid, pp. 155-160 and 173-174.

37 Tamimi, pp. 260-261.

38 Jensen, Michael Irving, *The Political Ideology of Hamas: A Grass Roots Perspective*, translated from Danish by Sally Laird, I.B. Tauris, London and New York, 2009, pp. 34-40.

39 "ISRAEL SIEZED WEAPONS SHIP 'VICTORIA'," *Global Jihad*, March 22, 2011, retrieved December 28, 2015, www.globaljihad.net/?p=4469.

40 Stern, Yoav and Issacharoff, Avi, "Egypt: Major Hezbollah Attack in Sinai Thwarted," *Haaretz*, April 10, 2009, retrieved February 24, 2012, www.haaretz.com/egypt-major-hezbollah-attack-in-sinai-thwarted-1.273842.

41 Chehab, Zaki, *Inside Hamas*, Nation Books, New York, N.Y., 2007, pp. 129-130.

42 Ibid, pp. 131-134.

43 Ibid, pp. 135-136.

44 Ibid, pp. 140-151.

45 Many considered this to be a very shortsighted Israeli policy which eventually backfired.

46 Eldar pp. 245-246.

47 Ibid, pp. 187-188.
"Fatah-Hamas Conflict," *Wikipedia*, retrieved July 20, 2011, en.wikipedia.org/wiki/Fatah–Hamas_conflict.
"Battle of Gaza (2007)," *Wikipedia*, retrieved July 20, 2011, en.wikipedia.org/wiki/Battle_of_Gaza_2007).
Ne'eman, "Palestinian Islamic Revolution – On the March," June 15, 2007, http://me-ontarget.org/pws/page!4962.

48 Eldar, pp. 196-197.

49 Ne'eman, "A Time of Reckoning," July 14, 2006, http://me-ontarget.org/pws/page!5049.
Ibid, "Seeking Total Victory" July 16, 2006, http://me-ontarget.org/pws/page!5048.
Ibid, "Cross-Cultural Misunderstandings," July 28, 2006, http://me-ontarget.org/pws/page!5036.
Ibid, "Hezbollah-Israel War: Comments," August 4, 2006, http://me-ontarget.org/pws/page!5033.
Ibid, "Hezbollah Victory," September 3, 2006, http://me-ontarget.org/pws/page!5011.
50 Eldar, pp. 218-230.
51 Wage, John, "Hamas Leaders Enforce Sharia Law in Gaza Strip," *CBN News*, October 12, 2009, retrieved August 16, 2011, http://www.cbn.com/cbnnews/world/2009/august/hamas-leaders-enforce-sharia-law-in-gaza-strip/?mobile=false.
"Hamas Parliament Votes for Sharia in Gaza," *Jihad Watch*, December 24, 2008, retrieved August 16, 2011, http://www.jihadwatch.org/2008/12/hamas-parliament-votes-for-sharia-in-gaza.
"Islamization of the Gaza Strip," *Wikipedia*, retrieved August 16, 2011, en.wikipedia.org/wiki/Islamization_of_the_Gaza_Strip.
52 Chehab, pp. 202-205.
53 Eldar p. 143.
54 For a sweeping perspective by an insider of Yasir Arafat's financial misdeeds see Uzrad Lew's work (in Hebrew), *Inside Arafat's Pocket*, Kinneret, Zmora-Bitan, Dvir, Israel, 2005.
55 Avishai, Bernard, "The Israel Peace Plan That Could Still Be," *The New York Times*, February 7, 2011, retrieved August 31, 2011, www.nytimes.com/2011/02/13/magazine/13Israel-t.html?pagewanted=all.
56 "World Bank: Real GDP Growth in West Bank and Gaza During 2008 - 2%," June 4, 2009, *Aldawaba*, retrieved August 31, 2011, http://www.albawaba.com/business/world-bank-real-gdp-growth-west-bank-and-gaza-during-2008-2.
57 "West Bank" and "Gaza Strip," *CIA World Factbook 2011*.
"Valley of Peace Initiative," *Wikipedia*, retrieved February 29, 2012, en.wikipedia.org/wiki/Valley_of_Peace_initiative.
For example, in 2010 West Bank unemployment was 16.5% and Gaza's 40%.
58 "Ending the Occupation, Establishing the State," August 26, 2009, retrieved March 1, 2012, www.mideastweb.org/palestine_state_program.htm.
Program by Palestinian PM Salam Fayyad to declare Palestinian independence by 2011. As is known, the Palestinian State initiative did not succeed in October 2011 at the UN. However the plan was presented and Fayyad began building infrastructure for the state entity. The Palestinian State received non-member observer state status at the UN on Nov. 29, 2012.
59 Ne'eman, "Hamas Dilemma," November 25, 2007, http://me-ontarget.org/pws/page!4945.
Ibid, "Gaza Humanitarian Issues in the Service of Islamists," January 29, 2008, http://me-ontarget.org/pws/page!4455.
60 Ibid, "Closing in on Hamas?" March 5, 2008, http://me-ontarget.org/pws/page!4449.
61 Ibid, "Coexistence or the Erez Crossroads as an Enemy," March 8, 2004, http://me-ontarget.org/pws/page!5984.
Ibid, "Karni Attack Directed at Palestinians, Not at Israel," January 15, 2005, http://me-ontarget.org/pws/page!5854.
Ibid, "Increasing Palestinian Suffering for Islamist Gains," April 10, 2008, http://me-ontarget.org/pws/page!4447.
"Main Terrorist Attacks Carried Out at Gaza Strip Crossings," *Israel Foreign*

Ministry, retrieved August 30, 2011, http://mfa.gov.il/MFA/ForeignPolicy/Terrorism/Palestinian/Pages/Main%20terrorist%20attacks%20carried%20out%20at%20Gaza%20Strip%20crossings%2016-Jan-2005.aspx

62 Wage, "Hamas Leaders Enforce Sharia" and "Hamas Parliament Votes for Sharia," December 24, 2008, *Jihad Watch*, retrieved August 30, 2011, www.jihadwatch.org/2008/12/hamas-parliament-votes-for-sharia-in-gaza.

63 Levitt, Matthew, *Hamas: Politics, Charity and Terrorism in the Service of Jihad*, Yale University Press, USA, 2007, pp. 17-18.

64 Ibid, pp. 107-108.

65 Ibid, direct quote, p. 110.

66 Ibid, p. 111-142.

67 Ibid, pp. 229-249.

68 "Gaza War," *Wikipedia*, retrieved August 30, 2011, en.wikipedia.org/wiki/Gaza_War.

69 Ibid.
Chodoff, Elliot, *Mideast on Target*, "Lessons of War," January 5, 2009, http://me-ontarget.org/pws/page!1605.
Ne'eman, "Egypt vs. Hamas (and the Moslem Brotherhood)," January 8, 2009, http://me-ontarget.org/pws/page!1608.
Ibid, "Hamas Perspectives and Options," January 10, 2009, http://me-ontarget.org/pws/page!1609.
Ibid, "Phase 3: Israel"s Military and Diplomatic Options," January 14, 2009, http://me-ontarget.org/pws/page!1611.
Ibid, "Demanded: End Game Scenario," January 16, 2009, http://me-ontarget.org/pws/page!1612.

70 "Casualties of the Gaza War," *Wikipedia*, retrieved February 29, 2012, en.wikipedia.org/wiki/Casualties_of_the_Gaza_War.

71 "Gaza War," *Wikipedia*.
Ne'eman, "End Game Scenario."
Ibid, "Hamas War and Cease-Fire" January 19, 2009, http://me-ontarget.org/pws/page!1613.

72 "Report of the United Nations Fact Finding Mission on Gaza Conflict," *United Nations Human Rights Office of the High Commissioner*, retrieved December 28, 2015, http://search.ohchr.org/results.aspx?k=report%20of%20the%20United%20Nations%20Fact-finding%20mission%20on%20the%20gaza%20conflict.
"United Nations Fact Finding Mission on the Gaza Conflict,"*Wikipedia*, retrieved December 28, 2015, https://en.wikipedia.org/wiki/United_Nations_Fact_Finding_Mission_on_the_Gaza_Conflict.
The document is commonly known as the "Goldstone Report."

73 Goldstone, Richard, "Reconsidering the Goldstone Report on Israel and War Crimes," *Washington Post*, April 1, 2011, retrieved February 29, 2012, www.washingtonpost.com/...goldstone.../AFg111JC_story.html.

74 Ne'eman, "Hamas War and Cease-Fire."
Ibid, "The EU Contains Obama's Initiative?" January 21, 2009, http://me-ontarget.org/pws/page!1614.
Ibid, "Anti-Semitism and the Liberal/Extreme," February 1, 2009, http://me-ontarget.org/pws/page!1617.

75 "President Obama's Cairo Speech Text," *USA Today*, June 4, 2009.

76 "Address by PM Netanyahu at Bar-Ilan University," text, June 14, 2009, *Israel Ministry of Foreign Affairs*, retrieved February 29, 2012, http://mfa.gov.il/MFA/PressRoom/2009/Pages/Address_PM_Netanyahu_Bar-Ilan_University_14-Jun-2009.aspx

77 Ne'eman, "Obama's Emerging Permanent Status Agreement," July 29, 2009, http://me-ontarget.org/pws/page!2317.

78 Akram, Fares, "At a Rally for Hamas, Celebration and Vows," *The New York Times*, December 14, 2011, retrieved January 20, 2016 from www.nytimes.com/.../on-anniversary-hamas-repeats-vows-on-israel-and-violence Speech in Gaza by Hamas PM Ismail Haniyeh, December 14, 2011, excerpts in *MEMRI*, retrieved March 1, 2012, www.memritv.org/clip/en/3247.htm.

79 "Islamization of the Gaza Strip," *Wikipedia*, retrieved January 12, 2012, en.wikipedia.org/wiki/Islamization_of_the_Gaza_Strip.

80 "Report of the Secretary-General's Panel of Inquiry of the 31 May Flotilla Incident," September 2011, retrieved January 12, 2012, http://www.un.org/News/dh/infocus/middle_east/Gaza_Flotilla_Panel_Report.pdf.

81 Ne'eman, "The Flotilla: Turkish Move to Lead the Muslim World," June 4, 2010, http://me-ontarget.org/pws/page!6111.
Ibid, "The New Ottomanism Taps the Palestinian Venue," September 6, 2011, http://me-ontarget.org/pws/page!8516.

82 "Gaza Strip Economy 2011," from the CIA World Factbook 2011, *Theodora*, retrieved January 31, 2012, http://www.theodora.com/wfb2011/gaza_strip/gaza_strip_economy.html.

83 Fayyad, "Ending the Occupation, Establishing the State."

84 Toameh, Khaled Abu and Keinon, Herb, "Fatah official says two-state solution is over," October 12, 2010, *Jerusalem Post*, retrieved Aug. 18, 2015, www.jpost.com/Israel/Fatah-official-says-two-state-solution-is-over.
   This is a continuing trend over the past five years, particularly as concerns the PA education and Fatah youth movement. The Greater Land of Israel settler movement, led by Jewish national religious ideologues and completely opposed to any compromise with the Palestinians, was possibly an unintended beneficiary of the hardening Fatah position and increasing Hamas influence. Hardliners on both sides reject the two-state solution.

85 "Joint Israeli-Palestinian poll – The Harry S. Truman Institute for the Advancement of Peace," March 17, 2010, retrieved January 31, 2012, http://www.geneva-accord.org/mainmenu/new-joint-poll-december-2011-support-for-the-geneva-initiative-significantly-increased.

86 Eldar pp. 236-241.

87 Ibid, p. 239.

88 "Hamas PM Ismail Haniyeh - We Are a Nation of Jihad and Martyrdom," *MEMRI*, November 15, 2010, retrieved August 20, 2015, www.youtube.com/watch?v=xRJc3NsFxF0.

89 Zahran, Mudar, "Is Jordan's King Losing Control over the Bedouin?" *Gatestone Institute International Policy Council*, June 20, 2011, retrieved August 20, 2015, www.gatestoneinstitute.org/2209/jordan-bedouin.

90 "Operation Pillar of Defense," *Wikipedia*, retrieved January 6, 2014, en.wikipedia.org/wiki/Operation_Pillar_of_Defense.

91 "General Assembly Votes Overwhelmingly to Accord Palestine "Non-Member State Observer State Status in United Nations," retrieved January 6, 2014, www.un.org/News/Press/docs/2012/ga11317.doc.htm.

92 al-Mughrabi, Nidal, "Egypt tunnel blockade takes toll on Gaza business," *Reuters*, December 9, 2013, retrieved January 7, 2014, http://www.reuters.com/article/palestinians-gaza-business-idUSL5N0JK1UZ20131209.

93 Yehezkeli, Zvi, *London and Kirshenbaum*, Israel Channel 10 Television, December 9, 2013.

94 Gunning, Jeroen, *Hamas in Politics: Democracy, Religion, Violence*, Hurst & Company, London, 2007, pp. 53-75.

95 Ibid, Chapter "Conclusion," pp. 263-274.

96 "Egypt Floods Gaza's Smuggling Tunnels," *Al-Monitor: The Pulse of the Middle East*,

February 19, 2003, http://www.al-monitor.com/pulse/originals/2013/02/egypt-floods-gaza-tunnels.html.

"Egypt's military says it destroyed 1370 Gaza smuggling tunnels," *The National World*, March 12, 2014, http://www.thenational.ae/world/palestinian-territories/egyptian-military-says-it-destroyed-1-370-gaza-smuggling-tunnels.

"Abbas backs Egyptian crackdown on Gaza tunnels," *Al-Arabiya*, December 12, 2014, http://english.alarabiya.net/en/News/middle-east/2014/12/12/Abbas-backs-Egypt-crackdown-on-Gaza-tunnels-.html.

All articles retrieved December 28, 2015.

97 "2014 Israel-Gaza Conflict," *Wikipedia*, retrieved October 6, 2014, https://en.wikipedia.org/wiki/2014_Israel–Gaza_conflict.

Ne'eman, "Some Gaza Conclusions," August 25, 2014, http://me-ontarget.org/pws/page!12724.

98 "Hamas: Ruling West Bank it could destroy Israel with speed that no one can imagine," *Palestine Media Watch*, e-mail release October 5, 2014, http://www.palwatch.org/pages/aboutus.aspx.

Gaza Strip

Ruled by Hamas since 2007

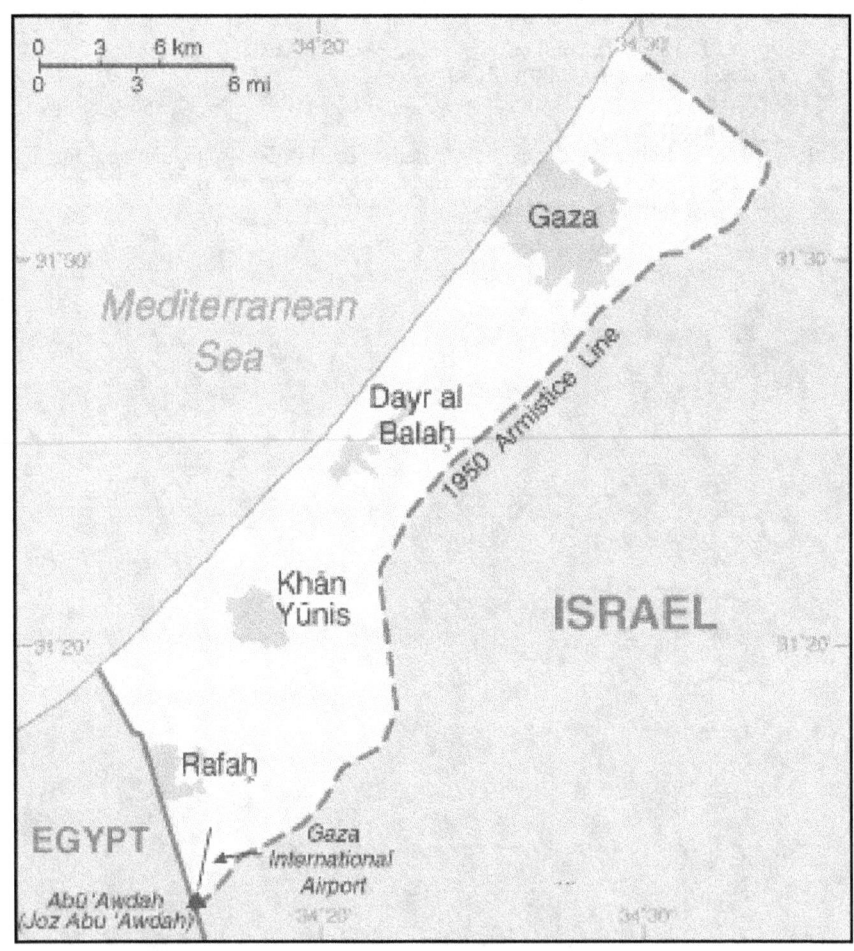

Credit: CIA World Factbook

VI

The Hamas Covenant Analysis

Islamism, Jihad, and Antisemitism

Overview

The purpose of this review and analysis of *The Hamas Covenant* is to expose its true sources of inspiration, in particular the anti-Jewish attitudes in the Koran and the Hadith. Beyond using certain quotes from the Koran, whether in context or not, *The Hamas Covenant* is largely dependent on the infamous and viciously antisemitic Czarist forgery *The Protocols of the Learned Elders of Zion* and its nefarious inspired text, Hitler's *Mein Kampf*, in particular Chapter XI, entitled "Nation and Race." A more in depth discussion of the Czarist-Nazi connection to Jihadi Islam is presented in Chapter VIII "Czarist-Nazi Integration." *The Hamas Covenant* demands a spirit of anti-Jewish behavior from all Muslims; action is sought against Christians and people of the Far East as well. Its modern Islamist influences are directly attributable to Sheikh Hasan al-Banna, the ideological father and founder of the Muslim Brotherhood, Sayyid Qutb, the most influential of the Islamist thinkers, and the Palestinian Jihadist Abdullah Azzam. Other Muslim Brotherhood ideologues contributed to the compilation of this Islamist and Jihadi statement of identity and intent, particularly those men present when assembling the *Covenant* in 1988.

What follows is not a critique of Islam, but a critical investigation of a text with wide-ranging ramifications felt at the end of the twentieth century and into the twenty-first. Although at one time seen as marginal, the spirit of *The Hamas Covenant* and Muslim Brotherhood continue to make global inroads as Islamic extremism sweeps through Pakistan, Afghanistan, Iran (although Shiite in belief), Yemen, Egypt, Syria, Iraq, Libya, Nigeria, the Horn of Africa, Turkey and among Muslim communities in Europe.

Let us recall that Hamas (the Islamic Resistance Movement or IRM) is only the Palestinian wing or faction of the Sunni Muslim Brotherhood, an organization advocating world domination as made clear in the *Covenant*. Lest we forget, Osama bin Laden's Al Qaeda is an extreme extension of Muslim Brotherhood thinking, which draws its strength from the more radical supporters of Islamic fundamentalism demanding no temporary compromises, but immediate action. As of mid-2014, the fanatical Jihadist Islamic State,

ISIS or ISIL (Islamic State of Iraq and Syria or Islamic State of Iraq and the Levant), began developing as a semi-state "Islamic Caliphate" entity in eastern Syria and western Iraq. We will refer to this group as the Islamic State.

This *Covenant* is the Hamas understanding of Islam, Jihad, and the need for world domination. Hamas does not represent the thinking of all Muslims, but it does represent a certain faction of that community of 1.5 billion believers. What exact percentage supports the ideals expressed in *The Hamas Covenant* is impossible to determine, but no doubt it is not a marginalized group. Over the years, estimates of support for Jihadi Islam of the Hamas type range between 10 to 15 percent of the world Muslim population, or some 200 million supporters in total. Others claim the number of adherents to be much higher, closer to one-third or half a billion Muslims. The Jewish population worldwide is approximately 13 million, meaning Jihadist support for the ideas expressed in the *Covenant* is 15 to 38 times the entire number of Jews in the world today.

As will be seen below, Hamas claims it is battling foreign influences in Islam. Their claim is only partially true. Hamas and the Muslim Brotherhood are solidly anti-Enlightenment, anti-egalitarian, anti-universal human rights, anti-secular Arab nationalist, as well as against all forms of religious equality, cultural expression and personal freedoms, unless they are in line with Hamas' form of Islam. Hence, Hamas condemns Western thinking, most notably the principle of equality.

Still, the Czarist (Russian) *Protocols* and fascism (in particular German National Socialism—or Nazism) clearly influenced Hamas. The *Covenant* is a throwback to the periods of extreme intolerance in Islam, and is now augmented ideologically with the two most reactionary regimes of recent times. One major historic common denominator between extremist Islamic regimes, Czarist rule and Nazi Germany, is their fanatical antisemitism. Previous Islamic regimes and Czarist rule oppressed Jews unbearably, yet Jews were still allowed to survive. The Nazis went a step further; achieving near-total Jewish extermination in areas under their control during WWII.

A careful reading of *The Hamas Covenant* places it between the oppressive Islamic regimes and the Russian Czars on one side, and the Nazis on the other, with a preference for the Nazi solution to "the Jewish problem." For Jews in particular, understanding *The Hamas Covenant* in this context is of paramount significance. The *Covenant* is the harbinger of "evil times" for Jews, as Czarism and Nazism were previously. Antisemitism is inherent in the total world order envisioned by the Muslim Brotherhood adherents of Hamas. We can call this new round of Jew hatred—Islamo-antisemitism.

Because of the dangerous reality we face in today's world, a greater comprehension is necessary to understand Hamas priorities and guidelines. *The Hamas Covenant* is rendered in full below in bold text. Commentary of each section follows, with bold italicized quotes from the Koran and Hadith when quoted within the context of *The Hamas Covenant*.

The Covenant of the Islamic Resistance Movement—Hamas[1]
[Also known as "The Charter of Allah"]

In the name of Allah the Merciful and the Compassionate Palestine, Muharram 1, 1409 A.H./August 18, 1988

In the name of Allah the Merciful and the Compassionate

***"You are the best nation that has been brought out for mankind. You command good and forbid evil and believe in Allah. If only the people of the Book [i.e., Jews and Christians] had believed, it would have been well for them. Some of them believe, but most of them are iniquitous. They will never be able to do you serious harm, they will only be an annoyance. If they fight you, they will turn their backs and flee, and will not be succored. Humiliation is their lot wherever they may be, except where they are saved from it by a bond with Allah or by a bond with men. They incurred upon themselves Allah's wrath, and wretchedness is their lot, because they denied Allah's signs and wrongfully killed the prophets, and because they disobeyed and transgressed"* (Koran 3:110-112).**

The "Introduction" to *The Hamas Covenant* makes the fundamental understanding of Muslim superiority over the entire world abundantly clear. Muslims are superior and the "best nation." Notice the included Koran verse 3:110–112 seemingly speaks to Muslims; however, it is also a statement against non-Arab Muslims. We can understand its interpretation in the full Arab Muslim Brotherhood light when reading the tract "Between Yesterday and Today," by Hamas spiritual mentor Sheikh Hasan al-Banna. In it, he unequivocally speaks of the spread of Islam by saying, "The transfer of authority to non-Arabs – i.e., Persians at one time, [Shiite Iranians, 945-1055] at another, Mamlukes [Egyptian slave dynasty, 1250-1517], Turks, and others – who had never absorbed genuine Islam, and whose hearts had never been illuminated with the light of the Koran because of the difficulty they encountered in trying to grasp its concepts, for all that they [the Arabs] read the words of Allah (Blessed and Almighty is He!)."[2] Al-Banna himself then quotes Koran 3:118 immediately after making the above statement.

"O ye who believe! Do not take as confidants those who are not of you; they will not fail to cast disorder among you; they are pleased by what troubles you. Hatred has been revealed out of their mouths; what their hearts conceal is yet greater. We have made the signs clear to you, if you would but understand" (Koran 3:118).

This attempt at Arab superiority through use of Koranic quotes is deceptive. In full context, the preceding verses, Koran 3:115–117, speak of relationships with the People of the Book—Jews and Christians—and unbelievers. The Koranic quote used is the conclusion that Muslims should only trust each other concerning belief, yet al-Banna turns it into an affirmation of Arab superiority over all others. Those Arab Islamists reading and acting by *The Hamas Covenant* in tandem with al-Banna's writings are absorbing two distinct yet integrated messages: First, Islam is the ultimate belief system and superior to all. Second, the Arab peoples are the only fully loyal messengers to proselytize Allah's revealed religion of Islam since all the others, "had never absorbed genuine Islam" and therefore are less than trustworthy. The Arab understanding of Islam is intrinsic; other people can never grasp the full meaning of the Koran. Conclusion: Most Jews and Christians are evil and certainly inferior to Muslims. However on the ethnic level there is a difference between Muslims of different backgrounds, at least according to the followers of al-Banna. They believe the Arabs are superior both spiritually and as a people. Such beliefs are racist.

We are already aware of the usual Islamic condemnation of Jews and Christians for their refusal to believe in Islam. The People of the Book are legally *dhimmi,* or protected second-class people, provided they swear full loyalty to their Islamic overlords. Most are "iniquitous" and "humiliation is their lot" along with receiving "Allah's wrath" and "wretchedness" because of their transgressions. And of course they are cowards in battle as "they will turn their backs and flee." In particular these beliefs set up the great contradiction of Israeli (Jewish) battlefield victories over the Arab Muslims. The need arises for the demonization of the Jews; otherwise, how can anyone explain the repeated Israeli "*dhimmi*" victories for close to seventy years?

> "Israel will exist, and will continue to exist, until Islam abolishes it, as it abolished that which was before it." [From the words of] The martyr, Imam Hasan al-Banna', Allah's mercy be upon him. [2]

Here we see the outright demand for the destruction of Israel as part of a worldwide Jihad. It should be understood that this statement refers not only to Israel the State entity, which was in existence for only nine months upon al-Banna's death, but the People of Israel or the Jews. In Arabic the terms "Jews" and "Israel" are often interchangeable (see al-Banna's understanding of Jihad in Chapter II "Ideologues").

> "The Islamic world is burning, and each and every one of us must pour water, even if it be a little, to extinguish whatever he can extinguish, without waiting for others." [From the words of] Sheikh Amjad Al-Zahawi, Allah's mercy be upon him.[3]

This quote, which ends the Introduction to *The Hamas Covenant (HC)*, paraphrases the Koran verse 5:64 in *HC* Article 32 (see below). Later, it is used in condemning the peace accords with Egypt and demanding Israel's destruction.

> **In the Name of Allah the Merciful and the Compassionate**
>
> **Preamble**
>
> Praise be to Allah. We seek help from Him, we ask forgiveness from Him, we ask Him for guidance, and we rely on Him. Prayer and peace be upon Allah's messenger and upon his family and companions, and those who are loyal to him and spread his message and follow his sunna [the Prophet's custom]. Prayer and peace be forever upon them as long as heaven and earth exist.
>
> Oh people, from the midst of great troubles and in the depths of suffering, and from the beating of believing hearts and arms purified for worship, out of cognizance of duty

and in response to Allah's command - thence came the call [of our movement] and the meeting and joining [of forces], and thence came education in accordance with Allah's way and a resolute will to carry out [the movement's] role in life, overcoming all of the obstacles and surmounting the difficulties of the journey. Thence came also continuous preparation, [along with] readiness to sacrifice one's life and all that is valuable for the sake of Allah.

Then the seed took form and [the movement] began to move forward through this stormy sea of wishes and hopes, yearnings and aspirations, dangers and obstacles, pains and challenges, both locally [in Palestine] and abroad.

When the idea ripened, and the seed grew, and the plant shot its roots into the ground of reality, away from fleeting emotions and improper hastiness, then the Islamic Resistance Movement [Hamas] [4] set out to play its role, marching onward for the sake of Allah. [In doing this, Hamas] joins arms with all those who wage jihad for the liberation of Palestine. [5] The souls of its jihad fighters meet the souls of all those jihad fighters who sacrificed their lives for the land of Palestine, from the time when the Prophet's companions conquered it until the present.

The covenant of the Islamic Resistance Movement (Hamas) reveals its face, presents its identity, clarifies its stand, makes clear its aspiration, discusses its hopes, and calls out to help it and support it and to join its ranks, because <u>our fight with the Jews</u> [underline added for emphasis - YN] is very extensive and very grave, and it requires all the sincere efforts. It is a step that must be followed by further steps; it is a brigade that must be reinforced by brigades upon brigades from this vast Islamic world, until the enemies are defeated and Allah's victory is revealed.

In the Preamble of *The Hamas Covenant* is the call for Jihad; the battle against the enemy identified as "the Jews." The war is "extensive and very grave" and must include "this vast Islamic world until the enemies are defeated and Allah's victory is revealed." Although Hamas is a Palestinian organization, its battle is not confined to Israel as the secular PLO/Fatah claims, but rather it is a universal struggle for the defeat of the Jews. It is in the Preamble we first see the all-encompassing hatred for the Jewish People, a common denominator with the *Protocols of the Elders of Zion* (mentioned in Article 32) and Hitler's *Mein Kampf.*

There are many who maintain *The Hamas Covenant* to be little more than an internal liberation ideology circular, not a serious policy. The lack of differentiation between "Jews" and "Israelis" is said to be mere semantics. Islamists define Israelis as part of an illegitimate nation state, and the true enemy, which is accused of instigating wars to capture lands in the Muslim Middle East. Sheikh Ahmed Yasin, the leading light in the establishment of Hamas, uses the words "Jew" and "Israeli" interchangeably, as do many Muslims. The implication is that Yasin was not antisemitic because when he says "Jew" he means only those living in Israel. Such a contention is completely false as we will see when reading through and analyzing the *Covenant.* More so, Yasin and the Hamas leadership are fully responsible for this document and understand the scope of Jewish-Muslim relations in its complete theological/historical light. Apologists claim *The Hamas Covenant* will either be canceled or changed,[3] yet this event has not taken place, nor should one expect such a decision in the future. *The Hamas Covenant* Preamble accurately reflects Divinely inspired Islamist thinking, which is both Jihadist and antisemitic.

> *This is how we see them coming on the horizon: "And after a time you will come to know about it"* **(Koran 38:88).**
>
> *"Allah has written: It is I and My messengers who will surely prevail. Allah is Strong and Mighty"* **(Koran 58:21).**
>
> *"Say: This is my way. I call on Allah with certainty, I and those who follow me, and glory be to Allah, I am not among the polytheists [idolaters]"* **(Koran 12:108).**

The first two quotes above from the Koran do not mention the Jews, but the third one (Koran 12:108) implies that the Jews are "polytheists." Jews were accused both in ancient Arabia, and in modern times, of aligning themselves with the polytheists against Islam and the word of Allah. Nowadays polytheism is defined as democracy, pluralism, liberalism, socialism, communism, capitalism and the nation state. Jews are accused of promoting these modern ideals with the intent of undermining and destroying Islam. This original accusation can be traced back to the Koran 5:82 (see Chapter II "Ideologues").

Traditionally, Jews are considered one of the Peoples of the Book and are allowed to survive under Islamic sovereignty. Yet, already in the *HC* Preamble Jews are downgraded to a questionable existence since polytheism ("*shirk*" in Arabic) is apostasy and cannot be forgiven. Islamic scholars differ as to whether it is punishable by death. Polytheistic or idolatrous religions have no right to exist once they are informed and reject the virtues of Islam. "When the sacred months are over slay the idolaters wherever you find them" (Koran 9:5). For a fuller text including conversion to Islam see Sura or Chapter "Repentance" found in Koran 9:3-7.

The step taken here in the name of eradicating "polytheism" is no longer traditionally Islamic, but is more similar to the Nazism of Hitler's Germany. The Nazis accused the Jews of the same secular polytheisms as mentioned in the Islamic context above. Under the Nuremberg Laws (September 1935) Jews first lost their rights when they were stripped of German citizenship and reduced to the status of "subjects." Thirteen other laws followed, making Jewish existence illegal.[4] The "Final Solution" was adopted for full implementation at the Wannsee Conference (January 1942), making the crime of Jewish existence punishable by death.[5]

Since *The Hamas Covenant* is not fully implemented due to Israeli sovereignty and the ability to fend off Jihadi attacks, it is still in the theoretical stage, untested. Hence we do not know for certain if Hamas jurists would demand capital punishment for Jews based on the charge of polytheism. It would not be unthinkable for Jihadis to insist on the death penalty (see Chapter II "Ideologues" Sayyid Qutb).

Chapter One:
Introduction to the Movement Ideological Premises

Article One

The Islamic Resistance Movement: Islam is its way. It is from Islam that it derives its ideas, concepts, and perceptions concerning the universe, life, and man, and it refers to Islam's judgment in all its actions. It is from Islam that it seeks direction so as to guide its steps.

Here Hamas claims Islam is the correct way. The only acceptable insight and understanding of Islam are those of the Islamic Resistance Movement or its Arabic acronym "HAMAS." Full explanations are rendered in the following articles below.

The Relation between the Islamic Resistance Movement and the Muslim Brotherhood

Article Two

The Islamic Resistance Movement is one of the wings of the Muslim Brotherhood in Palestine. The Muslim Brotherhood movement is a global organization and is the largest of the Islamic movements in modern times. It is distinguished by its profound understanding and its conceptual precision and by the fact that it encompasses the totality of Islamic concepts in all aspects of life, in thought and in creed, in politics and in economics, in education and in social affairs, in judicial matters and in matters of government, in preaching and in teaching, in art and in communications, in secret and in the open, and in all other areas of life.

Hamas is the Palestinian regional branch of the Muslim Brotherhood. The Brotherhood is the self-proclaimed world Islamic movement; the exclusive all-encompassing interpreter of the Islamic message, understandings, values and way of life. To be a true follower a believer must embrace the Brotherhood in its totality in all aspects of human existence as there are no other interpretations. Continuing after al-Banna, this is entirely in line with Sayyid Qutb's Islamist convictions as expressed in his work *Milestones*. Full obedience is reminiscent of all totalitarian regimes—whether religious or secular. We only need to recall all forms of fascism including its most radical expression—Nazism. On the left, totalitarianism was expressed through the communist movement turned dictatorial as implemented through Bolshevism (Stalinism) and Maoism.

Structure and Formation

Article Three

The Islamic Resistance Movement is founded upon Muslims who gave their allegiance to Allah and served Him as He ought to be served. *"I did not create jinns and men except that they should serve me"* **(Koran 51:56).**

[These Muslims] recognized their duty toward themselves, their families and their homeland, fearing Allah in all of this. They raised the banner of jihad in the face of the oppressors, in order to deliver the land and the believers from their filth, impurity and evil. *"We hurl the truth against falsehood and crush its head, and lo, it vanishes"* **(Koran 21:18).**

Hamas adherents serve Allah and are loyal to His demands, the foremost being the declaration of Jihad (Holy War) against all enemies who by definition suffer from "filth, impurity and evil."

Article Four

The Islamic Resistance Movement welcomes every Muslim who embraces its creed, adopts its ideology, is committed to its way, keeps its secrets and desires to join its ranks in order to carry out the duty, and his reward is with Allah.

Hamas is a religious activist movement receiving only those who accept its ideals and calls to action or "duty" meaning Jihad, especially against the Jews as was made clear in the Preamble.

The Islamic Resistance Movement— Dimensions of Time and Place

Article Five

The temporal dimension of the Islamic Resistance Movement - in view of the fact that it has adopted Islam as its way of

life - go back to the birth of the Islamic message and to the righteous early believers; Allah is its goal, the Prophet is its example to be followed, and Koran is its constitution.

Its spatial dimension: wherever there are Muslims who embrace Islam as their way of life, everywhere upon the earth. Thus, [Hamas] sends its roots deep into the ground, and it extends to embrace the heavens.

"Do you not see how Allah has given us a parable? A good word is like a good tree; its roots are firm and its branches extend to the heavens. It always bears its fruit at the right time in accordance with God's will. Allah recites parables to men so that they will take heed" **(Koran 14:24-25).**

The "Koran is its constitution" determination makes the central religious text in Islam a binding legal document for Hamas as a political movement—the same applies for the Muslim Brotherhood. The question becomes, "Who is interpreting the Koran?" The Hamas domain is all-inclusive, from "roots deep in the ground" to its "embrace the heavens" with all other interpretations invalid.

Distinctiveness and Independence

Article Six

The Islamic Resistance Movement is a distinct Palestinian movement that is loyal to Allah, adopts Islam as a way of life and works to raise the banner of Allah over every inch of Palestine. Under the wing of Islam, followers of other religions can all live safe and secure in their life, property and rights; whereas in the absence of Islam, discord arises, injustice spreads, corruption burgeons, and there are conflicts and wars. Allah bless the Muslim poet Muhammad Iqbal [6] who said: When faith is gone, there is no safety, And there is no life to him who has no religion. He who is content to live without religion has taken death as a consort of life.

The objective is clear, the Islamic imperative to conquer Palestine. Upon the Islamic conquest of Palestine other religions will be allowed to survive but must live under its banner. With "the absence of Islam" there is "discord," "injustice," "corruption," "conflicts and wars." For Jews and Christians who are "People of the Book" the Charter of Omar embodies legal discrimination or the *dhimma* status (see Chapter I "Negative Image of the Jew"). "Where faith is gone" refers only to the loss of the Islamic faith, otherwise why must all religions live "Under the wing of Islam?"

Paradoxically, Article 6 is quite "moderate" when only demanding legalized discrimination against the Jews. Previously the Jews/Israel were defined as the foremost enemy to be destroyed (al-Banna quote above).

The Universality of the Islamic Resistance Movement

Article Seven

Muslims who adopt the way of the Islamic Resistance Movement are found in all countries of the world, and act to support [the movement], to adopt its positions and to reinforce its jihad. Therefore, it is a world movement, and it is qualified for this [role] owing to the clarity of its ideology, the loftiness of its purpose and the exaltedness of its goals. It is on this basis that it should be regarded and evaluated; it is on this basis that its role should be recognized. Whoever denies its rights, refrains from helping it, becomes blind [to the truth] and makes an effort to blot out its role—he is like one who attempts to dispute with [divine] predestination. Whoever closes his eyes to the facts, intentionally or unintentionally, will eventually wake up [to find that] events have overtaken him and that the [weight of the] evidence has rendered him unable to justify his position. Precedence shall be given to those who come first [to the movement]. The iniquity of one's own relatives is more painful to the soul than the blow of a sharp sword. [7]

The leadership of the Islamic Resistance Movement or Hamas is Divinely self-evident. Hamas has an undeniable "predestination" or celestial victory at the End of Days. This victory is through world Jihad. Theologically there is a universal bond between Islamists, whether we are speaking of the regime in Sudan, Hamas in Gaza or the rising Islamic State regime of 2014 in Iraq/Syria. As seen in the quote below, Allah gave the Koran for all nations to adopt thereby eliminating any differences. Such is the test of loyalty while kinship is of little value.

> *"We have revealed to you the Book in truth, confirming the scripture that came before it and guarding it. Judge between them according to what Allah has revealed, and follow not their capricious will, turning away from the truth that was revealed to you. To each among you Allah has appointed a law and a way. If Allah had so desired, he would have made you a single nation. However, he desired to test you in all that he had given you. So vie with one another in good works. It is to Allah that you shall all return, and He will then reveal to you [the truth] about the matters in which you differed"* (Koran 5:48).

The Islamic Resistance Movement is one link in the chain of jihad in confronting the Zionist invasion. It is connected and linked to the [courageous] uprising of the martyr 'Izz Al-Din Al-Qassam and his brethren the jihad fighters of the Muslim Brotherhood in the year 1936. It is further related and connected to another link, [namely] the jihad of the Palestinians, the efforts and jihad of the Muslim Brotherhood in the 1948 war, and the jihad operations of the Muslim Brotherhood in 1968 and afterwards. Although these links are far apart, and although the continuity of jihad was interrupted by obstacles placed in the path of the jihad fighters by those who circle in the orbit of Zionism, the Islamic Resistance Movement aspires to realize the promise of Allah, no matter how long it takes.

The above paragraph is a partial recounting of the history of Jihad against the Jewish national movement beginning with 'Izz a-Din al-Qassam in the mid-1930s. The Muslim Brotherhood then takes credit for doing battle in 1948 through the Egyptian invasion of the newly proclaimed State of Israel even though the Brotherhood only fielded a small volunteer force. Speaking of Jihad from 1968 onwards is more fiction than fact since the secular Fatah and later the Palestine Liberation Organization (PLO) were the dominant groups from the late 1960s until the rise of Hamas in the 1980s. Notice there is even an admission that the Jihadist battle against Zionism was sporadic.

> **The Prophet, Allah's prayer and peace be upon him, says:** *"The hour of judgment shall not come until the Muslims fight the Jews and kill them, so that the Jews hide behind trees and stones, and each tree and stone will say: 'Oh Muslim, oh servant of Allah, there is a Jew behind me, come and kill him,' except for the Gharqad tree, for it is the tree of the Jews"* **(recorded in the Hadith collections of Bukhari and Muslim).**

The above quote is a clear call to kill Jews since Judgment Day, "the hour of judgment" or End of Days, cannot arrive until Muslims kill the cowardly Jews who hide behind stones and trees. According to Islamist thinking, Jewish existence prevents the Messianic End Time from arriving. So hideous is the Jew that nature itself will turn Jews over to the Muslim so the Jew can suffer his just fate of execution. For example, the trees Jews hide behind will reveal their location.

The implication is that traitors exist everywhere, even in nature. For example, the "tree of the Jews" is the Gharqad tree, and it will not reveal a Jew is hiding behind it. This is despite the fact that Allah created the world in a perfect state. Such thinking is reinforced by Qutb when he wrote, "The Jews have instilled men and regimes (in the Islamic world), in order to conspire against this (Muslim) Community."[6] And again with this statement, "This antagonistic force threatening the Islamic world today has a massive army of agents in the form of professors, philosophers, doctors and researchers—sometimes also writers, poets, scientists and journalists—carrying Muslim names because they are of Muslim descent!! And some of them are from the ranks of the 'Muslim religious authorities'!!"[7]

It must be emphasized that the above quote is not from the Koran, but as noted in the *Covenant* itself, is extracted from the Hadith collections. Furthermore there are those who argue that the quote is taken out of context from the Hadith. However we are speaking of its context within *The Hamas Covenant* where the authors chose the context accurately and effectively in their demand to destroy the Jews. Most significantly if there was any doubt as to the *dhimmi* status of the Jews and their right to exist with disabilities, here *The Hamas Covenant* determines the death sentence because Jews are an obstacle to the arrival of Judgment Day, when the good will be rewarded and the evil punished.

The word "Jews" is invoked in the above *HC* Article 7 passage, not the word "Zionists." This betrays the claim of the Islamist/Jihadist advocates that these activist Hamas Muslims are battling the Zionists, but have no claim against the Jews. Hamas says what it means without euphemisms or downplaying their exact intentions. The process begun is reminiscent of Nazi Germany. Here in Article 7 the war against the Jews is absolute, there is no compromise; the Jews must be exterminated.

The Motto of the Islamic Resistance Movement

Article Eight

Allah is its goal, the Prophet its model to be followed, the Koran its constitution, Jihad its way, and death for the sake of Allah its loftiest desire.

Hamas followers must show no fear. "Death for the sake of Allah" is the most praiseworthy desire for Muslims. Determined by an absolutist diocentrism, the ultimate goal of life is the unification with Allah brought about through a Jihadist's death. Speaking of "its most loftiest desire" means the Jihadist strives to become one with Allah. Allah is the almighty male Deity who is satisfied by worship and Jihadi actions. Just as Mohammed pleased Him previously, today's Jihadist continues in his footsteps. Allah is the stern Father returning his love conditionally, provided his son fulfills His wishes through actions—Jihad until death.[8] Unconditional or "motherly love" is completely absent. One's existence without the Jihadi lifestyle and action is of no value. Jihad allows the suicide-homicide bomber to achieve the ultimate in love and hate simultaneously by destroying one's enemy and fusing with Allah in the same action. This is a multiple human sacrifice—the infidels, whether *dhimmis*, heretics and/or pagans, alongside the "righteous" Jihadi warrior.

Chapter Two:
Causes and Goals

Article Nine

> The Islamic Resistance Movement has found itself in a period when Islam is absent from everyday life. Consequently, the balance has been disturbed, concepts have been confused, values have been altered, evil people have come into power, injustice and darkness have prevailed, the cowardly have become tigers, the homeland has been ravished, the people have been driven away and have been wandering in all the countries of the world. The rule of righteousness is absent, and the rule of falsehood prevails. Nothing is in its proper place. Thus, when Islam is absent, everything is transformed. These are the causes.
>
> As for the goals, they are to fight falsehood, vanquish it and defeat it so that righteousness shall rule, the homeland shall return [to its rightful owner], and from the top of its mosques, the [Muslim] call for prayer will ring out announcing the rise of the rule of Islam, so that people and things shall all return to their proper place. From Allah we seek succor.

Only Islam can bring about righteousness and a better world. The "homeland," referring to *waqf* lands, such as Palestine or any other territories ruled by Islam in the past, will be returned and there will be world order once again when Islam rules.

> *"If Allah did not ward off one group of people by means of another, the earth would certainly be in a state of disorder. Allah is most kind to all beings"* (Koran 2:251).

The Koranic quote above implies justice in Allah's "warding off one group of people by means of another" in obvious support for an Islamic military revival through Jihadist means and the great removal of people groups at the command of Allah to be implemented by the Muslim Brotherhood including Hamas. Islamic conquest is obligatory.

What is most interesting concerning Koranic quote 2:251 is that it is completely out of context since it refers to ancient the Israelites and later King David's victory over Goliath and the rout of the Philistines. The previous Koranic verse to the one quoted above states "David slew Goliath, and Allah bestowed on him [David] sovereignty and wisdom." When put in the Koranic context this quote justifies the Jewish return to the Land of Israel and the establishment of the Jewish State. The *Covenant* uses the verse out of context to justify Muslim conquests.

Article Ten

The Islamic Resistance Movement—while marching forward—offers support to all who are persecuted and protects all who are oppressed with all its strength. It spares no effort in upholding the truth and eradicating falsehood, in word and in action, here and in every place within its reach and its influence.

Hamas protects the "persecuted" and is the exclusive source of truth. In the spirit of *The Hamas Covenant* the persecuted are Muslims.

Chapter Three:
Strategy and Means

The Strategy of the Islamic Resistance Movement
Palestine Is an Islamic Waqf (Islamic Religious Endowment)

Article Eleven

The Islamic Resistance Movement maintains that the land of Palestine is Waqf land given as endowment for all generations of Muslims until the Day of Resurrection. One should not neglect it or [even] a part of it, nor should one relinquish it or [even] a part of it. No Arab State, or [even] all of the Arab States [together], have [the right] to do this; no king or president has this right nor all the kings and presidents together; no organization, or all the organizations together - be they Palestinian or Arab - [have the right to do this]

> because Palestine is Islamic Waqf land given to all generations of Muslims until the Day of Resurrection.

Palestine is Muslim land forever where no other people have a right to sovereignty. Muslims captured Palestine from the Byzantine Empire in 638 CE, upon which it became part of the "Islamic Waqf" and can never be relinquished to anyone. It is implied that to make a territorial compromise to achieve peace is unacceptable. Islamic ownership is eternal.

> This is the legal status of the land of Palestine according to Islamic law. In this respect, it is like any other land that the Muslims have conquered by force, because the Muslims consecrated it at the time of the conquest as religious endowment for all generations of Muslims until the Day of Resurrection. This is how it was: when the conquest of Al-Sha'm [8] and Iraq was complete, the commanders of the Muslim armies sent messages to the Caliph 'Umar b. Al-Khattab, asking for instructions concerning the conquered land - should they divide it up among the troops or leave it in the hands of its owners or what?

Any land ever conquered by Islamic forces becomes an indivisible part of the "Islamic Waqf" as a "religious endowment for all generations of Muslims until the Day of Resurrection." It is the obligation of all to liberate any territories ever having been possessed by Islam but henceforth held by foreigners. This is "defensive Jihad." "Offensive Jihad" is the implied next step for full realization of universal Islamic domination (see Chapter II "Ideologues").

The reference to the conquest Al-Sha'm (the Levant) and Iraq are instructive. The Islamic State (ISIL or ISIS) uses the same theological and historical message condoning conquest of these lands as a "religious endowment for all generations of Muslims." Palestine/Land of Israel is part of the Levant and appears on the Islamic State maps for future conquest. Any "land to be left in the hands of its owners" is subject to a Sharia court of law where those non-Muslim owners have no judicial standing.

> After discussions and consultations between the Caliph 'Umar b. Al-Khattab and the Companions of the Prophet, they decided that the land should be left with its [original] owners

to benefit from its crops, but the substance of the land, that is the land itself, should constitute Waqf for all the generations of Muslims until the Day of Resurrection. The tenure of the owners is only tenure of usufruct. This Waqf will exist as long as the heaven and earth exist. Any measure which does not conform to this Islamic law regarding Palestine is null and void. *"This surely is the very truth. Therefore, praise the great name of your Lord"* **(Koran 56:95-96).**

The land belongs to Islam forever, "as long as heaven and earth exist." The reference to "usufruct" is the right of the individual (in this case a non-Muslim) to work the land and enjoy its profits. This right is granted by Islamic rulers at their discretion as the land itself is *waqf* and belongs to Islam for eternity. Omitted from this section is any mention of taxes imposed on non-Muslims to be paid to the Muslim overlords. At the outset Muslims did not have to pay taxes.[9]

The present status of Palestine/Land of Israel is obviously not in line with Islamic law as the Jewish National Fund in the name of the Jewish People bought and owns large tracts of land, as does the Israeli State. Any such ownership of land by a national and/or state entity contradicts Islamic jurisprudence. Hamas considers the entire State of Israel (within the pre-1967 borders as well) to be illegally occupying *waqf* lands captured close to 1400 years ago in the name of Allah.

It must be pointed out that both the Crusaders (1099-1291) and the British (1917-1948) are seen to have illegally held these *waqf* lands, as Israel does today. All three entities are an affront to Allah's will of Muslim sovereignty over all. Conversely, the Islamic Arab conquest of these lands from the Christian Byzantines is Divinely justified.

Homeland and Nationalism as seen by the Islamic Resistance Movement

Article Twelve

Nationalism [9], as seen by the Islamic Resistance Movement, is part of the [Islamic] religious creed. There is nothing that speaks more eloquently and more profoundly of nationalism than the following: when the enemy tramples Muslim

territory, waging jihad and confronting the enemy become a personal duty of every Muslim man and Muslim woman. A woman may go out to fight the enemy [even] without her husband's permission and a slave without his master's permission.

There is nothing like this in any other political system - this is an indisputable fact. Whereas the various other nationalist [ideologies] are connected to physical, human and regional factors, the nationalism of the Islamic Resistance Movement is [likewise] characterized by all of the above, and in addition - and most importantly - it is characterized by divine motives which endow it with spirit and life, since it is related to the source of the spirit, and to Him who gives life. [The Islamic Resistance Movement] is raising the divine banner in the skies of the homeland, so as to firmly connect heaven and earth. When Moses came and cast down his rod both sorcery and sorcerer came to naught.

"The right way stands out clearly from error; therefore, whoever renounces falsehood and believes in Allah, he has indeed grasped the firmest handhold, which never breaks, and Allah is Hearing and Knowing" (Koran 2:256).

Islam, as religion and nationalism, are one and together of the Divine spirit and not of the physical or regional types. *Even* a woman or a slave need not seek their master's permission to join the Jihad to defend *waqf* Muslim lands. This is a "given" in Jihadist writings and is a personal obligation in the implementation of defensive Jihad (see Chapter II "Ideologues").

Peaceful Solutions, Initiatives and International Conferences

Article Thirteen

The initiatives, the so-called peace solutions, and the international conferences for resolving the Palestinian problem stand in contradiction to the principles of the Islamic Resistance Movement, for to neglect any part of Palestine is to neglect part of the Islamic faith. The nationalism of the Islamic Resistance Movement is part of its [Islamic] faith. It is in the light of this principle that its members are educated, and they wage jihad in order to raise the banner of Allah over the homeland.

"And Allah has full control over His affairs; but most people do not know" (Koran 12:21).

From time to time there are calls to hold an international conference in order to seek a solution for the [Palestinian] problem. Some accept this [proposal] and some reject it, for one reason or another, demanding the fulfillment of some condition or conditions before they agree to hold the conference and participate in it. However, the Islamic Resistance Movement - since it is familiar with the parties participating in the conference and with their past and current positions on the issues of the Muslims - does not believe that these conferences can meet the demands or restore the rights [of the Palestinians], or bring equity to the oppressed. These conferences are nothing but a way to give the infidels power of arbitration over Muslim land, and when have the infidels ever been equitable toward the believers?

No one except Muslims have the right to rule over lands captured by Islam, in this case Palestine, at any given period. The above declaration speaks of giving "infidels the power of arbitration over Muslim land," thereby nullifying any efforts as insincere by Jews, Christians or others willing to help in a negotiated settlement of the conflict. This follows the logic that the only solution is one where Islam controls all the land in Palestine. Implied is the condemnation of any Muslim state or authority agreeing to a compromise. Such a body is heretical for allowing compromise.

Specifically, as determined in the first paragraph, there are to be no negotiations with the Jewish State to achieve conflict resolution. To compromise is to "neglect" a "part of Palestine," which is akin to abandoning the "Islamic faith." Previously in the Preamble, Jews were relegated to polytheists; here in the last sentence they are "infidels." Under Sharia law, in either case, whether considered a polytheist or an infidel, a Jew's life is in jeopardy. The term "infidel" may also be read to mean those powers participating as brokers in the negotiations, meaning the Christian West and especially the US, the UN and the non-Islamist moderate Arab regimes. In the long run their fate cannot be expected to be much better than that of the Jews.

HC Article 13 reinforces the demand for Jihad against the Jews. There is to be no negotiation with the Jews, only war and destruction to solve the "problem" and restore "rights" in the Palestinian Muslim context. One does not negotiate with Jews nor accept any third party attempt to do so. Such absolutism is very reminiscent of Nazi racial attitudes toward the Jews.

> *"The Jews will never be content with you, nor will the Christians, until you follow their religion. Say: 'The guidance of Allah is the right guidance.' But if you follow their desires after the knowledge which has come to you, then you shall have no protector or guardian from Allah"* (Koran 2:120).

Dhimmis have no protection from Allah and they will never allow a Muslim to worship freely. All gain Allah's protection when the *dhimmis* live under Muslim domination.

The logical conclusion of Article 13 up to this point solidifies the injunction against Jewish sovereignty anywhere in Palestine. This reasoning follows that to allow Jewish rule leads Israeli Muslims to the acceptance of a *dhimmi* faith and punishment from Allah—the withdrawal of protection for the apostate, generally interpreted as capital punishment for those abandoning Islam. Advanced notice is also given to Muslims worldwide in the Koranic quote not to acquiesce to the sovereignty of the People of the Book or *dhimmis*.

> **There is no solution to the Palestinian problem except by jihad. Initiatives, proposals and international conferences are a waste of time and a farce. The Palestinian people is far too eminent to have its future, its rights and its destiny toyed with. As stated in the Hadith: "The people of Al-Sha'm are [Allah's] rod in His land. Through them, He wreaks vengeance on whomever He wishes among His servants. The hypocrites among them are not allowed to be superior to the believers among them, and they shall die in grief and distress." (Recorded by Al-Tabarani with a chain of transmitters to Muhammad, and by Ahmad [Ibn Hanbal] with an**

incomplete chain of transmitters to Muhammad which may be the accurate record, the transmitters in both cases being trustworthy - Allah alone is omniscient).

Jihad is the only answer to ensure the true destiny of the Palestinian Muslim people. Attempting to achieve peace is "a waste of time and a farce." The "hypocrites" involved in peace negotiations or conflict resolution with Israel "shall die in grief and distress."

The Three Circles

Article Fourteen

The problem of liberating Palestine involves three circles: the Palestinian circle, the [pan] Arab circle and the Islamic circle. Each of these three circles has its role in the struggle against Zionism and has its duties. It is a grave error and shameful ignorance to neglect any of these circles, for Palestine is an Islamic land. In it is the first of the two qiblas [directions of prayer] and the third most holy mosque, after the mosques of Mecca and Medina. It is the destination of the Prophet's nocturnal journey.

***"Praise be to Him who carried His servant by night from the most sacred mosque to the farthest mosque whose surroundings We blessed, so as to show him Our signs. He is the One who is hearing and seeing"* (Koran 17:1).**

The first two circles involve the local Palestinian and the pan-Arab; by implication both are Muslim. The third circle brings in the remainder of the Islamic world. Only Islam can defeat the Zionists and Jews (see Articles 34 and 35 below, and Chapter II "Ideologues").

Until 624 CE, the first "*qibla*," or Muslim prayer direction, was Jerusalem; at that time it was replaced by Mecca. The Jerusalem prayer direction is often attributed to the Prophet Mohammed's initial overtures to the Jews, which were rejected. The Prophet's "nocturnal" or night journey takes place at the "farthest mosque," the location of which Muslims identify as Jerusalem. Jerusalem is never mentioned in the Koran but gained major religious significance as the "Outer" or "Farthest Mosque"[10] during the Ummayad reign (661-750) in the region when its leadership was in serious conflict with the inhabitants of the Arabian Peninsula, who controlled the two holy cities of Mecca and Medina. Both the Dome of the Rock (691 CE) and Al-Aksa Mosque (711 CE) were constructed during the Umayyad era.[11]

> **This being the case, the liberation [of Palestine] is a personal duty of every Muslim, wherever he be. It is on this basis that one should consider the problem, and every Muslim should understand this. When the day comes and the problem is treated on this basis, and all the capabilities of the three circles are mobilized - the current circumstances will change and the day of liberation will draw near.**

All Muslims are obligated to battle the Jews to liberate the "farthest mosque," and consequently all of Palestine. Unity through the three circles is imperative for victory. As you can see below, the foolish Jews can deny Allah but they fear the Jihadist.

> *"You strike more fear in the hearts of the Jews than does Allah, because they are people who do not understand"* (Koran 59:13).

Jihad for the Liberation of Palestine is a Personal Duty

Article Fifteen

> The day the enemies conquer some part of the Muslim land, jihad becomes a personal duty of every Muslim. In the face of the Jewish occupation of Palestine, it is necessary to raise the banner of jihad. This requires the propagation of Islamic consciousness among the masses, locally [in Palestine], in the Arab world and in the Islamic world. It is necessary to instill the spirit of jihad in the nation, engage the enemies and join the ranks of the jihad fighters.

Once again these are the demands for a defensive Jihad defined as "a personal obligation of every Muslim" in "the face of the Jewish occupation of

Palestine." Every Muslim is obligated to go to war against the Jews in any manner he or she sees fit. The struggle is not only about battling an army, but, in essence, a war against all the Jews as the enemy deserving of death by definition (Preamble and Article 7 respectively). By implication all methods are acceptable.

The above mentioned "spirit of jihad" leads the entire nation in a communal death for Allah in the ultimate human sacrifice through destruction of the enemy and oneself, if necessary. Jihadists sacrifice themselves in love and dedication to the cause of world Islamization (see *HC* Article 8). Whether one be a suicide-homicide bomber or acting as a human shield for a Jihadist operation, all actions and deaths are in the name of Allah. The true Muslim will receive his or her reward in heaven. Such is the spirit of Jihad (see Chapter II "Ideologues").

> **The indoctrination campaign must involve ulama, educators, teachers and information and media experts, as well as all intellectuals, especially the young people and the sheikhs of Islamic movements. It is [also] necessary to introduce essential changes in the curricula, in order to eliminate the influences of the intellectual invasion which were inflicted upon them by the Orientalists and the missionaries. This invasion came upon the region after Salah Al-Din Al-Ayyubi defeated the Crusaders. The Crusaders then realized that it is impossible to vanquish the Muslims unless the way is first paved by an intellectual invasion that would confuse the [Muslims'] thinking, distort their legacy and impugn their ideals. Only after this [intellectual invasion] would there come invasion with troops. This [intellectual invasion] prepared the ground for the colonialist invasion, as [General] Allenby declared upon entering Jerusalem: "Now the Crusades have come to an end." General Gouraud stood at Salah Al-Din's tomb, and said: "Oh, Salah Al-Din, we are back!" Colonialism helped to intensify the intellectual invasion and helped it to take root. It still does. All this paved the way toward the loss of Palestine.**

Islamic consciousness and Jihad demand societal "indoctrination," most especially of the Islamic scholars (*ulama*), teachers, media and intellectuals. With this there is the "elimination of the influence of the intellectual invasion" of the West, Christianity and especially its rational Enlightenment thinkers. Islam was weakened to the point of defeat by this invasion of foreign values and perspectives and now insists on the elimination of the free flow of ideas and freedom of thought. This is Islamist totalitarianism or what is referred to at times as "Islamo-fascism."

Indoctrination and elimination of intellectual influence is the foundation of the totalitarian state as described by the Italian dictator Benito Mussolini, who coined the term "totalitario" in the 1920s when referring to his rule of state. He furthermore exclaimed, "All within the state, none outside the state, none against the state." It was absolute single party rule. Generally speaking, totalitarianism is the oppressive centralized state authority controlling all action through coercion, usually of the physical type.[12] Most often the people are swept up in euphoria and support such rule especially should the state be led by a popular charismatic figure such as Stalin, Mao or Hitler. The totalitarian state destroys or alters existing institutions and replaces them with monolithic goal-oriented ones. The infrastructure of state is directed toward centralized objectives, empowers and worships itself.

The Islamic totalitarian state would be something of a hybrid since certain objectives remain traditional in nature, such as the spreading of Islam worldwide. However, regime control where everyone works for the same ultimate purpose with no deviance and no freedom of maneuver or discourse is fully in the spirit of totalitarianism. Any form of dissent is crushed, even should it come from reformists inside the system. The monolithic state objective of spreading Islam through Jihad and the continual conflict status to which all Muslims are obligated must culminate in the universal totalitarian Islamic political entity of the future. The masses must conform to the will of the Jihadi state. Totalitarian states do not need one overriding powerful central figure, but more often than not this becomes the case.

As an example, Ayatollah Khomeini of Iran was moving in the direction of the dictatorial Islamic State by the time of his death in 1989, yet it was not quite totalitarian. The former Iranian President Ahmedinejad (2005-2013) was accused of attempting to overthrow the Islamic Republic in order to construct a hybrid Islamic-military totalitarian state where all loyalties were to him within the domain of a great Islamic leader.

Should a charismatic authoritarian leader arise in the image of Sheikh Hasan al-Banna he could take the reins of power in a Hamas or Muslim Brotherhood organization, unifying religion and state into one, and embody all decisions as Supreme Leader. Historically, such an eventuality can be seen as a reestablishment of the Caliphate even if it be more of a totalitarian system. The recent rise of the Islamic State of Iraq and Syria (ISIS) in June 2014 may take such a direction under the declared Caliph Abdul Bakr al-Baghdadi (original name Dr. Ibrahim Awwad Ibrahim).

Hasan al-Banna was critical of the German and Italian examples because their militarism was for the benefit of the secular nation and race, not for proselytizing the word of Allah and Islam.[13] Al-Banna still showed great respect for these regimes and favored them over the Allies in WWII. Forms of Arab racism in his writings exist as noted above. As usual, there must be the agreed-upon enemy or scapegoat to blame for all of society's ills. Like the Czarists, Hitler, Stalin and Khomeinist Iran, Hamas found that common enemy in the Jews. As for Christianity, the *Covenant* connects the Crusaders and the French-British alliance of WWI by linking Generals Allenby and Gouraud to the overall Western Christian onslaught against Islam as a continuing phenomenon. According to Hamas, only Islam has the exclusive right to conquer the world.

> **It is necessary to establish in the minds of all the Muslim generations that the Palestinian issue is a religious issue, and that it must be dealt with as such, for [Palestine] contains Islamic holy places, [namely] the Al-Aqsa mosque, which is inseparably connected, for as long as heaven and earth shall endure, to the holy mosque of Mecca through the Prophet's nocturnal journey [from the mosque of Mecca to the Al-Aqsa mosque] and through his ascension to heaven thence.**

For those with doubts concerning Jerusalem's holiness, *The Hamas Covenant* makes an inseparable connection between Jerusalem and Mecca by using the "Night Journey" (Koran 17:1). Islamic interpretation of the text identifies Jerusalem as the Al-Aksa, or "Outer" Mosque, therefore the city is Muslim, not Jewish or Israeli. As stated previously, Jerusalem is never mentioned by name in the Koran. The "Palestinian issue is a religious issue," which means it goes well beyond *waqf* territorial demands.

> *"Being stationed on the frontier for the sake of Allah for one day is better than this [entire] world and everything in it; and the place taken up in paradise by the [horseman's] whip of any one of you [jihad fighters] is better than this [entire] world and everything in it. Every evening [operation] and morning [the operation] performed by Muslims for the sake of Allah is*

better than this [entire] world and everything in it" (recorded in the Hadith collections of Bukhari, Muslim, Tirmidhi and Ibn Maja).

"By the name of Him who holds Muhammad's soul in His hand, I wish to launch an attack for the sake of Allah and be killed and attack again and be killed and attack again and be killed." (Recorded in the Hadith collections of Bukhari and Muslim).

The Hadith quotes are general declarations of the nobility of Jihad. Multiple deaths and martyrdom indicate an infinite self-sacrifice in the physical sense and a spiritual fusion, at-oneness and eternal love for Allah. The love of death becomes a value in itself.

Educating the Next Generations

Article Sixteen

It is necessary to educate the next Islamic generations in our region in an Islamic way, based on the performance of the religious duties, attentive study of Allah's book, study of the Prophet's sunna [custom], perusal of Islamic history and legacy based on reliable sources under the instruction of experts and scholars, and reliance on methods which will produce a wholesome outlook in thought and in faith. In addition, it is necessary to closely study the enemy and his material and human capabilities, to become familiar with his weaknesses and strengths, to recognize the powers that assist and support him. It is also necessary to be familiar with current events, follow new developments and study the analyses and commentaries regarding them. It is likewise necessary to plan for the future and to study each and every phenomenon, so that Muslims engaged in jihad will live with full awareness of their purpose, goal and way, and [with full awareness of] what is happening around them.

"Oh my son! If [a thing] is but the weight of a grain of mustard, though it be in a rock, or in the heavens, or on earth, Allah

will bring it forth. Allah discerns even the smallest thing; He is omniscient. Oh my son! Keep up prayer and enjoin the good and forbid the evil, and persevere whatever may befall you; surely this [behavior] is worthy steadfastness. Turn not thy cheek in scorn away from people; do not walk haughtily in the land. Allah does not love the arrogant and self-conceited" (Koran 31:16-18).

A full and thorough Islamic education as prescribed by Hamas is a prerequisite for victory. One must know the enemy (Jews, Israel, Zionism) and be fully aware of everyday events, small as they may be. Hamas focuses, interprets and directs knowledge to ensure the defeat of the enemy and to guarantee the Islamic future. Indoctrination is through "Islamic history and legacy based on reliable sources" approved by Hamas.

The Role of Muslim Women

Article Seventeen

The role of the Muslim woman in the war of liberation is no less important than that of the man, for she is the maker of men. Her role in guiding and educating the next generation is very important. The enemies have realized [the significance of] her role, and they believe that if they can educate her according to their wishes, guiding her away from Islam, they will have won the war. You find, therefore, that they continually make great efforts [to do this] by means of the media, the cinema and school curricula, through their agents who are incorporated in Zionist organizations that assume various names and forms such as the Freemasons, Rotary Clubs, espionage groups, etc. - all of which are dens of sabotage and saboteurs. These Zionist organizations have an enormous abundance of material resources, which enable them to play their game in [various] societies with the aim of realizing their purpose while Islam is absent from the scene and the Muslims are estranged [from their faith]. The followers of the Islamic movements [10] should fulfill their role in countering the schemes of these saboteurs. When Islam

is at the helm, it will totally eradicate these organizations, which are hostile to humanity and to Islam.

Muslim women have the all-important role of raising and educating their children to be the first to counter the Zionist organizations and their affiliates—the Freemasons, Rotary Clubs and other civic organizations seen as "espionage groups" with their "dens of sabotage and saboteurs." The danger is especially for Muslim women living in Western societies, where these groups are most active. The fear is that Muslim women might adopt liberal democratic ideals and pass them on to their children.

It is interesting to note the condemnation of "the media, cinema and school curricula." The average Muslim woman lives in a Muslim society, leading one to understand the Hamas assumption that the Muslim world has been infiltrated by the Zionists. The women themselves must make themselves immune to the influences of the above-mentioned organizations. The addition "etc." leads to conjecture over whether Hamas is using a general term to indicate condemnation of feminist and women's advocacy groups demanding equality with their male counterparts. Would they too be considered "dens of sabotage and saboteurs"?

Article Eighteen

The woman in the jihadist home and family, as mother or sister, has the primary role in managing the household, raising the children according to the moral ideas and values inspired by Islam, and teaching them to perform the religious duties in preparation for the jihadist role that awaits them. Hence, it is necessary to pay close attention to the schools in which the Muslim girl is educated, and to their curricula, so that she will grow to be a good mother, conscious of her role in the war of liberation. She should have adequate awareness and understanding concerning the management of domestic affairs, since economy and avoiding wastefulness in family expenses are among the requirements in the ability to persist in the current difficult circumstances. She should ever be aware that available funds are like blood that must flow only in the veins for life to continue in both young and old.

VI The Hamas Covenant Analysis

> *"Muslim men and Muslim women, the believing men and women, the truthful men and women, the persevering men and women, the humble men and women, the charitable men and women, the fasting men and women, the men and women who guard their chastity, and the men and women who remember Allah frequently - for them Allah has prepared forgiveness and great reward"* **(Koran 33:35)**.

The role of the Muslim woman is to successfully administer her household. This includes a Jihadi education for all the children and in particular a domestic education for her daughter in practical matters such as family expenses since financial donations to the Jihad cause are important for the overall battle. The role of women in fascism is one of homemaker and, most importantly, child bearer. This model was followed by fascist societies in Europe and, in particular, Germany. Male children are raised in preparation for war, or in this case Jihad.

A Jihadi education includes dying for Allah and, if need be, becoming a suicide-homicide bomber. Jihadi mothers prepare their children to become a martyr or "*shaheed*"— a human sacrifice in the battle against the enemy. Such child sacrifice is reminiscent of the ancient Phoenician religion where children became live burnt offerings to the god Ba'al Hammon in a ceremony known as "*tophet*," or incineration. Such was the custom, particularly in Carthage, and reviled throughout the ancient world.

Taking Articles 17 and 18 together one understands "the role of the Muslim woman," but without the shadow of Koranic authority we lack the entire picture. Mothers, daughters and sisters are honored yet they lack equality, falling under male authority. Koran 4:34 is clear on the topic, "Men have authority over women because Allah has made the one superior to the others, and because they spend their wealth to maintain them. Good women are obedient. They guard their unseen parts because Allah has guarded them. As for those from whom you fear disobedience, admonish them and send them to beds apart and beat them. Then if they obey you, take no further action against them. Allah is high, supreme."

And as for sex, the Koran 2:223 states clearly, "Women are your fields; go, then, into your fields as you please" however in the preceding verse men are not to go near women during menstruation (Koran 2:222). Women are often considered unclean and are not equal to men. They can be beaten if suspected of disobedience and are viewed as sexual property; the analogy given is that of a plowed field. Plowed, seeded fields produce crops; women produce children—the metaphor is clear. The role of the Muslim woman is to be subservient to her husband (see Chapter II "Ideologues"). For more on the status of women see Koran Chapters 2 (The Cow) and 4 (Women).[14]

The Role of Islamic Art in the War of Liberation

Article Nineteen

Art has rules and standards by which it is possible to determine whether it is Islamic or pagan. The Islamic liberation is in need of Islamic art that uplifts the spirit without subjecting one aspect of human nature to another, but rather uplifts all aspects in [perfect] balance and harmony. Man is a marvelous and unique creature made from a handful of clay and a breath of spirit. Islamic art addresses man on this basis, while pagan art addresses the physical body and gives dominance to the component of clay.

Books, articles, pamphlets, sermons, epistles, traditional songs, poems, [patriotic] songs, plays, etc. - when they have the characteristics of Islamic art, they are among the necessary means of ideological indoctrination. [They constitute] self-renewing sustenance for continuing the journey and refreshing the spirit, for the road is long, the suffering is great and the souls grow weary. Islamic art renews the energies, revives motion and awakens the soul to lofty ideals and wholesome conduct.

Nothing heals the soul when it is in retreat save moving from mode to mode.

"Pagan" art is a façade, a useless materialism while Islamic art is spiritual. This is a condemnation of classical Western art, such as the Greco-Roman and European art in general. Human statues or depictions are forbidden. One should consider the ramifications for Western and Far Eastern art, in particular sculpture. In Afghanistan the Taliban destruction of ancient Buddhas carved out of mountain faces in Bamiyan, eastern Afghanistan in March of

2001 was considered one of the greatest cultural crimes against world civilization.

Since June of 2014 the Islamic State (ISIS or ISIL) has perfected the destruction of "pagan art" in the name of "Islamic liberation"—to use Hamas terminology. The Nineveh Wall, Nimrud, Hatra and ancient works of art from the Mosul Museum in Iraq were damaged or demolished while the archaeological site of Palmyra and artifacts in al-Raqqah suffered a similar fate in Syria. The lead archaeologist working in Palmyra was beheaded. Most relics are from the ancient Assyrian era dating back thousands of years. Numerous Shiite mosques and Christian churches continue to suffer major damage and destruction. In Africa similar devastation hit Timbuktu in Mali when Islamic extremists condemned Sufi Muslim art and scripture in 2012.

Non-Islamic art is illegal and illegitimate. Pagan artists are identified with their work and both are to be destroyed. To destroy art is similar to book burning. We are reminded of Heinreich Heine who declared, "Where books are burned, people are next."[15] We know this is what happened in Nazi Germany. When quoting the above Article 19, "When works have the characteristics of Islamic art, they are among the necessary means of ideological indoctrination," we understand that art is in the service of Jihad and the Islamic totalitarian state.

On the seemingly more moderate side, we have art in the service of the secular totalitarian state. One sees its degradation by communism, whether of the Bolshevik Russian or Maoist Chinese type, and fascism—Nazi German, Italian, Eastern European, and Iberian in the twentieth century. Any form of dictatorship infringing on free will and free expression curtails artistic ability and brings about its elimination. But this pales in comparison to Islamic Jihadi artistic dictates.

An extreme response can be expected to any use of art to criticize Islam as evidenced by the Mohammed cartoons printed in Denmark (2005) which led to worldwide Islamic protests and rioting (2005-2006) and the *Charlie Hebdo* massacre in Paris (January 2015), an even more fanatical Islamist reaction to political cartoons relating to the Prophet.

All this is utterly serious and no jest, for the nation engaged in jihad knows no jest.

This last sentence sets the tone for Hamas society at large when stating, "the nation engaged in jihad knows no jest." Laughing, smiling, singing or a general light moment is forbidden in the Jihadi society. Enforcement of such puritanical laws and behavior began in Gaza in December 2008 (see Chapter V "Hamas Ideological Victory," subheading "The 'Cast Lead' Gaza War and Repercussions").

Social Solidarity

Article Twenty

Muslim society is characterized by solidarity. The Prophet, Allah's prayer and peace be upon him, says: "Blessed are the Banu Al-Ash'ar tribe. When they are afflicted with drought - whether in a town or on a journey - they would collect all that they have and divide it among themselves in equal shares."

This is the Islamic spirit that should prevail in every Muslim society. A society which is facing a wicked enemy with Nazi behavior, that does not distinguish between men and women, old and young, has an even greater need to grace itself with this Islamic spirit [of solidarity]. Our enemy employs the method of collective punishment, depriving people of their homes and possessions. He pursued them [even] in their places of exile, breaking bones, shooting women, children and elderly people with or without reason. He established detention camps to imprison thousands and thousands [of people] in inhuman conditions, all this in addition to destroying houses, orphaning children and unjustly convicting thousands of young people to spend the best years of their youth in the darkness of prison.

The Nazism of the Jews targeted both women and children. The terror they spread is directed at everyone. They fight people by destroying their livelihood, stealing their money and trampling their dignity. Their horrible treatment of people is like that of the worst war criminals. Deportation from one's homeland is a form of murder.

In the face of such behavior, we must have social solidarity among the people, and we must face the enemy as one body,

which, when one of its limbs is in pain, the rest of it reacts with sleeplessness and fever.

Article 20 is a call for solidarity, indicating it is seriously lacking. Much of Israel's success against Hamas is the result of intelligence information provided by Palestinian Muslims themselves, which resulted in detentions, arrests and street battles with Hamas Jihadists. Most notably, Hamas antisemitism borrows heavily from Nazism, particularly in Article 22 (see below) and engages in role reversal accusations.

The *Covenant* was published in August 1988 when Hamas was beginning to lose the upper hand in the battle against Israel, known as the "First Intifada," hence the accusations include that of Nazism. Under Nazi rule, populations dwindled as Jews were exterminated. The exact opposite took place from the twenty years after the 1967 Six Day War in the West Bank, Gaza and East Jerusalem.

The Palestinian population grew from some 586,000 in the West Bank in 1967 to some 858,000 by the end of 1987[16] when Hamas was officially established. This is an absolute population growth of over 46 percent in 20 years, quite a high level by anyone's calculations. This was accompanied by rapid economic growth. In 1967 the Gaza population was approximately 390,000 and increased to 633,000[17] twenty years later, an increase of 62 percent. The Palestinian Arab West Bank population as of 2013 was said to be some 2.5 million,[18] and Gaza some 1.7 million,[19] while East Jerusalem's was 372,000[20] up from 66,000 in 1967,[21] more than five-fold increase in four and a half decades. In summary, the Palestinian population was four and a half times larger in 2013 than it was in 1967 when the population was just over one million.

Equating Jews with Nazism is a phenomenon begun with the Soviet penetration into the Arab Middle East during the mid-1950s as Egypt, Syria and Iraq shifted toward the communist world. In particular the Baathist ideals of Assad's Syrian regime and Saddam Hussein's Iraq borrowed heavily from the Nazis.[22] Jews and Israelis were no longer being depicted as communists and Bolsheviks, as had been done previously. A composite caricature was developed in popular cartoons of long-nosed Jews in Nazi uniform with swastikas, their knives and/or teeth dripping with blood. The bloody long-nosed Jew is derived from Nazi publications pushing the accusation of the Blood Libel where Jews were accused of killing Christian children to use their blood for baking matzot for Passover. This new "Jew as Nazi" stereotype is used in the Arab world to the present day and is now part of the Arab/Islamist lexicon, enhancing the condemnation of Israel and the Jews.[23] Articles 31 and 32 follow suit, making the same accusation.

In the aftermath of WWII the Soviets believed everyone would forget their non-aggression pact with Nazi Germany (1939-41), just as the Arab nationalists and the Islamists hope their overtures, cooperation and support of the Nazis during WWII will be forgotten. In both cases wiping out the memory of their own alliances with Hitler is best served by eternalizing the false charge of Nazism against their Jewish victims. Public opinion does not recall the Soviet and Arab alliances with Nazism while the accusations against the Jews are of recent memory.

The accuser-perpetrator blames his victim for the heinous crimes he intends to impose or has imposed on his victim, thereby justifying his actions to himself, his people and others. Such behavior is in line with what is known as "The Big Lie" used by dictators and in particular by the Nazis. The bigger and more hideous it is, the more likely the constantly repeated lie will eventually be believed. Hitler lied when he accused the Jews of instigating WWII. Hamas follows in his footsteps knowing full well that the Grand Mufti of Jerusalem, Haj Amin el-Husseini, was the foremost Palestinian Arab-Nazi collaborator (see Chapter III "Jewish National Liberation" subsection "Haj Amin El Husseini" and Chapter VIII "Czarist-Nazi Integration into the Palestinian Islamist Jihad").

Through this role reversal "Big Lie" accusation the Arab Muslim world with Hamas in the lead works to convince everyone of the illegitimacy and pariah status of the Jewish State and Jewish existence. By exposing the "Big Lie" the Jewish State takes its place as equal to all other states and exercises the right to defend itself. Legitimate Israeli responses to ensure security are in accordance with the "Fourth Geneva Convention," August 1949:

- Article 5 allows for actions to be taken against a "spy and saboteur" or if a "person is definitely suspected of or engaged in activities hostile to the security of the State."
- Article 28 explains that the "presence of a protected person [civilian] may not be used to render certain points or areas immune from military operations." In other words, Hamas operatives may be apprehended or killed even if Israeli forces operate within a civilian environment.
- Article 49 determines that the "Occupying Power may undertake total or partial evacuation of a given area if the se-

- curity of the population or imperative military reasons so demand." Persons are to remain in the overall territory however.
- Article 54 deals with house destruction when considered "absolutely necessary by military operations." Once this was implemented, the question for Israel was whether it deterred terrorist activities. At times Israel ended the policy but more recently this course of action has been reinstated. The policy is legal but demands discretion.

Article Twenty-One

Social solidarity means offering help to everyone who is in need, be it material or moral, or joining with him to complete some work. Members of the Islamic Resistance Movement should regard the interests of the masses as their own, and they should spare no effort to achieve them and protect them. They must prevent reckless playing with matters affecting the future of the next generations or causing losses to their society. The masses are of them and for them, and the strength of the masses is strength for them; their future is theirs. The members of the Islamic Resistance Movement should be with the people on joyous occasions and at times of grief. They should espouse the demands of the masses and strive to serve the masses' interests, which are indeed their own. When this spirit prevails, friendship will deepen and there will be cooperation and empathy, unity will increase and the ranks will be strengthened to confront the enemies.

Unity with the masses and social cohesion on every level is the Hamas recipe for success. Hamas should not get involved in actions detrimental to their society, because their strength comes from mass support. This article could call into question popular support for attacks against Israel, which resulted in the defensive actions Cast Lead (2008-09), Pillar of Strength (2012) and Protective Edge (2014) operations and created extensive damage to Gaza infrastructure. It appears that popular backing for Hamas' Islamist ideals remain high and keeps them in tandem with "the demands of the masses" despite material suffering.

The Forces which Support the Enemy

Article Twenty-Two

The enemies have been planning expertly and thoroughly for a long time in order to achieve what they have achieved, employing those means which affect the course of events. They strove to accumulate huge financial resources which they used to realize their dream.

With money they have taken control of the world media - news agencies, the press, publishing houses, broadcasting services, etc. With money they sparked revolutions in various countries around the world in order to serve their interests and to reap profits. They were behind the French Revolution and the Communist Revolution and [they are behind] most of the revolutions about which we hear from time to time here and there. With money they have formed secret organizations, all over the world, in order to destroy [those countries'] societies and to serve the Zionists' interests, such as the Freemasons, the Rotary Clubs, the Lions, the Sons of the Covenant [i.e. B'nei B'rith], etc. All of these are organizations of espionage and sabotage. With money they were able to take control of the colonialist countries, and [they] urged them to colonize many countries so that they could exploit their resources and spread moral corruption there.

For an historical understanding of the accusations throughout Article 22, one can review the stereotypes portraying Jewish conspiracies as expressed by the Russian Czarist antisemites in *The Protocols of the Learned Elders of Zion* and in Adolf Hitler's *Mein Kampf.* In particular, Jewish money is said to lead to world domination. This is more fully discussed in Chapter VIII "Czarist-Nazi Integration with the Palestinian Islamic Jihad." Overall the Jews are the enemy, as stated in the Introduction-Preamble of *The Hamas Covenant.* Jewry is castigated for developing huge financial reserves to initiate their assault on the Muslim world. Hamas believes Jews initiated their conquests by capturing the media and continue to advance onward to achieve global domination.

The truth is that the Jews have little influence on the world media. The Arab/Muslim media throughout the Middle East is anti-Israel, pro-Palestinian and often pro-Islamist. *Al Jazeera* and *Al Arabiya* are the two most prominent satellite stations, with the former being accused by many in the Arab world of working for the rise of Islamism and the downfall of the Arab secular regimes. In particular, the Egyptian governments of former President Hosni Mubarak and current President Abdel Fattah a-Sisi have made the accusation against *Al Jazeera*. Media in Pakistan, Bangladesh, Afghanistan, Iran, Indonesia, Malaysia and the majority in Turkey are anti-Israel and often antisemitic. Quite a few are pro-Islamist.

Over the years, major Western media in Europe, including *Sky News,* the *BBC, France 24* and at times *CNN*'s European cable channel, are noted as quite critical of Israeli government policies and even as leaning pro-Palestinian, but not pro-Islamist. European newspapers and magazines are sympathetic to the Palestinians or considered "balanced." *CNN*'s European affiliate was considered so anti-Israel during the 2000-2004 Palestinian Low Intensity Conflict against Israel, commonly known as the Second Intifada, that Israelis demanded its removal from the general package offered by its two major cable TV suppliers. *CNN* in North America is considered "balanced," while *Fox News* is seen as pro-Israel. The *New York Times*, while supporting Israel's right to exist, is viewed as much more critical of Israel than it is of the Palestinians, while the *Wall Street Journal* leans more pro-Israel.

Most developing countries in Africa and Latin America lean heavily toward the Palestinian cause, though they are generally not pro-Islamist while Far Eastern countries such as China and Japan may be considered somewhat more balanced, but still tilting toward the Palestinians. India is inclined toward Israel, especially the present government which identifies more with the Jewish State because of its own continuing fears of Pakistani Islamist attacks. The above countries and media usually advocate a two-state solution. In the UN, however, these same member states constantly vote in favor of resolutions condemning Israel. Jewish influence in the media and the international community is negligible.

Only in the US and Canada is there more sympathy for Israel than for the Palestinians, both in governmental circles and the media. Many attribute this to Jewish activism in the "Israel Lobby," yet the largest group of pro-Israel support comes from Evangelical Christians.

Once again, non-Jewish international liberal community service organizations, such as the Rotary, Lions Club and the Freemasons, are blamed for serving Zionist and Jewish interests. To round out the international conspiracy, the French and communist revolutions as well as colonialism are blamed on the Jews.

> There is no end to what can be said about [their involvement in] local wars and world wars. They were behind World War I, through which they achieved the destruction of the Islamic Caliphate, reaped material profits, took control of numerous resources, obtained the Balfour Declaration, and established the League of the United Nations [sic] so as to rule the world through this organization. They were [also] behind World War II, through which they reaped enormous profits from commerce in war materials and paved the way for the establishment of their state. They [also] suggested the formation of the United Nations and the Security Council to replace the League of the United Nations [sic] and to rule the world through this [new organization]. Wherever there is war in the world, it is they who are pulling the strings behind the scenes. *"Whenever they ignite the fire of war, Allah extinguishes it. They strive to spread evil in the land, but Allah does not love those who do evil"* (Koran 5:64).

Jews are said to profit from involvement in all wars. Likewise they are alleged to have ruled humanity through the establishment of the League of Nations between the world wars. Most ludicrous is the claim of Jewish gain from WWII, where six million Jews perished in the Holocaust. As for the Jewish State, it almost did not come into being because the so-called "Jewish Problem" in Europe was virtually "solved" through annihilation. In 1948 the Arab world invaded the new-born State of Israel with the declared purpose of its extermination. As is documented, there are more UN resolutions castigating Israel than there are condemning any other state.

> The colonialist powers, both in the capitalist West and the communist East, support the enemy with all their might, both materially and with manpower, alternating one with the other [in giving support]. When Islam appears, all the forces of unbelief unite to oppose it, for all unbelief is one denomination.

The "capitalist West and the communist East" are accused of supporting the Jews. The *Covenant* was written in 1988, hence not only are the US and its allies said to be supporting Israel, but also the Soviet Union and the East Bloc. Lest one forget, Soviet Jews suffered persecution for decades and fought to leave the USSR because the situation was so dire. Many were jailed, suffered job loss and/or were denied exit permits. Finally succeeding in their battle for free immigration well over a million Jews immigrated to the West and Israel beginning in the 1970s.

> *"Oh you who believe, do not take as your intimate friends those outside your ranks, for they will spare no effort to harm you. They desire that which causes you suffering. Hatred has indeed come out of their mouths, but what they hide in their hearts is even worse. We have given you clear signs, if you understand"* **(Koran 3:118). It is not for nothing that the verse ends with His words** *"if you understand."*

The above schemes, plots and condemnations of the Jewish People originate in Czarist and Nazi antisemitism and are now blended into one entity with Jihadi Islam, best expressed by Hamas. Article 22 represents the full demonization of the Jews.

Chapter Four:
Our Positions on

A. The Islamic Movements

Article Twenty-Three

> The Islamic Resistance Movement looks on the other Islamic movements with respect and appreciation, for even if it is at variance with them in some given respect or thought, it agrees with them in many more respects or thoughts, and it views them - so long as their intentions are good and they are devoted to Allah - as falling under the rubric of legitimate opinion, that is, as long as their actions are within the bounds of the Islamic circle. Everyone who strives for truth receives his reward.

Theoretically and theologically the above paragraph could include the newly rising Islamic State. As of this writing Hamas and the global Muslim Brotherhood consider the Islamic State fanatical adversaries, yet they more often than not derive their ideological perspectives and incentives for action from the same Salafist sources (see Chapter V "Hamas Ideological Victory" and Chapter IX "Summary and Conclusions").

> **The Islamic Resistance Movement considers these movements as reinforcing it, and asks [Allah] to guide and direct us all. It never forgets to constantly raise the banner of unity and to strive assiduously to achieve unity in accordance with the Koran and the sunna.** *"Hold fast to Allah's rope, all of you. Do not be divided among yourselves, and remember Allah's favor to you. When you were enemies to one another, He brought your hearts together, and through his favor you became brothers. You were on the brink of a pit of fire, and He saved you from it. Thus Allah shows you His signs, so that you may follow the right way"* **(Koran 3:103). [11]**

Article 23 is a call for unity among all Islamic movements "in accordance with the Koran and sunna." Several years ago, Sunni Hamas expanded the interpretation of this clause to include the Shiites and therefore accept military and financial aid from Hezbollah and Iran to use in confronting Israel and the Jews. This change in policy took place after the death of Sheikh Ahmed Yasin (2004), who refused to work with the Shiites. As of 2011, there were disagreements with Tehran and reportedly certain ventures were suspended on the background of the recent Sunni Arab Islamic Awakening and the intervention of Turkish interests as directed by today's president but at the time, Prime Minister Erdogan. Joint efforts with Iran and Hezbollah can be resumed when deemed necessary.

Article Twenty-Four

> **The Islamic Resistance Movement does not allow to impugn or to blacken the name of individuals or groups, for [true] Muslims do not impugn or curse others. One should make a clear distinction between this and positions or behavior, for the Islamic Resistance Movement does have the right to**

expose error and to deter people from it and to strive to make the truth known and to adopt it in an impartial way in every given case. Wisdom is what the Muslim looks for, and he takes it wherever he finds it. [12]

Hamas has a mission to "expose error and deter people from it" while making "the truth known." Believers should steer others to the correct course of Islam as determined only by Hamas.

"Allah does not like it when people speak ill in public, except for those who have been wronged. Allah hears all and knows all. When you do good openly, or in secret, or forgive a wrong [done to you], surely Allah is Forgiving and Almighty" (Koran 4:148-149).

What of the *dhimmi* Jews and Christians? They have not been wronged? How should one view the fate what of those beneath the *dhimmi* status such as Zoroastrians, Bahai adherents and Hindus? Does Allah hear their voices or does he only hear the voice of the Islamist who oppresses the above mentioned groups?

B. The Nationalist Movements in the Palestinian Arena

Article Twenty-Five

[The Islamic Resistance Movement] respects them and appreciates the conditions that surround and affect them. It supports them as long as they do not pay allegiance to the Communist East or to the Crusader West, and it emphasizes to all the members and supporters [of these movements] that it is a jihadist, ethical movement, conscientious in its worldview and in its treatment of others. It abhors opportunism, wants only good for the people, both as individuals and as groups, and does not strive to attain material gain or fame for itself. It does not seek reward from people, and it goes forth with its own resources and what it has at hand - *"Muster against them all the force you can"* **(Koran 8:60) - in order to carry out the duty and win Allah's favor. It has no desire other than that.**

It reassures all the nationalist [groups] of all orientations that are operating in the Palestinian arena for the liberation of Palestine that it shall never be anything other than a support and an aid for them, in word and in deed, at present and in the future. It joins together and does not separate, preserves and does not scatter, unites and does not divide, it values every kind word, every sincere effort and every praiseworthy endeavor. It closes the door in the face of petty disagreements. It pays no heed to rumors and biased remarks, but is fully aware of [its] right to defend itself.

Anything that opposes or contradicts this orientation is fabricated by the enemy or by their lackeys in order to cause confusion, divide the ranks and create distraction with side issues. *"OH you who believe, if an evildoer brings you information [about any person], you should examine it carefully lest you hurt [innocent] people out of ignorance, and afterwards come to regret it"* **(Koran 49:6).**

Palestinian movements are forbidden any allegiance with foreign ideologies whether they are from the Communist world (atheism), as led historically by the Russians and Chinese, or the Crusader West of Christendom, represented by the US and Europe. Such alliances are viewed as "opportunism" and betraying the true "jihadist, ethical movement" of the Islamic Resistance. Any ideal besides that of Hamas only brings about a fracturing of the Jihadi movement.

In practice, Article 25 is an ideological attack against the PLO, whose greatest support came from the communist bloc until its collapse in Europe between 1989 and 1991. The PLO and its leader Yasir Arafat received some sympathy in Western Europe during this period, but on the diplomatic level they sought greater Western support in the 1990s, due to lack of choice. Alongside the collapse of communism, Arafat and the PLO lost a powerful patron in Iraq's secular Baathist leader Saddam Hussein as a result of his defeat at the hands of the Americans, Europeans and their Arab allies in the Desert Storm Gulf War of 1991. Arafat left Baghdad a day before the American air offensive in mid-January.

This clause is still fully relevant in the aftermath of the Oslo Accords (1993-95) and Wye Agreement (1998) which worked toward conflict resolution with Israel and were negotiated with the aid of the "Crusader West," in particular the United States.

Article Twenty-Six

The Islamic Resistance Movement - looking favorably as it does on the Palestinian nationalist movements - does not refrain from discussing new developments concerning the Palestinian problem in the local and international arena in an objective manner, so as to find to what extent [these developments] agree or disagree with the national interests in the light of the Islamic vision.

Article 26 is an invitation for the PLO to clarify its position in "light of the Islamic vision" of Hamas. Put bluntly the PLO is expected to reject secular nationalism and embrace the Hamas global Jihad initiative. Together the groups will merge under the Islamic banner and commence with the destruction of the State of Israel. The PLO/Palestinian Authority has not quite taken up the Jihadi challenge in the physical sense. Although in recent years there are many more calls for violence from the PLO and the ruling Fatah faction, the PA is working much more on the diplomatic front against Israel, in particular at the UN.

C. The Palestine Liberation Organization

Article Twenty-Seven

The Palestine Liberation Organization [PLO] is closest to the heart of the Islamic Resistance Movement. [We regard it as] a father, brother or friend, and a true Muslim does not spurn his father, his brother or his friend. Our homeland is one, our misfortune is one, our destiny is one and we share the same enemy.

Owing to the circumstances that surrounded the establishment of the PLO, and [owing to] the intellectual confusion which prevails in the Arab world as a result of the intellectual invasion to which it has been subject since

the defeat of the Crusaders, and which was intensified, and continues to be intensified by Orientalism and Christian missionary activities - the PLO has adopted the idea of the secular state, and we view [the PLO] accordingly. Secularist ideology stands in total contradiction to the religious ideology, and it is ideas which are the basis of positions, behavior and decisions.

There is expressed solidarity with the PLO as a blood relative and/or friend, and as one who shares a common destiny or fate. Here, Hamas extends sympathy and understanding because they share a common homeland and enemy. As well, Hamas is aware of the historic circumstances leading to the establishment of the PLO, which were heavily influenced by Western–secular ideas molding Arab and Palestinian nationalism. Still, Hamas holds to the idea that secularism is in complete contradiction to Islamic fundamentalism. In wishing to establish a secular state, Hamas believes the PLO is a victim not only of "Christian missionary activities," but of Orientalism, since advocating for a secular state is a Western invention.

"Orientalism" is the study of Eastern cultures by Western scholars. Arabs often accuse researchers of being insensitive through their study of the East as an "object." In particular, the Christian Palestinian scholar Edward Said is known for his attacks against Western academics and their criticisms of the Arab Middle East. Said wrote his famous book *Orientalism,* to condemn Western research and attitudes when investigating the Middle East. Hamas accuses the PLO of being part of the crime since their "Western" view of history is said to be analytical, rooted in deductive reasoning and devoid of Sharia law. On the other hand, Said showed allegiance to the PLO and believed in secular humanism, equal rights for all, advocated for democracy and sympathized with Jews as a persecuted group despite his anti-Israel stance. He bewailed the disappearance of "*ijtihad*" or critical thinking in Islam.[24] Hamas would certainly accuse Said of taking a secular *jahili* Christian missionary Orientalist approach.

The West's so-called "intellectual invasion" commenced in the nineteenth century when missionaries began their liaison with the Eastern Christian communities, particularly in the Beirut and Damascus urban centers. The contours of secular Arab nationalism began to develop and eventually brought in Muslim sympathizers, which overrode Sharia law. Hamas blames the secularization of the Arab world on Christian Europe and its Middle Eastern Christian co-religionists. The Hamas analysis is essentially correct. Had Enlightenment ideals such as "equality for all" not impacted Islam, Sharia law would continue to rule and European nation-state style governments in the region would be greatly curtailed or possibly not exist at all (see Chapter II "Ideologues" in particular the subsection on Abdullah Azzam).

> **Hence, with all our appreciation for the Palestine Liberation Organization and what it may yet become, and without belittling its role in the Arab-Israeli conflict, we cannot give up the Islamic identity of Palestine in the present and in the future to adopt the secularist ideology - for the Islamic identity of Palestine is part of our faith, and whoever is lax with his faith is lost.** *"Who spurns the religion of Abraham but one who has made himself into a fool?"* **(Koran 2:130).**

Hamas condemns the PLO for advocating a secular society, an obvious Christian European influence. Palestine must be Islamic—this is the one and only truth. Using the Koranic quote Hamas claims "the religion of Abraham" to be Islam.

> **When the PLO adopts Islam as its way of life, then we shall be its troops and the fuel for its fire that will burn the enemies. But until this time comes - and we pray to Allah that it be soon - the position of the Islamic Resistance Movement vis a vis the PLO is that of a son toward his father, a brother toward his brother or a relative toward his relative. He shares the other's pain when he is pricked by a thorn, and supports him in facing the enemy, and he wishes for him to find divine guidance and [follow] the right path. Your brother, your brother before all others! He who has no brother is like one who goes to war unarmed. Your cousin, you must know the strength of his wing, for how can the falcon rise up without wings? [13]**

The Hamas Covenant acknowledges appreciation for the PLO struggle, yet Hamas continues to emphasize their belief that the Islamic faith is the only acceptable identity for Palestinians. Despite expressing solidarity with the PLO as a father or brother, Hamas will only join them when the PLO "adopts Islam as a way of life."

D. Arab and Islamic States and Governments

Article Twenty-Eight

> **The Zionist invasion is a cruel invasion, which has no scruples whatsoever; it uses every vicious and vile method to achieve its goals. In its infiltration and espionage operations, it greatly relies on secret organizations which grew out of it, such as the Freemasons, the Rotary Clubs, the Lions and other such espionage groups. All these organizations, covert or overt, work for the interests of Zionism and under its direction, and their aim is to break societies, undermine values, destroy people's honor, create moral degeneration and annihilate Islam. [Zionism] is behind all types of trafficking in drugs and alcohol, so as to make it easier for it to take control and expand.**

Article 28 is a continuation of the conspiracy theories condemning secular civic organizations such as the Freemasons, the Rotary and the Lions, who in reality are all dedicated to bettering society. Hamas accuses the groups of advocating "Zionism" and in helping in the proliferation of the drug and alcohol markets to destroy society through "moral degeneration" and to "annihilate Islam." These organizations are liberal universalist in their concerns and the attacks against them for immoral behavior are preposterous. The accusations of "trafficking in drugs and alcohol" even go beyond certain slanders of the Czarist *The Protocols of the Learned Elders of Zion* against the Jews and Freemasons.

> **We demand that the Arab countries around Israel open their borders to jihad fighters from among the Arab and Islamic peoples, so they may fulfill their role and join their efforts to the efforts of their brothers - the Muslim brethren in Palestine. As for the rest of the Arab and Muslim countries, we demand that they facilitate the passage of the jihad fighters into them and out of them - that is the very least [they can do].**

All Arab countries must aid their Jihadist brethren, at least allowing them passage to attack Israel. Such a demand lines up with the minimum expected from neighboring Muslim countries during "defensive Jihad" (see Chapter II "Ideologues").

> **We should not fail to remind every Muslim that when the Jews occupied the Sacred place [i.e., Jerusalem] in 1967 and stood on the threshold of the blessed Al-Aqsa mosque, they shouted: "Muhammad is dead; his offspring are women." Israel with its Jewish identity and Jewish people is challenging Islam and the Muslims. May the cowardly know no sleep.**

To rally the Arab world, included is a reminder of the Israeli conquest of Jerusalem and the Al-Aksa mosque in 1967. To claim Israelis shouted "Muhammad is dead; his offspring are women," is absurd. Rather this article shows the Hamas contempt and misogynous attitude toward women, as they consider it an insult to say Muhammad's descendants are female. For the Islamic attitude toward women see Articles 17 and 18 above.

Upon the Jordanian surrender of Jerusalem close to fifty years ago and the symbolic "handing over of the keys," to the Temple Mount or Noble Sanctuary domain of the Al-Aksa Mosque and Dome of the Rock to Israel's Defense Minister Moshe Dayan, he returned the keys to those same Muslim religious *waqf* authorities who offered their submission. In so doing Israel showed the highest respect for Islam. From 1967 until what is known as the "Second Intifada" or Low Intensity Conflict-Terror Offensive in October 2000, there was freedom of visitation for all people to the mosques. Such open visitation was done in the spirit of the separation of state and religion. Hostilities have calmed down for the most part, yet since the year 2000 entry into the mosques is allowed only to Muslims and preferred "others" given special permission by the *waqf* authorities. Everyone else is allowed to visit the Temple Mount, but cannot enter the mosques. Due to everyday tensions and the separation between religion and state, Israel does not enforce the right of visitation by non-Muslims nor does Israel allow for Jews or Christians to pray on the Temple Mount. The right of prayer is exclusively Muslim, a "status quo" arrangement actively discriminating against non-Muslims.

The second-to-last sentence in the final paragraph "Israel with its Jewish identity and Jewish people is challenging Islam and the Muslims" proves Hamas cannot deny the fact that Israel is the Jewish State. That the Jewish

State has repulsed all Muslim attempts to destroy it contradicts the interpretations of the Koran by Islamic jurists over the centuries claiming the covenant with the Israelites/Jewish People was nullified. The fact that Hamas included this line in its *Covenant* is a challenge to Islam and Muslims who demand nullification of the Jewish State. Otherwise one could surmise that the continued existence of the State of Israel has Divine support, hence the almost hysterical demand for its destruction. As for the "cowardly," see Chapter II "Ideologues," most notably the subsection on Abdullah Azzam.

Nationalist and Religious Groups and Organizations, Intellectuals, and the Arab and Islamic World

Article Twenty-Nine

The Islamic Resistance Movement hopes that these groups will stand by it in every respect, help it, espouse its positions, back its activities and strive to enlist support for it, so that the Muslim peoples will be a support and a reinforcement for it, and [will provide] strategic depth on all levels: [in terms of] human and material resources, information, in every time and every place. [This should be done] by holding conferences, publishing ideological pamphlets, and by indoctrination of the masses with regards to the Palestinian issue - what is facing [the Palestinians] and what is plotted against [them]. [Likewise they should work to] mobilize the Islamic peoples, ideologically, educationally and culturally, so that they will play their role in the decisive war of liberation just as they did in defeating the Crusaders and routing out the Mongols, thus saving human civilization. This is not difficult for Allah.

"Allah has decreed: 'I and My messengers shall prevail'; Allah is strong and all-powerful" (Koran 58:21).

All cultural and educational groups in the Muslim world have an obligation to support Hamas in its struggle against their enemy the Jews, just as the Muslim world is said to have united to defeat the Mongols (1260 at Ein Jalut) and Crusaders (1265 and final destruction 1291) some seven to eight hundred years ago, "thus saving human civilization." This includes "indoctrination of the masses" to guarantee their mobilization in the decisive "war of liberation," better known as Jihad. The Egyptian non-Arab Mamluk leader Baybars is credited with the great victories of the 1260s yet as noted previously in the Introduction, Sheikh Hasan al-Banna defines the Mamluks as spiritually inferior despite their conquests in the name of Islam.

Article Thirty

Writers, intellectuals, media people, preachers in mosques, educators and all the various sectors in the Arab and Islamic world, are all required to fulfill their role and perform their duty. [This is necessary] due to the ferocity of the Zionist onslaught and the fact that it has infiltrated many countries and has taken control of the finances and media - with all the ramifications that follow from this - in most countries of the world.

Jihad is not limited to wielding arms and fighting the enemies face to face, for eloquent speech, persuasive writing, effective books, support and help - when [they are] performed with the sincere intention that Allah's banner will reign supreme - all constitute jihad for the sake of Allah.

[As the prophet said:] *"Whoever equips a warrior fighting for the sake of Allah is [himself] a warrior, and whoever supports the family of a warrior [who has set out to fight for the sake of Allah] is [himself] a warrior"* **(recorded by Bukhari, Muslim, Abu Da'ud and Tirmidhi in their Hadith collections).**

Whoever supports the military and takes orders from Hamas joins the Jihad. Aiding a warrior is another stipulation for Jihad (see Chapter II "Ideologues"). People of influence in the Arab/Muslim world—intellectuals, preachers, journalists and educators are called upon as loyal soldiers with no discretionary powers, but rather are "required to fulfill their role and perform their duty" in the battle against the "Zionist onslaught," who are once again accused of controlling the world's finances and media.

Article 30 is a restatement of Article 22 (see above) when claiming Zionist/Jewish control over world finances and the media. As noted in the commentary for Article 22, *Protocols* and Hitler's *Mein Kampf* made the same reprehensible accusations.

Followers of the Other Religions
The Islamic Resistance Movement is a Humane Movement

Article Thirty-One

The Islamic Resistance Movement is a humane movement which respects human rights and is committed to the Islamic tolerance toward the followers of the other religions. It is hostile only to those among them who display hostility toward it or stand in its way, hampering its activities and foiling its efforts. Under the wing of Islam, the followers of the three religions - Islam, Christianity and Judaism - can coexist in security and safety. It is only under the wing of Islam that safety and security prevail. Recent and ancient history provide the best evidence for this. The followers of the other religions should stop competing with Islam for sovereignty in this region, because, when they rule, there is nothing but carnage, torture and deportation, and they cannot get along with their own, let alone with followers of other religions. Both the present and the past are full of evidence for this.

Article 31 embodies both deceit and honesty. The Islamic Resistance Movement is not a "humane movement which respects human rights," as stated in the above and in previous articles. Throughout *The Hamas Covenant,* there are demands for Jihad and annihilation of all enemies, in particular the Jews. The key phrase to accurately describe Hamas' tolerance is "Islamic tolerance," which stipulates Jews and Christians are to live under the confines of the *dhimma,* or second-class status, and accept the full authority and superiority of Islam. The comment "they cannot get along with their own" refers to the various Christian sects who clashed with each other throughout the centuries in the Middle East. Hamas claims that if all Christians lived as *dhimmi* under Sharia law, there would be no conflicts between Christian denominations. It is possible that the Judaism referred to above relates to the tiny group of rabidly anti-Zionists, and at times the Holocaust-doubting Neturei Karta type and its affiliates—who have an overall population of several thousand, yet gain tremendous media attention, especially in the Arab Muslim world. They are well-known for supporting all enemies of the State of Israel, whether it is Iran's Ahmedinejad or Hamas.

Article 31 makes the claim that "only under the wing of Islam can safety and security prevail." This is partially true because of Hamas' self-proclaimed need for Jihad against any group with sovereignty, whether it be the Jews in Israel, Christian power-sharing in Lebanon or Coptic Christian equal rights in Egypt. The process of Jihad against these groups began in the seventh century when the Arab Islamic invasion overran the southern part of the Byzantine Empire and completely destroyed the Persians. The Turks finally wiped out the Byzantines during the sack of Constantinople in 1453—long after Jihad spread Islam to North Africa, Spain, Portugal and even to southern France. Eventually Christian forces retook the Iberian Peninsula in January 1492.

Omitted by Hamas are all of the inter-Arab, inter-Muslim wars plaguing Middle Eastern history from the time of the Arab Muslim invasion (638) to the present. Following are examples of the lack of "safety and security" when Islam prevails. Just to name a few: the Ummayad rise to power (661), the Abbasid defeat of the Ummayads (750), the massacre of Shiites at Karbala by the Sunnis (680), the numerous revolts against the Abbasids (750-1258), the Ismaili revolts in the Middle Ages, the Fatamid wars against the Ismailis, the Fatamid-Seljuk wars (900s-1000s), and the Mamluk territorial wars from the mid-thirteenth century until their defeat by the Ottoman Turks in the sixteenth century. During the Turks' four hundred year rule in the Arab world, from the early 1500s until 1917, there were numerous Muslim rebellions within the Ottoman Empire in addition to constant wars with the Persian Shiites. The twentieth century witnessed a civil war in Yemen where Egypt and Saudi Arabia squared off against each other in the mid-1960s, and the Iraq-Iran War (1980-1988) with well over a million dead and wounded. In both cases chemical weapons were used, first by the Egyptians and then by the Iraqis. Massacres in Sudan, Libya and Somalia accompanied by intra-Islamic Sunni versus Shiite strife in Iraq and Yemen continue to plague the Muslim world into the twenty-first century. Most recently the Islamic State (ISIS/ISIL) is terrorizing large swaths of the Middle East in particular in Iraq and Syria, the latter once again using chemical weapons, this time against its own population.

Hamas supporters constantly point out the supposed expansive liberal understandings of the Muslim Brotherhood toward Jews and Christians, always omitting the Charter of Omar and *dhimma* disabilities forced upon these populations, which would make their lives difficult at best and often quite miserable.

> *"They do not fight you together, except from within fortified villages, or from behind walls. They fight fiercely with one another. You consider them to be united, but their hearts are divided, for they are a people with no sense"* **(Koran 59:14).**

This quote from the Koran refers to the numerous battles Mohammed waged against the Jews. It is used in *The Hamas Covenant* to recall the 1948 Israeli defense line of fortified kibbutzim in the Negev who were given much credit for halting the Muslim Brotherhood and Egyptian army offensive directed toward Tel Aviv in Israel's War of Independence. Sayyid Qutb makes reference to this quote in his essay "Our Struggle With the Jews" Using it as a rallying cry for the Egyptian onslaught into the new born Jewish State. Jews and/or Israelis may have sharp disagreements with each other, but they are not involved in a civil war and certainly do not "fight fiercely with one another."

> **Islam accords to every person his rights, and prevents any infringement on other people's rights. The Nazi Zionist measures against our people will not succeed in prolonging the duration of their invasion, for the rule of injustice lasts but one hour, while the rule of the truth will last until the Hour of Resurrection.**

"Islam accords to every person his rights" is another deceitful statement. These rights are only according to Islamic Sharia law, and Hamas will be the arbiter of it when they rule. The simple conclusion is that the Zionist invasion (Israel) must collapse since it is not the "rule of truth." Hamas' great fear is that some form of negotiated agreement will take hold between Israelis and Palestinians. They do not want Muslims to continue to live in the Jewish State while holding Israeli citizenship and accepting Israeli sovereignty. To do so would undermine the Islamic "rule of the truth," which is said to "last until the Hour of Resurrection."

In the year 2000, the Israeli-Arab weekly newspaper *Kul al-Arab* conducted a poll to determine whether Arabs living in Israel would agree to living under the Palestinian Authority as part of a land swap between Israel and the

PA in order to ensure national and demographic consistency on both sides of the future border. The results showed that even in Umm el-Fahm, a Muslim Arab-Israeli city controlled by the Islamic Movement, well over 80 percent of the city's residents claimed they wanted to remain in Israel and retain Israeli citizenship in any peace accords with the Palestinian Authority.[25]

The Islamic Movement in Israel is theologically akin to Hamas and denies Israel's right to exist as a Jewish State. Anti-Israel as they may be, its members served and serve on Arab municipal councils and were elected to Israel's parliament, the Knesset. The best example of this seeming contradiction is former Knesset member Abdulmalik Dehamshe, who served from 1996-2006. Dehamshe was not only a member of the Islamic Movement in Israel, but he also believes in the destruction of the Jewish State. As of the time of this writing, in the mixed municipalities of Akko, Ma'alot and Misgav, coalitions have developed where Arabs serve as deputy mayors, although their populations are one-third or less than the total population. All of this begs the question: why would Arabs in Israel want to live and serve under a "Nazi Zionist" regime and how do Israeli Arab Muslims view this "rule of injustice?"

> ***"Allah does not forbid you to show kindness and act justly toward those who do not fight you on account of your religion, and do not drive you from your homes. Allah loves those who act justly"*** **(Koran 60:8)**.

Islamists are not forbidden "to show kindness and act justly toward those" who are not doing battle against Muslims; however, a Muslim is not obligated to behave in a positive manner toward such people. The conclusion here is, a believer is not commanded to show these people mercy either—they are not Muslims and not deserving of simple mercies. A Muslim is permitted, however, to take extra steps to treat non-Muslims with kindness.

The Attempt to Isolate the Palestinian People

Article Thirty-Two

World Zionism and the colonialist powers attempt, by clever maneuvering and meticulous planning, to pull the Arab States, one by one, out of the circle of the conflict with Zionism, so as to ultimately isolate the Palestinian people. It has already taken Egypt out of the circle of conflict to a large extent through the treacherous Camp David Accords [of September 1978], and it is trying to pull additional [Arab] countries into similar agreements so that they leave the circle of conflict.

In Article 32, Hamas accuses worldwide-Zionism and the Western powers of trying to make peace. Hamas fully condemns all conflict resolution as agreed upon in the "Israel-Egyptian Peace Treaty" and the "Framework for Peace in the Middle East," between Israeli Prime Minister Menahem Begin and Egyptian President Anwar Sadat in 1978-79 at Camp David, as well as what was achieved under the American auspices of President Jimmy Carter. The "Framework for Peace" obligates the parties to work for a "full and comprehensive peace," and was meant to include every state in the region putting an end to all conflict. Peace is founded on mutual recognition, secure, recognized boundaries and normalization.

Hamas condemned Egypt for leaving the conflict and would condemn Jordan for its peace agreement with Israel in 1994, fifteen years later. The Egyptian wing of the Muslim Brotherhood assassinated President Sadat in October of 1981. Thirty years earlier, in July 1951, a member of the Muslim Brotherhood assassinated King Abdullah I of Jordan after it became known that he secretly initiated peace talks with Israel in late 1950. Israel and Jordan were on the verge of ending the conflict. From Israel's inception the Muslim Brotherhood is consistent in its efforts to halt the peace process.

The Islamic Resistance Movement calls upon all the Arab and Muslim peoples to strive seriously and diligently to prevent this horrible scheme, and to alert the masses to the danger inherent in leaving the circle of the conflict with Zionism. Today it is Palestine, and tomorrow some other country

or countries, for the Zionist plan has no limits, and after Palestine they want to expand [their territory] from the Nile to the Euphrates, and when they finish devouring one area, they hunger for further expansion and so on, indefinitely. Their plan is expounded in *The Protocols of the Elders of Zion*, and their present [behavior] is the best proof for what we are saying.

Hamas lambastes the Camp David Accords as a "horrible scheme," citing supposed Zionist plans for expanding from the Nile to the Euphrates, all the while not explaining the apparent contradiction of Israel's second return of Sinai to Egyptian rule by 1982 as part of the 1979 Peace Agreement. The first time Israel returned Sinai to Egypt was in 1957 after the 1956 Sinai Campaign. As well, Hamas negates Israel's and Jordan's reconciliation and mutual recognition of each other's independence and territorial integrity. Geographically, Sinai sits between Israel's Negev Desert and the Nile River of Egypt, while the entire Jordanian kingdom is located between the Jordan River and Iraq, home of the Euphrates. The Camp David Accords and peace agreement with Jordan lay bare the outright lies and perfidy of Hamas claims on this issue.

Hamas used *Protocols* as their ultimate truth despite hard facts to the contrary. Article 32 shows the complete Hamas endorsement of all the false accusations made in *The Protocols of the Elders of Zion*. A supposed "Jewish super-state" is discussed (Protocol 5, Clause 11 and Protocol 9, Clause 3), but it does not define borders. Rather the Jews are accused of first devouring Europe over the ages and only finally coming to Jerusalem, which is found in the Introduction of the *Protocols*. Hamas transmits the same spirit of lies and deceit into the Middle East by claiming, "the Zionist plan has no limits" and using as supposed proof the desire to capture land from the Nile to the Euphrates.

Writing in *Mein Kampf*, Hitler claims that any Jewish State is "completely unlimited as to territory." He uses a racial explanation concerning the Jewish inability to live in a finite region as the reason for a borderless state that attempts to dominate the world. Hamas does not use racial terms but makes the same claim stating, "When they [the Jews] finish devouring one area, they hunger for further expansion and so on, indefinitely."

Following the Nazi lead, Hamas embraced *The Protocols of the Elders of Zion* as a fundamental text in its war for Jewish destruction. Whereas the Nazis succeeded in murdering six million Jews, Hamas is still in the declarative state of calling for the death of the Jews (Article 7) and a Jihad or Holy War to accomplish that mission (in particular Articles 8, 13, 15, 33 and quotes below).

> **Leaving the circle of the conflict with Zionism is an act of high treason; all those who do this shall be cursed.** *"Whoever [when fighting the infidels] turns his back to them, unless maneuvering for battle or intending to join another [fighting] company, he incurs Allah's wrath, and his abode shall be hell. Most unfortunate is his fate"* (Koran 8:16).

In this passage of *The Hamas Covenant*, anyone making a peace agreement with Israel is engaged in "high treason" and will be damned to hell. Logically, assassination of such traitors is not only legitimate, but obligatory.

> **All forces and capabilities must be pooled to confront this ferocious Mongol, Nazi onslaught, lest homelands be lost, people be exiled, evil spread on the earth and all religious values be destroyed. Each and every person should know that he is responsible to Allah.**
>
> *"Anyone who does a grain's weight of good shall see it, and anyone who does a grain's weight of evil shall see it"* (Koran 99:7-8).
>
> **In the circle of the conflict against world Zionism, the Islamic Resistance Movement sees itself as a spearhead or as a step forward on the road [to victory]. It joins its efforts to the efforts of all those who are active in the Palestinian arena. It now remains for steps to be taken by the Arab and Islamic world. [The Islamic Resistance Movement] is well qualified for the upcoming stage [of the struggle] with the Jews, the warmongers.**
>
> *"We have planted enmity and hatred among them [i.e., among the Jews] until the Day of Resurrection. Every time they kindle the fire of war, Allah extinguishes it. They strive to spread evil upon the earth, and Allah does not love those who do evil"* (Koran 5:64).

Hamas views the Jews and Zionism as the ultimate evil. The Arab Islamic world must aid in the destruction of the Jews and their state. Jihad is the only answer against Zionism and the Jews—notice once again there is no differentiation. The Koranic verse refers to the Jews and is a continuation of two previous verses (5:62-63), which accuse the Jews of being blasphemous and doing evil. Hamas attempts to demonstrate Divine proof of satanic-Jewish intrigues.

Article Thirty-Three

The Islamic Resistance Movement - proceeding from these general concepts which are in harmony and agreement with the laws of nature, and following the current of divine destiny toward confrontation with the enemies and jihad against them in defense of Muslims, Islamic civilization and Islamic sanctities, primarily the Al-Aqsa mosque - calls on the Arab and Islamic peoples and their governments, and on their NGOs and official organizations, to fear Allah in their attitude toward the Islamic Resistance Movement and in their treatment of it. They should act toward it as Allah wants them to, namely back it, support it, assist it and continuously reinforce it, until Allah's word is fulfilled. Then the ranks will all be united, jihad fighters will join other jihad fighters, and the masses all over the Islamic world will rush out and answer the call of duty, shouting: "Hasten to jihad!" This call will penetrate the clouds in the sky and continue to ring out, until liberation is accomplished, the invaders are defeated and Allah's victory is revealed.

"Allah surely helps whoever helps Him; Allah is strong and mighty" (Koran 22:40).

Article 33 is the Hamas self-proclamation of Divinity. Hamas flows with Allah's "laws of nature" as it declares Jihad against its enemies. As a summation of previous articles, this is yet another call for total unity among Arab Muslims and the Islamic world to support and join the Jihad against Zionism (Israel) and the Jews. Heavily inferred is the unity necessary for universal Jihad.

Chapter Five:
Historical Evidence Throughout the Generations Regarding Confrontation with Aggressors

Article Thirty-Four

Palestine is the center of the Earth and the meeting place of the continents; it has always been the target of greedy aggressors. This has been the case since the dawn of history. The Prophet, Allah's prayer and peace be upon him, points to this in his noble words with which he addressed his exalted companion, Mu'adh b. Jabal, saying: *"Oh Mu'adh, Allah will give you the land of Al-Sha'm after my death, from Al-'Arish to the Euphrates. Its men, women and handmaids will be [constantly] stationed on the frontier until the Day of Resurrection, for any one of you who chooses [to live in] some part of the coastal plains of Al-Sha'm or Bayt Al-Maqdis [i.e., Palestine], will be in a [constant] state of jihad until the Day of Resurrection."*

The aggressors coveted Palestine on many occasions. They attacked it with great armies in attempt to realize their greedy aspirations. The great armies of the Crusaders came there, bringing their religious creed and hoisting their cross. They managed to defeat the Muslims for a while, and the Muslims only managed to regain it when they fought under their religious banner, joined forces crying "Allah Akbar," and set forth in jihad under the command of Salah Al-Din Al-Ayyubi for nearly two decades, which led to a clear victory: the Crusaders were defeated and Palestine was liberated.

According to Islam, the Christian Crusaders lost to Jihad and now it is the Jew's turn to suffer defeat in the never-ending battle for an Islamic Palestine. "Palestine is the center of the Earth" and has known many wars and will continue to be a battlefield until the Day of Resurrection or the End Time. The Islamic invasion led by the Caliph Omar in 638 captured the Land of Israel/Palestine from the Christian Byzantines. This is further emphasized by the Prophet Mohammed's promise to give Mu'adh b. Jabal "the land from El

Arish to the Euphrates." Notice that the Prophet Mohammed himself refers to the Holy Land as "Bayt Al-Maqdis." This is a direct translation from the Hebrew "Bet Hamikdash," meaning "The House of the Holy," or the Temple. Here we have another full admission in Islamic literature of the Jewish connection to Jerusalem and the Land of Israel.

> *"Say to the unbelievers: You will surely be defeated and gathered in Hell. Most terrible shall be your resting-place"* **(Koran 3:12).**

By quoting Koran 3:12, *The Hamas Covenant* implies and accuses both Christians and Jews as "unbelievers." In this instance the verse is used out of context, as its true reference is to "Pharaoh's people" or idolaters/polytheists as cited in the previous two Koranic verses 3:10-11. This Koranic quote is purposefully taken out of context and used as "proof" of Christian and Jewish non-belief and condemnation to Hell.

The Hamas Covenant builds new theological contexts. Some refer to *The Hamas Covenant* as "The Charter of Allah," which is now holy writ. Koran Chapter 3 begins, "He has revealed to you the Book with the truth confirming the scriptures which preceded it; for he has already revealed the Torah and the Gospel for the guidance of men, and the distinction between right and wrong" (Koran 3:2-3). This Koranic quote is at least a partial endorsement of Judaism and Christianity, not a condemnation of these two religions as Hamas would have us believe. However, the Hamas text is full of hatred and Jihad and could not allow belief that any good could come from Christianity or Judaism.

> **This is the only way to liberation, and one cannot doubt the testimony of history. This is one of the rules of the universe and the laws of reality. Only iron can break iron, and their false, fabricated faith can only be overcome by the true faith of Islam, for religious faith cannot be attacked except through religious faith. And truth shall eventually triumph, for truth is the strongest.**

> *"We have already given Our Word to Our servants, the messengers, that they would be helped to victory and that Our army will triumph"* (Koran 37:171-173).

By innuendo, Hamas accuses Judaism and Christianity of being a "false, fabricated faith" to be defeated by the true devotion of Islam. The quote from the Koran concerns idol worshipers, as can be noted in the previous verses (37:158-170). Once again, *The Hamas Covenant* attempts to manipulate readers by quoting the Koran completely out of context to ensure condemnation of Christians and Jews.

Article Thirty-Five

The Islamic Resistance Movement studies the defeat of the Crusaders at the hands of Salah Al-Din Al-Ayyubi, the liberation of Palestine from them, as well as the defeat of the Mongols at 'Ayn Jalut and the breaking of their military strength at the hands of Qutuz and Al-Zahir Baybars, and the delivery of the Arab world from the Mongol conquest which destroyed all aspects of human civilization. [14] [The Islamic Resistance Movement] studies these [events] seriously, and draws lessons and examples from them. The current Zionist invasion was preceded by Crusader invasions from the west, and by Mongol invasion from the east. And just as the Muslims faced these invasions, made plans for fighting them and defeated them, they can [now] confront the Zionist invasion and defeat it. This is not difficult for Allah, providing that intentions are sincere and resolve is strong, and that Muslims draw benefit from the experiences of the past, shed off the influences of the intellectual invasion, and follow the ways of their predecessors.

Hamas believes the "Zionist invasion" will fail, as did the Christian Crusaders and the Mongols before them. Reading this far in *The Hamas Covenant*, one is expected to classify all invasions together as that of infidels to be destroyed by the Divine forces of the Islamic Jihad. The "intellectual invasion" is that of the secularists and Orientalists; these too will meet their demise at the hands of the Jihadists.

Conclusion:
The Islamic Resistance Movement-Soldiers [for the Cause]

Article Thirty-Six

The Islamic Resistance Movement, in its march forward, repeatedly emphasizes to all members of our people and of the Arab and Muslim peoples that it is not seeking fame for itself, or material gain, or social status, and that it is not aimed against any of our own people in an attempt to compete with them or to take their place - nothing of the kind. It does not oppose any Muslim, or any non-Muslim who is peaceful toward it, here [in Palestine] or elsewhere. It will always offer nothing but help to all groups and organizations that strive against the Zionist enemy and against its lackeys.

The Islamic Resistance Movement adopts Islam as its way of life. [Islam] is its creed and its law. [Any group that] adopts Islam as its way of life, here or elsewhere - be it an organization, association, state or any other group - the Islamic Resistance Movement will serve as its soldiers. We ask Allah to guide us, to guide [others] through us, and to judge between us and our people with truth.

"Oh, Lord, judge between us and our people with truth. You are the best of judges" (Koran 7:89).

At the end, we beseech: Praise be to Allah, Lord of the Universe.

Hamas poses as an altruistic movement claiming it seeks no fame in its Jihad, seeking only mutual aid and friendship to all those who do battle against Zionism. It does not oppose "any non-Muslim who is peaceful toward it," meaning those subservient, non-Muslims accepting Islamic religious law relegating them to the *dhimma* status and not asserting themselves in any independent manner—this being the condition for peace (see *HC* Article 31). Islam is the way of life for the Islamic Resistance Movement, Hamas. Anyone adopting Islam in the same manner as determined by Hamas will find support from Hamas as allies and soldiers.

Endnotes from Memri Translation

[1] Islamonline, http://www.islamonline.net/Arabic/doc/2004/03/article11.SHTML.

[2] Hasan al-Banna (1906-1949) founded the Muslim Brotherhood in 1928 and was its Director General until his assassination in 1949.

[3] Amjad Al-Zahawi was an Iraqi Sunni religious scholar who was affiliated with the Muslim Brotherhood and active in various initiatives to support the Palestinian cause.

[4] Hamas is the Arabic acronym for Islamic Resistance Movement (harakat al-muqawama al-islamiyya); it is also an Arabic word meaning enthusiasm, ardor or zeal.

[5] Due to the importance of the concept of Jihad in the Hamas ideology, we leave this term in Arabic wherever it appears in the text.

[6] The verses of Muhammad Iqbal (1873-1938), an Indian Muslim poet and religious thinker, are often used by both reformist and conservative Muslims in support of their opposite orientations.

[7] This is an oft-quoted verse by the famous pre-Islamic poet Tarafa.

[8] In medieval Islamic writings, Al-Sha'm usually refers to an area roughly corresponding to present-day Israel, Palestine, Lebanon, Jordan and Syria.

[9] The word "nationalism" here and elsewhere in this document is used as the equivalent of the term wataniyya, which is derived from the Arabic word watan ("homeland"), and in modern Arabic discourse signifies particularist territorial nationalism, in contradistinction to qawmiyya, which also means "nationalism," but is used to refer to pan-Arab nationalism.

[10] The phrase "the followers of the Islamic movements" is used here to translate the Arabic term al-islamiyyun.

[11] In the Arabic original the verse is mistakenly given as 3:102.

[12] This is a well-known Hadith (i.e., a saying attributed to the Prophet).

[13] These are two oft-quoted verses by the seventh century poet Miskin al-Darimi.

[14] Saif Al-Din Qutuz (d. 1260) was the Mamluk Sultan of Egypt from 1257 until his death. In 1260, the commander of his army, Al-Zahir Baybars (1223-1277), defeated the Mongols in the battle of Ain- Jalut.

Copyright© 2006 The Middle East Media Research Institute

Summary and Conclusion

The Muslim Brotherhood and its Palestinian Arab faction Hamas fully believe in the superiority of Muslims above all others. In their minds, Islam is the perfect religion and a calling all humanity should accept. A Holy War, or Jihad, is to be waged against all who deny such "truths" and the deniers will be defeated just like the Mongols, Crusaders and the People of the Book were in previous centuries. The People of the Book are expected to be granted the right to exist under Islamic rule provided they adhere to certain harsh stipulations known as the Charter of Omar (see Chapter I "Negative Image of the Jew").

Islamists such as those belonging to Hamas believe Jews and Christians should only be allowed *dhimmi* "advantages" as the People of the Book. Yet the contradiction exists in the more extreme punishments allocated for the Jews. In the Introduction to *The Hamas Covenant*, the reference to Israel by Hasan al-Banna—both the state and people—shows them slated for abolishment. Immediately afterward, in the Preamble, Hamas identifies the Jews as the overall enemy. Not only are Arab Muslims to defeat Zionism and take back Palestine, but they are to kill the Jews, as is explicitly declared in Article 7. At this point, Jews slide from *dhimmi* status to dehumanization. This downgrading follows the process, which began in Czarist Russia and culminated in the "The Final Solution" of Nazi Germany. In the finale the Jews are to suffer a total loss of rights and the vast majority is to be annihilated.

The Hamas Covenant is at least as much an antisemitic document par excellence as it is an Islamic call to action. Although extremist Islamist regimes existed in the past, the venom spewed here against Jews shifts closer in line with the *Protocols* and Hitler's *Mein Kampf* than it does to other Islamist works from the past. Hamas takes a major step in favor of the Nazi option of extermination through its constant demand for Jihad against the Jews. Only a tiny number of completely submissive Jews will be allowed to survive. As a window into the Muslim Brotherhood worldwide, *The Hamas Covenant* exposes the shift from traditional Islamic repression and institutionalized antisemitism, as legislated through the Charter of Omar and *dhimma* stipulations, to the adaptation of a grand policy of near total Jewish extermination.

As for Christianity, the West, the Far East and everywhere else around the globe, humanity will either convert to Islam or be subject to its universal government and legal system upon the inevitable Jihadist victory. Secular states populated by Muslims are to be destroyed and replaced by Islamist entities. Hamas is one cog in the wheel, using the Israel/Palestinian front as their sector in the universal Jihad. Focusing on Israel and Palestine gives Hamas the veneer of a territorially-based regional nationalism, downplaying the organization's continual claim of fulfilling its role in world Islamic conquest.

The most dangerous aspect of *The Hamas Covenant* is that it is defined as the word of Allah, indisputable, never to be changed. The Hamas authors believe in a Divine decree to destroy Jews. The world is to face Jihad until there are no challenges anywhere to the absolutist Hamas Islamist doctrines or "truths."

Endnotes

1. Translation of *The Hamas Covenant,* MEMRI – Middle East Media Research Institute Special Dispatch - No. 1092, February 14, 2006, No. 1092. Usage permission granted by Yigal Carmon through e-mail communication on June 23, 2014. Full credit given for English translation by MEMRI.
2. al-Banna, Hasan, *The Five Tracts of Hasan al-Banna*, translated by Wendell, Charles ed., University of California Press, Berkeley, 1978, Tract "Between Yesterday and Today," p. 19
3. Tamimi, Azzam, Hamas, *The Inside Story*, Olive Branch Press, Northampton, Mass, USA, 2007, pp. 147-152.
4. Shirer, William, *The Rise and Fall of the Third Reich*, Simon and Schuster, New York, 1960, p. 233.
5. Ibid, p. 965.
6. Qutb, Sayyid, "Our Struggle With the Jews," in Nettler, Ronald, *Past Trials and Present Tribulations: A Muslim Fundamentalist's View of the Jews*, Vidal Sasson International Study of Antisemitism, Hebrew University, Jerusalem, Israel, 1987, p. 75.
7. Ibid, p. 76.
8. Both the concept of fusion with the beloved and satisfying the demanding, conditional fatherly love are developed by Erich Fromm, the famous philosopher psychoanalyst in his renowned work *The Art of Loving*, 1956.
9. Lewis, Bernard, *The Arabs in History*, Harper and Row Publishers, New York, 1966, p. 71.

10 *Koran*, Chapter or Sura 17 entitled "The Night Journey" opens with this passage. From verses 17:2-8 there is a brief theological discussion concerning the First and Second "Israelite" Temples.
11 Ben Dov, Meir, Jerusalem, *Man and Stone*, Modan Publishing House, Tel Aviv, Israel, 1990, p. 282.
12 "Totalitarianism," *Encyclopedia Britannica*, retrieved November 3, 2014, www.britannica.com/topic/totalitarianism ..
13 al-Banna, Hasan, Tract "Our Mission," pp. 109 and 113.
14 "What does the Koran say about women?," *Freethought Nation*, retrieved October 29, 2015, freethoughtnation.com/what-does-the-koran-say-about-women.
15 Exhibit in Historic Museum, short film "Book Burning," *Yad VaShem Holocaust Museum and Documentation Center*, Jerusalem, Israel.
16 Benvenisti, Meron and Khayat, Shlomo, *The West Bank and Gaza Atlas*, The Jerusalem Post, Jerusalem, 1988, p. 27.
17 Ibid, p. 109.
18 Hasson, Nir, "Demographic Debate Continues // How Many Palestinians Actually Live in the West Bank?," *Haaretz*, retrieved October 1, 2014, http://www.haaretz.com/israel-news/.premium-1.532703.
19 Abu Toameh, Khaled, "Palestinian Population in W. Bank, Gaza, about 4.5 million," *The Jerusalem Post*, November 7, 2013, retrieved October 1, 2014, http://www.jpost.com/National-News/Palestinian-population-in-W-Bank-Gaza-about-45-million-319569.
20 "East Jerusalem by the Numbers," *Association for Civil Rights in Israel*, May 7, 2013, retrieved October 1, 2014, www.acri.org.il/en/2013/05/07/ej-figures.
21 "East Jerusalem," *Wikipedia*, retrieved October 1, 2014, en.wikipedia.org/wiki/East_Jerusalem.
22 Patterson, David, *A Genealogy of Evil, Anti-Semitism from Nazism to Islamic Jihad*, Cambridge University Press, New York, 2010, Chapter 7, "Secular Offshoots: The Baath Party and the PLO."
23 Lewis, Bernard, *Semites and Anti-Semites*, W. W. Norton and Co., New York and London, 1999, see "Afterword to the Norton Paperback Edition," pp. 260-272.
24 Said, Edward, *Orientalism*, Vintage Books, New York, (1978) 2003 edition, Preface to the Twenty-Fifth Anniversary Edition.
   Previously, in his book *Covering Islam* (1981), Said accused the Western news media of enormous bias while he presented Islam as much more liberal and equality based. The Jihadi Islamists would not agree but we do not know what Said's stand would be in 2015 as he died in Sept. 2003. Politically Edward Said served on the Palestine National Council of the PLO from 1977-91, but resigned when he felt the PLO under Arafat was conceding too much to Israel in negotiations.
25 Dershowitz, Alan, *The Case for Israel*, John Wiley and Sons Inc., Hoboken, NJ, 2005, p. 223.

West Bank

Joint Israeli-Palestinian Authority rule since 1994.
Jewish settlements not indicated.

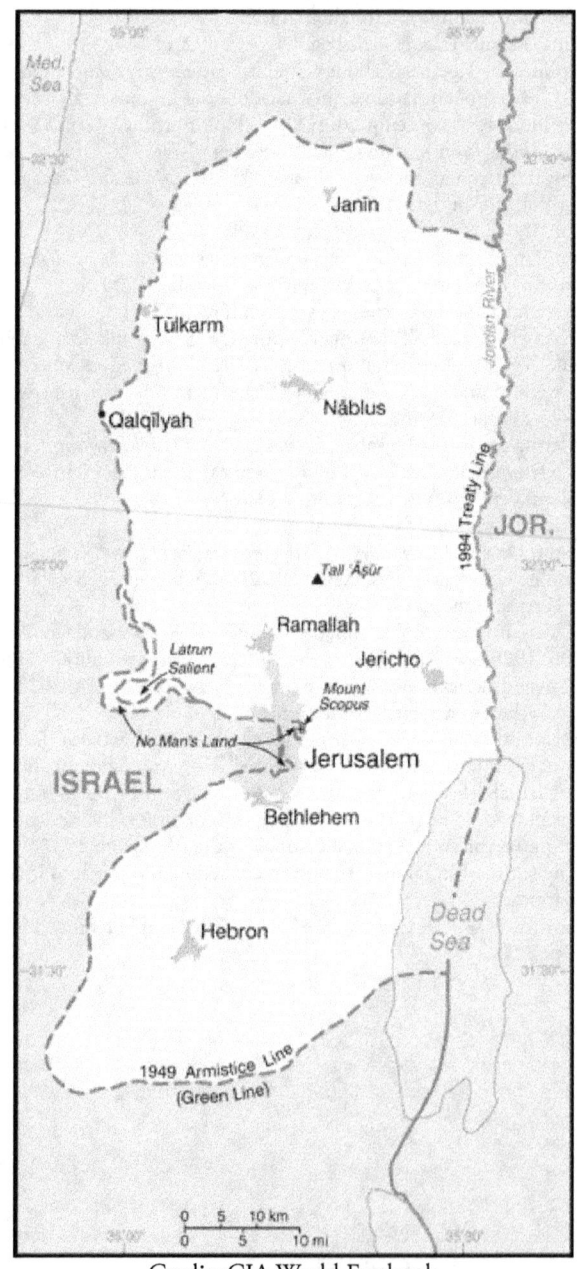

Credit: CIA World Factbook

VII

A Comparative Analysis

The Palestinian National Charter and The Hamas Covenant

Overview

Although the secular Palestine Liberation Organization (PLO) and Hamas are often considered bitter enemies, when examining their seminal doctrines we see the two have much in common. *The Palestinian National Charter (PNC)* of the Fatah-led PLO and *The Hamas Covenant (HC)* both call for the destruction of Israel and the liberation of Palestine through armed struggle. At the conclusion of the Wye Accords in late 1998, the PLO was obligated to amend all references to Israel's destruction from the *PNC*, in order to stay within the Oslo Accords framework for conflict resolution between Israel and the Palestinians. To date the Palestinians have not changed the verbiage in the *PNC*, despite letters of intent and declarations of promise by PLO officials and the late Yasir Arafat. *The Hamas Covenant,* considered "The Charter of Allah," will never change. It must remain in its original form—rock solid in its intent, an everlasting religious document. The two documents share a similar spirit and worldview when calling for Israel's destruction. When relating to Jews the PLO embodies a concealed antisemitism while the Hamas position is overtly manifest. Islam most definitely influenced the PLO, while secular Palestinian nationalism impacted Hamas actions and rhetoric. The major difference between the two documents is their scope of conflict. *The Hamas Covenant* advocates world Jihad and an almost universal destruction of the Jews, while neither world Jihad nor Jewish annihilation are declared objectives of *The Palestinian National Charter (PNC)*.

The Palestinian National Charter

Below is the full text of *The Palestinian National Charter (PNC)* with explanations and commentary.[1] The *Charter* is the foundational document of the Palestine Liberation Organization (PLO). It expresses the organization's identity, objectives and strategy in its struggle against Israel. Below each section are comparisons between *The Hamas Covenant (HC)* and the specific *PNC* Articles indicated. Relevant commentaries from Professor Yehoshafat Harkabi's book, *Palestinians and Israel*[2] are cited for a fuller understanding of secular

Palestinian nationalism in its original 1960s and 1970s context. The analysis and conclusions presented when comparing the two texts are fully my own.

THE PALESTINIAN NATIONAL CHARTER
Resolutions of the Palestine National Council, July 1-17, 1968

PNC, Article 1: **Palestine is the homeland of the Arab Palestinian people; it is an indivisible part of the Arab homeland, and the Palestinian people are an integral part of the Arab nation.**

The *PNC* uses the term "Arab" to emphasize Arab nationalism as the secular and cultural definition. "Palestinian" indicates the specific region in the Middle East ranging from the Jordan River to the Mediterranean Sea corresponding to the "Land of Israel" as defined by Jews and Judaism. Historically, the Jihadi Islamic conquests of the seventh century CE imposed the Arab cultural and linguistic heritage on the Middle East. Hence, even the term "Arab" insinuates a Muslim cultural acquiescence although non-Muslims are included, such as Christians living in the region. The *PNC* makes their Arab declaration for the Western audience, who view the term "Arab" as referring exclusively to language and culture while obscuring the Muslim aspects of the "Arab" identity.

HC Comparison:

The Palestinian branch of the Muslim Brotherhood emphasizes its identity as first and foremost Islamic (*HC*, Article 2). Arab and Palestinian identities are secondary and there is no room for Christians or others. In the *HC*, secularism is the antithesis to religious ideology. Palestinian national movements are evaluated based on their attitude toward Islam. When the PLO adopts Islam as its identity and way of life, Hamas envisions unification between the two movements (*HC*, Article 27).

PNC, Article 2: **Palestine, with the boundaries it had during the British Mandate, is an indivisible territorial unit.**

Palestine is an indivisible territorial unit between the Jordan and the Mediterranean Sea including the Negev Desert. The *PNC* does not allow for a two-state solution option whereby Arab and Jewish States exist side by side in "peace and security" as is often stated by those advocating conflict resolution.

HC Comparison:

The *HC* addresses the issue of "Palestine" as important, even if artificial, in order to sound more palatable to the general Arab and specifically Palestinian Arab public. The overall battle for Palestine is overwhelmed by the demand for its full incorporation into the Islamic *waqf* or "endowed lands" belonging to Islam for eternity (*HC,* Article 11). Hamas envisions conquering all land worldwide. In the Islamist struggle, Hamas is only one unit in the global Jihad army whose responsibility lies on the narrow front called "Palestine."

> ***PNC, Article 3:* The Palestinian Arab people possess the legal right to their homeland and have the right to determine their destiny after achieving the liberation of their country in accordance with their wishes and entirely of their own accord and will.**

The *PNC* emphasizes the Palestinian Arab people's legal and national right "to their homeland." Palestinian rights and self-determination are understood to be absolute by definition. The *PNC* declared an inalienable "right" to their homeland, which requires implementation. Although secular because the people determine their own "destiny," one hears Islamic religious overtones as only Palestinian Arabs have a right to the land.

HC Comparison:

Palestine is *waqf* land belonging to Islam forever. Legal rights originate from *Sharia* law—the law of Allah (*HC,* Article 11). Islam *always* determines the people's destiny.

> ***PNC, Article 4*: The Palestinian identity is a genuine, essential, and inherent characteristic; it is transmitted from parents to children. The Zionist occupation and the dispersal of the Palestinian Arab people, through the disasters which befell them, do not make them lose their Palestinian identity and their membership in the Palestinian community, nor do they negate them.**

As pointed out by Harkabi, this *PNC* article is a Palestinian Law of Return, similar to the secular laws pertaining to Jews immigrating to the State of Israel. Once a Palestinian, always a Palestinian; a nationality passed on from one generation to the next. This is a secular definition.

HC Comparison:

Secular Palestinian identity is not the beginning. One's identity as a Muslim is far more important (*HC*, Articles 1 and 2). Hamas does not emphasize issues concerning refugee status, as hinted at in the *PNC*, Article 4, since the conquest of Palestine is a pan-Islamic responsibility and not just for those directly affected by the conflict. It makes little difference if there was displacement or not. Islam must regain Muslim *waqf* lands. Such action is a universal Islamic responsibility (*HC*, Article 7, paragraph 1).

Neither the PLO nor Hamas admit any responsibility for the "disasters which befell" the Palestinians although conflict resolution was at hand in 1947-48 if the two-state solution Partition Plan (UN Res. 181 – Nov. 29, 1947) was accepted. The Arab and Muslim world, including Palestinians, reject all compromise.

> ***PNC, Article 5:*** **The Palestinians are those Arab nationals who, until 1947, normally resided in Palestine regardless of whether they were evicted from it or have stayed there. Anyone born, after that date, of a Palestinian father - whether inside Palestine or outside it - is also a Palestinian.**

As shown by Harkabi, *PNC* Article 5 is a continuation of Article 4. Only Arabs are Palestinians, Jews are excluded. The Palestinian identity is passed down through the father, a tradition originating in Islam. Despite the "secularism" of the *PNC*, no reference is made to the status of the mother.

Through UN resolutions and the establishment of the United Nations Relief Works Agency (UNRWA), Palestinian refugee status became eternal and is passed down to the next generation, whether through the mother or father, as opposed to the general refugee status of other people in similar circumstances. Other refugees worldwide have done their best to begin life anew. The Palestinians use their refugee status to their political and diplomatic advantage. Such a claim is possible because it is made against a non-people, the Jews, as we will see below in *PNC*, Article 20. Implementation of full Palestinian refugee return translates into the destruction of the Jewish State, nullifying the two-state solution.³ UN Resolution 194 (December 1948) allowed for "compensation" to replace the demand for refugee return.

HC Comparison:

Islam is the determining factor as to whom the land belongs, not whether an individual was a Palestinian or the descendant of a Palestinian. Reference to exile is made in *HC* Article 20.

In real time Hamas demands full Palestinian refugee return as the major condition for even a limited *hudna* of several years, this to ensure Israel's demise. Such insistence solidifies support for Hamas among refugees and their descendents. Whenever the PLO considered options other than full return, such as compensation, Hamas insisted that such policies are a betrayal of the Palestinian Muslim cause.

> ***PNC, Article 6:*** **The Jews who had normally resided in Palestine until the beginning of the Zionist invasion will be considered Palestinians.**

Harkabi points out that the "Zionist invasion" or the implementation of Jewish national rebirth is considered to have begun with the Balfour Declaration in 1917 by the Palestinians. The above Article 6 states immigrating Jews cannot be considered "Palestinians" and presumably can be expelled. Jews who lived in Palestine before 1917 and their descendants will be considered citizens in the future Palestinian State. According to *PNC* Article 5, only Arabs are Palestinians. Hence, we see a contradiction in the *PNC*. How can these "acceptable" Jews ever achieve full and equal status in a Palestinian Arab State? This inconsistency appears intentional, denying Jews full equality in the Arab world. Questions arise as to how to trace the lineage of each and every Jew and who will determine which Jews can or cannot live in a Palestinian State. How will denial of citizenship and expulsion be implemented for those not meeting the PLO criteria?

HC Comparison:

The Hamas Covenant advocates eliminating the Jews, for they are the enemy. Zionism is seen as the secular state extension of overall Jewish perfidy. Technically, Jews can live under Jihadist–Islamist rule accepting the *dhimma* status. In reality, only the tiny rabidly anti-Israel ultra-orthodox Jewish Netura Karta sect and their allies would willingly accept the *dhimmi* status. The Jewish date of arrival in Islamic Palestine would be of no importance, but rather the Jews' willingness to prostrate themselves before their Muslim overlords (see Chapter VI *"The Hamas Covenant* Analysis," Article 31). All other Jews would conceivably be annihilated.

> *PNC, Article 7:* **That there is a Palestinian community and that it has material, spiritual, and historical connections with Palestine are indisputable facts. It is a national duty to bring up individual Palestinians in an Arab revolutionary manner. All means of information and education must be adopted in order to acquaint the Palestinian with his country in the most profound manner, both spiritual and material, that is possible. He must be prepared for the armed struggle and ready to sacrifice his wealth and his life in order to win back his homeland and bring about its liberation.**

Insisting Palestinians are raised, educated, and informed in "an Arab revolutionary manner" while engaging in the "armed struggle" is a concept parallel to the Hamas understanding of Jihad. The *PNC* expects Palestinian individuals to make sacrifices. The *PNC* uses the term "Arab" as opposed to "Muslim" or "Islamic" (*HC* Article 15, paragraph 1 and Article 16). Palestinians are connected to the land in every manner, the same way Muslims are connected to the *waqf* lands (*HC,* Article 11). The PLO demands Palestinian Arabs must be informed and educated concerning the struggle and "ready to sacrifice his wealth and his life" to liberate the homeland. Here the *PNC* uses more nationally focused language since it does not speak of world Jihad or in secular terminology the "armed struggle" in a religious sense, as does *The Hamas Covenant.* Still, the *PNC* requires the same course of action involving ultimate material and personal sacrifice through overall commitment.

HC Comparison:

The direct line between the *PNC* and *HC* originates with the concept of "defensive" Jihad (see Chapter II "Hamas Ideologues," subsections—al Banna and Azzam). In the *HC* Articles 12, 15 and 16, similar sacrifices of wealth and life are demanded for the national cause. In *HC* Article 19 Islamic art makes its own specific contribution. Hamas condemns secular nationalism as lacking the full spiritual commitment Islam demands. The *HC* explains the Jihad imperative as far superior to all others—a hint that shows Hamas' opinion that the secular PLO is not serious. Furthermore, the *HC* states the major role of the Muslim woman is to impart a Jihadi education to her children (*HC*, Articles 17 and 18). We do not find a role particular to women in the *PNC*.

> *PNC, Article 8:* **The phase in their history, through which the Palestinian people are now living, is that of national *(watani)* struggle for the liberation of Palestine. Thus the conflicts among the Palestinian national forces are secondary, and should be ended for the sake of the basic conflict that exists between the forces of Zionism and of imperialism on the one hand, and the Palestinian Arab people on the other. On this basis the Palestinian masses, regardless of whether they are residing in the national homeland or in diaspora *(mahajir)* constitute - both their organizations and the individuals - one national front working for the retrieval of Palestine and its liberation through armed struggle.**

Article 8 of the *PNC* is a call for unity to fight Jewish nationalism (Zionism) and imperialism, or Western influence as described in the developing world's national and "liberationist" literature of the 1960s and 1970s. At the time, factionalism set in between the larger more mainstream Fatah, and the left wing Popular Front for the Liberation of Palestine (PFLP) and Democratic Front for the Liberation of Palestine (DFLP) groups. All Palestinians are called upon to fight for the "liberation of Palestine through armed struggle," whether they live in the "homeland" or abroad.

HC Comparison:

PNC Article 8 is similar to the Hamas call for world Islamic unity (*HC*, Article 23). The difference is *The Hamas Covenant* expands into condemnations of both the East and West (*HC*, Articles 25 and 26), an implied criticism of the PLO. There is a fundamental difference in scope as the PLO is only addressing Palestinians in the Palestinian-Israeli (Jewish national) arena while Hamas sees the Palestinian Islamic struggle as one link in the chain of an overall battle to conquer the world. Jihad, at first defensive, transforms to offense and continues until universal victory.

> *PNC, Article 9:* **Armed struggle is the only way to liberate Palestine. Thus it is the overall strategy, not merely a tactical phase. The Palestinian Arab people assert their absolute determination and firm resolution to continue their armed**

struggle and to work for an armed popular revolution for the liberation of their country and their return to it. They also assert their right to normal life in Palestine and to exercise their right to self-determination and sovereignty over it.

Although built into a secular nationalist setting this strategy of the "Armed struggle" is the same as the Hamas call to Jihad. There are no compromises. Calls for "self-determination," "sovereignty," and "the right to normal life in Palestine" mean Arab domination and the full denial of any Jewish national rights. These words are euphemisms for Israel's destruction. Such was the leftist rhetoric of the 1960s, focusing on the armed struggle by the "*fedayeen*" or resistance fighters (Harkabi). Their function is the same as the "*mujahideen*" guerilla groups, which at times were not necessarily religious or even Muslims, as condemned by Abdullah Azzam (see Chapter II "Ideologues"). The word *mujahideen* originates from the root "j-h-d" and defines such armed groups as Jihadists in the popular understanding. Calls for the armed struggle of the *fedayeen* are couched in secular terms but have roots in total victory as advocated by Jihadism.

HC Comparison:

These absolutist ideals began as Islamic and mutated into an all-encompassing secular Arab Palestinian nationalism, and then reappeared in *The Hamas Covenant*. *HC* Article 3 demands Jihad in the general cleansing sense while *HC* Article 8 defines Jihad as its "path," "way," or "methodology," depending on which translation is used. Jihad is the strategy and not simply a tactic. Hamas calls all Muslims to Jihad (*HC* Articles 13, 15, 30, 32, and 33), while the *PNC* confines "their armed struggle" and "armed popular revolution" to Palestinian Arabs. These *PNC* concepts are similar to Hamas' Jihad, but in a more focused context. The PLO plans to re-emerge in its pure state once again via the calls for the destruction of Israel, calls similar to those in *The Hamas Covenant*. The solution is the annihilation of the State of Israel (Harkabi) and its replacement by a Palestinian Arab State.

For Hamas, there is little difference between destruction of the Jewish State and the destruction of Jews. On the theological level, the Israeli State is to be replaced by an Islamic one. Hamas goes well beyond the *PNC* with vicious antisemitism. The concept of self-determination, sovereignty and "normal life in Palestine" are expressed in Islamic domination (*HC*, Articles 6 and 31), demanding second-class *dhimmi* status for Jews and Christians.

> ***PNC, Article 10:*** **Commando action constitutes the nucleus of the Palestinian popular liberation war. This requires its escalation, comprehensiveness, and the mobilization of all the Palestinian popular and educational efforts and their organization and involvement in the armed Palestinian revolution. It also requires the achieving of unity for the national (*watani*) struggle among the different groupings of the Palestinian people, and between the Palestinian people and the Arab masses, so as to secure the continuation of the revolution, its escalation, and victory.**

"Commando action" is a deceptive term because it includes terror attacks against all Israeli Jews, civilians as well, defined as the Zionist enemy. Engaging in commando action is a step prior to becoming a suicide-homicide bomber. Both types of attacks are often against civilians; however in commando action, the attacker plans to escape. In any case, all Palestinians and Arabs are expected to unify in participation to achieve victory.

HC Comparison:

In the *HC* a suicide-homicide bomber sees death as his or her escape. Bombers believe they fuse with Allah by implementing the "Divine" desire to kill the Jews (*HC,* Article 7, 8 and 28 last paragraph). Hamas upgrades from the *PNC's* command to kill Zionists, to the killing of Jews in general. The *HC* echoes the *PNC's* call for mobilization and national unity among all Palestinian Muslims. Hamas specifies and intensifies the demand for full Islamic unity in battling Israel and the Jews. *HC* Articles 17 and 18 define the role of women in the mass Jihadi struggle as opposed to no specific clause as such in the *PNC*. The above *PNC* Article 10 probably inspired *HC* Articles 15, 16 and 30 with their specific demands for total participation in the destruction of Jews and Zionism.

> ***PNC, Article 11:*** **The Palestinians will have three mottoes: national (*wataniyya*) unity, national (*qawmiyya*) mobilization, and liberation.**

For the PLO "national unity" is a love of nation or a fusing of the national being into one—unity similar to other national movements. On the theoretical level, liberation is national and physical for the secular PLO, as the *PNC* speaks of "national mobilization, and liberation." This is the secular version of unification with Allah—the ultimate goal for believing Muslims.

HC Comparison:

The Hamas motto is found in *HC*, Article 8.

> **Allah is its goal,**
> **the Prophet its model to be followed,**
> **the Koran its constitution,**
> **Jihad its way,**
> **and death for the sake of Allah its loftiest desire.**

The Hamas objective to fuse with Allah by way of a Jihadist death is considered one's "loftiest desire." The Prophet Mohammed is their ultimate example. For Hamas, the Koran is the legal document or "constitution," which lays out the mobilization leading to the Holy War/Jihad or religious war of "liberation" in the name of Allah (see Chapter II "Ideologues" and Chapter VI *"The Hamas Covenant* Analysis.")

Hamas demands the physical liberation of all Palestine for Islam, while invoking spiritual and religious terminology and seeking the same land-oriented objectives of total conquest over Palestine: the destruction of the State of Israel.

The mottoes appear to have a different emphasis but they serve similar objectives in reference to land. The difference is in loyalty and "way of life." The *PNC* is loyal to the secular nation while Hamas is loyal to Allah and the Islamic nation. These are the ideological roots of the clash between the PLO and Hamas (*HC*, Article 27).

> *PNC, Article 12:* **The Palestinian people believe in Arab unity. In order to contribute their share toward the attainment of that objective, however, they must, at the present stage of their struggle, safeguard their Palestinian identity and develop their consciousness of that identity, and oppose any plan that may dissolve or impair it.**

The Palestinians are one part of the puzzle of secular Arab unity or Pan-Arabism. Perpetuating the Palestinian identity within an overall Arab unity is vital to achieve that all-encompassing national goal. Due to their struggle, the special characteristics of Palestinian identity contribute to Pan-Arab unity. Palestinians insist they are part of a greater whole, but fear being fully assimilated into an overall Arab nation, a position they cannot outwardly admit, because to do so would expose their suspicions as to the ulterior motives of their Arab allies. There is both cooperation and tension between the specific Palestinian Arab identity and the inclusive Pan-Arabism. This is particularly acute as concerns Syria, which traditionally claimed Palestine, Lebanon and Jordan as part of what is called "Greater Syria."

HC Comparison:

In the *HC*, the identity issue is much simpler. Muslim identity and universal Islamic solidarity are expected, but Hamas emphasizes a specific Palestinian Arab identity because of the PLO/*PNC* influence. For Hamas, Palestinian uniqueness is superficial. The Palestinian Arab identity is secondary to the greater Islamic nation, as constantly expressed in *The Hamas Covenant* and Muslim Brotherhood understandings. Hamas rejects the specifics of a secular Palestinian Arab identity (*HC,* Article 25, 26 and most importantly 27). The *PNC* emphasizes Pan-Arabism (Harkabi) while Hamas demands Pan-Islamism.

> *PNC, Article 13:* **Arab unity and the liberation of Palestine are two complementary objectives, the attainment of either of which facilitates the attainment of the other. Thus, Arab unity leads to the liberation of Palestine, the liberation of Palestine leads to Arab unity; and work toward the realization of one objective proceeds side by side with work toward the realization of the other.**

The PLO expects the Arab world to participate in the liberation of Palestine as a function of Arab unity. Article 13 became a challenge to the Arab world in the mid-twentieth century, in particular to Egypt's Nasser and the Syrian Baath leadership. If they could not liberate Palestine, then the struggle for Arab unity is questionable. Harkabi shows the liberation of Palestine to be a Pan-Arab unifying point. It would be the "Big Bang" leading to the singularity of identity and purpose in the Arab world. To fail on the Palestinian front is to fail in Arab unity.

HC Comparison:

The *HC* presentation of unity is exactly the same as the *PNC* when it comes to confronting the Muslim world and challenging it to step up for Palestinian Islamic liberty. The *HC* translates the conquest of Palestine as the continuing proof of cohesiveness and Islamic unity. Like Pan-Arab nationalism mentioned above, participation in universal Jihad for Palestine facilitates union and tests true loyalties (*HC*, Articles 14, 15, 28, 29, 30 and 32). For Hamas, the successful liberation of Palestine is only one more step in world Islamic conquest. Pan-Arabism claims a specific part of the globe—the Arab world. Hamas, as part of the Muslim Brotherhood, lays claim to the entire world—one front at a time after Muslim lands are recovered (see Chapter II "Ideologues" subsection on Abdullah Azzam).

> **PNC, Article 14:** **The destiny of the Arab nation, and indeed Arab existence itself, depend upon the destiny of the Palestine cause. From this interdependence springs the Arab nation's pursuit of, and striving for, the liberation of Palestine. The people of Palestine play the role of the vanguard in the realization of this sacred *(qawmi)* goal.**

Article 14 is a continuation of Article 13, and clarifies the "vanguard" role of Palestinian Arabs in the realization of not only their own destiny, but of the overall Arab cause and Arab unity itself. To fail is to threaten the continuation of the Arab existence (Harkabi). The Palestinians have appointed themselves the commandos of Arab unity through their own front-line struggle; a confrontation they believe belongs to the entire Arab world.

HC Comparison:

Substitute the word "Muslim" for the word "Arab," and *PNC* Article 14 becomes the Hamas Islamist position in principle. Failure on the Palestinian front is equal to a worldwide Islamic setback. Palestinian Muslims may be the "vanguard," but all others are obligated to aid in "defensive" Jihad as stated in *HC* Articles 14, 15, and by inference Article 3 (see Chapter II "Ideologues," subsections on al Banna and Azzam).

The Hamas Covenant raises a question of concern for Arab national existence in the event of failure to "liberate" Palestine. The root of this question comes from the Jihadi diatribes expressed by Qutb against the Jews and their nation state plot—the State of Israel. For both Arab nationalists and Islamists, the Jewish State is situated in the center of their national and religious homeland. They see Israel as an occupying force trying to spread evil. Evil is defined as Jewish perfidy originating in the seventh century or a Jewish form of imperialism, as detractors define Zionism. Islamists determine both as Jewish evils. This is arguably what is known as a "zero sum game," or defined as "winner takes all." Israel's existence contradicts Arab nationalism (Harkabi), just as Jewish continuity nullifies the Islamic essence for Hamas (*HC,* Article 28, last paragraph).

> ***PNC, Article 15:*** **The liberation of Palestine, from an Arab viewpoint, is a national** *(qawmi)* **duty and it attempts to repel the Zionist and imperialist aggression against the Arab homeland, and aims at the elimination of Zionism in Palestine. Absolute responsibility for this falls upon the Arab nation - peoples and governments - with the Arab people of Palestine in the vanguard. Accordingly, the Arab nation must mobilize all its military, human, moral, and spiritual capabilities to participate actively with the Palestinian people in the liberation of Palestine. It must, particularly in the phase of the armed Palestinian revolution, offer and furnish the Palestinian people with all possible help, and material and human support, and make available to them the means and opportunities that will enable them to continue to carry out their leading role in the armed revolution, until they liberate their homeland.**

The *PNC* obligates the "Arab nation" throughout the world to aid the Palestinians in every way possible in their struggle to eliminate Zionism (the State of Israel).

HC Comparison:

Hamas makes parallel demands of Muslims to participate in the struggle against Israel and world Jewry. Palestine is a Pan-Arab confrontation line or, in the case of Hamas, a Pan-Islamic Jihadist front line (*HC,* Articles 28, 32 and 33).

Most significantly, the *HC* emphasizes the integration of the three circles: The Palestinian, the Arab, and the Islamic, with the latter having the greatest influence. *HC* Article 14 specifically insists on this point. It is through the Palestinian and Arab overlap that Hamas constructs the conduit whereby *PNC*/PLO understanding and loyalty can be channeled to the Islamists themselves. The *PNC* emphasizes the Arab/Palestinian identity, but as they call on the Arab world there is the latent Islamic understanding. This overlap of loyalties facilitates the shift of secular Palestinian nationalist supporters to Hamas.

> **PNC, Article 16:** **The liberation of Palestine, from a spiritual point of view, will provide the Holy Land with an atmosphere of safety and tranquility, which in turn will safeguard the country's religious sanctuaries and guarantee freedom of worship and of visit to all, without discrimination of race, color, language, or religion. Accordingly, the people of Palestine look to all spiritual forces in the world for support.**

Only secular Palestinian Arab rule will end the conflict and bring about the perfect solution of equality, freedom of religion, "safety and tranquility." Let us recall that Jews were forbidden from entering the secular Arab Hashemite Kingdom of Jordan and therefore had no access to holy sites in the West Bank under Amman's jurisdiction from 1949-1967. Never were Palestinian voices raised against such discrimination. Both the secular PLO and Hamas claim everyone will enjoy freedom of religion, but we must think of "freedom of religion" from an Islamic standpoint—not a Western one. Yet the PLO presents the future Palestinian State as an open liberal society and calls upon "all spiritual forces in the world for support" for a policy with no realistic basis for implementation. Under Israeli jurisdiction from June 1967 until the Second Intifada or Low Intensity Conflict in 2000 there were full visitation rights for everyone at holy sites, regardless of religious affiliation. From October 2000 and continuing to the present, the Palestinian Muslim religious authorities have as a general rule refused to let Jews, Christians or other non-Muslims enter the Al-Aksa Mosque or Dome of the Rock structures on the Temple Mount in Jerusalem.

HC Comparison:

The *HC* claims all religions will live in perfect harmony under Islamic rule. This means Sharia law and the second-class *dhimmi* status for Jews and Christians (Articles 6 and 31), and no promises of access to holy sites. Under Hamas rule we can expect everyone except Muslims to be barred from Islamic holy sites.

VII A Comparative Analysis

> *PNC, Article 17:* The liberation of Palestine, from a human point of view, will restore to the Palestinian individual his dignity, pride, and freedom. Accordingly the Palestinian Arab people look forward to the support of all those who believe in the dignity of man and his freedom in the world.

The Israeli (Jewish) victory was and continues to be a terrible humiliation for the Palestinians and Arab world as a whole. Victory, meaning the destruction of Israel, will return "dignity" to the Palestinians.

HC Comparison:

The *HC* understanding of the Palestinian humiliation is theological. Palestine is *waqf* land consecrated for Muslim ownership until Judgment Day (*HC*, Article 11). Muslim dignity, pride and sovereignty will be restored with the re-conquest of *waqf* land. The End of Days or Judgment Day is delayed from arrival due to the existence of the Jewish State and the survival of the Jewish People within a sovereign state entity. According to the *HC*, not only is dignity at stake with the existence of a Jewish State, but world redemption is delayed. Victory through Jihad will eliminate all obstacles (*HC*, Article 7, 8 and 9).

> *PNC, Article 18:* The liberation of Palestine, from an international point of view, is a defensive action necessitated by the demands of self-defense. Accordingly, the Palestinian people, desirous as they are of the friendship of all people, look to freedom-loving, and peace-loving states for support in order to restore their legitimate rights in Palestine, to re-establish peace and security in the country, and to enable its people to exercise national sovereignty and freedom.

The PLO deems Israel an illegal entity (Harkabi); therefore any attack against her is legal. Israel embodies implied original sin, thereby relegating any defensive actions illegal. By definition anyone supporting the Palestinian cause of the destruction of the State of Israel desires peace and freedom. In the *PNC*, the expression "defensive action" is used instead of making clear PLO demands to initiate conflict. The PLO defines its supporters as "freedom-loving" and "peace-loving," regardless of whatever their nature might be. States invited to support the Palestinian cause may be Muslim or non-Muslim.

HC Comparison:

Attacking Israel is viewed as "defensive" Jihad, since Palestine belonged to the Islamic world until the end of WWI. Allah gives Hamas' its "legitimate rights," yet as an organization and political movement they depend on the Islamic world and in particular the Muslim Brotherhood for friendship and support (*HC*, Articles 2, 28, 29, 30, 32 and in particular 33). The *HC* does not seek alliances with non-Muslims although in certain cases temporary cooperation can be considered.

> ***PNC, Article 19:*** **The partition of Palestine in 1947 and the establishment of the state of Israel are entirely illegal, regardless of the passage of time, because they were contrary to the will of the Palestinian people and to their natural right in their homeland, and inconsistent with the principles embodied in the Charter of the United Nations, particularly the right to self-determination.**

Beginning with the problem of the "international point of view" mentioned in Article 18 and continuing through the supposed illegality of the Partition Plan (UN Resolution 181) and the establishment of the State of Israel, Article 19 is as deceitful as the previous one. The Palestinian birthright is one twin in the two-state resolution passed by the UN in November 1947. The international body recognized both Jewish and Palestinian Arab nationalism at the same time. The vote was more than the necessary two-thirds majority: 33 in favor, 13 against, 10 abstentions and one no-show. The abstentions and no-show are not included in the final tally as per UN voting rules. Apparently the member states did not see themselves as violating their own UN Charter.

HC Comparison:

The *HC* offers a simpler explanation. Islam is the only answer and Allah is the only adjudicator, therefore Palestine as an Islamic *waqf* is part of the Muslim world and no humans have the right to give it to others or split any part of it (*HC*, Articles 11 and 13). The illegality of Israel is a theological issue, not one given to discussion in human forums. People who accept the existence of the Jewish State thereby abandon the struggle against Zionism and incur Allah's condemnation (*HC*, Articles 13 and 32).

PNC, Article 20: **The Balfour Declaration, the Mandate for Palestine, and everything that has been based upon them, are deemed null and void. Claims of historical or religious ties of Jews with Palestine are incompatible with the facts of history and the true conception of what constitutes statehood. Judaism, being a religion, is not an independent nationality. Nor do Jews constitute a single nation with an identity of its own; they are citizens of the states to which they belong.**

PNC, Article 20 is a continuation of Article 19. The British issued the Balfour Declaration and then the League of Nations, the predecessor of the United Nations, incorporated it into the Palestine Mandate. The Palestine Mandate is an internationally drafted and recognized legal commitment. The British ruled Palestine and implemented the Jewish National Home policy through the Palestine Mandate authority granted them by the League of Nations. The *PNC* nullifies these documents.

The *PNC* outright denies Jewish religious, historical and national claims to Palestine (the Land of Israel). Secular Palestinians decided Jews are not a nation, but are only members of a faith, Judaism. Here too lies a contradiction, as Judaism does claim the same specific land as the Palestinian Arabs. All of Article 20 is a complete falsehood. If through the secular national eyes of the PLO, religious affiliation is not a criterion for claiming territory, then by extension, Islam cannot make claims to any land mass. They are only members of a faith, after all. If secular Arab nationalists recognize Islamic land-claims, then Jews would have just as strong a case for asserting their rights over the Land of Israel.

The secular Palestine Arab national movement not only strips the Jewish People of its national memory, but eradicates its geographic roots and religious foundations through the institutionalized nullification of the concept of the "People of Israel," a term consistently used in the *Tanakh* (Hebrew Scriptures). Here in Article 20 we find "Erasement Theology" embodied in the secular *PNC* as a form of holy writ. The Palestine national movement is not demanding the physical elimination of the Jews, but determines the historic, cultural and religious death of the Jewish People as a "given." If memory is deleted, then one is only a Jew in name, Judaism an empty shell, and the eradication of the Jewish People as a group is just a matter of time. Palestinian antisemitism hides behind supposed Enlightenment liberalism of individual rights rooted in equality for all. Yet, as made clear in previous *PNC* articles, one must be a Palestinian Arab to enjoy such rights. Jews are legally shorn of memory as well as national rights.

This is similar to the Soviet attitude toward their own Jews from the 1950s through the 1980s. Jews were a community but had little access to religious, cultural or national texts and were forbidden the study of Hebrew. Such is the policy of "forced assimilation" bringing about the dissolution of a people.

HC Comparison:

By basing its insights on the Koran, Hamas has a complete, even diametrically opposed, understanding of the Jewish connection to the Land of Israel. Since the Koran is its constitution (*HC* Article 8) it is clear they believe everything written in the Koran is true. The Israelites/Jews did have a covenant with Allah, received a promised "blessed land" (Koran 7:137), had two Temples destroyed in the Farther Temple (Jerusalem), and therefore certainly did live in Palestine or the Land of Israel (Koran 17:1-17:8). Most importantly, by keeping the covenant, the Israelites/Jews guarantee their own redemption in the afterlife by returning to the Land (Koran 17:104) as discussed in Chapter IX "Islamic Abrogation."

According to the Jihadists, the Muslim Brotherhood and Hamas, the Jews were expelled from the Land for violating the covenant with Allah and for doing evil. Their greatest evil was not accepting Mohammed as the Prophet, doing battle against him and not accepting his revelations. All of the above statements classify the Jews as a "people" —albeit disgraced and nefarious. For Hamas, the Jews are not just a religious community. The Koran clearly views Jews as more than a faith group, and there is no denial of Jewish nationhood in *The Hamas Covenant*. Paradoxically, Hamas can be accused of theological denial when confronted with the establishment of the State of Israel and the ingathering of the Jewish People into the Land of the Covenant. The abrogation clause (Koran 2:106) may be invoked, but rings false in light of reality (see Chapters II "Ideologues" and IX "Abrogation"). By declaring Judaism only a religion and denying Jewish nationhood the PLO is not caught in the contradiction, yet they are denying the Divine truth of the Koran.

We are familiar with the "Replacement Theology" of certain Christian theologians relegating Judaism to the trash heap of history and replacing it with Christianity. Later, Islam adopted the same concept toward both Judaism and Christianity. Hamas follows in these theological footsteps by declaring Islam the only true religion. Judaism and Christianity are replaced through Divine supersession. The *dhimma* status suffered by the "People of the Book" involved the eradication of national and religious memory. Islam reduces both Jews and Christians to an inferior community status.[4] Despite all, the Koran and Hamas view the Jews as a people who had a homeland.

Certainly Hamas theology is physically more deadly, but *PNC* ideals are not far behind in the danger they present in their total denial of the collective Jewish being. *PNC* antisemitism is comparable to the Soviet brand (1917-1991) of cultural and religious denial, mentioned above.

> *PNC, Article 21:* **The Arab Palestinian people, expressing themselves by the armed Palestinian revolution, reject all solutions which are substitutes for the total liberation of Palestine and reject all proposals aiming at the liquidation of the Palestinian problem, or its internationalization.**

Here we have the complete rejection of a two-state solution from a secular perspective, and support for a military struggle similar to Jihad. The *PNC* rejection of compromise comes from the Palestinian Arab nation, while the Hamas rejection is founded on what are deemed Islamic principles and "the will of Allah."

HC Comparison:

In the *HC,* rights to Palestine/Land of Israel are based on the *waqf* endowment and belligerency toward Jews as sanctified in the Koran's Sura or Chapter 9. *PNC* Article 21 is written in the same non-compromising spirit as the *HC* Articles 11, 13 and 32.

> *PNC, Article 22:* **Zionism is a political movement organically associated with international imperialism and antagonistic to all action for liberation and to progressive movements in the world. It is racist and fanatic in its nature, aggressive, expansionist, and colonial in its aims, and fascist in its methods. Israel is the instrument of the Zionist movement, and geographical base for world imperialism placed strategically in the midst of the Arab homeland to combat the hopes of the Arab nation for liberation, unity, and progress. Israel is a constant source of threat vis-a-vis peace in the Middle East and the whole world. Since the liberation of Palestine will destroy the Zionist and imperialist presence and will contribute to the establishment of peace in the Middle East, the Palestinian people look for the support**

> of all the progressive and peaceful forces and urge them all, irrespective of their affiliations and beliefs, to offer the Palestinian people all aid and support in their just struggle for the liberation of their homeland.

In the *PNC's* Article 22, Jewish nationalism is elevated to a universal evil as a "racist," "fascist," "aggressive, expansionist and colonial" power. As Harkabi and others pointed out, this was the agenda of extremist elements emanating from the New Left in the 1960s. The *PNC* and *The Hamas Covenant* borrow from each other, most notably in singling out Israel as a world menace aligned with the most reactionary, repressive powers and as part of a universal conspiracy against the Palestinians. Although the *HC* was written later, its ideological underpinnings were understood by the 1950s. The Hamas and PLO reasons for hating Israel are identical. Israel is seen as the ultimate satanic nation state by both. The *PNC* does not condemn the Jews as its enemy, only their "illegitimate" nationalism—Zionism. The PLO/*PNC* views Israel or Jewish nationalism as an extension of international imperialism.

HC Comparison:

The *HC* condemns the Jewish State as an appendage of a worldwide Jewish conspiratorial evil. In both the *HC* and *PNC*, Israel is a small part of a much larger insidious international plot. Hamas is more vicious than the PLO, as noted in Articles 20, 22 and 32, by blending Islamic, Czarist and Nazi antisemitism not only in the condemnation of Israel and the requirement for its liquidation, but in its demands for annihilating Jews worldwide (*HC*, Articles 7 and 32). For Hamas, there is an all-encompassing conflict of eradication against Israel, Judaism and the Jews (*HC*, Article 28, last verse).

The *PNC* stops short of the Hamas accusation identifying the Zionists/Jews as Nazis (*HC*, Articles 20, 31 and 32). The secular Palestinian condemnation of Zionism is a major stepping-stone in the direction of Hamas demands for Jewish destruction. If Jewish nationalism is "racist," "fascist," etc., then Jews have these same attributes.

> ***PNC, Article 23:*** The demand of security and peace, as well as the demand of right and justice, require all states to consider Zionism an illegitimate movement, to outlaw its existence, and to ban its operations, in order that friendly relations

among peoples may be preserved, and the loyalty of citizens to their respective homelands safeguarded.

The PLO requests foreign states to consider Zionism or Jewish nationalism "an illegitimate movement," outlawed internationally in the name of security, peace and justice. Article 23 is in contradiction to itself when discussing "security and peace" because it was the leaders of the Palestine national movement and Arab States who rejected the Partition Plan designed to ensure "friendly relations among peoples," in particular Jews and Arabs. In 1947, had the Palestinians accepted the Partition Plan, it would have allowed for full implementation of individual human rights in both the Arab and Jewish States, replacing the British Mandate. Furthermore, these two conflicting nations were to be recognized as equally legitimate.

The matter of "loyalty of citizens to their perspective homelands" is aimed at Jews supporting the State of Israel while living in other countries instead of fully focusing their loyalty toward the country of their residency. As Harkabi points out, people living in democratic nation states have other loyalties as well, not just the "narrow, formal nationalistic approach." Nullifying one's rights to secondary loyalties of religion, culture and ethic/national identity is the ultimate in dictatorial secular nationalism, the former Soviet Union being the case in point.

Continuing with this logic, Muslim Arabs outside of the Arab world—those living in Europe, North and South America—would have to forgo loyalty to their nation of ethnic origin. Forbidding outside loyalties would criminalize people if they contributed to any cause associated with their previous ethnic and/or religious homeland. To take this a step further, any Israeli Arabs who call themselves "Palestinian Arabs with Israeli citizenship," as many do nowadays, would be disloyal and could be charged with treason. But obviously such a move against "disloyal" Israeli Arabs by the Israeli authorities would be deemed illegal because the Jewish State is an illegitimate entity according to the *PNC*.

HC Comparison:

The Introduction and Preamble of *The Hamas Covenant* makes clear Israel and the Jews are the foremost enemy and must be killed (*HC,* Article 7). They believe Zionism is an evil Jewish plot, by definition illegal (see Chapter II "Ideologues," subsection on Qutb), and to be battled in every possible way (*HC,* Articles 7, 15, and 17). At the same time, they see Israel and Jews as having powerful, and mythical, allies in the capitalist West, the communist

East, and in international institutions (*HC*, Articles 22, 25, and 28). Hamas allies are Arab/Islamic organizations (*HC*, Introduction-Preamble and Articles 29, 32 and 33).

> ***PNC, Article 24:*** **The Palestinian people believe in the principles of justice, freedom, sovereignty, self-determination, human dignity, and in the right of all peoples to exercise them.**

All people have rights, especially when it comes to sovereignty and self-determination, except for the Jews who are not a people (Article 20). Article 24 sounds harmless enough but when put in the context of the *PNC* reveals outright discrimination and antisemitism, Stalinist-Soviet style once again. The Jewish People are left bereft "of justice, freedom, sovereignty," and "self-determination" once they are denied their heritage and bond to the ancient homeland (*PNC*, Articles 19, 20 and 21). The secular PLO goes so far as to ignore the Koran when they nullify Jewish peoplehood and nationality. Jews must find their "human dignity" as individuals within the rights of other peoples. The *PNC* excludes Jews from the family of nations.

HC Comparison:

Using the Koran as its constitution (*HC*, Article 8), Hamas recognizes the Jews as a "people," one destined for elimination. All other peoples and a few surviving Jews are designed to be subjects living under Islam.

> ***PNC, Article 25:*** **For the realization of the goals of this Charter and its principles, the Palestine Liberation Organization will perform its role in the liberation of Palestine in accordance with the Constitution of this Organization.**

The PLO's Constitution and the *PNC* were written at the same time in July 1968. The Constitution echoes the *PNC* but is a technical document relating to the general principles, national assembly and the executive branch of the PLO. The PLO Constitution provides organizational structure to facilitate the implementation of PLO objectives. Overall, the PLO Constitution is another call for the destruction of the State of Israel, similar to that of Hamas.

The PLO Constitution is not the same document as the *Constitution of Palestine* drawn up by the Palestinian Authority (PA) in 2003, which does not specifically outline borders in the future Palestinian State. The PA Constitution of 2003 declares Jerusalem as the capital of Palestine (Introduction and Article 3). It also invokes the Palestinian right of refugee return (Introduction). Borrowing from the PLO Constitution the PA is to develop executive, legislative and judicial branches of government. At the same time, the PLO retained its status as the "sole and legitimate representative of the Palestinian people," a contradiction if one thinks in terms of elections and democracy. The *PNC* and the PLO Constitution of 1968 are secular documents, but the future Palestinian State, according to the 2003 PA Constitution, declares Islam as its official religion and Sharia law as the basis for judicial decisions (Constitution of Palestine, Article 4).[5] The Palestinian Authority moved closer to the Hamas state ideal.

HC Comparison:

The Islamist influence, apparently by way of *The Hamas Covenant*, is overwhelming. In practice, the supposedly secular PLO and Palestinian Authority began fusing with Hamas when it came time to implement the principles and legislation necessary for building a concrete Palestinian State as illustrated by the 2003 Constitution. For sure Hamas applauds the inclusion of Islam and Sharia law as religious and judicial pillars in the future Palestinian State.

On the other hand, Hamas is at variance with this same Constitution for declaring the PLO as the only "legitimate representative of the Palestinian People" and for not outlining its borders as ranging from the Jordan River to the Mediterranean Sea.

> ***PNC, Article 26:*** **The Palestine Liberation Organization, representative of the Palestinian revolutionary forces, is responsible for the Palestinian Arab people's movement in its struggle - to retrieve its homeland, liberate and return to it and exercise the right to self-determination in it - in all military, political, and financial fields and also for whatever may be required by the Palestine case on the inter-Arab and international levels.**

In the 1960s, the PLO became the self-appointed liberation organization to represent the Palestinian People. In October 1974, the seventh Arab Summit Conference officially designated the PLO as the "sole, legitimate representative" of the Palestinians. They remained so until Hamas challenged their authority in the late 1980s. The PLO and Fatah gained legitimate representative status in the 1996 parliamentary and presidential elections; Hamas did the same in the 2006 parliament.

The PLO takes full responsibility for "retrieving," "liberating," and "returning" any land held by Israel. "Liberation" and "self-determination" are euphemisms for the destruction of Israel. The PLO's role is to destroy Israel in the "military, political, and financial fields" for the sake of the Palestinians, and responsible for inter-Arab relations on an international level.

HC Comparison:

In 1988 Hamas also took the self-appointed leadership role, making the same claim as the PLO but this time with "Allah's blessings." Hamas took a similar approach when representing Palestinian Muslims as part of a broader Islamic front working for world Islamic conquest (*HC,* Articles 7 and 8). The Hamas call for Israel's destruction is identical to that of the *PNC,* but is based on the Islamic right to all *waqf* lands, and rule by Sharia law (*HC,* Article 11). Hamas sees itself as the only legitimate organization representing Palestinian Muslims.

> ***PNC, Article 27:*** **The Palestine Liberation Organization shall cooperate with all Arab States, each according to its potentialities; and will adopt a neutral policy among them in the light of the requirements of the war of liberation; and on this basis it shall not interfere in the internal affairs of any Arab State.**

PLO policies toward the Arab world are a result of the support they receive from those Arab countries in the "war of liberation." Understanding their own precarious position in building alliances and intending to focus on the destruction of Israel the *PNC* declares they will not intervene in other Arab countries' affairs so as not to alienate any potential allies. Hence the PLO declares a neutral policy toward inter-Arab rivalries.

HC Comparison:

Hamas has a more radical view when it comes to judging Arab countries' support of Jihad against Israel and the Jews. Islam trumps all other interests, especially secular nationalism. All Muslims are to battle Israel. Hamas does not accept neutrality in the long term (see Chapter II "Ideologues"). One either supports Jihad or is against it, becoming either a friend or an enemy.

> *PNC, Article 28:* **The Palestinian Arab people assert the genuineness and independence of their national *(wataniyya)* revolution and reject all forms of intervention, trusteeship, and subordination.**

Since 1993 the Palestinian Authority draws its legitimacy from agreeing to work with Israel to jointly implement the Oslo Accords. Yet, according to the above article, the PLO is independent and rejects all mediation or compromises leading to less-than-full national sovereignty. By agreeing to the Oslo Accords (1993), the PLO/Fatah crossed the line and contradicted their own *Charter* when they established the Palestinian Authority (PA) to negotiate with Israel. This may be a tactical stage (or not) in a multi-step approach to lull Israel into complacency and eventually destroy the Jewish State. In retrospect, one could ask whether PA Chairman Yasir Arafat had a subtle destruction plan in mind during the Palestinian Low Intensity Conflict (LIC) and terror offensive waged against Israel from 2000-04.

HC Comparison:

The *HC* advocates the same uncompromising policy stated in *PNC* Article 28 and denounces any attempt at compromise, specifically the 1978-79 Camp David Accords between Israel and Egypt (*HC,* Articles 11, 13, 32). Hamas also consistently condemns the Oslo Accords of the 1990s.

> *PNC, Article 29:* **The Palestinian people possess the fundamental and genuine legal right to liberate and retrieve their homeland. The Palestinian people determine their attitude toward all states and forces on the basis of the stands they adopt vis-a-vis to the Palestinian revolution to fulfill the aims of the Palestinian people.**

Article 29 is a continuation of Article 27, clearly going beyond neutrality. The PLO reciprocates the foreign policies of other states toward the "Palestinian revolution." Those supporting the revolution will have the Palestinians as an ally and those who undermine or do not support them will be treated as enemies. As Harkabi explained it, anyone showing friendship toward Israel becomes an enemy. More so, the PLO views itself as a government in exile, and therefore behaves as a state, at least on the declarative front.

HC Comparison:

As stated before, Hamas views everyone within the prism of Islam as interpreted by the Muslim Brotherhood, and determines its policies toward others in the same reciprocal manner as the PLO. Previous commentary on *PNC* Articles 27 and 28 clarify their stand. The Hamas expectation is for Muslims with secondary Palestinian and Arab identities to subjugate themselves to the Islamic demand for Jihad and the "liberation of Palestine" (*HC*, Article 14).

> *PNC, Article 30:* **Fighters and carriers of arms in the war of liberation are the nucleus of the popular army which will be the protective force for the gains of the Palestinian Arab people.**

The *PNC* views people who fight and carry arms as the heart of the true "popular army," meant to include not just Palestinians, but all Arabs. There is also no prohibition against any non-Arab willing to join the cause.

HC Comparison:

The mention of a "protective force" can be considered parallel to the "defensive" Jihad explained in the commentary on *PNC* Article 14, in tandem with Hamas (*HC*, Articles 3, 7, and 10). The issue of non-Muslims joining the Jihad against Israel and the Jews is not clearly stated in the *HC* (see Chapter II "Ideologues").

> *PNC, Article 31:* **The Organization shall have a flag, an oath of allegiance, and an anthem. All this shall be decided upon in accordance with a special regulation.**

The PLO is preparing for a Palestinian State with a flag and an oath. Hamas has a flag and an oath of allegiance, but as of now there appears to be no official Hamas anthem. Hamas could easily use the PLO anthem without facing contradiction. The PLO anthem lends a veneer of Palestinian nationalization to one's overall Islamic identity. The PLO belongs to the Arab world, while Hamas is the Palestinian component of the Muslim Brotherhood. Below is the PLO National Anthem calling for sacrifice and war, the same actions advocated by Hamas.

Fida'i - Fedayeen Warrior

My country, my country (or "warrior, warrior")
My country, the land of my grandfathers
My country, my country (or "warrior, warrior")[6]
My country, my nation, the nation of eternity
With my determination, my fire and the volcano of my revenge
The longing of my blood to my land and home
I have climbed the mountains and fought the wars
I have conquered the impossible, and crossed the frontiers
My country, my country, the nation of eternity
With the resolve of the winds and the fire of the guns
And the determination of my nation in the land of struggle
Palestine is my home, Palestine is my fire, Palestine is my revenge and the land of eternal
My country, my country, the nation of eternity
I swear under the shade of the flag
To my land and nation, and the fire of pain
I will live as a guerrilla, I will go on as guerrilla, I will expire as guerrilla until I will be back
My country, my country, the nation of eternity[7]

PNC, Article 32: **Regulations, which shall be known as the Constitution of the Palestinian Liberation Organization, shall be annexed to this Charter. It will lay down the manner in which the Organization, and its organs and institutions, shall be constituted; the respective competence of each; and the requirements of its obligation under the Charter.**

As mentioned in the commentary on *PNC* Article 25, the PLO Constitution is a technical document. PLO ideals were transferred to the Constitution of the quasi-governmental Palestinian Authority of 2003. It was their first step as an autonomous but not-fully-independent regime striving for statehood. To repeat, the PA Constitution guarantees democratic rights and freedoms while power is exercised through the executive, legislative and judicial branches. Arabic is the official language, Islam the official religion and the principles of Islamic Sharia law form "the main source of legislation."[8] The PLO administered the Palestinian Authority since 1994 and won the parliamentary elections in 1996. The 2006 legislative ballots, however, resulted in a national unity government for a year and a half.

Hamas overthrew the PA in Gaza in June 2007 and took full control. In response, Fatah established an emergency government in the West Bank. Both Hamas and Fatah administered their own rules in their respective territories. Neither regime was, or is democratic, nor were elections held. Ideologically the PLO Constitution echoes the *PNC*. The Constitution of the Palestinian Authority is meant to become the basis for all law under the PA and the independent Palestinian State in the future. It must be emphasized that PLO/Fatah leaders make it clear their first loyalty is to the PLO/Fatah. Their secondary commitment is to the quasi-state Palestinian Authority. This leaves open questions of loyalty to the future independent Palestinian State as opposed to allegiance to a political faction.

HC Comparison:

The Koran is the Constitution of Hamas (*HC*, Article 8). Like the PA, Hamas is loyal to itself first and to a state framework second. Proof of loyalty to its organizational ideology became clear after the Hamas coup against the PA in Gaza in 2007.

> ***PNC, Article 33:*** **This Charter shall not be amended save by [vote of] a majority of two-thirds of the total membership of the National Congress of the Palestine Liberation Organization [taken] at a special session convened for that purpose.**

A two-thirds majority can amend the *PNC*. Allowing amendment or correction is an anthropocentric ideal where humans rule, can make decisions and reverse them. War and peace are in the hands of "the people" and their representatives after elections.

It is within *PNC* Article 33 that we see the possibility of conflict resolution should a two-thirds majority decide to change the *Charter*. The PLO voted to amend the *PNC* by removing the offending anti-Israel passages already in 1996 by a vote of 504 to 54 and then again two years later after the Wye Plantation Accords when President Clinton came to Gaza to witness another vote in December 1998. The revised *Charter* was meant to pave the road for a two-state solution.[9] Eighteen years later (2016) nothing has changed nor are there any concrete proposed draft changes. As stated above, *The Palestinian National Charter* remains in full effect despite declared intentions to the contrary. One can only conclude that "secular" Palestinians prefer to preserve the *PNC* as is, despite their declared obligation to change it.

HC Comparison:

From the Hamas viewpoint, the *Covenant* or "Charter of Allah" is unchangeable; its essence is diocentric, of Divine inspiration. Allah is everlasting, does not change His mind and accordingly does not suffer human intervention in His dictates. Allowing amendment is a major point where *The Hamas Covenant* and *The Palestinian National Charter* differ.

Conclusion

The Palestinian National Charter and *The Hamas Covenant* are similar in many ways. In calling for the destruction of the State of Israel, neither minces their words. The PLO speaks of a war of liberation on a secular nationalist level; Hamas insists on Holy War or Jihad. The PLO borrowed Jihadist concepts prior to the official establishment of Hamas and secularized them. Hamas retrieved the veiled Islamist lexicon from the PLO and returned it to its original Islamic context—Jihad.

The PLO demands a secular Arab State to replace Israel. Some Jews would be allowed to stay with supposed equal rights, but they would expel most of the Jews who would survive the Palestinian Arab victory. Hamas demands an Islamic State and would slaughter any Jews who refused to accept the *dhimmi* inferior status. For Hamas it is of no significance when Jews arrived (until 1917, after 1948, etc.), but rather it is their acceptance or non-acceptance of the *dhimma* strictures that would determine if they live or die (*HC,* Article 31).

On the declarative level, the PLO is battling the State of Israel and Zionists—meaning all supporters of Jewish nationalism. Hamas is battling Israel, Judaism and the Jewish People worldwide (*HC,* Article 28 last paragraph). Hamas calls for the overall annihilation of Jews (*HC,* Article 7), except for the few who would choose to accept full Muslim sovereignty and the humiliating *dhimmi* arrangement.

While both the *PNC* and *HC* are exterminationist toward the State of Israel, for Hamas the Jews are the ultimate conspirator, working to undermine the Islamic world on every level. The PLO finishes its combat mission with Israel's destruction. Hamas views the destruction of Israel as only one victory against universal Jewish conspiracies. Hamas is clear in its antisemitic policies while the PLO denies any enmity toward Jews. Still, the PLO denies the Jewish right to national memory and an independent national existence in the ancient homeland.

There are major concerns about the PLO and the secular Palestinian nationalist attitude toward Jews. To single out Jews as the only people whose nationalism is to be denied, is to separate them from the rest of humanity. Their approach is simple prejudice, specifically antisemitism. On the level of written intent, secular Palestinian nationalism is not annihilationist toward Jews worldwide. Still, we must keep in mind that Haj Amin el-Husseini, the Grand Mufti of Jerusalem held in such great esteem by both Hamas and the PLO was a staunch Nazi ally during World War II. He outwardly demanded the extermination of the Jews and sent Bosnian Muslim troops to help facilitate the Holocaust (see Chapter III "Jewish National Liberation," Chapter VI "*The Hamas Covenant* Analysis" Article 22 and specifically Chapter VIII "Czarist Nazi Integration").

This leaves us with the last dilemma of whether the "anti-Zionism" of secular Palestinian nationalism is a veneer for an overall universal antisemitism, or do the secularists say exactly what they mean? Behavior and comments by Palestinian Authority officials originating in the PLO from the mid-1990s to the present, lead many to believe that claims they make denying antisemitism ring hollow.[10] Article 4 of the PA's Constitution of Palestine sets Islam as the official religion of their territories and state-to-be in the future. Sharia law is the "main source of legislation." Islam and Sharia law lean on the Koran as its pillar of legitimacy. The Koran is extremely ambiguous toward the Jews, and is often exceedingly hostile. Sharia law does not advocate Jewish equality but rather the *dhimmi* status extrapolated from the Koranic verse 9:29 and implemented over the centuries. More so, there are numerous antisemitic quotes by PLO and PA leaders as noted previously in this work.

The PLO/*PNC* denial of Jewish peoplehood and nationalism goes a step beyond "Replacement Theology" reviewed earlier. Today's *PNC* is the political expression of "Palestinianism" as discussed by Bat Ye'or,[11] and not only a Palestine national document asserting the rights of the Arab population. As we have noted previously, "Replacement Theology" gives way to a secular "Erasement Theology" cloaked in the guise of liberal, secular, individual human rights, whereby Jews will find national expression in other cultural and state entities. Supposedly, only Zionism or Jewish nationalism is seen as illegitimate and detrimental to universal stability.

If the PLO were to finally change the *PNC* to acknowledge Israel's right to exist, as demanded by the Wye Accords, it would prove secular Palestinian nationalism to be neither antisemitic nor anti-Zionist. Until then, there is much in common between the PLO and Hamas foundational documents and their respective political movements in relation to Israel, Jewish nationalism and Jews even if the *PNC* does not explicitly call for Jewish destruction.

While *The Hamas Covenant* advocates world conquest through Jihad, *The Palestinian National Charter* does not make any such demand. The PLO/PA may be brought closer to such Jihadi thinking in the future due to the increasing Islamization of secular Palestinian nationalism and the 2003 Constitution declaring Islam the official religion.

Endnotes

1. Kadi, Leila S. (ed.), "Basic Political Documents of the Armed Palestinian Resistance Movement," *Palestine Research Centre*, Beirut, December 1969, pp. 137-141.
2. Harkabi, Yehoshafat, *"Palestinians and Israel,"* Keter Publishing House, Jerusalem, 1974, pp. 49-69.
3. "Refugee," *Wikipedia*, retrieved June 20, 2011, en.wikipedia.org/wiki/Refugee.
   "Palestinian Refugee," *Wikipedia*, retrieved June 20, 2011, en.wikipedia.org/wiki/Palestinian_refugee.
   Pipes, Daniel, "[UNRWA] The Refugee Curse," *Daniel Pipes Middle East Forum*, August 19, 2003, retrieved June 20, 2011, www.danielpipes.org/1206/unrwa-the-refugee-curse.
   Note on the Applicability of Article 1D of the 1951 Convention relating to the Status of Refugees to Palestinian refugees, at unispal, retrieved December 31, 2015, http://www.unhcr.org/4add88379.pdf

   If one follows Article 1D as mentioned above eventually everyone on the planet may be recognized as a Palestinian refugee.
4. Bat Ye'or, *Islam and Dhimmitude Where Civilizations Collide*, translated from the French by Miriam Kochan and David Littman, Fairleigh Dickinson University Press, Teaneck NJ, USA, 2002, Chapter 10, "The Politics of Dhimmitude in Europe," and Chapter 11 "Conclusion," pp. 305-400.
5. "Constitution of Palestine," *Wikisource*, retrieved June 3, 2015, http://en.wikisource.org/wiki/Constitution_of_Palestine.
6. "Fida'i," *Wikipedia,* retrieved June 3, 2015, https://en.wikipedia.org/wiki/Fida%27i.
   The words "warrior, warrior" appear as the translation of *"fida'i, fida'i."*
   "Palestinian National Anthem - Fida'i," *Lyrics Translate*, retrieved June 3, 2015, http://lyricstranslate.com/en/site-search?query=Palestinian+National+Anthem&op=Search#gsc.tab=0&gsc.q=Palestinian%20National%20Anthem&gsc.page=1.
   The word *"biladi"* or "land" appears in the first line while *"fida'i"* or "warrior" appears in the third.
7. "Palestine National Anthem," *Middle East Facts*, retrieved June 3, 2015, http://www.middleeastfacts.com/middle-east/palestinian-national-anthem.php.
   The overall translation is taken from this site where the word *"biladi"* meaning "land" or "country" is used. The Palestine Affairs Council also emphasizes the word "country."
8. "Constitution of Palestine," *Wikisource*.

9   Rubin, Barry and Rubin, Judith Colp, *Yasir Arafat, A Political Biography*, Oxford University Press, New York, 2005, pp. 167-168.
10 For a comprehensive review of Fatah, PLO and Palestinian Authority (PA) antisemitism, see "Palestine Media Watch," for foreign language translation, numerous articles and primary source research of Palestinian media, both written and electronic, http://www.palwatch.org/pages/aboutus.aspx.
11 Bat Ye'or, pp. 366-371.

VIII

Czarist-Nazi Integration into Palestinian Islamic Jihad

Overview

To understand how Czarist/Nazi attitudes toward the Jews developed within the Palestinian Muslim domain we must begin with a synopsis of Haj Amin el-Husseini's pro-Nazi policies as the Palestinian Mufti of Jerusalem, his Nazi activism in WWII, and conclude with the 1948 War. Next, we will review Jewish community loyalties to their nations of origin during both world wars, thereby debunking the slanders concerning traitorous Jewish behavior. Often Jewish communities were "rewarded" for their loyalty with persecution, as was the case in Europe preceding and during WWII. Finally there will be an in depth review of *The Hamas Covenant* Article 22 with textual comparisons between the Czarist *The Protocols of the Learned Elders of Zion* and Hitler's *Mein Kampf,* as concerns the Jews. Context will be provided for other accusations, particularly as pertains to Middle Eastern history.

The Grand Mufti Haj Amin el-Husseini and the Nazis

The Palestinian Mufti of Jerusalem, Haj Amin el-Husseini, visited the German General Consul in Jerusalem at the end of March 1933, claiming to represent Muslims everywhere. Consul Wolff sent a telegram to the Nazi leadership relating the Mufti's greetings, congratulations on the Nazi rise to power, and a wish for an alliance to spread fascism while destroying Jewish influence and democracy. That summer the Arab Nazi movement came to light. Those who would later become secular Baathists insisted they were racists and great admirers of Nazism.

Throughout the 1930s, Arab and Islamic parties not only sent congratulations to the Nazis on the passage of the viciously antisemitic Nuremberg Laws, but used the Nazi fascist model for building political parties, youth movements and paramilitary organizations. These political factions influenced people at the time such as the Baathists Michel Aflaq and Salah a-din al-Bitar, and future Egyptian Presidents Gamal Abdul Nasser and Anwar Sadat.[1]

As for the Mufti, he acquired funds from the Nazis both for al-Banna's Egyptian Muslim Brotherhood and for his own 1936-39 revolt against the British. Simultaneously, he worked with the Brotherhood to liquidate the internal pro-British Palestinian Nashashibi family opposition.[2] In the words of David Patterson, the Mufti Haj Amin el-Husseini was the cultivator and the "fertilizer" necessary to ensure the growth of the new hybrid Nazi/Jihadi evil unified into one species.[3]

Fleeing the British in 1937 after the onset of the revolt, Haj Amin el-Husseini hid out on the Temple Mount in Jerusalem, escaped to Lebanon and then made his way to Iraq. Nazi propaganda and influences were very heavy in Iraq in the late 1930s and increased in the early 1940s. The Mufti's influence expanded significantly after the pro-Nazi Rashid Ali al-Kilani and his Golden Square conspirators overthrew the pro-British Iraqi government in early April 1941.

The Mufti Haj Amin el-Husseini worked with Rashid Ali to solidify an alliance with the Nazis and organize battle plans for joint efforts to defeat Britain. Much to his dismay, the Nazis were preoccupied in the Balkans and unable to offer immediate assistance. Haj Amin had already written a letter to Hitler in January urging support and coordination for the Arab cause against the British and the Jews.[4] When the British began their counter-offensive to retake Iraq a month later on May 2, Haj Amin and the Muslim clergy declared Jihad against the British in a special communique to the Iraqi people. Rashid Ali demanded immediate aid, especially for aircraft and Axis military advisors. Haj Amin, for his part, explained that with German-Italian commitment assured, insurgency actions would commence in both Transjordan and Palestine. All this was expected to dovetail nicely with the solidification of Nazi power in the Balkans, an Axis reinforcement of the fascist Vichy government in Syria, the sidelining of any possibility of Turkish opposition and the overall German desire to control the eastern Mediterranean.[5]

But the Germans could not fully commit, as they were already involved in Eastern Europe preparing for Operation Barbarossa and the invasion of the Soviet Union that June. Still, they were able to provide some arms, ammunition, aircraft and financial aid, whereby the Mufti is said to have received $25,000. He also procured Italian help, especially with funding, but most assistance arrived too late. The British and Transjordanian Arab Legion defeated the Iraqi Nazis by late May. Rashid Ali's government and the Mufti fled while the British recaptured the country in early June.[6]

British success led to an outburst of fury directed against the Jewish population of Baghdad who were celebrating the Shavuot, or Pentecost holiday, on June 1-2. The attack resulted in the pillage, plunder and slaughter of 180 Jews in the pogrom known as the "Farhud." The Islamists saw the Jews as beneficiaries of the British presence due to their now legally equal status in Iraq, and resented the British mandate in Palestine calling for a Jewish National Home. More importantly, the Iraqi nationalist and Islamist groups, which Haj Amin helped, fully identified with the virulently antisemitic Axis message and took action against the Jews despite Rashid Ali's defeat.[7] This fit in well with the Mufti's adherence to Hitler's racial theories and his demand that Arab countries maintain the equivalent right to deal with their Jewish minorities "by the same method applied to solve this problem on the territories of the Axis power."[8]

A few months later the Mufti sought a letter of commitment for full German involvement in the Middle East when he met the Nazi foreign minister von Ribbentrop and Hitler on November 28, 1941. He stressed the importance of the Arab-German friendship in the face of the British, Jewish and Bolshevik enemy. Hitler hesitated in making any sort of pledge, first wanting to secure the southern Caucasus Mountains as a springboard into western Asia. Although Hitler was adverse to a public declaration of unity between the Nazis and the Islamists, Haj Amin was assured of a top Arab leadership position in the emerging alliance with the Nazis once Hitler achieved his goals. Hitler, as "Fuhrer" or guide, did not want to make empty declarations and feared another Iraqi failure. Kilani received a promise of German support as the future premier of Iraq and the Italians recognized the Mufti as the leader of the northern tier of Arab countries. Despite offering to build a joint fighting force, the Mufti failed to convince the Germans to recognize his pan-Arab efforts to build a unified Greater Syrian entity.[9]

The record of Hitler's commitment from the above November meeting with Haj Amin el-Husseini is quoted in the following memorandum:

> 1. He (the Fuhrer) would carry on the battle to the total destruction of the Judeo-Communist empire in Europe.
>
> 2. At some moment which was impossible to set exactly today but which in any event was not distant, the German armies would in the course of this struggle reach the southern exit from Caucasia.
>
> 3. As soon as this had happened, the Fuhrer would on his own give the Arab world the assurance that its hour of liberation had

arrived. Germany's objective would then be solely the destruction of the Jewish element residing in the Arab sphere under the protection of British power. In that hour the Mufti would be the most authoritative spokesman for the Arab world. It would then be his task to set off the Arab operations which he had secretly prepared. When that time had come, Germany could also be indifferent to [Vichy] French reaction to such a declaration.[10]

Regarding the above Articles 2 and 3, once German forces moved south out of the Caucasus mountain region, the Mufti was destined to become the leading Nazi in the Arab world, responsible not only for Arab liberation, but for Jewish extermination. In the spring of 1942 the Grand Mufti Haj Amin el-Husseini and Rashid Ali al-Kilani secretly exchanged letters with the Nazis. Arab leaders pledged to help the Germans fight their common enemies. The letter assured assistance to Arabs living under British rule and recognized their right to independence while speaking of Arab political unity, should that be the desire of the Arabs themselves. The Jewish National Home in Palestine was slated for destruction as General Erwin Rommel's German Afrika Korps moved across North Africa into western Egypt. In early May 1942, Haj Amin and Rashid Ali publicized the contents of the letter exchange and agreement on Axis radio.[11] The threat ended when Rommel was defeated at the Battle of El Alamein a half a year later.

The Mufti advocated for an Arab legion to serve as a semi-independent force in the German army, in particular focusing on Egypt and North Africa. The following March he pushed for an overtly pro-Arab German declaration.[12] Haj Amin worked to build an independent pan-Muslim fighting brigade made up of men from Crimea, the North Caucasus region, Azerbaijan, central Soviet Asia and India. In the end, the Balkan Muslims—mostly Bosnians—were close at hand. The Mufti prevented the immigration of 4,500 Balkan Jews to Palestine, which included 4,000 children. He intervened with the Bulgarian, Italian, Romanian and Hungarian governments and convinced them to prohibit their Jews from immigrating. Instead, they arranged to have these Jews shipped to Poland to ensure they would be "supervised properly," a euphemism for their destruction. The Mufti was building a pan-Islamic international policy for a Jihadist/Nazi victory on all fronts.

Haj Amin collaborated closely with the Nazis in the implementation of Hitler's "Final Solution" of Jewish extermination as evidenced by his diary entry after his November 1941 meeting with Hitler.[13] Envious of the Einsatzgruppen success in exterminating one and a half to two million Jews in Eastern Europe, the Mufti offered his services. Being on intimate terms with mass murderers Adolf Eichmann and Heinrich Himmler, he too demanded a part in the Final Solution. He visited the concentration-extermination camps at Auschwitz, Majdanek, Mauthausen, Theresienstadt and Bergen Belsen, while befriending their commandants. He organized pro-Nazi military units consisting of tens of thousands of Bosnians in the Handschar Division, and Albanians in the Skanderberg Division. He prepared Muslim volunteers and made radio broadcasts urging German-Arab victory and Jewish destruction. His speeches constantly condemned the Jews as enemies of Islam and accused them of dominating Britain, the US and international communism. As Patterson clearly states, "in keeping with the thinking shared by Nazis and Islamic Jihadists, al-Husseini espoused the view that the Jews fall outside the circle of redemption or rehabilitation. The *only* way to handle the Jewish Question is through extermination."[14]

The Mufti was known for his vicious annihilationist diatribes against the Jews on various occasions. His most well-known declaration was on March 1, 1944 when he called for the universal Islamic massacre of Jews as he addressed Bosnian Muslim Nazi troops. "Kill the Jews wherever you find them. This pleases God, history and religion. This saves your honor. God is with you."[15] In preparation, he issued his infamous pamphlet *Islam and Judaism* to his Muslim Balkan Nazi troops in 1943. He began by declaring, "It is unworthy for us Moslems to pronounce the words Islam and Judaism in one breath, when Islam stands so high above its deceitful opponent." Giving a short theological and historic survey, he explained the roots of the conflict as the "treachery of Jewry and its enmity to the founder of Islam, the Prophet Mohammad." He reminded them of the history that some Jews converted to Islam only to return to Judaism and condemn Islam as a "lie." Others tried to kill the Prophet Mohammad by flinging rocks at his head. In the end, the Jews of Medina were expelled, fled to Kaybar and would be punished again for not accepting defeat. The pamphlet goes on to describe the attempt to poison Mohammed by a Kaybar Jewess "Zaynab, Salam Ibn Mishkam's wife." It continues to say, "We Moslems must incessantly think about the Khaybar feast. If the Jews at that time betrayed Muhammad in this way, they would surely persecute us with their deceit, to destroy us." Mohammed suffered health problems ever since the Khaybar feast, the inference being that in the end he may have succumbed to the poisoning. He accused the Jews of "immorality" and of plotting to kill Moses after being "expelled" by the Egyptians. The Jews were and are locusts he declared, and accused them of intentionally spreading diseases in Syria and Palestine during the Roman period. This explains why the Jews are still referred to as "microbes" in Arabia. The Mufti continued to say Jews sowed strife among the different Arabian tribes.

Returning to the present moment, the Mufti said the Jews were "excrement" because they "succeeded in buying land from the poorest people and from unscrupulous landowners" a deed inferred to have begun with evil Jewish business practices since the time of Mohammed. The Jews were further denounced for having aligned themselves with the pagans and for telling them that paganism is superior to Islam, even though the Jews professed to believe in God. El-Husseini quoted Koran 5:82, "You shall surely establish that the greatest enmity toward those who believe is entertained by Jews and pagans."[16] He next quoted Mohammed, "It shall never be possible to see a Moslem and Jew together, without the Jew having the secret intention of destroying the Moslem."[17]

The final quote in the pamphlet *Islam and Judaism* is the constantly repeated notorious Hadith quote, but in a slightly different form, as found in *The Hamas Covenant* Article 7. "The Day of Judgment will come only when the Moslems completely destroy the Jews, when every tree behind which a Jew is hiding, will say to the Moslem: 'A Jew is standing behind me, kill him!' Only the Gharqad tree [thorn bush], which is a small shrub with many thorns and grows around Jerusalem, will not participate in this, for it is a Jewish tree!"[18]

Following WWII, Haj Amin fled prosecution by the Yugoslav authorities for war crimes, and was welcomed as a hero in Cairo. He provided the fertile ground for Jewish extermination as an acceptable doctrine among the Jihadists when he arrived in Egypt in 1946 and met up with Muslim Brotherhood leaders Hasan al-Banna and Sayyid Qutb.[19] The Mufti saw himself as a firm Muslim Axis ally. The Reich Ministry for Foreign Affairs quoted Haj Amin as declaring, "The victory of the Axis powers will also be the victory of the Islamic nations."[20] As further noted by Patterson, "the Ikhwan [Muslim Brotherhood] becomes a hybrid of Nazism and Islam to form Islamic Jihadism, making the ultimate extermination of the Jews not just a political or territorial aim, but a fundamental and defining element of their worldview: one cannot be part of the Brotherhood, just as one cannot be a Nazi, without espousing the extermination of the Jews."[21] Haj Amin el-Husseini was planning for Jewish annihilation and Nazi victory in Palestine.[22]

Haj Amin continued his efforts organizing along with the Egyptian Muslim Brotherhood to send volunteers to attack the Jewish community in mandated Palestine beginning in early 1948, even before the declaration of independence by the newborn Israeli State. The Brotherhood took part in the 1948 War but lacked recruits and was unsuccessful; the newborn state survived the conflict. Despite the dark shadow of all these revelations, Haj Amin el-Husseini remains a highly respected leader in the Arab world today. More significantly, the Mufti's Islamo-Nazi hatreds toward the Jews were passed on to Hamas and, in part, to the secular PLO/Fatah. By the 1980s the Mufti's legacy of hatred for the Jewish People and its nation state became fully integrated into much of the Palestinian national mindset, especially as seen in Hamas and its *Covenant*. For decades the secular PLO/Fatah was considered anti-Zionist, but not antisemitic; however, in recent years the Mufti's antisemitic imprint is becoming more manifest as evidenced in numerous statements and articles originating with Palestinian Authority officials[23] (see Chapters III, VI, and VII – "Jewish National Liberation," *"The Hamas Covenant"* and "Comparative Analysis," respectively).

Jewish Involvement in WWI, WWII, and the Bolshevik Revolution

Antisemites, especially of the Czarist, Nazi and Jihadi types, constantly accuse the Jews of betraying their home countries, especially in times of war. The historic record defeats these accusations and shows that Jewish communities demonstrated unswerving loyalties to the European powers involved in WWI. These powers included the German, Russian, French, British, Austro-Hungarian and Ottoman Empire militaries. Jews suffered the same casualties, if not more per capita, than non-Jews. Prior to WWII, most Jews lived in Eastern and Central Europe. Below are some statistics as relates to this period of time.

After the start of WWI, in 1915 Czarist forces expelled about 600,000 Jews from much of Eastern Europe, mainly Poland. Jews living under the German occupation suffered the usual hardships of war, including 70,000 of their people forcibly transferred to Germany proper for exploitation in labor battalions. About 650,000 Jews served in the Russian army during WWI (with an estimated 100,000 killed), 320,000 Jews served in the Austro-Hungarian army (40,000 killed) while 100,000 Jews fought in the German army (12,000 killed).

Throughout the Bolshevik Revolution (1917-1921), Czarist forces massacred Jews. In particular, the Czar's Ukrainian allies were responsible for between 50,000 to 200,000 deaths with countless others wounded. The exact numbers are not known.[24] In general, Jews favored the Soviet revolution due to the promise of equality, but were deeply disappointed when such promises remained declarative more than anything else. Antisemitism was in full bloom under Stalin and the Bolsheviks by the late 1920s and reached its height in the purges of 1936-38, just prior to WWII, where many Jews were imprisoned and executed as enemies of the revolution. This of course would not stop them from taking up arms in June 1941 to defend the Soviet Union against Hitler's invasion.

While the Palestinian Mufti of Jerusalem Haj Amin el-Husseini actively worked with the Nazis, one and a half million Jews fought on the Allied side to defeat Germany and its Axis allies. Some 200,000 Soviet Jewish soldiers were killed, 50,000 died in the Polish army and over 12,000 died serving in the American military.[25] These numbers represent the highest casualty rates, though Jews died serving in other armies as well. As is well known, six million Jews died in the Holocaust. Any claims concerning Jewish gain in WWI, the Bolshevik Revolution or WWII as stated in *The Hamas Covenant*, Article 22, (see below) are outright lies.

After the Holocaust, Stalin continued active persecution of the Soviet and Eastern European Jews who survived. In 1952 he orchestrated the antisemitic Slansky Trial in Czechoslovakia where Jews were accused of betraying the nation and the Communist Party. Simultaneously, he planned the destruction of Soviet Jewry under the guise of what became known as the "Doctors' Plot" where mostly Jewish doctors were arrested and accused of plotting and assassinating the Soviet leadership. Stalin accused the entire Jewish community of betrayal, in order to gain support for total Jewish annihilation. Stalin's death in early 1953 ended persecution of this type, as his successors did not follow through with his plans.[26] However, for over thirty years continuing into the 1980s Jews suffered from economic, educational and cultural discrimination under Stalin's successors.

Czarist/Nazi Antisemitism in *The Hamas Covenant*

As previously discussed in Chapter VI, "*The Hamas Covenant* Analysis," there are several Articles accusing the Jews of Nazism, while many others use Czarist/Nazi accusations in their condemnation of the Jews. Articles 20 and 31 accuse Israel of Nazism through "accusation reversal." Hamas knew and knows full well of the Muslim Brotherhood alliance and support for Nazism in the past and its continuous support through spokesmen such as the Egyptian cleric Sheikh Yusuf al-Qaradawi, at present. Article 28 links back to the Czarist/Nazi conspiracy theories accusing the Jews of achieving world conquest through proxy civic organizations such as the Freemasons and Rotary Clubs. Article 30 condemns Zionism for its dominance of the world media and finance, echoing the Czarist/Nazi libel against the Jews. Most importantly our focus will be on Article 22, the heart of the Hamas vilification of the Jews.

Virtually everything written in *The Hamas Covenant* Article 22 was influenced by the notorious antisemitic Czarist forgery, *The Protocols of the Learned Elders of Zion*,[2728] published in 1903. By demonizing the Jews and blaming them for all Czarist Russia's troubles, the attitudes expressed in *Protocols* helped set the stage for violent attacks against the Jews in 1903, especially the Kishinev Pogrom. The document is said to be the twenty-four "Protocols" of a secret universal Jewish leadership conspiracy plotting world conquest. The Russian secret police invented these "Protocols" to prove "Jewish evil" and in the 1920s these diatribes served as the foundation for Chapter XI of Volume I, in Hitler's *Mein Kampf*, entitled "Nation and Race."[29]

In the Introduction to *Protocols,* under the heading "Elders" and preceding the main body of the forgery, the "Jewish plot" for world conquest is said to be "exposed." Supposedly, the Jewish scheme began in 929 BCE with King Solomon and continues to this day. Jewish leaders, in particular the Zionist leaders Theodor Herzl and Chaim Weizmann, are accused of seeking world domination, not just a national home in Palestine. To quote part of the Introduction to the *Protocols:* "The desire for a National Home in Palestine is explained as only camouflage and an infinitesimal part of the Jew's real objective. It proves that the Jews of the world have no intention of settling in Palestine or any separate country and that their annual prayer that they may all meet 'Next Year in Jerusalem' is merely a piece of their characteristic make-believe. It also perpetuates the idea that the Jews are now a world menace, and that the Aryan races will have to domicile them permanently out of Europe." With the publication of *The Protocols*, Czarist antisemites demanded Jewish expulsion from Europe. Later, Hitler picked up the terms "Aryan races" and "race and nation" from *Protocols*, and the Nazis sought Jewish extermination.

The outline below shows the undeniable similarities between the deadly antisemitism of *The Hamas Covenant* Article 22, the *Protocols of the Elders of Zion* and Hitler's *Mein Kampf* Chapter XI. Quotes are as found in the original translations.[30] Each section's sub-header identifies the area of accusation the three documents have against the Jews. Bullet points are verbatim examples and/or represent a summary of what *The Hamas Covenant*, the *Protocols,* and *Mein Kampf* respectively say about the issues. Quotes from *The Hamas Covenant* Article 22 are in bold while quotes using capital letters are in the original.

Accumulation of Financial Resources

- *The Hamas Covenant* claims that Jews must "**accumulate huge financial resources**" for their plots to succeed.

- In Protocol 1, Clause 7, the Czarists accused the Jews of plotting to acquire the world's gold resources to facilitate global conquest through the "despotism of capital." Protocol 22, entitled "Power of Gold," once again reminded everyone of Jewish greed and their quest to control world finance.

- Hitler concurred and went a step further: "Finance and commerce have become his [the Jew's] complete monopoly," and continues, "the Jewish influence on economic affairs grows with terrifying speed through the stock exchange. He becomes the owner, or at least the controller, of the national labor force." The Jewish objective is to dominate the world economy.

Take-Over of the Media

- *The Hamas Covenant* states: "**With money they [the Jews] have taken control of the world media - news agencies, the press, publishing houses, broadcasting services, etc.**"

- In Protocol 2, under "Destructive Education" in Clause 5, Jews are accused of planning and gaining domination over the press. This plot is further discussed in Protocol 7 under "Universal War," Clause 5, where the conclusion is: "THE PRESS, WHICH, WITH A FEW EXCEPTIONS THAT MAY BE DISREGARDED, IS ALREADY ENTIRELY IN OUR [Jewish] HANDS." Protocol 12 in its entirety is dedicated to Jewish domination of the media as a means of controlling the world.

- Likewise, Hitler exclaimed: the powerful "weapon in the service of the Jews—the press. With all his perseverance and dexterity he seizes possession of it. With it he slowly begins to grip and ensnare, to guide and to push all public life, since he is in a position to create and direct that power which, under the name of 'public opinion,' is better known today than a few decades ago."

VIII Czarist-Nazi Integration into Palestinian Islamic Jihad   365

Responsibility for Revolutions Worldwide

- *The Hamas Covenant*: "With money they sparked revolutions in various countries around the world in order to serve their interests and to reap profits. They were behind the French Revolution and the Communist Revolution and [they are behind] most of the revolutions about which we hear from time to time here and there."

- Phantom "Jewish" authors described the Jewish responsibility for the French Revolution in Protocol 3, Clause 14. "The secrets of its preparations are well known to us for it was wholly the work of our hands." This statement comes after the Jewish conspiracy entitled "Liberty, Equality, Fraternity," found in Protocol 1, Clause 26, under the heading "WE SHALL END LIBERTY." This section is said to lay bare the supposedly Jewish rabble-rousing slogan to overthrow the aristocracy. Protocol 3, Clause 7, under the heading "WE SUPPORT COMMUNISM," states, "We [the Jews] appear on the scene as alleged saviors of the worker from this oppression when we propose to him to enter the ranks of our fighting forces – Socialists, Anarchists, Communists – to whom we always give support in accordance with an alleged brotherly rule (of solidarity of all humanity) of our SOCIAL MASONRY."

- *Mein Kampf* views universal social process and revolution as a Jewish plot for world domination. After discussing the Zionist "swindle," the use of democracy and labor unions to ensure Jewish power and the Jewish Marxist plot, Hitler stated that now "begins the great last revolution. In gaining political power the Jew casts off the few cloaks that he still wears. The democratic people's Jew becomes the blood-Jew and tyrant over peoples."

Formation of Secret Organizations

- *The Hamas Covenant* continues: "With money they have formed secret organizations, all over the world, in order to destroy [those countries'] societies and to serve the Zionists' interests, such as the Freemasons, the Rotary

Clubs, the Lions, the Sons of the Covenant [i.e. B'nei B'rith], etc."

- The *Protocols* attack Freemasons for functioning in the name of the Jews as "a screen for us and our objects" (Protocol 4, Clause 2) and once again when stating that the, "SECRET MASONRY WHICH IS NOT KNOWN TO, AND AIMS WHICH ARE NOT EVEN SO MUCH AS SUSPECTED BY, THESE 'GOY' CATTLE, ATTRACTED BY US INTO THE 'SHOW' ARMY OF MASONIC LODGES IN ORDER TO THROW DUST IN THE EYES OF THEIR FELLOWS" (Protocol 11, Clause 7). The Protocols accuse Freemasons of being an arm of the Jews serving as, "THE AGENTS OF INTERNATIONAL AND NATIONAL POLICE" (Protocol 15, Clause 4), that will eventually meet their end when liquidated by the Jews (Protocol 15, Clause 9).

- *Mein Kampf* attacks the supposed Jewish conspiracies of communism, the Freemasons and the press at once, declaring, "Thus there arises a pure movement entirely of manual workers under Jewish leadership, apparently aiming to improve the situation of the worker, but in truth planning the enslavement and with it the destruction of all non-Jewish Peoples. The general pacifistic paralysis of the national instinct of self-preservation begun by Freemasonry in the circles of the so-called intelligentsia is transmitted to the broad masses and above all to the bourgeoisie by the activity of the big papers which today are always Jewish. Added to these two weapons of disintegration comes a third and by far the most terrible, the organization of brute force. As a shock and storm troop, Marxism is intended to finish off what the preparatory softening up with the first two weapons has made ripe for collapse."

Against Democracy

- Hamas speaks of the **"capitalist West"** as a euphemism for liberal democratic societies. Although democratically elected to power, just like the Nazis were, Hamas used its armed forces to overthrow the state building framework of

VIII Czarist-Nazi Integration into Palestinian Islamic Jihad

the Palestinian Authority in the Gaza Strip in June 2007 and instituted an Islamist dictatorship. Following in Hitler's footsteps, Hamas condemns both the **"imperialist forces of the capitalist West and the communist East"** as Jewish puppets begging their own destruction by their Jewish overlords. For Hamas, democracy is used as a means to an end, the achievement of Islamic power. Secular rule is forbidden as is the people's will or democracy. What is not in line with Sharia Law is prohibited.

- The *Protocols* discuss how the Jews will exploit a "SO-CALLED LIBERALISM" in the name of "political freedom" in order to weaken and conquer the non-Jewish nations of the world (Protocol 2, Clause 6). The Jews will use constitutions to undermine the monarchies and eventually bring about the overthrow of all state institutions to ensure universal conflict, chaos and poverty (Protocol 3, Clauses 2-4). Such democracy and "'freedom'" will be used to deceive the foolish masses and will become a "brute force which turns mobs into bloodthirsty beasts" (Protocol 3, Clause 20). The "chaotic license of liberalism" will allow for Jewish control of all institutions of society including elections, the press, the law and in particular education (Protocol 9, Clause 9) which in turn will lead to mob support for Jewish dominance (Protocol 10, Clause 4). The "Poison of Liberalism" and "Constitutionalism" will be used to ensure Jewish "DESPOTISM" and world sovereignty (Protocol 10, Clauses 8-20).

- *Mein Kampf*, states it this way, "His [the Jew's] ultimate goal at this stage is the victory of 'democracy' or, as he understands it: the rule of parliamentarianism. It is most compatible with his requirements; for it excludes the personality – and puts in its place the majority characterized by stupidity, incompetence, and last but not least, cowardice." Hitler passed the Enabling Act of March 23, 1933,[31] giving the Reich cabinet and Fuhrer full powers of decision-making and taking the first major step in ending German democracy.

Supporting International Conflict

- Below is the continuing Hamas message of Jewish world domination through massive financial resources and the Jewish plot to spark global conflict. The similarities to Hitler's speech are obvious (see below). The consequences for the Jews are meant to be identical as well. Once again the Jews are accused of instigating universal turmoil. *The Hamas Covenant,* Article 22: **"There is no end to what can be said about [their involvement in] local wars and world wars. They were behind World War I, through which they achieved the destruction of the Islamic Caliphate, reaped material profits, took control of numerous resources, obtained the Balfour Declaration, and established the League of the United Nations [sic] so as to rule the world through this organization. They were [also] behind World War II, through which they reaped enormous profits from commerce in war materials and paved the way for the establishment of their state."** Most of these accusations originate in *The Protocols of the Elders of Zion* and are repeated in Hitler's *Mein Kampf.*

- *Protocols* accuse the Jews of fostering economic wars (Protocol 2, Clause 1), working for universal conflict of all against all (Protocol 3, Clause 3), and advocating "universal war" (Protocol 7, Clause 3). As noted above the Jews are said to conspire to conquer the world through communist agitation and the control of world finance.

- In *Mein Kampf* Hitler explains that the Jew seeks wars where all fight each other. The Jew implements such a policy through his control of the media, economy, state apparatus, the press and a false concern for the workers. To succeed in world conquest the Jew "weaves a net of enemies, thanks to his international influence, incites them to war and finally, if necessary, plants the flag of revolution on the very battlefields." Such thinking culminated in his infamous speech to the Reichstag on January 30, 1939 when Hitler showed the same irrational antisemitic conspiracy theories found woven throughout *The Hamas Covenant*. He condemned Jews to death for the imagined crimes of international finance (capitalism) and Bolshevism (communism), two entirely opposite political/economic systems. To quote Hitler, "Today I will once more be a prophet: if the international Jewish financiers in and outside Europe should succeed in plunging the nations once more into a world war, then the result will not be the Bolshevizing of the earth, and thus the victory of Jewry, but the annihilation of the Jewish race in Europe!"

The Downfall of the Islamic Caliphate & Worldwide Autocracy

- Hamas explains that **"They [the Jews] were behind World War I, through which they achieved the destruction of the Islamic Caliphate,"** thereby placing blame on the Jews for both events. Such supposed Jewish behavior can be traced back to Koran 5:82, accusing the Jews and pagans of having "enmity to the faithful." The final downfall of the Caliphate occurred under Mustafa Kemal Ataturk, the secular leader who created the modern Turkish state in the 1920s. Islamists, particularly Sayyid Qutb and his followers, charged Ataturk with working for the Jews, leading to questions posed by Islamists as to whether Ataturk may have been a Jew himself. Many Islamic fundamentalists see the collapse of the Caliphate as part of an overall Jewish conspiracy to destroy Islam. This parallels the Czarist antisemitism expressed in *Protocols* accusing the Jews of plotting to destroy the Russian aristocracy and Christian clergy and Hitler's determination that the Jews are destroying all peoples and religions (see below). Hamas falsely accuses the Jews of obliterating the Muslim institution of autocratic rule, the Islamic Caliphate.

- In *Protocols*, the Czarist forgers accuse the Jews of destroying aristocracy in general, and more specifically working to destroy the "RUSSIAN AUTOCRACY" (Protocol 15, Clause 3), the clergy and even the Papacy through "FREEDOM OF CONSCIENCE" to ensure the "COMPLETE WRECKING OF THAT CHRISTIAN RELIGION" (Protocol 17, Clauses 2-3).

- In *Mein Kampf* Hitler states, "Hence today I believe that I am acting in accordance with the will of the Almighty Creator: *by defending myself against the Jew, I am fighting for the work of the Lord.*"[32] Hitler sees himself as a Christian whose mission it is to destroy the evil Jewish race who he accuses of attempting to conquer the world. He does not believe the Jews adhere to a religion as they have no ideals and only seek practical gain. For Hitler the Jews were a racial "menace" working to destroy all peoples and religions. Hitler like the Czarists condemns the Jews for having "conscience" as noted in the following quote, "With our movement the intervening age, the middle age, has come to its end. We terminate a wrong course of mankind. The tablets of Mount Sinai have lost their validity. Conscience is a Jewish invention." Here we note Hitler's barbarism to which he fully admits. Jewish conscience when adopted by Germans undermined their rights as the stronger to conquer and commit genocide against the weaker peoples. Such thinking impeded the German will to conquer the world. Here is another reason the Jews would need to be exterminated - to eliminate conscience.[33]

The Issue of Conscience

"Conscience" is not discussed directly in *The Hamas Covenant*. In Koran 5:82 it states plainly that "the most implacable of men in their enmity to the faithful are the Jews and pagans," who nowadays are accused of spreading the ideals of democracy, liberalism, socialism and secularism (see Chapter II "Ideologues"). Nor are Christians absolved as the verse continues "and that the nearest in affection to them are those who say: 'We are Christians.'" These socio-economic and political frameworks are very much the result of "conscience" as expressed by the Enlightenment thinkers of the early modern period in Europe. Matters of "conscience" led to the overthrow of the aristocratic and religious autocracy in Europe. The Hamas Jihadists see themselves facing the same enemies and in particular the Jews, whom they likewise accuse of destroying the Islamic Caliphate. The above secular Enlightenment concepts deny Sharia law and infer equality for all members of humanity, both infractions being an affront to the Hamas faithful.

"Conscience" leads to compassion, a trait considered a weakness by dictatorial and totalitarian regimes, most notably the Czarist, Nazi and Jihadi Islamist types. All chapters in the Koran begin with the verse, "In the Name of Allah, the Compassionate, the Merciful" except for Chapter 9 entitled "Repentance." Here the believers are urged to "slay the idolaters wherever you find them," while the *jizya* tax and hellfire branding are invoked against Jews and Christians. Unrelenting Jihad is the answer to non-compliance with Islamic dictates, no matter how peaceful one may be (Koran 9:1-14). Allah may be referred to as "compassionate" in parts of the Koran but in the Jihadi world view compassion and mercy are only granted to the believers, such attitudes toward all others were and are abrogated.

Jews Accused of Global Manipulation

To repeat, according to Hamas, the Jews **"reaped material profits, took control of numerous resources, obtained the Balfour Declaration, and established the League of the United Nations [sic] so as to rule the world through this organization. They were [also] behind World War II, through which they reaped enormous profits from commerce in war materials and paved the way for the establishment of their state."**

In addition to exposing the blatant antisemitism expressed in the Hamas quote above, the historical record must be set straight. In 1917, the British issued the Balfour Declaration supporting a Jewish National Home in Palestine. London was already engaged in the McMahon Correspondence with Sharif Hussein of Mecca in 1915 to establish an Arab kingdom with the halt of hostilities at the end of WWI. The fact that the British only followed through in part had little to do with Jews or Zionism and had everything to do with the Sykes-Picot Agreement and the French. These two European powers divided the Middle East into mandates, thereby preventing the establishment of a unified Arab Kingdom. The Zionist leader Chaim Weizmann met with Sharif Hussein's son Faisal to discuss Arab-Jewish national issues, joint interests and cooperation in January 1919. The idea was to work together for Jewish immigration and Arab development in Palestine. Faisal's condition for implementation was that the British fulfilled all of their promises to the Arabs, which they did not.[34] As a result the Weizmann-Feisal Agreement was never carried out.

The world organization between the two world wars was the League of Nations. Jewish groups were not involved in assembling the League, which consisted of state entities. This fact alone illustrates Hamas' ignorance when addressing international issues and accusations of the Jews' involvement with the League of Nations. The Hamas accusation only makes sense if one believes in the conspiracy theories of Jewish world dominance. The League fell apart by the late 1930s when international efforts were most needed to save Jews from Hitler's Nazi persecutions, highlighted best by the massive German state-sponsored Kristolnacht (November 1938) pogroms. This Nazi state-driven terror came a few months after the failed Evian Conference, which sought to find German Jewry safe havens in other nations. The League of Nations proved itself a complete failure in saving Jews.

In 1948, Israel was born in destitution despite the Holocaust—not as a result of it. Much of the world, including the United States and especially Britain, considered a form of trusteeship designed around the Morrison-Grady Plan (1946)[35] as a remedy for the Palestine Mandate. Should it have succeeded, Britain would have retained its strategic interests, eliminating all possibilities for a Jewish State. Both Jews and Arabs rejected the proposal. It was not until November 29, 1947 that the United Nations presented Resolution 181, commonly known as the "Partition Plan," and the world considered the two-state solution whereby Jewish and Arab States were to live side by side as two political entities sharing a common economy. The Arab world, including the Palestinians, rejected the Partition Plan, but the Jews accepted it.

Palestinian Arabs initiated attacks against Palestinian Jews on November 30. Later, five Arab countries invaded Israel: Egypt, Jordan, Iraq, Syria and Lebanon. Haj Amin el-Husseini directed his irregular "Jihad Army" from the outside, but without any success, while another Nazi collaborator, Fawzi Qawuqji, joined the invasion from Syria, hoping to finish Hitler's work. Both failed (see Chapter III "Jewish National Liberation"). Article 22 of *The Hamas Covenant* is the logical continuation of Haj Amin el-Husseini and Fawzi Qawuqji's legacies.

Certain Hamas apologists admit the *Protocols* to be false and concede incorrect selective usages of Koranic quotes. They expect Article 32 and others to be stricken from the *Covenant* at a later date or at least changed as not to be antisemitic. Article 22 completely demonizes the Jewish People, going beyond the traditional Islamic anti-Jewish attitudes and more fully adopts the Czarist Russian and Nazi approach. The logical conclusion when dealing with demons is to destroy them. As for any expected change in the *Covenant*, this is most improbable since Hamas followers commonly refer to it as the "Charter of Allah," a binding Islamic document Divinely inspired by Allah Himself.

VIII Czarist-Nazi Integration into Palestinian Islamic Jihad 373

Endnotes
1. Lewis, *Semites and Anti-Semites*, W.W. Norton and Co., New York and London, 1999, pp. 147-150.
2. Patterson, David, *A Genealogy of Evil, Anti-Semitism from Nazism to Islamic Jihad*, Cambridge University Press, 2010, pp. 111-112.
3. Ibid, p. 108.
4. Hirszowicz, Lukasz, *The Third Reich and the Arab East*, Rutledge and K. Paul, London, 1966, pp.108-109.
5. Ibid, pp. 134-166.
6. Ibid, pp. 166-172.
7. Rejwan, Nissim, *The Jews of Iraq*, Weidenfeld and Nicolson, London, 1985, pp. 217-224.
8. Lebel, Jennie, *The Mufti of Jerusalem, Haj Amin el-Husseini, and National Socialism*, Cigoja Stampa Publishers, English translation, Paul Munch, Belgrade, 2007, pp. 81 and 84.
9. Hirszowicz, pp. 218-225.
10. Laqueur, Walter and Rubin, Barry, eds., *The Israel-Arab Reader, A Documentary History of the Middle East Conflict*, Penguin Books, New York and London, 1995, p. 71.
    Hitler, the Mufti, Foreign Minister von Ribbentrop and Minister Grobba attended the meeting. Protocol of the meeting taken from *Documents of German Foreign Policy 1918-45*, in Laqueur and Rubin, pp. 68-72.
11. Hirszowicz, pp. 225-228.
12. Ibid, pp. 257 and 309.
13. Ibid, p. 312.
14. Patterson, pp. 115-119.
15. "WHAT IF THE JEWS LOST ANY WAR," 4th quote, *Peace for Our Time*, retrieved July 22, 2011, www.peaceforourtime.org.uk/page146.html.
16. Lebel, *The Mufti of Jerusalem*, Appendix "Islam and Judaism" pamphlet by The Great Mufti of Jerusalem" (Zagreb 1943), pp. 311-319. Quotes and explanations are taken from the pamphlet.
17. Ibid, p. 318.
18. Ibid, pp. 318-319.
19. Patterson, p. 121.
20. Lebel, p. 175.
21. Ibid, p. 126.
22. For a full review see *Nazi Palestine: The Plans for the Extermination of the Jews of Palestine*, by Klaus-Michael Mallman and Martin Cuppers, translated by Krista Smith, 2010.
23. For more information, see *Palestine Media Watch*, www.palwatch.org.
24. Barnavi, Eli, "World War I and the Jews," *Jewish History, 1914 to 1948*, www.myjewishlearning.com.
    "World War I," *The YIVO Encyclopedia of Jews in Eastern Europe*, www.yivoencyclopedia.org/article.aspx/World_War_I.
    "Epilogue: Relevant Commentaries on the Lists," *DIE JUDISCHEN GEFALLENEN (German Jewish Soldiers)*, www.germanjewishsoldiers.com/epilogue.php. Budnitskii, Oleg, "Russian Jews Between the Reds and Whites 1917-1920," *University of Pennsylvania Press*, translated by Timothy J. Portice, www.upenn.edu/pennpress/book/toc/14908.html, all sources retrieved May 27, 2015.
25. Exhibit "Jewish Soldiers in World War II," *Latrun Armor Museum*, Israel.
26. For an overall review of Russian and Soviet Jewry, see Salo W. Baron's *The Russian Jew Under Tsars and Soviets* (1987).
    Also for persecutions specifically under Stalin's regime, see Louis Rapoport's work

Stalin's War Against the Jews (1990).

For an in depth study done as these major post WWII persecutions were unfolding, see *The Jews in the Soviet Satellites,* by Peter Meyer, Bernard Weinryb, Eugene Duschinsky and Nicolas Sylvain, 1953.

27 All quotes in this section are taken from "The Protocols of the Learned Elders of Zion," with revised 1923 Introduction, *Solar General,* retrieved January 25, 2016, http://solargeneral.org/wp-content/uploads/library/protocols-of-zion.pdf.

Note: this site is antisemitic.

Protocols of the Elders of Zion originally accessed under "The Winds," July 20, 2001.

For a dissection of the plagiarisms, half truths, outright lies and antisemitic slanders, see "The Protocols of the Elders of Zion," *Wikipedia,* https://en.wikipedia.org/wiki/The_Protocols_of_the_Elders_of_Zion, and "Protocols of the Learned Elders of Zion," *Britannica Encyclopaedia,* www.britannica.com/topic/Protocols-of-the-Learned-Elders-of-Zion.

For official Palestinian Arab usage of the *Protocols,* see "The Protocols of the Elders of Zion," *Palestine Media Watch,* http://www.palwatch.org/STORAGE/OpEd/Protocols_of_the_Elders.pdf.

All sites accessed August 30, 2015.

28 Often the word "Learned" is removed from the title and the forgery is referred to as *The Protocols of the Elders of Zion.* Both terms are correct.

29 Hitler, Adolf, *Mein Kampf,* originally published 1925/1926, accessed online at *Hitler Historical Museum,* retrieved January 25, 2001, http://www.hitler.org/writings/Mein_Kampf.

30 All quotes concerning Hitler are from Chapter XI in *Mein Kampf* unless otherwise noted.

Protocols quotes in all capital letters were written in capitals in the original document.

31 Shirer, William, *The Rise and Fall of the Third Reich,* Simon and Schuster, New York, 1960, p. 198.

32 Walker, Jim, "Hitler's religious beliefs and fanaticism," *NoBeliefs.com for freethinkers,* retrieved October 6, 2015, www.nobeliefs.com/hitler.htm.

33 "Hitler quotes from Adolf Hitler," *SimpleToRemember.com Judaism Online,* retrieved October 7, 2015, www.simpletoremember.com/articles/a/hitler-quotes. Heinsohn, Gunnar, "Shofar FTP Archive File: people/h/heinsohn.gunnar/why-auschwitz," *The Nizkor Project,* September 26, 1996, retrieved January 5, 2016, www.nizkor.org/ftp.cgi?people/h/heinsohn.gunnar/why-auschwitz.

The quote from Hitler is found within the text written by Gunnar Heinsohn.

34 "Agreement Between Emir Feisal and Dr. Weizmann," *MidEast Web Historical Documents,* retrieved January 15, 2010, www.mideastweb.org/feisweiz.htm.

35 "Morrison-Grady Plan (1946)," *Encyclopedia.com,* retrieved Nov. 11, 2015, www.encyclopedia.com/doc/1G2-3424601875.html.

Sachar, Howard, *A History of Israel from the Rise of Zionism Until Our Time,* Alfred A. Knopf, New York, USA, 2007, pp. 270-275.

IX

Conflict Resolution in the Shadow of Islamic Abrogation

Constructive Perspectives Toward Jews, Christians and Others

Overview

Islam has the tools to curtail Jihadi demands from within. Applying these tools will have a much greater impact on world peace than any attempt to force Islam into a democratic framework, or replace it with secular nationalism. Perhaps surprisingly to non-believers, the solution to the world Jihad offensive and Islamic antisemitism is found in the Koran itself. Having survived fourteen centuries, Islam has the internal strength and latent flexibility to avoid a devastating clash with other cultures and religions. These qualities are critical not only for the future of Islam but for global self-preservation.

Islam was overwhelmingly successful in spreading its message for about a millennium, beginning with its inception in the early 600s. Full faith in Allah, along with commitment to Sharia law and military Jihad were accredited with the continual victory. The end of Islamic dominance may be dated as early as the close of the sixteenth century, or several generations later with the failed siege of Vienna in 1683. The first major setback was the "*Reconquista*" in Spain, which ended in 1492. That was viewed at the time as a temporary check and Christendom was expected to suffer defeat; however, to date, no successful Islamic counteroffensive has ensued. For the past four centuries the Christian West has landed one blow after another on their Muslim rivals, culminating with the break-up of the Ottoman Empire after WWI, the European occupation of large swaths of the Middle East and the development of nation state frameworks in these regions by the mid-twentieth century.

I will make the argument that Islam, like all major religions, is given to numerous interpretations. Bernard Lewis, in his study *What Went Wrong?*, sees a lack of freedom, discrimination and prejudice within Islamic societies as leading the Muslim world to the failures, frustrations and extremism experienced today. If Muslims intend to improve their own societies, then lashing out at and blaming the West and other cultures for their misfortunes is futile. The suggestion is for Muslims to look inward seeking answers. The current Jihadi solutions feed off of societal failure in present day Islam. In Lewis' conclusion, he admits his perspective is one of a Western observer. Lewis implies the need for a more Western, secular value-based society, admitting this to be a long and difficult path.[1]

Edward Said accused Lewis and other Western scholars of "Orientalism" in the 1970s and beyond. Said condemned these researchers as Westerners studying the East, or Orient, as an "object" and, in particular, upbraided such academics for seeing Islam as inferior and inflexible.[2] Without delving into the academic debate that followed, let it be stated clearly here that Islam is central to its adherents' identity and that it can modify its direction, but such change can only come from within their core of belief and understanding. Foreign pressure affects all people and cultures and Islam is no exception. For true believers, Western coercion forcing an Islamic re-evaluation of its attitude toward non-Muslims will not stand the test of time when compared to change formulated from within firmly established Islamic traditions of interpretation. Devout Muslims define secular nationalism as Western-induced and forced upon Muslim societies under duress.

Conflict resolution must include pious Muslims, thus an Islamic framework is imperative. This chapter will review relevant Koranic quotes and analyze the dangers of continuing on the present path. I will present proposals for using alternative sacred verses to achieve a peaceful outcome to the Islamist clash with the West, Judaism, Christianity and the non-Abrahamic religions and philosophies, particularly those originating in the East. I will make it clear that no one needs to accept Jihadi dictates, regardless of how liberal or "multi-cultural" one's perspective may be. Where issues of morality and ethics are considered we are obliged to seek out the best solution on the universal plane and eliminate demands for war, whether considered holy or not. The liberal, democratic West must stand its ground firmly while simultaneously suggesting peaceful Islamic alternatives to help bring conflict resolution. Such a move could alter specific Jihadi aspirations, including those of Hamas.

Changes must be undertaken while dealing with the realities of the Western Christian world's economic and technical superiority. Worse yet for Islamists, the existence of the State of Israel for close to seven decades remains a festering sore defying Allah's Divine will. Today's Islam lives within a contradiction of theological challenges that can either be constantly contested through Jihad, or accepted as bequeathed by Divine forces. Islamic leadership and jurists have discretion in decision making but need religious texts to support any judgments handed down. Politics, religion and self-preservation all mix as judicial judgments (*fatwas*) and are seen as crucial in maintaining the continuation of an Islamic society rooted in Sharia law. In our nuclear age, the threat of mass destruction is a powerful motivation for both the West and

IX Conflict Resolution in the Shadow of Islamic Abrogation

the Muslim world to find answers for survival and an alternative to Jihadi Islam.

The unorthodox route suggested here is based on quotes from the Koran and will seek an alternative to continued conflict. Moderate statements in the Koran toward non-Muslims and pro-Israelite/Jewish comments abound, yet few vestiges of peaceful intent are noticeable among today's Muslim fundamentalists. There are two primary reasons for this. First, Islamism worldwide is wracked by internal conflict in the battle against the moderates and secular nationalists. The clash is between peaceful Muslims and Jihadi-inspired groups such as the Muslim Brotherhood, which includes the Palestinian Hamas, the Taliban, Al-Qaeda, the Islamic State (ISIS or ISIL) and similar types. Jihadists accuse peaceful Muslims in their home societies of betraying Islam by accepting the present world order. Demands for war are part of the internal debate; demands for violence often increase public support. Second, Jihad activism against non-Muslim societies leads to never-ending war with expected Divine victory over the Western world and other non-Islamic societies. Although this conflict jeopardizes all of humanity, the Jihadists base their behavior on a specific interpretation of Islam whose ideological roots reach back to the seventh century. Today this interpretation is considered "radical" or "fundamentalist," but over the initial millennia of Islamic expansion from the Iberian Peninsula to India and beyond, such a Jihadist perspective was considered quite legitimate and mainstream. As in previous eras, today's Jihadists use "abrogation" or negation to invalidate peaceful verses toward Jews, Christians and other non-Muslims and instead institute Holy War or Jihad in the name of Allah to spread Islam. The key clause is the neutral Koranic verse 2:106 which remains open for interpretation by the acknowledged Divinely inspired Islamic jurists and commentators:

> If We abrogate any verse or cause it to be forgotten We will replace it by a better one or one similar. Do you not know that Allah has power over all things?

This verse contains the origin of a solution. A moderate, conciliatory, co- or multi-existence approach exists side-by-side with the Jihadi interpretation. The moderate approach also has roots in the seventh century, although this understanding is considered a minority opinion held by very few at present.[3] We must consider the possibility that this theological perspective can be re-legitimized and re-introduced through a conscious effort directed toward the

Muslim faithful and Islamic scholars. To achieve a more peaceful world, a re-emphasis on positive Koranic quotes relating to Judaism, Christianity and others is necessary to override Jihadi claims. This specific theological perspective through the use of "reverse abrogation" contributes to conflict resolution between Islam and others, provided an effort is made by the accepted Islamic authorities to do so. Achieving such a reversal is an uncertain prospect at best. It will require an intensely focused effort, but without a shift in emphasis toward inclusive universalism and away from Jihad, the conflict between Islam and the rest of humanity can be expected to continue. The non-Muslim world needs a moderate ally from within the Islamic clerical elite to abrogate Jihad, achieve conflict resolution and benefit humankind.

One could believe there is no hope for peace since Jihad and Islamic extremism grew out of the "abrogation" of conciliatory verses in the Koran while emphasizing conflict-oriented dictates. Yet, extremism can be reversed. Not all Islamic scholars, either in the distant past or in the modern period, follow "abrogation." Since 9/11 the debate has intensified.[4] Most significantly Chapter 9 in the Koran is belligerent and warlike in the extreme, invoking punishment against the People of the Book and others deemed as enemies. Chronologically it appears toward the end of the Koran and abrogates the previously prescribed positive characteristics attributed to and peaceful commands given regarding Jews and Christians. The suggestion here is to engage in conflict resolution through the emphasis on peace-seeking passages and a re-interpretation of the Koran. Being guardedly optimistic, it is my belief that the overwhelming majority of the 1.5 billion Muslims worldwide do not seek a war of infinite duration with all other peoples. Still extremist Islamic interpretations are making the most inroads within peaceful Muslim societies. One can be led to believe that no religiously acceptable alternative exists to the Jihadist perspective, since none is truly being provided from within the Islamist world.

Unfortunately, pro-Western secular regimes in the Middle East are often corrupt and repressive toward Islam as a way of life and therefore do not present an alternative for the devout. The average Muslim finds secular regimes offensive, and although he may not be a Jihadist at the outset, he becomes open to such ideals after feeling the threat of foreign and particularly Western influences. Any challenge to the Jihadi, antisemitic, anti-female, anti-Western and xenophobic worldview pressed by Islamist radicals must come from the Koran, which is the ultimate Islamic source. The West can only provide a partial solution through secularization, material well-being and the temporary military containment of Jihadism. None of these efforts alter basic understandings within Islam, but rather they lead to increased conflict. In times of societal stress, Jihadism will return unless an Islamic theological solution is provided at its source. Only a reformed pluralistic Islam, accepting of other world cultures and religions can defeat Jihadism. The winner of the ideological battle in the arena of Islamic thinking today will be the policy makers in tomorrow's Muslim states. Change must come from within, through a victory of the reformers.

The solution lies in reinstating and highlighting the previously abrogated conciliatory or positive Koranic verses toward Jews, Christians and others. Using the same tools, warlike interpretations must be downplayed, ignored and finally nullified or abrogated. Conflict resolution must begin with the Muslim recognition of the rights of the Jewish People to life and equality. Viewed universally, we are not speaking of a narrow nationalist conflict between Palestinians and Israelis in the eyes of diocentric, devout Muslims. The conflict is perceived as a global zero-sum game, one with Islam pitted against Judaism and world Jewry only as a first step. The true struggle continues with the subjugation of all humanity. For the Jihadists, the Jews must be destroyed and the world subjected to Islamic rule (see Chapter II on "Ideologues" sub text on Sayyid Qutb).

Chronologically, the earlier chapters written in the Koran are favorable toward the Israelites/Jews. They date from the early period of Mohammed's revelations from 610-624 CE, prior to the outright Jewish rejection of Mohammed as a prophet. Originally Mohammed and his followers prayed toward Jerusalem and sought alliances with the Arabian Jewish tribes. Only after his disappointments in dealing with the Jews did the early Muslims become anti-Jewish (see Chapter I "Negative Image of the Jew in the Arab Muslim word"). Yet the Koran remained ambivalent and positive statements toward the Jews were neither edited nor removed. Even in the later chapters there are reminders of previous positive insights of Allah's relationship with the Israelites/Jews. To re-emphasize my argument, the abrogation clause is fully relevant, allowing for human interpretation and intervention to be achieved through either conflict/conquest or reconciliation. In particular regarding Hamas and its *Covenant*, today's Muslims face a dilemma when making a choice between war and peace with the Jewish People, whether in Israel or in the Diaspora. The Islamic world can choose peace, reinforcing such a policy decision with verses from the Koran.

Positive Attitudes Toward Israelites/Jews in the Koran[5]

Islamic Endorsement of the Israelite Covenant at Sinai

If previous Islamic scholars decided on Jihad and battle against the People of the Book while eliminating all positive mention of them in the Koran, today's Islamic scholars can decide against Jihad and subjugation of the People of the Book in favor of conflict resolution based on Koranic passages, in particular the pro-Jewish, pro-Sinai covenant verses. One may ask why this was not done. One reason is that Jews were seen in a negative light; to acknowledge the legitimacy of Jewish nationalism in the Middle Eastern heartland is to deny a 1,400 year-old tradition. Israel, Jews and the West continue playing convenient scapegoat roles when Muslims seek answers in connection with their own societal dysfunctions. In traditional Islam supporting or defending the Jews would be akin to treason. It would involve a significant change in perspective, but Israel's existence today, after numerous wars, could be interpreted as the realization of Divine will.

The Israelite Exodus from Egypt plays a major role in the Koran, and is mentioned numerous times in many contexts. Most often the Koran follows the *Tanakhic* (Hebrew Scriptures or Old Testament) narrative closely, both in its positive and critical attitudes toward the Children of Israel. For instance, the Koranic verses 10:90-93, 28:43-49, 46:12, 45:16-17 and 2:122 are positive as shown below:

> We led the Israelites across the sea, and Pharaoh and his legions pursued them with wickedness and hate. But as he was drowning, Pharaoh cried: 'Now I believe that there is no god save the God in whom the Israelites believe. To Him I give up myself.'
>
> 'Now you believe!' Allah replied. 'but before this you were a rebel and a wrongdoer. We shall save your body this day, so that you may become a sign to all posterity: for most men give no heed to Our signs.'
>
> We settled the Israelites in a blessed land and provided them with good things. Nor did they disagree among themselves until knowledge was given them. Your Lord will judge their differences on the Day of Resurrection (Koran 10:90-93).

IX Conflict Resolution in the Shadow of Islamic Abrogation 381

The Egyptians drown when pursuing the Israelites after Moses split the Red (Reed) Sea as recounted in much greater detail in the Book of Exodus Chapter 14. Allah then "settled the Israelites in a blessed land" the reference being to the "Promised Land" or "Land of Canaan" and later referred to as the "Land of Israel." Entering the Promised Land corresponds to the Book of Joshua in the Hebrew Bible.

Continuing onward, there are the parallels of the holiness of God's word as revealed in both the Torah[6] and the Koran as quoted below:

> And after we had destroyed the previous generations We gave Moses the Scriptures as a clear testimony, a guide and a blessing for men, so that they might give thought.
>
> You [Mohammed] were not present on the western side of the Mountain when We charged Moses with his commission, nor did you witness the event. We raised many generations after him whose lives We prolonged. You did not dwell among the people of Midian, nor did you recite to them Our revelations; for We sent forth to them other apostles.
>
> You were not present on the Mountain-side when We called out to Moses. Yet We have sent you forth as a blessing from your Lord to forewarn a nation to whom no apostle has been sent before, so that they may take heed and may not say, when evil befalls them on account of their misdeeds: 'Lord, had You sent us an apostle, we should have obeyed your revelations and believed in them.'
>
> And now that they have received from Us the truth they ask: 'Why is he not given the like of what was given to Moses?' But do they not deny what was given to Moses? They say: 'Two works [the Torah and the Koran] of magic supporting one another!' And they declare: 'We will believe in neither of them' (Koran 28:43-49).

God gave Moses the Scriptures, or Torah, on Mount Sinai. The Koran makes clear Mohammed was "not present on the Mountain-side when We called out to Moses." Revelation was first given to the Children of Israel through Moses well before Mohammed appeared on the scene. In Judaism the "Sinai motif" is particularly significant as Divine revelation is not only awarded to an individual such as Moses or the Patriarchs before him, but rather is passed down as a mass event obligating the entire People of Israel.

Allah sent Mohammed "to forewarn a nation to whom no apostle has been sent before." He is the Messenger to the Arabs, many of whom did not follow him. Rather the pagan Arabs declared their disbelief that both the Torah and the Koran were revealed from God. Mohammed insisted the Koran was the Muslim interpretation of the Torah verses. The same pagan deniers condemned both the Torah and the Koran as "two works of magic supporting one another!" And next:

> Yet before it [the Koran] the Book of Moses was revealed, a guide and a blessing to all men. This Book confirms it. It is revealed in the Arabic tongue to forewarn the wrongdoers and to give good news to the righteous (Koran 46:12).

The Torah was received before the Koran and once again there is a positive correlation between these two Divinely revealed texts. The awarding of the Koran to the Arabs is compared with the giving of Torah at Sinai, both monumental events in the annals of humankind according to the above Koranic verse. To make clear the importance of the "Sinai motif," the Children of Israel have been "exalted" above the other nations as seen below in verses 45:16-17 and 2:122:

> We gave the Scriptures to the Israelites and bestowed on them wisdom and prophethood. We provided them with good things and exalted them above the nations. We gave them plain commandments: yet it was not till knowledge had been vouchsafed them that they disagreed among themselves from evil motives. On the Day of Resurrection your Lord Himself will judge their differences (Koran 45:16-17).

> Children of Israel, remember that I have bestowed favours upon you and exalted you above the nations. Fear the day when every soul shall stand alone: when neither intercession nor ransom shall be accepted from it, nor any help be given it (Koran 2:122).

The Israelites were given the Scriptures, prophecy, wisdom and the commandments. Although disagreements erupted among the Israelites themselves, often due to "evil motives," it is not the place of men to judge them. Allah will judge them on the Day of Resurrection.

The recognition that Allah "exalted them above the nations," is clear reference to the definition of the Hebrew "*sgula,*" meaning a "treasure" or "virtuous" people as recounted in Exodus 19:5. The Israelites were chosen and chose to accept the covenant with God. In essence, by using the Israelite example, Allah is urging the Arabs to accept his covenant with them through the Koran. This Islamic covenant recognizes the agreement between the Israelites and Allah, and does not nullify it or allow for human judgment as for whether the Israelites are adhering to their part of the agreement or keeping God's laws.

Here is a further emphasis on the firm connection between the previous revelation of the Torah to the Israelites and the new revelation of the Koran to the Arabs, as recounted in 40:53-54, 29:46 and 32:23-25.

> We gave Moses Our guidance and the Israelites the Book to inherit: a guide and an admonition to men of understanding. Therefore have patience; Allah's promise is true (Koran 40:53-54).

> Be courteous when you argue with the People of the Book, except with those among them who do evil. Say: 'We believe in that which is revealed to us and which was revealed to you. Our God and your God is one. To Him we surrender ourselves' (Koran 29:46).

> We gave the Scriptures to Moses (never doubt that you [Mohammed] will meet him) and made it a guide for the Israelites. And when they grew steadfast and firmly believed in Our revelations, We appointed leaders from among them who gave guidance at Our bidding. On the Day of Resurrection your Lord will resolve for them their differences (Koran 32:23-25).

The Israelites received the "Book" or Torah as a guide, just as at a later time Muslims received the Koran to show them the correct direction. Muslims believe Allah is the same God of Israel, and are obligated to accept the Israelite

Torah in addition to their own holy scriptures, the Koran. Honest disagreements between Jews and Muslims must be dealt with in a courteous manner, showing the proper respect. Once again, should the Israelites disagree among themselves, "On the Day of Resurrection your Lord will resolve for them their differences." As mentioned above, it is Allah's domain to judge or settle disagreements among the Israelites, by inference adherents of Islam have no part in God's relationship with the Jews.

Upon accepting the Torah and sealing the covenant at Sinai, the trek continues with both blessings and aid from God alongside His condemnation of the Israelite failure to fully show faith. The Koranic quotes below are in tandem with the account given in the Torah.

> Children of Israel! We delivered you from your enemies and made a covenant with you on the right flank of the Mountain. We sent down manna and quails for you. 'Eat the wholesome things with which We have provided you and do not transgress, lest you should incur My wrath, We said. He that incurs My wrath shall assuredly be lost, but he that repents and believes in Me, does good works and follows the right path, shall be forgiven. But, Moses, why have you come with such haste from your people?'
>
> Moses replied: 'They are close behind me. I hastened to You so that I might earn Your pleasure' (Koran 20:80-84).

Next we will read Koranic scripture verifying full Muslim acceptance of the "covenant" between God (Allah) and the Children of Israel. The covenant promises the Land of Israel, or Canaan, to the Israelites. The covenant was first made with Abraham, then Isaac, and Jacob, until it was solidified at Mount Sinai not only with Moses, but with all of the Children of Israel. The covenant is everlasting as God (Allah) does not go back on His word.

> Bear in mind the words of Moses to his people. He said: 'Remember, my people, the favours which Allah has bestowed upon you. He has raised up prophets among you, made you kings, and given you that which he has given to no other nation. Enter, my people, the holy land which Allah has assigned for you. Do not turn back, or you shall be ruined.'

'Moses,' they replied, 'a race of giants dwells in this land. We will not set foot in it till they are gone. Only then shall we enter.'

Thereupon two God-fearing men whom Allah had favoured said: 'Go in to them through the gates, and when you have entered you shall surely be victorious. In Allah put your trust, if you are true believers.'

But they replied, 'Moses, we will not go in so long as they are in it. Go, you and your Lord, and fight. We will stay here.'

'Lord,' cried Moses, 'I have none but myself and my brother. Do not confound us with these wicked people.'

He replied: 'They shall be forbidden this land for forty years, during which time they shall wander homeless on the earth. Do not grieve for these wicked people' (Koran 5:20-26).

This story is similar to the one it claims to recount in the Torah, found in Numbers 13-14, though it is greatly abridged. The twelve tribal leaders were sent to spy out the land, only two, Joshua and Caleb came back with a good report. The other ten had no confidence of victory when entering "the holy land which Allah has assigned for you," to quote the Koran. As punishment, the Israelites were then forced to wander in the desert for forty years before entering the Land. During that time the older defeatist generation would die out and be replaced by those born in freedom.

Israel and the Promised Land in the Koran

Continuing onward, not only are we reminded of the destruction of Pharaoh's forces in the Red (Reed) Sea, but the Israelites are instructed to live in the appointed land. There is a continuation of the promise, or covenant, as it is prophesied all of Israel will assemble together in the Promised Land in the "hereafter," - meaning an "ingathering of the Exiles" at the End of Days and in the afterlife as related in Koran Chapter 17 "The Night Journey" paralleling Exodus 30:5 in the *Tanakh* or Hebrew Scriptures.[7]

> Pharaoh sought to scare them out of the land: but We drowned him, together with all who were with him. Then We said to the

Israelites: 'Dwell in this land. When the promise of the hereafter comes to be fulfilled, We shall assemble you all together' (Koran 17:103-104).

And then, as quoted below in Chapter 7 the persecuted Israelites were rewarded "dominion over the eastern and western lands which We had blessed," a clear reference to the Holy Land both east and west of the Jordan River, thereby fulfilling the word of the Lord.

> We gave the persecuted people dominion over the eastern and western lands which We had blessed. Thus your Lord's gracious word was fulfilled for the Israelites, because they had endured with fortitude; and we destroyed the edifices and towers of Pharaoh and his people (Koran 7:137).

And finally there are the initial passages in the "Night Journey" Chapter 17 discussing the First and Second Temples in Jerusalem, both of which were destroyed, the former by the Babylonians and the latter by the Romans. This is the solid affirmation in the Koran of the existence of both sanctuaries.

> Glory be to Him who made His servants go by night from the Sacred Temple to the Farther Temple whose surroundings We have blessed, that We might show him some of Our signs. He alone hears all and observes all.

> We gave Moses the Scriptures and made them a guide for the Israelites, saying: 'Take no other guardian than Myself. You are the descendants of those whom We carried in the Ark with Noah. He was a truly thankful servant.'

> 'Twice you shall commit evil in the land. You shall become great transgressors'.

> And when the prophecy of your transgression came to be fulfilled, We sent against you a formidable army which ravaged your land and carried out the punishment with which you had been threatened.

> Then We granted you victory over them and multiplied your riches and your descendants, so that once again you became a numerous people. We said: 'If you do good, it shall be to your own advantage; but if you do evil, you shall sin against your own souls.'
>
> And when the prophecy of your second transgression came to be fulfilled, We sent another army to afflict you and to enter the Temple as the former entered it before, utterly destroying all that they laid their hands on.
>
> We said: 'Allah may yet be merciful to you. If you again transgress, you shall again be scourged. We have made Hell a prison-house for the unbelievers.'
>
> This Koran will guide men to that which is most upright. It promises the believers who do good works a rich reward, and threatens those who deny the life to come with a grievous scourge (Koran 17:1-10).

Mohammed made his "Night Journey" from the Sacred Temple (Mecca) to the Farther Temple (understood to be Jerusalem) thereby connecting the two in holiness according to the first verse. The central narrative of ancient Israelite-Jewish history followed immediately, as noted above. Moses received the Scriptures and the covenant at Sinai. The Israelites arrived in the Land and built the Temple, only to see it destroyed twice, due to evils committed by the people. This is perfectly in line with Jewish understanding. The First Temple was destroyed by the Babylonians and rebuilt after the return from exile as noted in the *Tanakh* - Books of Ezra and Nehemiah, while the Second Temple was destroyed by Rome and never rebuilt. The Muslim Dome of the Rock was erected on the site of the Temples and stands to this day; however, the Koran never mentions Jerusalem by name. The Koranic claim of a prophecy concerning the destruction of both the Temples is not found in the *Tanakh* or Hebrew Scriptures, although the destruction of the First Temple is recounted. The Koran speaks of such prophecies, but for sure well after the fact since the Koran was compiled only in the seventh and eighth centuries CE.[8]

What is significant, and in line with reward and punishment, is found in the fifth Book of Moses, or in the Torah, "Deuteronomy" Chapters 4-11 and

28. The Koran reaffirms this perspective in the last verses of the above quote. With righteousness there is a full redemption for the Israelite descendants as seen below in Deuteronomy 30:1-5.

> And it shall come to pass when all these things are come upon thee, the blessing and the curse, which I have set before thee and thou shalt call them to mind among all the nations, into which the Lord thy God has driven thee, and what return to the Lord thy God, and shalt obey his voice according to all that I command thee this day, thou and thy children, with all thy heart, and with all thy soul; that then the Lord thy God will turn thy captivity and have compassion upon thee, and will return and gather thee from all the nations, amongst whom the Lord thy God has scattered thee. If thy outcasts be at the utmost parts of the heaven, from there will the Lord thy God gather thee, and from there will he fetch thee: and the Lord thy God will bring thee into the land which thy fathers possessed, and thou shalt possess it; and he will do thee good, and multiply thee more than thy fathers.

The choice between good and evil continues to be laid out before the Israelites in the next few verses, Deuteronomy 30:6-20, with the people being given free will. To reiterate, the promise of resurrection for the Israelites/Jews in Deuteronomy 30:5 and a return to the Promised Land is reinforced by the Koran 17:103-104 as quoted previously. Righteousness is to be rewarded and evil will be punished. The Koran fully concurs in 17:10 (above) and adds the need for rabbinic guidance below in Koran 5:44.

> There is guidance and there is light, in the Torah which We have revealed. By it the prophets who surrendered themselves to Allah judged the Jews, and so did the rabbis and the divined; they gave judgment according to Allah's scriptures which had been committed to their keeping and to which they themselves were witnesses (Koran 5:44).

Allah judges the Jewish People by their adherence to Torah. The prophets, rabbis and Divined surrendered themselves to the Torah. When viewing modern Israel it certainly would appear that the Zionist leadership over the past century or so provided the correct guidance as the State of Israel was established in the Land of Israel through a Jewish return to the Holy Land. When considering the numerous above quotes the Israelites/Jews are keeping the covenant both according to the Torah and the Koran. The re-establishment of Jewish sovereignty in the Land of Israel can be taken as proof that the Jews have found favor with Allah.

More Exodus narratives exist in Koranic verses 20:9-94, 7:103-160 and 2:47-63, although they do not strictly follow the *Tanakh*. Koranic recollections concerning the Israelite prophets as well as Kings David and Solomon are found in Koran 38:17-35 and 27:15-44.

These texts bring forth the question: how can a believing Muslim deny Allah's involvement in physically redeeming the Jews in their ancient homeland, the Land of Israel, today in the modern period? The Jews must have been righteous to earn redemption and an "Ingathering of the Exiles," otherwise Allah would not allow for such an eventuality as He is by definition in Islam, the ultimate "Good."

Changing Islamic Attitudes Toward Christians and Christianity

Although not discussed in depth in the previous chapters, the Koranic perspective toward Christianity is also ambivalent. Christianity is under attack due to the doctrine of the Trinity, and deemed as less than monotheistic. Jesus is seen in a positive light, a prophet or holy man, but certainly not as the Son of God. Like the Jews, Christians are People of the Book and all restrictions directed by the Charter of Omar, payment of the *jizya* tax from Koran 9:29 and Jihadi exhortations apply. Because Christianity was virtually non-existent in the Arabian Peninsula during the seventh century, the early Muslims had little contact and no wars with Christians until after Mohammed's death in 632. The Arab invasion of the Byzantine Empire marked the beginning of the clash with Christendom. The early affinity, admiration and respect shown for the Children of Israel/Jews were not extended to the Christians. On the other hand, the direct clash and feeling of betrayal Mohammed and the early Muslims felt toward the Jews after the Battle of the Trench in defense of Medina have no corresponding historical event in the Muslim-Christian relationship. With Christianity, Muslims have less identification and reduced disappointment.

In the end Christians and Jews were grouped together as People of the Book, but the most intense positive and then reactive negative feelings were reserved for the Jews. From both sides of the spectrum these same emotions were greatly moderated toward Christians, making the ambivalence less sharp but extant nonetheless. Negative stereotypes began with the Jews and were later passed on in a more subliminal manner to the Christians. Any Jewish rehabilitation based on Koranic verses will certainly aid the Muslim relationship with Christendom. Add to this an emphasis on positive comments from the Koran and the same nullification or abrogation will work to the benefit of Christians as well as Jews.

As discussed above, negative comments about Christians in general revolve around the Islamic rejection of the Christian belief that Jesus is the Messiah and Allah is only one part of the Trinity; see examples below in the Koran Chapter 4 and two quotes from Chapter 5:

> People of the Book, do not transgress the bounds of your religion. Speak nothing but the truth about Allah. The Messiah, Jesus the son of Mary, was no more than Allah's apostle and His Word which He cast to Mary: a spirit from Him. So believe in Allah and His apostles and do not say: 'Three.' Forbear, and it shall be better for you. Allah is but one God. Allah forbid that He should have a son! His is all that the heavens and the earth contain. Allah is the all-sufficient Protector. The Messiah does not disdain to be a servant of Allah, nor do the angels who are nearest to him. Those who through arrogance disdain His service shall all be brought before Him (Koran 4:171-172).

> Unbelievers are those who declare; 'Allah is the Messiah, the son of Mary.' Say: 'Who could prevent Allah from destroying the Messiah, the son of Mary, together with his mother and all the people of the earth? His is the kingdom of the heavens and the earth and all that lies between them. He creates what He will and has power over all things' (Koran 5:17).

> Unbelievers are those that say: 'Allah is the Messiah, the son of Mary.' For the Messiah himself said: 'Children of Israel, serve Allah, my Lord and your Lord.' He that worships other gods besides Allah shall be forbidden Paradise and shall be cast into the fire of Hell. None shall help the evil-doers.

> Unbelievers are those that say: 'Allah is one of Three.' There is but one God. If they do not desist from so saying, those of them that disbelieve shall be sternly punished (Koran 5:72-73).

In addition, verses 17:111, 18:1-5, 19:88-92 and others admonish Christians for believing Allah has a son. However, there are also positive comments about Christians, in particular Muslim theological support for belief in the Gospels. This does raise a contradictory point, however, since the Gospels are the basis of the Christian claim that Jesus was and is the Messiah. On the other hand, such a juxtaposition of two opposites coming from Allah in holy writ may allow for some form of acceptance of the Christian understanding of Jesus' Messianic role. Allah can never be wrong therefore there must be some validity to the New Testament claim that Jesus is the Christ, born of a human mother, the Virgin Mary. See the quotes below from Koranic Chapters 5, 3 and 6:

> If the People of the Book accept the true faith and keep from evil, We will pardon them their sins and admit them to the gardens of delight. If they observe the Torah and the Gospel and what is revealed to them from Allah, they shall be given abundance from above and from beneath (Koran 5:65-66).

The need to "observe" the Torah and Gospel is repeated again in 5:68. Jews and Christians can receive their just reward if they follow their Holy Scriptures.

> Say: 'We believe in Allah and what is revealed to us; in that which was revealed to Abraham and Ishmael, to Isaac and Jacob and the tribes; and in that which Allah gave Moses and Jesus and the prophets. We discriminate against none of them. To Him We have surrendered ourselves (Koran 3:84).

> We gave him [Abraham] Isaac and Jacob and guided them as We guided Noah before them. Among his descendants were David and Solomon, Job and Joseph and Moses and Aaron (thus are the righteous rewarded); Zacharias, John, Jesus and Elias (all are upright men); and Ishmael, Elisha, Jonah and Lot. All these We exalted above Our creatures, as We exalted some of their fathers,

their children, and their brothers. We chose them and guided them to a straight path (Koran 6:84-87).

Note that positive comments about Jesus and the Gospels are usually accompanied by accolades for the Torah, the Patriarchs, Hebrew prophets and/or kings. In 5:82 discussed in earlier chapters Christians may be seen as adversarial, but are looked upon as behaving much better than the Jews. "That is because there are priests and monks among them; and because they are free from pride."

Chapter 19 entitled "Mary" is most important for understanding the Koran's rendition of the birth of Jesus. First John was born to Zacharias as recounted in the Gospels, while simultaneously there is the "annunciation" of the birth of Jesus to the Virgin Mary, as seen in Luke Chapter 1 of the New Testament. In the Koran, Mary gives birth next to the trunk of a palm tree, eats dates and drinks water from a brook; a story not told in the Christian scriptures. She (Mary or Miriam) is also said to be the sister of Aaron, showing confusion between the two separate characters of the Bible. Islamic scholars do not take the verse literally but rather see it as indicating Mary's righteousness. There is no theological clash up to this point, but there is afterward as recounted in 19:34-19:38.

> Such was Jesus the son of Mary. That is the whole truth, which they are unwilling to accept. Allah forbid that He Himself should beget a son! When he decrees a thing He need only say: 'Be,' and it is.
>
> Allah is my Lord and your Lord: Therefore serve Him. That is the right path.
>
> Yet the Sects are divided concerning Jesus. But when the fateful day arrives, woe to the unbelievers! Their sight and hearing shall be sharpened on the day when they appear before Us. Truly, the unbelievers are in the grossest error (Koran 19:34-38).

Later in "Mary," verses 19:88-92 emphasize the crime of believing Jesus to be the Son of Allah, "Those who say: 'The Lord of Mercy has begotten a son,' preach a monstrous falsehood, at which the very heavens might crack, the earth break asunder, and the mountains crumble to dust. That they should ascribe a son to the Merciful, when it does not become Him to beget one!"

The issues with Christians in today's world are not only over matters of theology. The continuing battles against Christendom and the West are a continuation of the Muslim invasion of Spain and southern France in 732 and the later penetration through the Balkans with the attempt to capture Vienna, the last time in the seventeenth century. The Crusader invasion in the Middle Ages is seen as the commencement of the Western onslaught against Islam, continuing into the twentieth and twenty-first centuries and presently resulting in the secularization, and thus weakening of Muslim societies. Islamic Jihadism places itself in never-ending world conflict with the Christian West in what truly can be termed a "Clash of Civilizations," while the Jews are accused of being involved in eternal conspiracies and subterfuge.

To prevent universal conflict, Muslim leadership must emphasize points of agreement concerning Christianity, in particular the above quotes calling on Christians to adhere to their Gospels. This is somewhat different than finding common ground with the Jews and the Sinai motif, the covenant and the centrality of the Land of Israel. A Koranic argument can be made for the covenant and Jewish claim to the Land of Israel while the Gospels are Allah's word, even if Jesus is not considered the Messiah.

Non-Western, Non-Abrahamic Faiths

Finding common ground with polytheists, atheists and idolaters is much more difficult for Muslims than finding a shared understanding with Jews and Christians. Non-Western, non-Abrahamic faiths are often viewed only in a negative light. Hinduism, Buddhism, Confucianism, Taoism and other Eastern religions and philosophies are not mentioned in the Koran. The attitude toward them as non-believers is adversarial. Still, one quote exists which may be helpful. In verse 2:256 the Koran begins, "There shall be no compulsion in religion." This verse is extremely useful in nullifying Jihad against the non-Abrahamic faiths, even though idol worship is condemned in the immediate following lines of 2:256, "True guidance is now distinct from error. He that renounces idol worship and puts his faith in Allah shall grasp a firm handle that will never break. Allah hears all and knows all." And then these non-believers are condemned to hell.

The question arises as to what are "faith in Allah" and the definition of "idol worship?" If Islamic jurists determine "faith in Allah" as belief in the same one central power of creation of all beings, then this definition would prove to be inclusive of a great number of religions and philosophies. With an acceptance of Jews and Christians as legitimate equals, such a moderate and liberal understanding may be projected upon other groups as well. Unfortunately, in earlier centuries, Hindus lived under Muslim rule in the Indian subcontinent for hundreds of years and in many cases suffered miserably.[9] There is further support for a tolerant position if one engages in *"ijtihad,"* defined as Islamic independent thinking or reasoning. Such use of *ijtihad* is demonstrated when considering this passage from Koranic verse 4:135 even if the topic at hand does not refer to those of other faiths:

> Believers, conduct yourselves with justice and bear true witness before Allah, even though it be against yourselves, your parents, or your kinsfolk. Whether the man concerned be rich or poor, know that Allah is nearer to him than you are. Do not be led by passion, lest you should swerve from the truth. If you distort your testimony or refuse to give it, know that Allah is cognizant of all your actions.

According to the self-proclaimed Muslim *"refusenik"* Irshad Manji and Professor Khaleel Mohammed, the above verse can be universal in the acceptance of others.[10] Lest one forget, Allah metes out punishment in the next world, as noted below, and if the Prophet Mohammed's task "is only to give warning" then any demand to conduct Jihad due to religious differences is neutralized.

> Know that Allah is stern in retribution, and that He is forgiving and merciful. The duty of the Apostle [Mohammed] is only to give warning. Allah knows all that you hide and all that you reveal (Koran 5:98-99).

> He knows the visible and the unseen. He is the Mighty One, the Merciful, who excelled in making of all things (Koran 32:6-7).

Although not particularly explicit, one finds non-coercive and conciliatory verses toward all humanity in the Koran, where each individual has his own reckoning before Allah. Allah created everyone and everything, meaning the entire world. He is infallible and therefore His creations are excellent, but humans must choose the correct direction. This begs the question: what mortal has the right to judge another, whether he is Muslim or not? Islamic attitudes toward the People of the Book can be quite accepting while the attitudes toward the non-Abrahamic religions need not be hostile.

Decision Making in Political Islam and Hamas

Whether Islamic jurists decide to advocate for Jihad or decide in favor of universal peace and acceptance of other religions and philosophies is the central question today. For believers, decisions about interacting with the world are seen as dictates or inspiration from Allah, but in reality these decisions are simply policy determination and implementation. War or peace is the choice of Islamic leadership throughout the world. As can be seen, both options exist in the Koran. Ultimately Muslims will have to resolve if they desire eternal Jihad, or if Islam is willing to take its place as one of the world's peace-loving religions. To avoid physical conflict, Muslims interested in attracting new adherents must follow a policy of persuasion, not coercion. Irshad Manji explains that Muslims have free will to interpret the Koran, a text "at war with itself" due to its never-ending contradictions. As she puts it, "The decisions that Muslims make are ours alone. They cannot be laid at God's feet."[11] Such a broad statement refers to attitudes toward Jews, Christians, pagans, women, homosexuality, and, most significantly, world Jihad with dire consequences for everyone if the Holy War option is chosen.

In order to avoid misunderstanding concerning the non-Islamic West, Far East, Russia and many others, it must be made crystal clear to all Muslims that choosing the Jihadi path will only lead to death and destruction on a level comparable to the German–Japanese defeat of World War II. Such a threat will not deter the true Jihadists, but it may very well undermine further support for their cause. Those considering the Jihadi direction may decide on a more moderate path for reasons of self, national and religious preservation. It is in their interest not to have a Jihadi-led society.

Hamas is part of this picture. The antisemitic, Jihadi *Hamas Covenant* is viewed as an alliance with Allah, a sacred document. But if theological emphasis by Koranic commentators can be redirected and parts nullified or abrogated without violating Islamic legal dictates, then *The Hamas Covenant* becomes increasingly irrelevant. Muslims will view Jews and others in a more favorable light and Jihad could become a policy of the past. However not all issues will be resolved through Islamic abrogation or reinterpretation. Too much of the *Covenant* adopts Czarist and Nazi antisemitism, revealing solutions similar to both. A rejection of European antisemitism by Muslim jurists as "foreign" or "in contradiction" to Islam will be necessary. The *Covenant*, also known as the "Charter of Allah," would become insignificant and ready for burial. First there must be a broadly accepted abrogation of Jihad and discrimination, then in the aftermath the death of *The Hamas Covenant*. Conflict resolution between the Israeli/Jewish People and Palestinians/Muslims will take a giant step forward. Unfortunately such a scenario is quite optimistic and apparently depends more on Western policies in the global struggle against Jihadi Islam than any conscious move by Muslims to change the emphasis to different verses from the Koran.

Western Tolerance of Jihadi Islam

To the Jihadist mindset, the West is weak and has a credibility problem, because it defends Islamist intolerance through its obsession with "multiculturalism."[12] Multiculturalism is a liberal attitude where everyone shows tolerance for everyone else. The issue is the definition of "tolerance" and its limitations. Tolerance is defined in *Webster's Dictionary* as "sympathy or indulgence for beliefs or practices differing from or conflicting with one's own" while "toleration" is defined as "the act of allowing something."[13] Such "tolerance" by definition leads to a broad accommodation of everyone else's culture and beliefs, even if certain people, religions or groups are considered to have customs or encourage behavior considered inferior or completely unacceptable by these same Western rationalist thinkers. We are familiar with the terms "universalism" and "relativism" emanating from the West. Universalism asserts all religions and cultures have shared positive values and behaviors, while relativism insists there are many religions and cultures, each deserving equal respect for their truths. As Manji points out[14] the only tolerance mainstream Islam demands is for itself and not for others. Jihadi Islam is infinitely worse, using understanding and tolerance for its behavior as a weapon against the West.

I will take this a step further. Western scholars identified with the supposedly "liberal" multi-cultural and anti-Orientalist schools can be broken down into two camps. The first are those who demand not only understanding but also support for Jihadi ideals, often defined as some form of "liberation," while being fully cognizant of the fact that Jihadi Islam has no tolerance for anyone except other Jihadis. In my mind, these intellectuals, who invariably consider themselves "liberal" in their worldviews, are hypocritical and deceitful, claiming "tolerance," but believing themselves superior to all others because of their "openness" and denial of moral/ethical boundaries when confronting "evil." Any such determinations are considered prejudicial value judgments. They rarely condemn the Jihadi demands and actions discussed in this work. These liberal Jihadi sympathizers prefer living in the West, far from such Islamist societies. They are known to make a visit or two to show support. They often explain Jihadi Islam as a reaction to Western imperialism or provocations. Such types in the West forgive Islamist exhortations demanding the destruction and/or subjugation of other people, while excusing brutal behavior as disconnected from Islamist interpretations. This intellectual group does not suffer under the heavy hand of Jihad. They blatantly hide behind the right to freedom of speech and the "liberal" banner while allowing the Islamist program for the destruction of Western civilization, beginning with Israel and quite often Jews in general.[15]

In an effort to prevent its own annihilation, the West must avoid situations where it is expected to tolerate and even promote another society's intolerance—be it Jihadist Islam or any other ideology. If the entire West were to argue for cultural or moral relativism to promote tolerance of Jihadi Islam, the result would be collective suicide.

The second camp are those Western intellectuals who adopt the attitude of a superior father figure, interpreting Jihadist behavior as "posturing," as if one were dealing with a child who really does not mean what he is saying. These academics hold universalism and relativism in holy esteem. The Islamists' believe Allah ordains their position, and hence there is no room for negotiation. Yet these Western scholars project their own values onto the Islamic Jihadists, convinced of their own righteousness and superiority, usually of the secular type. They often explain Jihadi demands such as those held by Hamas as negotiating positions.[16] This specific group of scholars does not accept or even consider that other people and cultures may think and view the world differently but constantly equate Western and especially Israeli security responses with Jihadi Islamist offensive actions.[17] It is difficult for these Westerners to accept that Islamists truly believe Allah has called them to spread the message of Islam by force if necessary. Enshrining non-violence as a transcendent value, these generally secular, liberal intellectuals are convinced that virtually everyone, including proclaimed Jihadists, do not truly mean to use violence to impose their religion, culture and values. In essence these left wing intellectuals tolerate Jihadi violence while condemning the right of Israel and the West to self defense. Liberal non-violent ideals are completely rejected by Hamas and Jihadi Islam as pertains to their own mode of operation. However Jihadis fully embrace these leftist understandings when demanding full surrender by their non-Muslim victims. Their potential prey is expected to accept a self imposed *dhimmization* or face the wrath of Allah's Jihad. Without using violence or the threat of violence, the Islamists cannot succeed. For Jihadis there are no mistaken civilian casualties, only the intent to destroy and/or subjugate all others. As the ardently secular philosopher Sam Harris makes abundantly clear, tolerance toward Jihadis is immoral.[18] It must be said outright, such "tolerant" liberalism is just one more form of Orientalism where Western scholars view "Easterners" as an "object" and impose their secular, Enlightenment values on the adherents of Islam. Jihadist Islam will only adopt the persuasive approach toward converting others to Islam as a first step when inviting non-Muslims to accept their said-to-be "true religion." Let us recall that to convert out of Islam is punishable by death. Upon refusing the invitation to convert to Islam, coercion through Jihad is the answer (see Chapter II "Ideologues" and Koran 9:4-14). There is no dialogue in Jihadi Islam.

By accepting the "multi-cultural" or "cultural relativist" perspective of the need to accept or even engage with the Jihadi ideals contained in the Muslim Brotherhood-Hamas doctrines we are led to the Islamic State (ISIS/ISIL) atrocities. What began as an Islamist ideal will end with its physical implementation. Yes, the road is often indirect and Jihadi outcomes are also influenced by other factors, but these are minimal in comparison to the Jihadi commitment. We must face reality—if the Jihad concept did not exist there would be no Islamic State. Jihad should not be confused with national territorial disagreements, which may erupt in violence but are limited in scope. These do not involve world conquest or extermination of others. Anyone willing to show "understanding," in the sense of empathy, and engage in dialogue with the non-negotiable demands for Jihad (the *dhimma* status and the forced conversion of infidels) gives legitimacy to Jihadi ideals. Those showing empathy encourage extremism and are accomplices to the crimes committed in the name of Islam through Jihad.

Such "liberal" behavior is akin to the appeasement policies of the British government and especially the Labor party's efforts at non-judgmental attitudes and even outreach toward Adolf Hitler and the Nazi regime during the 1930s. True pacifists, the British Laborites attempted to take "the moral high ground" against total evil and the ultimate immorality. Often condemning the British and French governments, they tried to fully "understand" the Nazis and even came close to justifying German aggressive actions until the eve of WWII.[19] The existential issue here is that such hypocritical, self-destructive perspectives have a chance of becoming policy in the West today. By taking the supposed moral high ground while invoking rational superiority, this ultra-universalist approach expresses sympathy for dictators, murderers and Jihadist policies following the misguided principle that "no one has the right

IX Conflict Resolution in the Shadow of Islamic Abrogation

to judge another." At best we have a dangerous amoral policy here, lacking all foundations for ethics and morality.

Hence, no moral high ground exists when there are no red lines forbidding wanton murder of pagans/idolaters and for those *dhimmis* allowed a conditional survival to suffer the destruction of their essential human rights, such as freedom and equality. What does exist is a superiority complex and an attempt at secular godliness whereby all are equally loved, cared for and defended regardless of behavior. Appeasement only leads to annihilation at the hands of one's adversary, or a late realization that battle is unavoidable, as was the case in WWII. If an individual or political party advocated murder or extermination of a group or a society, Westerners would consider it incitement. Yet, when emanating from Islam these demands are not only "understood," they are included within the acceptable domain of cross-cultural and interfaith exchanges. It takes a group such as the Islamic State to cause most Westerners to see Jihadist ideals as falling outside the realm of tolerance and multi-cultural dialogue.[20] Even here there are exceptions. Unfortunately the viciously anti-Western pro-Islamist said-to-be "scholars" in Middle Eastern studies academia continue to lay all responsibility for Islamic terrorism on the American government and people. Their constant excuses include "workplace" violence, alienation, cultural prejudice and the pressures of Islamophobia against Muslims as catalysts for even the most sadistic, inhuman Islamic State inspired attacks against innocent civilians as was the case in Paris and San Bernardino.[21] As shown in previous chapters, extremist Islamic behavior is embodied within Muslim Brotherhood and Hamas doctrine, yet the West "tolerates" it. Hence, any dialogue or conversation between the people of the globe and Islam must begin with the demand to jettison Jihadi world conquest and all laws concerning the *dhimma* status. The acceptable avenue to achieve such a change is to engage in "reverse abrogation" and to seek the answers which are found in the Koran.

The United States has been unsuccessfully bogged down battling Jihadi Islam for well over a decade. Washington has few on-the-ground allies in its battles against the Taliban in Afghanistan, and those who support the US send only small troop contingents. Pakistan is at times engaged in operations against the Taliban, yet it does its best to keep America at arm's length. Many say "collateral damage," the accidental killing of civilians, within Pakistan exacerbated the rift between Washington and Islamabad. It is well known that Jihadists and Taliban sympathizers who opposed the secular regime of President Zardari infiltrated the Pakistani ISI intelligence service. The rift with the US exists for decades. In 2011, the ISI apparently knew of Osama bin

Laden's whereabouts long before US Special Forces killed him.[22] By 2013, the Islamist-dominated regime led by Prime Minister Nawaz Sharif became an even less reliable ally in battling Islamic terrorism. This is a clash of theology and ideology between Islam and the West. Collateral damage is unfortunate, but serves only as an excuse by Islamic regimes, even moderate ones, not to cooperate with the West in curtailing the Jihadists.

Most recently, the supposed success in containing the Iranian march toward nuclear weapons and Middle East domination by the six world powers[23] and the EU only emboldened Tehran and other Islamists to continue their battle against liberal democracy. When the West fails to relay a sharp anti-Jihadi message, it encourages more Jihadi aggression. Unchecked Jihadi ideals lead to their manifestation. With the present July 2015 agreement strongly endorsed by President Obama,[24] Iran's nuclear program is said to be on hold for the next decade, but what of their imperial interests throughout the region? Iran projects power through its forces and proxies in Iraq, Syria, Hezbollah in Lebanon and the Houthis in Yemen, despite certain limited new sanctions imposed to halt these activities. With the cancellation of the main body of sanctions due to the nuclear deal, Iran can be expected to fund conflict throughout the Middle East and simultaneously spread Khomeinist ideological demands for the destruction of Israel, the US and eventually the West in general. Military conquest or Jihad goes hand in hand with ideological indoctrination throughout the region. There are those who define American policy fantasies of peace with Iran as "delusions."[25] We must be brutally honest, in the long run no one should expect the Iranians to keep their agreement with the West to limit nuclear development.

One only needs to recall the crisis experienced in the 1930s when the Japanese captured Manchuria and the coastal areas of China, while the Germans annexed Austria and shortly afterwards the British and French allowed the Nazis to move into Czechoslovakia. Eastern and even Western European countries found themselves under fascist and pro-Nazi regimes. Once German victories were consolidated, governments in Slovakia, Hungary, Romania and Bulgaria allied themselves with Hitler against the Soviet Union. In France, the pro-Nazi Vichy regime opposed the British and Americans. Such fanaticism was and is contagious. In the end, the Allies were victorious, but not before the massive destruction incurred during World War II and the Holocaust. The cost was enormous for all states involved, in particular Imperial Japan and Nazi Germany.

Such a war must only be a last resort. It involves the complete defeat of the previous society in a "total war" whereby the devastation is so overwhelming the erstwhile power elite and their ideals become discredited beyond rehabilitation. The victors then impose their values and way of life on the defeated. World suffering would have been reduced dramatically had Nazi and Japanese Imperial ideologies been discredited from the outset. Unfortunately the West chose "tolerance" and the policy of "appeasement." Throughout the post-war years of 1945-90, capitalist democracy flourished in West Germany and, before reunification, Stalinist communism dominated the East. After Hiroshima and Nagasaki, Japan was relegated back to the beginnings of democratic government under American tutelage. Thanks to the Marshall Plan for Europe and reconstruction projects in the Far East, Japanese and German "militarism" were converted into the obsession to rebuild their economies, most specifically in the construction of industrial might. All this occurred only in utter defeat when no options remained.

World Islam may very well face the prospect of total destruction should Jihadism win the day. Devastation would ensue, but none can promise the rebuilding of those societies. For years the West had serious disagreements over coordinating sanctions policy against President Ahmedinejad and the Ayatollah regime in Iran, despite the clear evidence pointing to the Iranian development of nuclear weapons and the continuing threat to exterminate the State of Israel and dominate the Persian Gulf. For twelve years, there were sanctions and although there is now an agreement the question is, who truly trusts Tehran to hold up its part of the bargain? The recently elected "moderate" President Rouhani and extreme right wing Supreme Leader Ali Khamenei continue to make clear the Iranian intent to keep all nuclear facilities and mobilize for Israel's obliteration. Unfortunately, the mere act of negotiating with Iran recognizes and even legitimizes the Iranian theological demand for Israel's obliteration and world Shiite Islamic domination. The West faces a dilemma in deciding whether to increase sanctions, not knowing whether Iran is a potential friend or continuing foe. As of this writing, the world powers have decided to work with Iran and hope to curtail Tehran's plans for regional domination.

Due to the high cost in lives lost and mangled, societal disruption and material damage, it is best not to become entangled in wars to contain Jihadi Islam. However, the military option must always remain on the table. No doubt a preferable approach is to foster the rise of a moderate Islam based on interpretation emphasizing the Koran's conciliatory verses toward Jews, Christians and, by extension, the world. Many will dismiss the idea of moderate Islam as wildly unrealistic or, at best, overly optimistic. Yet it defies logic to avoid the internal conflict within Islam while simultaneously applying external pressure to induce change. Most importantly, Muslims themselves must be convinced that a peaceful resolution of their conflicts with the West is the only way to a better future for themselves. Muslims must ask themselves why Islam is in such crisis worldwide. Are they defying Allah's will through the previous abrogations in Islam? Working within Islamic interpretation is not a Western cultural *jahili* invasion, but rather an authentic way to arrive at a modus vivendi to guarantee the survival of all peoples and religions, including Islam. With the continually expanding disasters upending the Islamic world, Muslims need to ask themselves whether some form of "reverse abrogation" may be an acceptable answer to their woes.

Endnotes

1. Lewis, Bernard, *What Went Wrong?*, Harper Perennial, New York, USA, 2002.
2. Said, Edward W., "Afterward," *Orientalism,* Vintage Books, New York, USA, 1994, Twenty-Fifth Anniversary Edition, 2003.

    Said was damning in his criticism of Western research of the Orient. In particular when referring to the Middle East he declared, "Orientalism, which is the system of European or Western knowledge about the Orient, thus becomes synonymous with European domination of the Orient" and thereby dismisses most Western study of the region as being jaded, prejudicial and racist at times. Said has valid points but can be charged with very much overstating the case. Said accuses Westerners of overestimating the influence of Islam, while many of his critics believe he down played the significance of religion in the Arab/Muslim world.
3. Prof. Khaleel Mohammed of San Diego State University, the Egyptian intellectual Tawfik Hamid, the Canadian Indian Muslim professor Salim Mansur, the Italian imam and professor Abdul Hadi Palazzi and the independent thinking Muslim "refusenik" Irshad Manji can be seen as representatives of this group.
4. Bukay, David, "Peace or Jihad? Abrogation in Islam," *Middle East Quarterly,* Fall 2007, retrieved November 14, 2011, www.meforum.org/1754/peace-or-jihad-abrogation-in-islam.
5. All quotes are taken from the Penguin Classics *The Koran* with translation and notes by N. J. Dawood, reprinted in 1977. Here the Koran's chapters or suras obviously retain their original numbering but are organized in chronological order.
6. Torah is the term used for the first five books of the Hebrew Scriptures or *"Tanakh."* Other terms for Torah include the Pentateuch or the Five Books of Moses. It is here the Israelite Exodus from Egypt is recounted.
7. The Hebrew Scriptures or *"Tanakh"* used for references and quotes are from *The Jerusalem Bible,* English text revised and edited by Harold Fisch, published by Koren Publishers Jerusalem, Ltd., Jerusalem, 1969.
8. The Christian Scriptures (New Testament) discuss the prophecy of Temple destruction, however there are numerous disagreements as to when these were recorded in the Gospels or even when the Gospels were written—before or after 70 CE. Such a discussion is beyond the scope of this work.
9. For numerous examples, see Bostom, Andrew, (ed.), *The Legacy of Jihad: Islamic*

IX Conflict Resolution in the Shadow of Islamic Abrogation

Holy War and the Fate of Non-Muslims, Prometheus Books, Amherst, New York, USA, 2005.

10 Manji, Irshid, *The Trouble with Islam Today*, 2003, St. Martin's Griffin, New York, 2003, Introduction by Khaleel Mohammed and p. 1.

Irshid Manji, quote on p. 3, "I am a Muslim Refusenik. That doesn't mean I refuse to be a Muslim; I refuse to join an army of automatons in the name of Allah. I take this phrase from the original refuseniks - Soviet Jews who championed religious and personal freedom… Over time, though, their persistent refusal to comply with the mechanisms of mind control and soullessness helped end a totalitarian system."

11 Ibid. pp. 36-38.
12 Ibid. pp. 199.
13 "Tolerance," *Merriam Webster*, retrieved July 9, 2015, www.merriam-webster.com/dictionary/tolerance.
14 In Manji's work, *The Trouble with Islam Today,* this is the overriding theme.
15 Such pro-Islamist "scholars" include Steven Salaita, Mark LeVine, Asad Abu Khalil, Joseph Massad, Noam Chomsky and John Esposito among others. This group is anti-Western, viciously anti-Israel and in many cases outwardly antisemitic. Through their actions and positions taken in justifying hard-line radical Islamic interpretations, such as those of the Muslim Brotherhood, Hamas, Hezbollah and extremist Islam, this author believes they have lent credibility to Islamic State atrocities. All these organizations work off the same Sharia law principles including the need for Jihad.
16 Among these are also the "Mainstream Protestant" churches which deplore the "Palestinian plight" and blame all ills on Israel. In general they also believe in "Replacement Theology," whereby the Jews and the covenant are replaced by Christianity. Islam has the same ideology replacing Judaism, Christianity and their scriptures. This group adopts a secular, liberal attitude to Arab/Islamic violence; one of forgiveness, especially when perpetrated against Israel. On the other hand, Israel comes in for general condemnation while there is understanding and even outreach toward Hamas and Hezbollah, despite the fact that these organizations do not see Christians as their equals. Other Islamist groups with similar ideals persecute Christians worldwide, the Islamic State being the most extreme example.
17 Herf, Jeffrey, "A Pro-Hamas Left Emerges," *The American Interest*, retrieved August 5, 2015, www.the-american-interest.com/…/a-pro-hamas-left-emerges.

In addition see Chapter V of *Hamas Jihad,* and the positions taken by Gunning, Jensen and Nepp as examples.

18 For a fuller discussion of this issue, see Sam Harris, *The End of Faith,* W.W. Norton, New York, USA, 2004, in particular Chapter 4, "The Problem with Islam."
19 For a more comprehensive rendition of how liberals approach the issue of war see Michael Howard, *War and the Liberal Conscience,* Oxford University Press, England, 1978.
20 The Islamic State is regularly dismissed as not representing Islam. The Obama Administration's policy statements often fall into this mentality. Others dismiss ISIS as "insane." Europeans are in a dilemma because of successful Islamic State recruitment on their continent and the rising suspicions that such vicious behavior is connected to certain Islamic ideals. Furthermore, the appeal of the Islamic State, al-Qaeda and other fanatical groups is spreading, and not only among Muslim youth. The Paris and the San Bernardino, California attacks in late 2015 may bring a re-evaluation of the situation whereby Western politicians and security officials will be forced to publicly define such extremism as originating from a specific interpretation of Islam. This is yet to be seen as of this writing.

21 Stillwell, Cinnamon, "Academia on San Bernardino Attack: No Jihad Here," *The Middle East Forum: Promoting American Interests*, December 11, 2015, accessed December 29, 2015, http://www.meforum.org/5712/us-professors-san-bernardino.
22 "Inter-Services Intelligence in Afghanistan," *Wikipedia*, retrieved December 29, 2015, https://en.wikipedia.org/wiki/Inter-Services_Intelligence_activities_in_Afghanistan.
Reidel, Bruce, "Pakistan, Taliban and the Afghan Quagmire," *Brookings Institute*, August 24, 2013, retrieved August 5, 2015, http://www.brookings.edu/research/opinions/2013/08/26-pakistan-influence-over-afghan-taliban-riedel.
23 The countries involved are the USA, Russia, China, Britain, France and Germany.
24 Whether the agreement with Iran was ever to be ratified by the US Congress was of no real significance. Russia, China, Europe as a whole and the United Nations can all be expected to lift sanctions and do business with Tehran. Tens of billions of dollars will flow into the Iranian economy with or without the US.
25 Karsh, Ephraim, "Obama's Middle East Delusions," *Middle East Quarterly*, Winter 2016, retrieved December 29, 2015, http://www.meforum.org/5685/obama-middle-east-delusions.

X

Summary and Conclusion

Analysis of Hamas Revolution and World Islamism

Overview

Hamas gained popular support in Gaza beginning in the 1970s and by 2007 physically overthrew the Palestinian Authority (PA) quasi-state structure. From the outset, Hamas emphasized Islam to the detriment of secular Palestinian nationalism and took a non-compromising hostile position toward Jews and Israel. On the political level, the organization projected honesty and unity of purpose alongside of social programs that demanded a narrowing of gaps between socio-economic classes in Palestinian society. In subsequent years, Hamas Islamist ideals permeated Gazan society. The revolution was not intended only for Gaza, but was directed at West Bank Palestinians and Muslims throughout the Middle East and beyond, as attested to in *The Hamas Covenant*. The primary target populations beyond their immediate constituencies continue to be the Palestinian Diaspora, Jordan, Egypt and Syria. Hamas and/or the Muslim Brotherhood have directly influenced each of these groups and their adherents are inspired to export Hamas revolutionary ideals to impact the global Muslim community. As the Palestinian branch of the Muslim Brotherhood, Hamas will always be a faction of the greater Islamist movement.

An analysis of the Hamas Revolution is in order. Did Hamas go through stages similar to those of the French and Russian Revolutions, or was it more of a revolt against the PA? The template used by many historians and political analysts looks at distinct reasons for rebellions and transitional periods. These are accompanied by specific stages and plateaus of political/military action as the group achieved its ideological objectives. As we will determine, the Hamas Revolution succeeded only in part. Whether the revolution will go forward or recede is impossible to predict. Still, we know Hamas revolutionary actions played a part in the Muslim Brotherhood's overall influence on the Islamic Middle East in recent years.

Discontent, Revolutionary Stages, and Hamas

Many commentators claim Hamas instituted a revolution at the will of the people. Those who maintain Hamas is a democratic force point to their 2006 electoral victory in the Palestinian Legislature. The question is whether in areas where Hamas rules, was their rise to power a revolution advocating democracy and the will of the people? In the wider Middle Eastern sense, the Islamic Awakening or supposed "Arab Spring" is far from its finale as a completed revolution. The Islamic regimes are not fully settled, and some countries like Libya, Syria and Iraq are more failed states than organized ones. A prime example is how the Syrian civil war achieved new milestones of brutality and societal breakdown. In a much less violent example, Egypt's Islamic Revolution came crashing down when the military overthrew the Muslim Brotherhood regime in the summer of 2013, although the generals did have a fair amount of popular support. As of this printing, it is unclear whether the Egyptian people support the secular nationalist military government of General Abdul Fatah a-Sisi, or the freely elected but deposed Islamist president, Mohammed Morsi. The game is far from over.

There is little doubt Hamas brought about a revolution in the Gaza Strip. But the Hamas rise to power is incomplete in the West Bank, halted by the continuing Israeli military presence in Areas B and C, as well as by the activity of the US sponsored Palestinian Authority police force under US General Keith Dayton's leadership. It may be reasonable to assume that barring these two outside factors, Hamas would also rule in the West Bank.

To put the Hamas role into context I will uphold the traditional theoretical understanding of rebellion and revolution. Revolutionary theorist and historian Crane Brinton[1] compares revolution to a fever beginning slowly, gaining force and rising to a burning delirium better known as a "reign of terror" at its height. It is in the "fever" of revolution that the disease dies; in other words, evil people lose power. Afterward, the fever subsides and although there is change, society returns to a new norm, no longer in the clutches of a religious/ideological tyrannical regime. In the early stages, the dysfunctional old regime collapses in its own mired corruption. Members of the old order doubt their own moral and ethical right to rule, while their adversaries declare them downright illegitimate. The revolutionaries declare that an overthrow of the illegal government will bring the ideal "heaven on earth." The "all powerful force" is embodied within the revolution, thereby guaranteeing victory. Purity will be victorious and virtue will reign. It should be pointed out that religious terminology is often used in describing revolutions, regardless of whether the revolution is in the name of God or in the name of humankind. Revolutionaries project the absolutism of purpose and objectives as true and righteous.

In the first stage of a revolution, the old regime is unable or unwilling to raise the force necessary to stop the revolt. They suffer from a lack of popular loyalty, a collapsing tax base and disintegrating organization. Realizing they are losing power the regime attempts reform in a sort of transitional period, but fails. Loyalty of both intellectuals and the security forces begins to shift, at least partially. The shift increases as the old regime collapses. The revolutionaries are ideologically driven, non-compromising, well organized and in the process of developing an armed force. Stage two is a transition period where the old regime is neither resolute in declaration, nor action. The die is cast. Shifting into third gear, the rebels enjoy mass support, sweep away the old regime and replace it with a broad-based coalition. Due to the higher ideals and the emphasis on rights and freedom, this era is known as the "rule of the moderates." The dissatisfied, more extremist factions rise up and declare a lack of ideological purity on behalf of the "compromising" moderates, forcing them from power.

The extremists in stage four comprise a very small, highly disciplined, "self-deified" ideologically committed group. An election, or plebiscite, may be called, but this is most likely done only once to ensure support. A second election is not called, which might allow loyalties to shift in other directions. The democratic system or "will of the people" is far from the ideological objective and would only serve to allow backtracking or reforms that thwart the full revolutionary realization. This group may be referred to as "philosopher killers" and are known for the demonization of their enemies. These extremists suffered oppression in the past and are experienced ideologues; their supporters often see them as saints. They attempt to exterminate any fanatics challenging their rule, and augment such policies with puritanical laws.

For the average person there is what can be called a "religious satisfaction" with the revolution, but the people as an entity suffer from its harshness. There is some wealth distribution, demands for monastic behavior, a form of morality police is established and revolutionary determinism is society's new leader. Many revolutionary ideals may have been imported, but the Messianic fervor or success leads to revolutionary export. On the local level, heaven is attained with forms of sainthood and martyrdom. The revolutionaries revel in their "chosenness" and lead in the name of their superior power, or God, to bring a universal adherence to their values. Revolutionaries obtain world conquest through "manifest destiny," and close all gaps between heaven and earth. In essence, a universal utopia is the ultimate objective. Extremist rule is often associated with a reign of terror which the revolutionaries identify as a reign of virtue. Brinton speaks of secular revolutions, but the actions of revolutionaries and their followers' interpretations of the radical events taking place is often apocalyptic, contained within a virtually religious realm.

The last stage, known as Thermidor (a term from the French Revolution), settles in once the height of fervor calms down. Usually led by autocrats and dictators, nationalism re-enters the Messianic spirit and universalism takes a back seat to everyday necessities. Revolutionary export implies constant fervor and clash, a far too intensive lifestyle for any given society over an extended period of time. Many prefer to return to the less tense period before the upheaval. The people want to indulge themselves and even seek the small vices of yesteryear. Summarized, this is the thrust of Brinton's theory, one fairly well accepted.

The Hamas Islamist revolution, like all revolutions, has its roots in deep discontent and feelings of deprivation. Ted Gurr advanced the theory of "relative deprivation" as a motivational factor in revolt against existing conditions and authority.[2] Overall "relative deprivation" is defined as the perceived discrepancies "between value expectations and value capabilities," or in other words the difference between what one believes he should be able to attain and what is actually achieved.[3] These gaps between expectations and reality lead to frustrations, violence, rebellion and possible revolution. The more intense the deprivation, the sharper and longer the violence will continue. Relative deprivation (RD) relates to material gaps such as concerns the economy, status within a society, rights of a group or of individuals, and/or political participation usually defined as national and religious. Often there are multiple causes, and in general the ultimate goal is "freedom." Besides being "perceived," which means RD is real or imagined (most often the former) such expectations are often handed down as legends, historic givens and/or rights. The larger the RD, the greater the justification will be for political violence. A regime deemed "illegitimate" because it does not represent the people will be the focus of popular discontent.[4]

Rebels make gains through violence and terror attacks. Full-blown revolutions consolidate a unity of ideals and purpose, eventually leading to demands for a Utopian order to right all wrongs. When working for the advancement of social cohesion there is an expectation of godly intervention against the hated common enemy. As opposed to those in power, people express anger, discontent and an effective reordering of the world through alternative media and communications including verbal, print, and electronic venues. Actions

X Conclusion: Hamas Revolutionary Impact

are directed against the "oppressor" whether in the framework of a civil war or an outside power. When dealing with a civil conflict, questions of regime loyalty on the part of the military and police play a major role, as the security forces may switch sides thereby determining the outcome. Creating multiple security forces indicates doubts of loyalty to the regime. This is often enhanced by a recent defeat in conflict by an outside power.

In general, the volunteer insurgents have very high levels of loyalty, while pressured regime forces are less committed. The rebels succeed in obtaining institutionalized support from within and without. They gain territory on the ground with military training and arms acquisition. Frustration and hostility increase cohesion, forming a group identity encroaching on the specifics of personal identification. The revolutionary leaders combine the best of the old and new worlds as they package the ideal vision for mass consumption. Deep discontent fosters new religions, sects, ideals and Messianism, all of which can be diocentric or anthropocentric in perspective. A successful revolution means this rising elite, or in many cases the highly educated suffering from RD, will become the new leaders who now must redistribute resources, eliminate discrimination and alleviate inequalities and abuse while committing to progress.[5]

The revolutionaries understand they must convert discontent into violence against the regime, while justifying their actions through their ideals. They will vilify the regime, use violence and/or terrorism, invite retaliation to increase mass suffering and exploit the increasing discontent to force even more regime attacks and possibly outright warfare. In general, the discontented seek "remedial action" from the regime, not violent solutions. But when peaceful action proves impossible and the RD gap becomes unbridgeable, they take physical action. Most applicable is the following quote taken from Gurr in explaining the susceptibility of different societies to violence, "Not all new beliefs provide justifications for violence, and most that do are derived from peoples' own cultural and historical experience rather than alien sources."[6] In the case of Hamas, one can easily understand a rebellion to reaffirm commitment to Islam while forcing the unpopular PA/Fatah regime out of power. Furthermore, Israel is certainly defined as an outside adversary of the worst type, acting as a direct or indirect occupying power supporting the corrupt exploitive PA/Fatah regime. For Hamas, Israel and the PA must be defeated together.

Hamas Revolution Disrupted by Outside Forces

The above stages of revolution and theoretical analysis of rebellion are applicable as general guidelines for understanding the Hamas uprising and Gaza takeover. Of course, the intensity of each stage varies from one revolution to another since political and social theories are neither science nor math. During revolutions we note that "small determined, ideologically driven minorities are those who change history and the world while the satiated and corrupt will eventually feel their wrath and be overthrown."[7] Hence, the surprise when the far inferior Hamas force deposed the "criminal" Fatah-led PA regime in Gaza within a week of fighting. Hamas attributes its success to a higher power, since they launched the revolt in the name of Allah, Islam and Sharia law.

The old Fatah/PLO-controlled Palestinian Authority regime represented secular Palestinian nationalism, solidified in the 1960s and supposedly holding Palestinian loyalties for forty years. Hamas supporters saw PA negotiations with Israel, under Yasir Arafat, as betraying Islamic principles by agreeing to discuss what was perceived as a two-state solution in the 1990s under the Oslo umbrella. Islamists particularly viewed the PA as repressive and corrupt, and the regime lost the people's support despite its involvement alongside Hamas in the 2000-04 Low Intensity Conflict (Second Intifada) against Israel. A system of "dual sovereignty" developed by the late 1970s when people began showing loyalty to the Islamic authorities and later the armed underground and not to the Fatah/PLO operatives deemed "the sole legitimate representative of the Palestinian People."[8] The process of transition intensified in the 1990s, deepening alienation between the people and PLO leadership. Corruption was rife, especially as pertained to Arafat and his associates in the PA ruling elite. They stole hundreds of millions, if not billions, of dollars.[9] Fatah won the elections in 1996, but when their four-year term of office expired in the spring of 2000, there was no further balloting. In light of the failure of the Camp David negotiations that summer, clashes with Israel ensued and lasted four years.

There was to be a double Jihad, against Israel and the corrupt secular PA.[10] Victory over both would defeat evil, bring a heavenly future in the name of Allah and reward his virtuous chosen. The beginning of the end of the old regime started with Arafat's death in November 2004, and continued with the Hamas parliamentary electoral victory of January 2006. PA attempts at re-imposing its authority were unraveling and hence the moves to establish a national unity government (NUG) to include Fatah, Hamas and independents. It was a very insecure transition. President Mahmoud Abbas spoke of national unity, yet continued to work with Israel and the US to undermine and outmaneuver Hamas, especially in the first half of 2007. Here is the seeming contradiction: To join a NUG, Hamas needed to acknowledge the Fatah/PA government as legitimate in some form. Fatah, the Palestinian organization advocating national liberation, was certainly legitimate, yet according to Hamas, Fatah's secularism was illegitimate (*HC,* Article 27). By participating in the 2006 parliamentary elections, Hamas implied recognition for the Oslo Accords, but never admitted as such. Hamas used the Oslo framework for elections to gain representation and international political legitimacy on the Palestinian front, and as a way of showing popular discontent with the Fatah/PA government. But recognition of Israel would not be forthcoming. The week-long civil conflict of 2007 was a rebellion and overthrow of the government, but not a full revolution, hence, we will not use the term "war."

Ideologically Fatah/PLO and Hamas have a fair amount in common. Their similarities are obvious when comparing *The Palestinian National Charter (PNC)* and *The Hamas Covenant (HC)*, both demanding Israel's destruction (see Chapter VII "Comparative Analysis") The two documents draw deep inspiration from the Grand Mufti of Jerusalem Haj Amin el-Husseini, embracing not only his antisemitism but his views concerning Arab identity and Islam. Haj Amin demanded full unity of the Arab and Muslim worlds, viewing the Arab world as a Muslim domain. Working with the Muslim Brotherhood in Egypt from 1945 onward, the Mufti helped facilitate the establishment of branches of the organization in the Palestine Mandate during the late 1940s. In later years Haj Amin was a personal hero to the more astute and diplomatic Yasir Arafat. Although not caught making blatant anti-Jewish remarks, Arafat fully admitted Mufti Haj Amin el-Husseini was his icon and role model. Both leaders were a mix of Islam and Arab nationalism embodying anti-Western and anti-democratic understandings.

The post-2000 secular Palestinian Arab nationalism was becoming more of a veneer and less of an influence for Muslims in the region. Hamas sought a revision from an anti-Zionist stand as expressed in *The Palestinian National Charter*, into an outright antisemitic policy platform, begging implementation and Jewish demise worldwide. As a reminder, the *PNC* still holds the anti-Zionist positions that were never corrected; the Fatah/PLO was expected to make revisions to the *PNC* to reflect the goal of an agreed upon two-state solution with Israel, the foundation of the Camp David 2000 negotiations. Lest one forget, by the end of his life Arafat publicly shifted much closer to

Hamas' behavior and objectives rejecting the two-state solution. In his heart he may have already been there years before.

The difference between Article 20 of the PLO's *PNC* denying Jews a homeland and history, and the *HC*, which calls for universal Jewish destruction (Article 7), is the gap between the denial of Jewish cultural and national identity (*PNC*), and physical extermination (*HC*). Ideologically, and through both words and deeds, the legacy of Haj Amin is fully in tandem with the *HC* and Jewish physical destruction.

During the ensuing battles in Gaza, the Fatah/PA forces melted away in the face of the highly motivated Hamas fighters (see Chapter V, "Hamas Ideological Victory"). When integrating Arab nationalism into the Islamist identity, as represented by Haj Amin el-Husseini and apparently Arafat as well[11] but to a lesser degree, one realizes that to shift allegiance from the Fatah/PA to Hamas is not a totally revolutionary undertaking. It certainly is a revolt against existing circumstances, in particular corruption and oppression. "Relative deprivation" existed on all levels: political, economic and social. The Hamas majority in parliament, heavy representation in the 2006 NUG and control of ministries raised expectations, but the resulting influence was disappointing. Feeling cheated by the secular-led Palestinian Authority, the expectation gap widened and Hamas fighter motivation soared. When the opportunity arose, the Hamas armed forces named for the Islamist hero "Izz a-Din al-Qassam" quickly determined the military outcome. The PA regime collapsed in Gaza, but Israel held the West Bank in check until US General Keith Dayton organized a new security apparatus to enforce Fatah rule.

From June 2007 to the present (2016), Hamas continues to rule Gaza. Many viewed Prime Minister Ismail Haniyeh's government as moderate because, despite increasing border tensions with Israel and a partial blockade for the next year and five months, there were semi-successful cease-fires until November 2008. Once Haniyeh lifted the cease-fire, Israel suffered from heavy shelling, and responded in kind, launching the Cast Lead operation at the end of December. Hamas instituted Sharia law throughout Gaza several days prior to the Israeli military operation. Unsuccessful military engagements with Israel forced Hamas back from the brink of increased extreme Islamist behavior, both inside Gaza and externally against Israel.

There was no continual Jihad because Israel delayed Hamas from further action. Desperately needing aid from the Arab/Muslim world to counteract the continuing Israeli naval and land blockade, Hamas was obliged to retain a working relationship with Hosni Mubarak's fairly hostile secular Arab nationalist regime in Egypt in order to kept the tunnels functioning and allow for the flow of smuggled goods, while garnering international support. Plans for greater revolutionary purity were postponed to guarantee material and diplomatic support from outside sources. Patience and prudence were necessary to rebuild, consolidate resources, and acquire support. Hamas could not tout a harsh anti-Egyptian policy. Only in the aftermath of regrouping could ideological purity take the lead once again.

By 2012, in the aftermath of Mubarak's overthrow, the atmosphere fostered enormous hope for Hamas concerning revolutionary continuity and export. Egypt's newly elected Muslim Brotherhood President Morsi allied with Hamas and tunnel traffic increased between Sinai and Rafiah, demonstrating Egyptian support and bringing the two Islamist partners into closer cooperation. Hamas was shifting toward its most intense stage, but was losing control to the more radical factions in the Gaza Strip. Most specifically, Ahmed Jaabari, who led the Hamas military wing "Izz a-Din al-Qassam," challenged the Hamas political leadership. He was high on Israel's elimination list. Jaabari's political intentions were not clear, but his behavior ruled out the possibility of attaining a cabinet position at some point in the future. Jaabari was killed by an Israeli air strike at the outset of the Pillar of Defense operation in November 2012. A week later, a greatly weakened Hamas struggled to re-impose its sovereignty after Israeli air strikes destroyed most of its weapons and ammunition. With no real opposition, Haniyeh's Hamas government in Gaza persevered. In July 2013, Egypt's Morsi government was overthrown, partially by popular will but more so by the Egyptian military and General a-Sisi in what is seen as a counter-revolution.

Going into 2014, Hamas was isolated in the Gaza Strip, a complete reversal of fortunes from the prior year and a half. General a-Sisi's government in Cairo was bent on halting the smuggling and closing down the illicit tunnel traffic in commercial goods and military contraband. As noted previously, the crackdown against smuggling also brought about fuel shortages, effectively paralyzing Gaza's power plant. More significantly, Egypt's military rulers feared Islamist activists and especially Salafist and Al Qaeda supporters who were freely making their way from Gaza to Sinai, and into the Egyptian urban landscape of Cairo and Alexandria. Weakened economically and militarily, Hamas agreed to join a NUG on June 2, 2014, and Fatah's Rami Hamdallah replaced Ismail Haniyeh as prime minister. Beating a tactical retreat, Hamas and Gaza were paying a heavy price for their previous Islamist victory. Still, even membership in the NUG would not stop Hamas activism.

Looking back on 2012, the Hamas revolt was readying for the full implementation of its ideological demands as it reached the heights of revolutionary Islamic fervor. The shift was delayed but not canceled by the previous Cast Lead clash. As well, the Pillar of Defense operation could not nullify the Hamas march to Jihadi victory. General a-Sisi's overthrow of Morsi and the Brotherhood was a much more serious challenge. Previous encouragement from the Tunisian and Libyan revolutions faltered, as both countries were in disarray. The bloody civil war in Syria pitted the Assad regime ally against the increasingly fractured Sunni Jihadis. This left Hamas to choose between its former Iranian/Hezbollah patrons and the wholehearted Turkish support they had since the *Marmara* incident in May 2010. It was increasingly difficult for Hamas to reconsolidate and retain power when the Arab/Muslim world was in deepening turmoil. Haniyeh's "victory lap" as an honored guest throughout the Arab/Muslim world in January 2012 and speeches from the winter of 2011-12 indicated a harshening of tone and return to building a total Islamist society with a full commitment to Jihad. Events had moved far too quickly for anyone to recall the Hamas hero status from just a few years prior.

Israel's July 2014 Protective Edge operation came as an immediate response to shelling and tunneling by Hamas and other Islamist organizations, launched after the murder of three Israeli teenagers. Hamas was in economic and political distress because of the increasingly effective Israeli blockade now augmented by Cairo, a result of Egyptian regime change. In addition, President Abbas and the Fatah/PA were making a comeback. Gaza sustained major physical damage, but Hamas and its allies redeemed themselves politically. Hamas popularity was restored through resistance to Israel and Fatah/PA despite participation in the NUG in the West Bank and Gaza. In the aftermath of Operation Protective Edge, due to the need for reconstruction funds from a spectrum of donor nations including the US and EU, Hamas could not capitalize on its renewed political support. The 2014 NUG arrangement with Fatah remains a major obstacle to Hamas' direct control over Gaza's destiny since funds for reconstruction must be channeled through the PA in Ramallah. Since they continue to enjoy popular support, Hamas' setback may only be temporary.

Hamas seized power in Gaza in 2007, yet the height of revolutionary fervor was disrupted (beginning in late December 2008) by Israel's three military operations, and a-Sisi's rise to power in Egypt. The extremist stage is incomplete as of this writing. Ideological commitment and intensity remain high but cannot be kept so indefinitely. The most radical phase of ideological adherence and Jihad will be contingent upon events in other Arab/Muslim countries, especially Egypt. Presently in Egypt, the secular nationalists are in

X Conclusion: Hamas Revolutionary Impact   415

power; in the future the Muslim Brotherhood may return with even greater support. Hamas and the Jihadis require increasing Middle Eastern instability and the reinstatement of the Muslim Brotherhood regime in Egypt in order to be victorious.

Without their temporary setback, would Hamas have gone beyond the Islamism of Haj Amin el-Husseini? It is worth recalling that in the 1930s Izz a-Din al-Qassam and the Black Hand represented the violent Muslim Brotherhood policies even more than the Grand Mufti Haj Amin el-Husseini. Within the parameters of a traditional society, Hamas initiatives, whether peaceful or violent are more of an Islamist revolt and demand for a return to fundamentalism than a full revolution. Prime Minister Haniyeh's "political Hamas" was considered moderate while the military wing under the command of Jaabari was deemed radical. Jaabari defied the politicians, but he did not force them from power. It is not a stretch to compare the Haj Amin and al-Qassam symbiosis with that of Haniyeh and Jaabari respectively. Both Haj Amin and Haniyeh are known for political roles, while al-Qassam and Jaabari were terrorist military leaders eliminated by outside adversaries, the former by the British and the latter by Israel.

Political Hamas preached neither democracy nor national unity with Fatah. Meanwhile, the radicals of the Izz a-Din al Qassam Brigades were content with their virtual monopoly of force. We may be seeing a return to aspects of a pre-revolutionary society resulting from the crushing physical defeat suffered at Israeli hands during the November 2012 Pillar of Defense operation, and even more so with the July 2014 Protective Edge bombardments and ground incursion. During the intervening year and a half, the Gaza front was one of the quietest in recent memory. At the time, the Hamas Revolution was stunted or in abeyance, mainly due to outside events, influences and intervention. Pre-2007 Gaza was a heavy mix of Islam and Arab nationalism. Taking the cutting edge off of militant Islam and re-emphasizing Arab identity will do Hamas little damage and may aid in shoring up support. This was particularly critical when the donor nation summit met in Cairo in the autumn of 2014 and agreed to appropriate $5.4 billion for Gaza reconstruction. The funds are to be funneled through President Abbas and the PA, bypassing Hamas.

It appears Hamas was on the threshold of an extremist stage, but today they face dire threats in the form of radical Salafists further to the right than themselves, al-Qaeda and Islamic State-types seeking to replace them. During the second half of 2015 Hamas was in military conflict with these fanatical groups and still retained the upper hand, but was exercising less-than-full control over the Gaza Strip. Hamas may not have the breathing space it needs to recuperate. To retain local support for its Islamist regime in Gaza, Hamas must unify, rearm and reassert its power sooner rather than later. No doubt if the Islamic Awakening retakes the offensive we can expect Hamas participation. Unauthorized rocket attacks against Israel initiated by fanatics may push Hamas to re-engage in battle earlier than planned. Hamas would prefer to confront Israel only after securing regional power support, be it Turkish, Iranian or Egyptian Islamist. Israel can be expected to retaliate and there may be frequent "restarts" possibly triggered by tunneling activity into Israel. Internally, Hamas will work to purge anyone opposed to their regime.

In the interim, we can foresee a remission in the direction of Palestinian Arab nationalism especially concerning participation in a NUG. The Palestinian national essence, although less significant than Islam, is viewed as a limited territorial commitment in the *The Hamas Covenant*. To broaden its appeal, we can expect Hamas to return to the Islamist/nationalist messages of Haj Amin el-Husseini. Palestinian nationalism will be played up, Islam downplayed a trifle and Jewish perfidy continually emphasized. The Fatah/PA will suffer further condemnation for betrayal and delegitimization by negotiating with Israel and accepting the two-state solution concept, at least in name.

If suffering under unbearable pressures, Hamas could accept a *hudna*-induced two-state arrangement for a limited period of time. The Hamas conditions for the *hudna* would necessitate a full Israeli withdrawal and Palestinian refugee return, nothing less. The *hudna* will end when Hamas feels strong enough to at least achieve political and military gain, if not overall victory. Of course these *hudna* conditions are theoretical since they are totally unacceptable to Israelis, including those on the furthest reaches of the Zionist left.

Revolutions lead to shifts and new baselines. In the case of Hamas, in comparison to the Fatah/PA regime, there will be more Islam, less secular identity and even less tolerance. One can foresee an intensified antisemitism harking back to seventh century Islam in content, yet highly integrated with Czarist/Nazi motifs and ideals. Jihad, necessary for cleansing, will not be far behind. Anti-Western and anti-Christian attitudes can be expected to continue. Certainly pagans and Far Eastern non-Muslim peoples have no place in the Hamas world. On the local level Hamas will labor to spread its ideals to the West Bank, whether through some form of NUG arrangement or by way of a more militant initiative. As a member of the Muslim Brotherhood, Hamas demands a Jihadi outcome of revolutionary export to the world. The *HC* will remain the "Divine" guideline for future rule.

Islamic World Domain, Background, Message and Ramifications

The Muslim Brotherhood/Hamas draws its legitimacy and strength from the birth of Islam and continues to see its primary enemy, the Jews, in the Islamic historic theological setting. Jews are in alliance with idolaters and/or polytheists and pagans (Koran 5:82), which nowadays means material well-being and man-made social systems such as democracy, socialism and capitalism. The Jews are the adversaries and in the broad context "modernity," meaning liberty, the free flow of ideas and humankind ruling itself, is evil. Sayyid Qutb, the foremost Islamist thinker of our time, made this specific understanding the focal point of the Muslim clash with the West.

Irshad Manji points out that a narrow-minded fundamentalist Islam emerged in the sixteenth century Ottoman Empire when clerics destroyed an observatory, which sought knowledge about the heavens. By the mid-eighteenth century, the printing press was banned. A century later, Mecca's leading cleric canceled three Ottoman reforms directed at eliminating slavery, freeing women from the veil and allowing non-Muslims to live in the Arabian Peninsula. These events took place on the background of the crystallization of the Saudi kingdom in a deal between Mohammed ibn Saud and Mohammed ibn al-Wahhab in the mid-1700s. The former forged political unity, while the latter solidified an extreme conservative religious reform highlighting a return to the Jihadi ruthlessness of Islam's first century. These Wahhabist ideals continue to be acted upon through the Islamic State destruction of world heritage sites deemed "pagan" or "un-Islamic."

A reactionary reformation set in, rooted in the ideas of the thirteenth and fourteenth century cleric Ahmed ibn Tamiya who denounced the Mongols (Tatars) for integrating their own secular laws into government despite their conversion to Islam. For ibn Tamiya, state and religion were one, and only the rule of Sharia law acceptable. In the mid-twentieth century, Qutb took inspiration from ibn Tamiya in his demand for the rule of Sharia law, and his fanaticism led Egypt's Nasser to execute him. Ibn Tamiya railed against the Mongol invaders no less than Osama bin Laden condemned the American presence in the Muslim Arab world. In later years, Qutb's exiled brother Mohammed taught Osama bin Laden in Saudi Arabia. In Islamist eyes, today's Americans are yesterday's Christian European Crusaders. Today's materialist Jewish-influenced New York City is the modern cosmopolitan pre-Islamic *jahili* Mecca of the 620s. The eternal resurgent pristine Islam is destined to capture America. The Islamist predictions are that New York will fall to Islam just as Mecca did close to 1,400 years ago.[12]

In theory, the above phenomena are in line with *The Hamas Covenant*. The superiority of Islam and the need for Sharia law are emphasized throughout the *HC*, but most emphatically in the Introduction-Preamble, Articles 2, 3, 8 and 11. Jews are the ultimate enemy, as seen in the Introduction-Preamble, Articles 7, 15, 20, 22, 28 and 32, and aligned with them are the Western capitalists and Eastern communists (Articles 22 and 25). The *HC* lists the previously defeated as Tatars/Mongols (Articles 29 and 35) and the Crusaders (Articles 15, 29, 32, 34 and 35) who are now part of the Christian West and allies of the Zionists/Jews. Hamas claims full membership in the world Islamist and Jihadist movement (Articles 1-5), claiming the same theological rights, history, previous and present enemies. The Jews and Israel are the most immediate adversaries slated for destruction. Hamas as part of the Muslim Brotherhood will battle the others at some future date.

The traditional stereotype of the Jews as a discarded, despised, cowardly and laughable *dhimmi* community gave way to the twentieth century contradiction of the re-assertive nationalist Jew claiming rights in his ancient home, the Land of Israel. The success of the State of Israel was humiliating in the face of previous Islamic historic and religious understandings. The only way to explain Jewish strength when confronted with Islamic "righteousness" was through theological demonization. Jewish insidiousness was in alignment with all the forces of evil, including the devil himself. Today the Jews, Christianity and the secular West are viewed as one entity—all anti-Islamic. Jews are singled out for special hatred even more so nowadays because they are the people who dared to establish an independent entity of their own on *waqf* lands in the Islamic-dominated Middle East. Czarist and traditional Islamic antisemitism gave way to embracing Nazi accusations and solutions before the final replacement of the Divine by the Jihadi determinists who will now act as Allah Himself. Jews worldwide are to pay the ultimate price of Islamist hatred.

Furthermore, wittingly or not, the rise of secular Arab nationalism is accused of aiding the Jewish success in building a state entity. Secular Arab nationalism in the Middle East is charged with betraying Islam by not invoking an eternal Jihad against the Jews. The anthropocentric leadership of these Muslim nations showed discretion and took it upon themselves to make either war or peace. In diocentric Islamic thinking such considerations are forbidden; devout Muslims have no right to make such determinations. The singular path must be that of Jihad, with the *hudna* Islamic cease-fire enacted only when absolutely necessary. Agents of betrayal come in the form of King Farouk, Gamal Abdul Nasser and Anwar Sadat of Egypt, King Hussein of

Jordan, the Shah of Iran, the Turkish leader Mustafa Kemal Ataturk and those Palestinians who engaged in the Oslo process with Israel, such as Yasir Arafat. Some of these Arab Muslims made peace with Israel, while others were involved in constant war, but because they were deemed secular and made decisions invoking secular logic, they could never defeat the Jews. Secularism was and is the eternal cardinal sin of the betrayers. Hamas believes in a future total reform when Islam recaptures these bastardized societies. On the Palestinian front, Hamas must defeat Fatah (*HC,* Articles 26 and 27) and impose Sharia law throughout Palestine/Land of Israel, allowing for no territorial or theological compromises (*HC,* Articles 11, 12, 13 and 32).

As a small piece in the world-Islamic puzzle, Hamas followed the trends begun in the mid-twentieth century when revolting against the Western imposition of its values. The Western liberal argument in the post-WWII era claims everyone would benefit from adopting Western values and the accompanying democratic form of government, whether through persuasion or coercion. Still, much of the Muslim world as well as parts of the Far East continue to reject the Western ideals of individual human rights and democracy.[13]

Urging readers to take a broad view of the historic understanding, Samuel Huntington's *The Clash of Civilizations and the Remaking of the World Order* (1996) dedicates a significant amount of research to understanding what is titled "The Islamic Resurgence." Western Enlightenment values such as human rights, liberalism, secular understanding and democracy are perceived as the universal panacea to be adopted by all people and cultures to ensure the advancement of humanity. Yet by the end of the twentieth century, the Muslim world justified its resistance to such values, as the West began declining on the economic, demographic and military fronts.

Islamic movements advocated the destruction of secular and material (capitalist and/or socialist) ideals in favor of a return to the true source of Islam as led by the Prophet Mohammed and the first caliphs. Nationalism, communism and socialism are viewed as doomed to failure, being empty secular materialist replacements for a Divinely given Islam. By the end of the twentieth century, Western-influenced regimes throughout the Muslim and Arab world indeed failed, in particular in supplying social services to their people. Most failed countries had corrupt former military power elites exploiting the people for their own narrow interests. In Tunisia and Egypt, mass revolutionary activism led to elections where the popular will of Islam began to rule by the winter of 2011-12, less in the former and overwhelmingly in the latter.

A reconstructed, purified Islam, as defined by the Muslim Brotherhood and other Islamist groups, provided the much craved for education and social welfare demanded by the masses. Islam was not to be modernized; rather modernity was to be Islamicized.[14] Islamic revivalism is two pronged. First there is domestic improvement for the Muslim community reaching down to the individual and second is the demand to crush all Western influences. Once rejuvenated, the Muslim community is to take offensive action, destroy Western influences and replace them with the culture determined by Allah, Islam and Sharia law. A reinvigorated Islam will embrace modernity and in particular its technological advancements while advocating the obliteration of Western values and societal norms. Social and universal justice coupled with rule on earth must be Islamic, not Western Enlightenment.

The slogan "Islam is the Solution," sums up opposition to nation states in favor of the Divine perfection of heavenly dictates. Dismissing secularism, law and education are to be Islamic and thereby permeate every aspect of society. Demand for social change may be seen as mainstream while the political movements of Islamism or Islamic fundamentalism embody extremism. According to Huntington, the Islamic awakening is similar to the European Reformation in its demand for sweeping reform and reevaluation of everything. The differences are in scope, the European model was limited to northern Europe while the "Islamic Resurgence" affects one and a half billion Muslims. Today's Islamists are well-educated, urbanized, lower middle class youth. For them, the power of religion is obvious; the Pope toppled Communism and Ayatollah Khomeini overthrew the Shah of Iran.[15]

Although democratic elements and Islamism may be the two opposition forces in the Muslim world used to challenge secular dictatorships, they are only temporary allies since Islamic culture and society are steadfast enemies of Western liberalism and the democratic process. Liberal democrats are easily repressed by nationalist dictatorships. However, when outlawed due to security concerns, Islamists can take refuge within religious organizational frameworks such as mosques and charities while continuing educational indoctrination. The democratic thinkers not only lack such institutions, but have shallow roots within the Muslim world. Furthermore, secular dictatorial regimes in the Muslim world identify more with Islam on the cultural level and are willing to integrate aspects of Islamic law, values and symbols within the secular national system. This forms a political/cultural hybrid in the hope of co-opting moderate Islamists.[16] We saw this clearly in the Egyptian Revolution of 2011 begun by the secular Facebook and Tweeter generation who initiated the massive demonstrations that toppled the Mubarak regime. In the aftermath, the Islamists represented by the Muslim Brotherhood and

Salafists won the elections by a landslide vote of over 70 percent. The Muslim vote was 80 percent in favor.

As of this writing, on the demographic level, the West and Christendom have had no population growth while Islamic populations are increasing at several percent a year. From an economic perspective, the West has material comfort while the majority of the Muslim East lives in relative poverty. Europe's population is aging at a time when the Muslim world is enjoying a "youth bulge." Still, too many young people with too few opportunities to achieve the promises of a secular world can become a major destabilizing factor, not only in their own lands but worldwide, and particularly in relation to the declining West.[17] This trend may be slowed, but continues into the second decade of the twenty-first century.

Islam expects Muslim loyalty to be exclusively to Allah and the Koran. Family, clan, tribe and culture may be pieces in the puzzle, but they must come second to the eventual unification of everyone in the global Islamic empire. There is no room for democracy, the nation state, or the separation of religion and the supposed "secular" authority, as exists in the West where one differentiates between "church and state." The Prophet Mohammed and caliphs represent the power and glory of past leadership, which held both religious and political power, in contrast to the Western Christian artificial duality.[18]

Writing in the 1990s, Huntington touted the possibilities for a multi-cultural, multi-religious world once the "resurgence" calmed down. He believed a "core state" such as Turkey would arise and unify the Muslim world, representing Islam to all others and seeking nonconflictual arrangements. He expected a balance to be reached.[19] It is here that Hamas and the universal Muslim Brotherhood ideals exhort all believers to crush any attempt at splitting sovereignty into a multi-cultural and multi-religious world. Although a temporary ceasefire or *hudna* may be called, the Western interpretation of a cease-fire to achieve conflict resolution is unlikely. Ideologically, the Muslim Brotherhood-Hamas were and are ordained to overthrow any "core state" regime (Turkey or otherwise) attempting to eternalize a world balance of power preventing Islam from obtaining universal conquest. The lure of peace, stability, and material-well-being for their own societies are obstacles in the path of the command to spread the word of Allah to everyone and Islamicize humanity. Hamas uses the Palestinian-Israeli conflict as a cover for principles concerning Islamic world conquest aspirations; the *HC* is clear on this point in Articles 11, 13, 14 and 32.

The West has a major problem by remaining in denial. Prior to 9/11 Huntington's perspective appeared unnatural despite the warning signs. In the early 1990s, President Bill Clinton maintained that the battle was against violent Islamic radicals, despite evidence pointing to overall Islamic rejection of Western values and forms of government. As seen previously, President Barack Obama in his June 2009 Cairo speech almost eight years after the al-Qaeda attacks on America, went a step further in naiveté by declaring, "Hamas does have support among some Palestinians, but they also have to recognize they have responsibilities, to play a role in fulfilling Palestinian aspirations, to unify the Palestinian people, Hamas must put an end to violence, recognize past agreements, recognize Israel's right to exist."[20] Hamas will reject any attempt at engagement or compromise, with complete disdain (*HC,* Articles 11, 13 and 32). What is shocking is that Obama never recognized Hamas, nor the Muslim Brotherhood, for who and what they claim they are. Failing to recognize Islamists as "Divinely inspired" is the greatest of insults, and only makes Americans and Europeans a more implacable and foolish enemy.

In 1990, Bernard Lewis, the eminent Western scholar on Islam, explained the rumblings in the Muslim world to be "no less than a clash of civilizations – that the perhaps irrational but surely historic reaction of an ancient rival against our Judeo-Christian heritage, our secular present, and the worldwide expansion of both."[21] Lewis made the comment once the Khomeini Revolution in Iran, the Soviet banishment from Afghanistan and the pro-Islamist coup by Zia ul-Haq in Pakistan, were in the past. Complementary trends showed the strengthening of extremist Islam in Sudan under Omar al-Bashir and Hezbollah's Khomeinism gaining strength in Lebanon. While it is said that Muslim commentators spoke of the coming clash, others mentioned above were already engaged in their anti-Western mission. Hamas has no pretenses as to its role and views itself as a "cog in the wheel," just part of the overall Jihad apparatus.

The West is understood to be "materialist, corrupt, decadent and immoral," devoid of religion and belief. Secular liberals are most despised as "arrogant" in their attempt to analyze Islam and Muslim societies. Western Christianity, because it is a recognized *dhimmi* belief system, is seen in a more positive light. The greatest clash is between the Western "universality" of culture verses Islamic religious global conquest.[22] The democratic West feels an obligation to extend its secular liberalism to the Muslim world, while Islamists live under the Divine obligation to Islamicize the planet. Islamists see the West as culturally self-righteous, and there are those in the West who see Islamists as religiously self-righteous. Islamists condemn the West as conceited for its attempt to implement man-made laws, as opposed to bending a knee to Allah and taking upon themselves Islam and Sharia law.

Islamists view the Soviet Afghan war of the 1980s as a Jihadi victory, and not a victory of the democratic West over the communist East. The American-led West is next on the Jihadi agenda, each superpower being destroyed in turn. The early 1990s saw US and European involvement in the Gulf by confronting Saddam Hussein after the Iraqi invasion of Kuwait. In retrospect the entire West is portrayed as the enemy despite being called in by the Saudis, Kuwaitis and others to halt Saddam Hussein's murderous imperial policies.[23] The US finally toppled Saddam in 2003, eternalizing his hero status in the Arab/Muslim world. Islamic unity was solidified despite Saddam's pro-Baath secular policies and harsh anti-religious repression leading to the deaths of hundreds of thousands.[24] The West is condemned to the fate of the Byzantine Empire, crushed by Islam in 1453.

Although seen as a "fault line" or "civilizational" wars, there is no doubt of the absolutism of the split between *Dar al-Islam*, the abode of Islam, and *Dar al-Harb*, the abode of war. Other civilizations accept faiths and cultures different from their own, but according to Jihadists, Muslims heralding compromise and a multi-cultural existence or any acceptance of these principles, are deemed betrayers. There is the rationalist expectation that exhaustion as a result of continual conflict will lead to moderation.[25] But as long as the fundamentalist Jihadist understanding is unchanged, any halt in conquest will only be a temporary delay or *hudna* necessary to regroup and reactivate the conflict. To continue with the geological metaphor from the Islamic Jihadi perspective, peace will only arrive when plate tectonics and fault lines are concepts from the past. The earth is destined to be one single Islamic plate.

"Islam's borders *are* bloody," much more so than any other religions or civilizations of recent memory. Already twenty years ago over 50 percent of wars involved Muslims, as did two-thirds to three-fourths of inter-civilizational wars.[26] As of late and just prior to the 2011 Islamic Awakening the number rose to 80 percent of all wars involving Muslims.[27] Internal Muslim bleeding from the 1990s continues into the 2000s as we witness wars in Libya, Somalia, Yemen, Iraq, Syria, the Palestinian territories, Afghanistan and Pakistan. Forms of civil conflict continue to plague Egypt, Tunisia, Algeria and Sudan. The estimate is that from 2006 to 2008 some 98 percent of all those killed in terror attacks by al-Qaeda were themselves Muslims. This was a serious increase from the already astounding 85 percent of Muslim victims at the hands of al-Qaeda, from the more inclusive 2004 to 2008 statistics.[28] Despite these casualties, overwhelming majorities in most Muslim countries and significant minorities in Western Muslim communities support extremist Islamist groups such as Hamas, Hezbollah, al-Qaeda and the Taliban. Support may be limited for the Islamic State, however tens of thousands of Muslims including many Westerners have joined the emerging caliphate and participate in its brutalities. There are very few wars worldwide that do not involve Muslims (the Russian-Ukrainian conflict being the most notable). This translates into a not insignificant approval of terror attacks against Western targets, including suicide bombings, the full implementation of Sharia law, rape of non-Muslim women and quite notably female honor killings.[29] It is not clear how much support the fanatical Islamic State enjoys. The Pew Research Center recently reported very limited global Muslim support for Islamic State actions adding up to several percentage points, while on the other hand a May 2015 on line poll by the *Al Jazeera* Arabic language satellite station based in Qatar showed an overwhelming 81 percent approval rating in the Arab world for Islamic State goals and behavior.[30]

Nowadays when pitting Islamists against their secular nationalist kin, repression by the former highlights everyday life in Ahmedinejad's or even Rouhani's Shiite Iran and in general wherever Sharia law rules in Arab countries or entities. Following this trend, it appears that even the Islamic State may solidify rock bottom support in a small yet deadly activist sector of the Sunni Muslim world. That is not to say the secular nationalist regimes are not dictatorial, they usually are; however, they do allow for non-Islamic thinking, therefore the possibility for democratic reform exists. The confrontation continues and it matters little which specific Middle Eastern region is under discussion.

For Palestinians, this is the heart of the Hamas-Fatah clash. Hamas rules in Gaza, and Fatah holds the upper hand in the West Bank. Hamas is not only part of the Islamic Resurgence but takes its place as those in the universal vanguard despite its lack of physical military power and dependence on a very narrow land base in the Gaza Strip. Hamas was a forerunner, taking the overall Palestinian legislative elections in January 2006 in Gaza and the West Bank, and overthrowing the PA NUG arrangement with Fatah in Gaza in June 2007. Islamists worldwide were inspired. It is believed and most likely quite accurate to claim that only Israeli and American support for Fatah's President Mahmoud Abbas and then Prime Minister Salam Fayyad prevented a similar outcome in the West Bank. Hamas continues to confront its Fatah rivals through an underground threat in the West Bank, and by way of embracement in a NUG, a panacea constantly advocated as a way to end Palestinian disunity.

X Conclusion: Hamas Revolutionary Impact

Politically the question of the necessity of Palestinian national unity, or whether there was to be a Hamas Islamic victory sometime in the very distant future, was overridden by the events of the Arab "Islamic Awakening" of 2011, ushering in revolution and changing the Middle East forever. The pace quickened and several of the previous Arab nationalist regimes are now gone; others are collapsing. Even the supposedly "popular" Egyptian military coup against the Muslim Brotherhood may be of short duration, Islam remaining a major political force. Any Palestinian unity government will prove Fatah the weaker party, and a redirection toward the Hamas-Muslim Brotherhood ideals will be in order. Peace with Israel will not be an option but instead only a limited *hudna* implemented for practical purposes, leaving the confrontation with the Jewish nation state and world Jewry to be delayed to some future date.[31] Hamas acknowledges that the decision to go to war is preordained. This complements the overall Jihadist policy toward the rest of the world with its emphasis on the destruction of the secular West. So the question is whether a NUG is preferable where Hamas will work to weaken Fatah from the inside, or whether remaining on the outside and applying constant pressure will bring about the same result of Islamic domination.

Hamas built popular support, defeated Fatah politically and militarily, mobilized much of world opinion against Israel and acted as a brake on US and EU liberal secular democratic influences on Palestinian Muslims, particularly in Gaza but also in the West Bank. Today Hamas is recalled as one of the pioneers of Islamist victories in the pre-2011 phase. Hamas projects ideological power and defiance toward all enemies of Jihad. In the pre-revolutionary period the Muslim Brotherhood took control of Sudan, the Khomenists overthrew the Shah of Iran, and Afghanistan fell to the Taliban. Afghanistan will fall again when the Americans leave, as did much of Iraq to the Islamic State. Hamas is one link in the Islamic chain. Just recently, Egypt became the newest and largest link following the Hamas lead even if it only lasted two years. Hamas is a symbol and icon to be emulated as evidenced by the tumultuous welcomes afforded Prime Minister Ismail Haniyeh when visiting the Arab and/or Muslim world.

Hamas is vindicated in the Arab Muslim world and beyond, whether through the recent revolutions or by Turkish Prime Minister Erdogan's increasingly anti-Israel, pro-Islamist policies. Outmaneuvering Iran, Ankara became the primary Hamas patron since the *Mavi Marmara* event of May 2010. This is an enormous feat considering that the Turkey of Ataturk was still the "model" bastion of a secular state in the Muslim world at the turn of the past century. Erdogan's Islamist Justice and Development Party is curtailing secularism in Turkey step-by-step, particularly in the military, and clamping down on journalistic freedoms. Straddling Islam in the Middle East and democracy in the West, the Turks are encountering little criticism and remain NATO members. Gaza and Ankara have much in common even if at the moment the Turks retain diplomatic relations with Israel and speak of a Palestinian State within the 1967 borders.

With Erodgan's increasing Islamic activism and retention of popular support as evidenced by his party's victory in the November 2015 elections, Hamas has a Sunni role model which can be viewed as just passing through the democratic stage on its way to a more perfect Sharia-ruled, Islamic society. This does not contradict any sort of positive relationship with the Iranians, as proven by the warm welcome Prime Minister Haniyeh received during his February 2012 visit. Whether a slow steady Islamization at the expense of democracy is a realistic option in the Turkish framework is yet to be seen, but it certainly gives hope to Hamas and the universal Muslim Brotherhood. When it appeared the Turks would not go far enough in Islamization, it was probable the Egyptians would replace them in the post-2011 era. Cairo may be a more comfortable benefactor, but in the meantime the Egyptian military acts contrary to Hamas interests, and Erdogan retains his influence. As much as Hamas needs a patron, because of its small size and military weakness, it is an inspiration to be emulated throughout the Muslim world. Hamas is not only fully embedded in the movement, but is a leader in the Islamic Resurgence.

As of this writing, Hamas is not engaged in a direct clash with the West, even if crossing swords with Israel quite often. Hamas is cognizant of the need for overall Jihadi victory and the use of tactical *hudnas* when necessary to achieve the goal of Islamic triumph.[32] In the meantime, the Muslim Brotherhood is working to regroup and consolidate its internal victories throughout the Islamic world. Hamas can be expected to do likewise. By late 2012 the Muslim Brotherhood was on the verge of victory throughout the Middle East. Seen through an Islamic lens, this was comparable to the period just prior to Mohammed's victorious entrance into Mecca in early 630 CE. Recent setbacks may be compared to the Medina years of the 620s when the Prophet Mohammed was still building an offensive force. Consolidation and the defeat of adversaries in the Muslim world and patience in taking on non-Muslims are the key factors in the prerequisite to the "export" of the Islamic Revolution on a universal scale, better known as "offensive Jihad." In essence the "export" began in earnest with the Islamic State Caliphate policy decision to attack Paris in mid November 2015 in a well coordinated operation. The Paris attacks may signify the beginning of offensive Jihad against the West. Although viewed as fanatical, Islamic State actions are the logical continuation of Muslim Brotherhood and Hamas policy objectives. Hamas and the Brotherhood are simply more cautious, fearing for their own existence. In practice Hamas is more "moderate" at the moment because the timing may not appear prudent, but theologically Hamas beliefs are very much in tandem with those of the Islamic State. This begs the question as to whether Hamas loyalists will consider shifting their allegiance to the Islamic State and participating in global Jihad while not abandoning the front against Israel. An Islamic State military and/or terror initiative against Hamas should not be ruled out.

Hamas Future within the Palestinian and Arab/Islamic World

The Hamas objective is to reunite with all Muslims in a reconstituted Islamist Middle East and beyond (*HC,* Articles 5, 7, 9, 11 and 33). They dream about a renewed Caliphate and one supreme Muslim ruler. Palestinian Muslims are to play their part as one of the regional actors in this "New Middle East," no longer separated by nation state boundaries or vying for outside patrons, whether American, European or Russian. The Muslim world is pre-destined to become a superpower and spread the Islamist message.

Well before such a victory, Hamas must capture the Palestinian Authority apparatus either through elections or violent overthrow. Should elections be won, future open balloting should not be expected. Elections will either be Iranian style, with a limited amount of acceptable Islamic candidates with somewhat more liberal or conservative views, or the Palestinian Legislature can vote itself out of existence. It does not seem reasonable to expect Hamas to allow for Fatah or any form of Western-influenced secular comeback unless faced with no options. In that case, Hamas will consolidate itself in the opposition until achieving power. The Hamas ultimate objective is Islamic power, and they can play the democratic game as a tactic for as long as necessary.

Hamas is intent upon defeating Fatah (*HC,* Article 27) and once that is accomplished, Israel is next. In the aftermath of the 2014 Protective Edge repercussions, taking on Israel is best not done alone with little chance of battlefield victory although periodic eruptions to release internal social, political and economic pressures can be expected. In the meantime, Hamas speaks of and implements a *hudna*, never intending to accept Israel's right to exist. Militarily defeating Fatah in the West Bank means conflict with Israel since a Hamas-controlled West Bank is an existential threat to the Jewish State. Presently taking on the IDF would be counter-productive; however, undermining Fatah support through a dual sovereignty approach will supply a slow and steady erosion provided the PA does not reform itself and make its rule beneficial to the masses.

On the material level, Fatah can deliver a higher standard of living and security for PA citizens but to do so they must partner with Israel, the US and Europe. Should the Palestinian Authority in the West Bank accrue popular support for a two-state solution and Palestinian independence, such a move would be a defeat for Hamas (*HC,* Article 11, 13, 15 and 32). All outstanding issues need to be resolved with Israel, including mutual recognition, borders and security, a "non-militarized" Palestinian State, Jerusalem and refugees. It is very unlikely such an agreement can be reached with the sides so far apart on all of the issues, especially over the status of Jerusalem and refugee return. Most of the Israeli right/religious do not support a two-state solution, but rather advocate more settlements and a continuation of the status quo. Most have no faith in the PA and many would like to see it collapse.

But even should the Israeli political left prevail, for a Palestinian leader to drum up popular support for an "End of Conflict" agreement with Israel appears hallucinatory, especially in light of the weakening of secular Arab nationalism and the continuing struggles of the 2011 Islamic Awakening. For many Palestinian Muslims, an alliance with the West is to the detriment of life in a society dominated by Islam and is totally unacceptable. Politically, serious peace overtures by the Barak and Olmert governments were spurned in the past when Jordan and especially Mubarak's Egypt stood behind such initiatives. Today the gap has widened further.

At present, a wave of Islamism continues to engulf the West Bank even if seemingly deflected by Western and Israeli influences. The PA will be undermined should it make full peace or even settle for an interim accord with Israel. The Hamas long-term future depends on the level of success of the Islamic sweep throughout the Middle East. Left on its own, Fatah/PA would suffer the same fate as the other secular Arab regimes. But President Abbas has an advantage because Hamas must defeat Israel to wrest control of the West Bank from Fatah, which is highly unlikely. To defeat Israel, a coordinated Jihad led by the surrounding Arab nations and worldwide volunteers would be necessary. The Hamas challenge is to muster allies in its Jihad against Israel as made clear throughout their *Covenant* in Articles 28-30, 32 and 33.

Were free and open elections held, Hamas would be the winner. One must question the effect on the PA security services and their coordination with the IDF and Israeli intelligence. The scenario of unhindered rocket attacks from the West Bank hitting Haifa, Tel Aviv, Jerusalem and Beersheva must be taken seriously. The shorter range Qassams would continue to be fired from Gaza on Sderot less than a kilometer away, while longer range Grads threaten Beersheva, Tel Aviv, central Israel and Jerusalem. Attacks of all types against Jewish settlements in the West Bank would be a matter of routine. The Israeli response could include a reoccupation of the West Bank, essentially putting an end to the Oslo Accords, the Palestinian Authority and the on-ground interim arrangements. The ramifications of Israel being forced into a virtual one-state solution confronted by Hamas as the unchallenged representative of Palestinian Muslim loyalties are deep and far-reaching, especially as concerns security. Such a discussion is beyond the range of the topic at hand.

Hamas can be expected to be true to its ideology and *Covenant*, making few, if any, limited temporary concessions even when on the defensive. Once successful, an uncompromising implementation can be expected. *The Hamas Covenant* was and is the overall vision of history, victory, the Islamist worldview and strategy to achieve the holy objectives as dictated by Allah through his faithful. For an Islamist, any reversal is as fleeting as Israel's existence. History is understood to be linear—Islam will be victorious. It is only a matter of time; one must be patient. We can expect recurring peaks of Jihadism as the direct result of the Islamic Awakening.

For those who expected a liberal, democratic form of Islam after the 2006 elections they will continue to be disappointed. Palestinians were free to vote but freedom of thought and criticism outside of Sharia law were not and are not part of the Hamas agenda. Nor would non-Muslims or women be equal under a Palestinian Islamic regime. One does not openly criticize or demonstrate against Hamas unless willing to face the consequences.

Revolution to overthrow an unjust regime is positive but replacing it with a more repressive one, even if freely elected, only increases suffering. It is easy to claim that the Palestinian people have a responsibility for the Hamas rise to power and that they can demand Hamas step down in the name of a more liberal regime. Revolutions in the name of human rights and democracy overthrew dictatorships in the past and will do so in the future. As concerns Hamas, this is not the case. Hamas controls Gaza and can call elections at will, but for the past nine years has declined to do so. Hamas continues to solidify power and barring massive outside intervention can be expected to reign for years into the future. Most significantly, Hamas continues to enjoy popular support; the Palestinian people do not want a liberal democratic regime. They support Islamist governance. The only serious challenges to Hamas rule are the Salafist Islamic State fanatical types. As a counter weight, the Egyptian counter-revolution is applying pressure to Hamas and acts as a temporary brake on the spread of Hamas influence both within and outside Palestinian circles.

On the other hand, Hamas may enter into elections should the April 2014 agreement with Fatah prevail, but the Fatah leadership will nix any balloting should they believe Hamas will emerge the victor. Theologically powerful, but physically weakened, elections or a national unity arrangement would only be acceptable to Hamas as a carefully devised policy of self-preservation. As of early 2016, Hamas enjoys majority support making such a move by the Fatah/PA semi-suicidal. Hamas can exploit participation in a NUG before elections to Islamicize Fatah (*HC,* Article 27) and transform itself into the ultimate victor with or without balloting. Both organizations draw from the same mid twentieth century ideals pursued by Haj Amin el-Husseini. As shown previously, *The Hamas Covenant* and *The Palestinian National Charter* overlap in most places and therefore Fatah would have little problem reverting to Islamist terminology.

The issue of "reverse abrogation" discussed in the previous chapter, allowing for theological acceptance of others, appears unlikely in the foreseeable future even though it is the best chance for conflict resolution between Islamists and the rest of humanity. Such a move demands a universal Islamic initiative to be open and accepting of others. At present the Islamic agenda is diametrically opposed to these Enlightenment values. Returning to today's realities, even should Hamas suffer temporary setbacks, as is the case at present, the anti-democratic, antisemitic, Jihadi Hamas Revolution and its legacy, although not complete, will continue to reverberate throughout the Arab world. Specifically, Hamas may absorb Fatah, first ideologically and then physically.

Endnotes

1. Brinton, Crane, *The Anatomy of Revolution,* Vintage Books, New York, USA, Expanded Edition, 1965.

    Four major revolutions are compared: The English, American, French and Russian. Brinton's theories are derived from the historical parallels. He defines these revolutions as "popular," "democratic," and advocating "freedom." They are placed in the modern period of Western world history from the 17th to 20th century. This author believes the revolutionary steps analyzed by Brinton for the most part apply to Hamas as well.

2. Gurr, Ted, *Why Men Rebel,* Woodrow Wilson School of International Affairs, Princeton University Press, New Jersey, 1970.

    Drawing on the research of many other scholars and bringing his own conclusions to the table Gurr's work is an overall analysis of why violence and revolution break out. The major theory is "relative deprivation" and those gaps between expectations and reality. In particular, Gurr deals with society's or the people's perspective of the intensity, scope, unmet expectations, frustration, shifting loyalties and legitimacy (or lack) of the power elite and those revolutionaries challenging the existing order.

3   Ibid, pp. 13-21.
4   Ibid.
    This is the main theoretical argument made by Gurr.
5   Ibid.
    Gurr's analysis of revolutionary implementation.
6   Ibid, pp. 357-358.
7   Ne'eman, Yisrael, "Palestinian Islamic Revolution – On the March," *Mideast on Target*, June 15, 2007, http://me-ontarget.org/pws/page!4962.
8   The PLO was recognized as the institutionalized leadership of the Palestinian People at the Seventh Arab League Summit Resolution in Morocco on October 28, 1974. They declared the PLO the "sole legitimate representative of the Palestinian People." Information retrieved August 26, 2015 from *Palestine: Information with Provenance* (PIWP database), cosmos.ucc.ie/cs1064/jabowen/IPSC/php/event.php?eid=486.
9   See Uzrad Lew's book *In Arafat's Pocket* (Hebrew). Lew estimates Arafat stole at least $300 million. This number is derived only from what Lew was able to fully document. The numbers most likely go much higher. Lew, as an activist Israeli dove advocating peace and the two-state solution was employed as a senior financial advisor to Arafat and hoped that by putting the PA financial resources in order he would be contributing to the peace process. As he writes in his book, he was sorely disappointed by rampant PA corruption driven by Arafat and his advisor Mohammed Rashid.
10 Yousef, Mosab, *Son of Hamas,* Tyndale House Publishers, Inc., USA, 2010, pp. 52-53.
11 Arafat was playing a double game, leading the secular Fatah while shifting to a more Islamist position.
12 Manji, Irshad, *The Trouble With Islam Today,* St. Martin's Griffin, New York, 2003, pp.144-150.
13 Huntington, Samuel, *The Clash of Civilizations and the Remaking of the World Order,* Simon and Schuster, New York, 1996, pp 81-91.
14 Ibid, p. 96.
15 Ibid, pp. 109-114.
16 Ibid, pp. 114-115.
17 Ibid, pp. 117-121.
18 See Sayyid Qutb's writings in *Milestones.*
    See Chapter II "Ideologues," *Hamas Jihad.*
19 Huntington, pp. 174-178.
20 Text of Obama speech in Cairo, June 4, 2009, *USA Today,*, retrieved January 20, 2016,
    www.usatoday30.usatoday.com/news/world/2009-06-04-Obama-text_N.htm
21 Lewis, as quoted by Huntington, p. 213.
22 Huntington, pp. 213-214 and 218.
23 Ibid, pp. 246-249.
24 Ne'eman, "Arab World Heroes," July 31, 2006,
    http://me-ontarget.org/pws/page!5035.
25 Huntington, pp. 291-297.
26 Ibid, pp. 256-258.
27 Williams, Armstrong, "WILLIAMS: The endless wars of Islam," *Washington Times*, September 12, 2010, retrieved December 16, 2015, http://www.washingtontimes.com/news/2010/sep/12/williams-the-endless-wars-of-islam/.

28 Mayer, Alexander, "Muslims account for 85% of casualties in al-Qaeda attacks," *Long War Journal,* December 9, 2009, www.longwarjournal.org/.../muslims_account_for_85_percent.
Musharbash, Yassin, "Surprising Study on Terrorism: Al Qaeda Kills Eight Times More Muslims Than Non-Muslims," *Spiegel,* http://www.spiegel.de/international/world/surprising-study-on-terrorism-al-qaida-kills-eight-times-more-muslims-than-non-muslims-a-660619.html. Both retrieved March 2, 2012.

29 Muslim Opinion Polls "A Tiny Minority of Extremists?" *The Religion of Peace,* retrieved October 14, 2014, www.thereligionofpeace.com/pages/opinion-polls.htm.

Presented are wide-ranging studies, in particular Pew Research and Surveys. There are extensive links to global surveys of Muslim public opinion both in the West and Arab/Muslim nations where Islam is the dominant religion.

30 Constantine, Tim, "Eighty-One Percent Support Islamic State and Its Barbaric Acts," *Washington Times,* May 28, 2015, retrieved August 19, 2015, www.washingtontimes.com/news/2015/may/28/tim-constantine-islamic-state-and-their-barbaric-a/.

This survey was done by the *Al Jazeera* Arabic language satellite station covering 38,000 respondents much more than the usual representative number used in public opinion polls. The question posed was, "Do you support the organizing victories of the Islamic State in Iraq and Syria (ISIS)?" 81% voted "Yes" in support of the Islamic State and 19% were opposed. The survey received global coverage.
Poushter, Jacob, "In nations with significant Muslim populations, much disdain for ISIS," November 17, 2015, retrieved December 16, 2015, www.pewresearch.org/.../in-nations-with-significant-muslim-populations.

31 Elliot Chodoff's theory of "delayance," unpublished 2009.

32 Akram, Fares, "At Rally for Hamas Celebration and Vows," December 14, 2011, *The New York Times,* retrieved December 28, 2015, www.nytimes.com/2011/12/15/world/middleeast/on-anniversary-hamas-repeats-vows-on-israel-and-violence.html?_r=0.

Glossary

Significant Names

Abbas, Mahmoud/Abu Mazen – Current leader of Fatah. Chairman and/or president of the Palestinian Authority (PA) since his election in January 2005 after the death of Yasir Arafat. Served briefly as prime minister under Arafat in 2003. Actively rules only in the West Bank.

Abdullah II – King of Jordan since 1999 after the death of his father, King Hussein.

Arafat, Yasir – Leader of Fatah beginning in mid 1960s. Took the helm of the PLO in 1969 and combined the Palestinian political and military organizations under one umbrella group. Signed the Oslo Accords with Israeli Prime Minister Yitzhak Rabin in 1993 and was the first Chairman of the Palestinian Authority. Was elected Palestinian president in 1996. Participated in failed Camp David negotiations with Israeli Prime Minister Ehud Barak in 2000. Led the Second Intifada or Low Intensity Conflict (2000-04) against Israel. Died November 11, 2004.

Azzam, Abdullah – Palestinian Islamist and Jihadist ideologue. Believed in world Jihad, fought in Afghanistan against the Soviets in the 1980s. Wrote *In Defence of the Muslim Lands* and *Join the Caravan*, urging universal Jihad by all Muslims. Assassinated in November of 1989.

al-Baghdadi, Abdul Bakr/Ibrahim Awwad Ibrahim – Leader or "caliph" of the Islamic State (ISIS or ISIL) beginning in 2014. Was arrested and released by American forces in Iraq in 2000s.

al-Banna, Hasan – Egyptian Islamist, Jihadi ideologue and political leader who established the Muslim Brotherhood in 1928. Pro-Axis during WWII, involved in assassinations of pro-British Egyptian officials during the 1940s.

Essay *On Jihad* is well known. Assassinated by King Faruk's agents in February 1949.

Barak, Ehud – Labor Party leader and elected Israeli prime minister 1999-2001. Army chief of staff 1991-95. Failed to achieve peace with the Palestinian leader Yasir Arafat at Camp David 2000. Served as defense minister (2007-09) under Prime Minister Ehud Olmert and continued in Benyamin Netanyahu's government 2009-13.

Begin, Menachem – Israeli prime minister 1977-83, right wing ideologue and leader of the Likud until 1983. Believed in the "Greater Land of Israel" and developed settlement policies in the West Bank. Signed Camp David Accords in 1978-79 with Egyptian President Anwar Sadat. Returned Sinai for peace with Egypt and agreed to autonomy for the Palestinians, but no state. Died in 1992.

Dayton, Keith – American general who built and directs the Palestinian Authority security forces since 2007.

Erdogan, Recep Tayyip – Turkish leader of the Islamist Justice and Development party. Served as Turkish prime minister 2003-14 and president of Turkey 2014 to the present.

Haniyeh, Ismail – Hamas activist and leader. Appointed prime minister of the Palestinian Authority in 2006 after leading Hamas to electoral victory over Fatah. Dismissed by President Abbas after Hamas military takeover of the Gaza Strip in June 2007. Effectively became prime minister of the unrecognized Gaza mini-state.

Herzl, Theodor – Zionist visionary. Wrote *The Jewish State* in 1896 and convened the First Zionist Congress in 1897. Set the foundations for the establishment of the State of Israel. Died in 1904.

Hussein, King – King of Jordan 1953-99. Fought against Israel in the 1967 Six Day War, in which he lost East Jerusalem and the West Bank. Made peace with Israel in October of 1994. Died in 1999.

el-Husseini, Amin – Known as Haj Amin el-Husseini. Became the leading cleric of the Muslim community as Grand Mufti of Jerusalem in 1922 under the British Mandate. Clashed with the Zionists and British in the 1920s. Established the Arab Higher Committee in the 1930s, rebelled against the British Mandate, fled to Iraq and Nazi Germany during WWII where he actively worked for Jewish destruction during the Holocaust. Failed in his war efforts against the Jewish State in 1948. Died in 1974.

Jaabari, Ahmed – Hard line Jihadist from Gaza. Trained and developed the Hamas military force Izz a-Din al-Qassam. Kidnapped Israeli soldier Gilad Shalit. Was known to challenge "political" Hamas led by Ismail Haniyeh. Killed by Israeli air force in Gaza in November 2012 at the outset of the Pillar of Defense operation.

Khomeini, Ruhollah – Iranian Shiite Muslim Grand Ayatollah who overthrew the Shah of Iran in February 1979. Revered as the Supreme Leader and Jurist-Theologian. He gave full support to the Lebanese Shiite Hezbollah and demanded Israel and Jewish destruction worldwide. Died in 1989.

Mashal, Khaled – The overall Palestinian leader of Hamas worldwide. Living abroad, his influence is much greater outside of Gaza and the West Bank. Often deals in international forums. Survived assassination attempt by Israeli Mossad in 1997, moved Hamas headquarters from Jordan to Syria in 1999 and relocated to Qatar in the wake of the Syrian Civil War in 2012.

Mohammed, Prophet – Central Prophet and Messenger in Islam. Through him Islam and the Koran are revealed to the world. Captured the Arabian Peninsula and began the spread of Islam worldwide.

Morsi, Mohammed – Egypt's fifth president, leader of the Muslim Brotherhood's Freedom and Justice Party. He was freely elected in June 2012 and served for a year until overthrown by popular demonstrations and intervention by the army. He was jailed by the military in July 2013 where he has been tried and may face the death sentence.

Moses – In *Tanakh* (Hebrew Scriptures or Old Testament) led the Israelites from slavery to freedom in the Exodus from Egypt across Sinai. Gave the Israelites the Law contained in the Pentateuch or "Five Books of Moses." Did not enter the Promised Land.

Nashashibi Family – Palestinian Muslim family who opposed the el-Husseinis and Haj Amin el-Husseini in particular, during the British Mandate period. Were known to be moderates and pro-British.

Nasrallah, Hasan – General Secretary of the Lebanese Shiite Muslim political party and militia Hezbollah. Close associate of the ayatollahs in Iran.

Nasser, Gamal Abdul – Leader of the Free Officers who overthrew King Faruk and the Egyptian monarchy in 1952. President of Egypt 1954-70. Unsuccessful battling Israel in the 1967 Six Day War. Died 1970.

Netanyahu, Benyamin – Leader of the conservative Likud party. Prime minister 1996-99 and again from 2009 to the present. Signed the Wye Accords with the Palestinians in 1998. Served as Finance Minister 2003-05.

Olmert, Ehud – Led Kadima centrist faction and was Israeli prime minister 2006-09. Mayor of Jerusalem 1993-2003 when still in the Likud. Was prime minister during the Second War in Lebanon in the summer of 2006. Convicted on corruption charges and presently serving time in prison.

Peres, Shimon – Israeli president 2007-14. Labor party prime minister 1984-86 and once again from late 1995 to mid 1996 after Yitzhak Rabin's assassination. Served as defense minister in Rabin's first government 1974-77 and varying stints as foreign minister and other functions from the 1980s until the mid 2000s. Was the major architect of the 1993 Oslo Accords.

al-Qaradawi, Yusuf – leading Muslim Brotherhood Egyptian cleric. Well known in the Arab/Muslim media and leading antisemite continually demanding the destruction of the State of Israel and the Jewish People.

al-Qassam, Izz a-Din – Muslim Brotherhood activist and Jihadist in Mandated Palestine during the 1930s. Led the Black Hand guerrilla band attacks against the British, Jews, Christians and moderate Muslims. Challenged the Palestinian Arab leadership, including Haj Amin el-Husseini. Killed in a shoot-out with the British in November 1935.

Qutb, Sayyid – Egyptian Muslim Brotherhood ideologue, Islamist intellectual and advocate of world Jihad. Most noted for his 18 volume religious commentary *In the Shade of the Koran*. Wrote *Milestones*, a pro-Jihadi, anti-Western tract declaring the need for the destruction of all non-Muslim forces and composed the viciously antisemitic essay "Our Struggle With the Jews." Known for his opposition to secular Arab nationalism. Jailed by Egyptian President Nasser and executed in 1966.

Rabin, Yitzhak – Israeli prime minister 1974-77 and again 1992-95. Served as chief of staff of the Israeli Army (IDF) 1964-68 and ambassador to the US 1968-73. Was Defense Minister in two national unity governments 1984-90. Signed the Oslo Accords in 1993. Assassinated by a religious right wing Jewish extremist on Nov. 4, 1995.

Sadat, Anwar – Egyptian President 1970-81. Initiated 1973 October (Yom Kippur) War with Israel. Retrieved Sinai in 1978-79 in Camp David Accords with Israel. Shifted Egypt out of the Soviet and into the American orbit. Assassinated by the Muslim Brotherhood in October 1981.

Shamir, Yitzchak – Israeli Likud prime minister 1983-84 and again 1986-92. Considered very hard line and rejected the two-state solution. Believed fully in the "Greater Land of Israel."

Sharon, Ariel – Israeli prime minister 2001-2006. General in the Israeli Army, well known for victories in 1967 and 1973. Instrumental in building the right-wing Likud party in the mid 1970s. Active in settlement building as agriculture minister 1977-81 and again in the early 1990s as housing minister. Was defense minister during the First War in Lebanon in 1982 and forced to resign in the aftermath. Later foreign minister 1998-99 and replaced Benyamin Netanyahu as party leader. In a comma for eight years before his death in 2014.

a-Sisi, Abdul Fatah – Egypt's sixth President since 2013 after overthrowing Muslim Brotherhood President Mohammed Morsi. Served as chief of staff of the Egyptian military and defense minister 2012-13. Banned the Muslim Brotherhood and was "elected" president in 2014.

Weizmann, Chaim – President of the State of Israel 1949-52. Leader of the Zionist movement during WWI in Britain, and later worldwide. At times considered an extreme moderate. Many questioned his full commitment to a Jewish State despite his activism as regards Jewish development in the Palestine Mandate. Died 1952.

Yasin, Ahmed – Leader of the Palestinian Muslim Brotherhood and considered first leader of Hamas. Charismatic spiritual leader confined to a wheelchair. He was seen as directing Hamas terrorist operations and was killed by an Israeli airstrike in March 2004.

Yousef, Hassan – One of several current ideological leaders of Hamas, residing in the West Bank Ramallah region. Father of Mosab Yousef.

Yousef, Mosab – Hamas operative until arrested by Israel. Worked for the Israeli General Security Service and helped break the 2000-04 Intifada or LIC. Converted to Christianity and resides in the US today. Son of Hassan Yousef.

Terms

Aliya – In Hebrew literally "to ascend." Term used to describe the immigration of Jews to Israel.

Anthropocentric – Secular, human centered understandings.

Antisemitism – Dislike, prejudice, discrimination and/or persecution of Jews.

Arab Spring or Islamic Awakening 2011 – Uprisings against the Arab military regimes beginning with Tunisia followed by Egypt, Libya, Syria and in part Iraq and Yemen. Originally thought to be in the name of liberalism, democracy and human rights. Activism was replaced by Muslim Brotherhood objectives and demands for Islamic, Sharia law administered state entities.

Due to deep Islamic loyalties held by most Muslims in the Arab Middle East and superior organization by the Muslim Brotherhood, this author believes the "Arab Spring" was more of a Western media invention leaning heavily on "wishful thinking" and comparisons to European history than a realistic definition of the events on the ground. What has taken place in the last five years is an "Islamic Awakening."

Avariz – In Islam a special tax, often collected in the event of an impending war.

Glossary 439

Covenant, Israelite/Sinai – Agreement between God and the Israelites. The Israelites/Jews are to keep God's laws as outlined in the Torah or the first five books of the Hebrew Scriptures *(Tanakh)* and they will live in the Land of Israel or "Promised Land." If the covenant is broken the Israelites/Jews will be expelled.

Dajjal – In Islam the devil, who is said to be Jewish.

Dar al-Harb – In Islam, the realm of war.

Dar al-Islam – In Islam the realm of Islam, inferring peace and prosperity.

Dawa/Dawah – In Islam the "calling" to both social service to the community and service in the Holy War or Jihad.

Dhimma/Dhimmi – Refers in particular to Jews and Christians as "People of the Book" meaning their scriptures (Torah and Gospels) have positive value. According to Islam, either these were not followed correctly in the case of the Jews or were not totally correct as in the case of Christians. Overall, these two communities are allowed to follow their traditions provided they are loyal to their Islamic overlords. They suffer from special disabilities and must pay the special *jizya* tax.

Diocentric – God or Allah centered society (author's term).

Effendi – Upper class in the Arab world. Often own much land.

Erasement Theology – Term formulated by the author of this book to describe the *dhimmi* phenomenon whereby Jews and Christians were pressured to forget their histories. This is a similar policy followed by the former Soviet Union when obliterating all knowledge of Jewish history. Where a people's history and identity is eliminated, their physical existence is called into question as well.

Fard Ayn – A compulsory duty for all Muslims, such as prayer.

Fard Kifaya – A command to all, but may only be fulfilled by some.

Fatwa – A legal opinion or interpretation handed down by Islamic jurists.

Fay – Spoils taken from the infidels, or *dhimmis,* in war.

Fellah – A peasant or farmer in the Arab world.

Fitnah – When used by Jihadi Islamists and based on early Islamic understandings, is defined as "disbelief" or reverting back to one's previous non-Islamic ways.

Hanafi – The largest school of religious thought in Islamic jurisprudence. Considered the most tolerant and broad minded.

Hejira – The Prophet Mohammed's flight from Mecca to Medina in 622 CE with his first Muslim adherents. This was the result of persecution by the rulers of Mecca.

Houri – Pure, beautiful, dark-eyed women in heaven awaiting the righteous Muslim. In recent decades, the term is associated with the 72 "dark-eyed virgins" promised to a *shaheed* or Islamic martyr.

Hudna – Islamic cease-fire achieved to gain a truce to rearm, retrain, reload and restart conflict against the enemy and achieve victory. The concept is the opposite of the Western understanding of a cease-fire designed to attain conflict resolution through mutual recognition by the warring parties. A *hudna* can be agreed to for several years, but broken at will by the Muslim side to achieve victory over the infidel adversary.

Ijtihad – An Islamic legal term denoting independent thought or reasoning. Questions arise as to whether such thinking is influenced by modern reformist understandings or is a call for reverting to what is perceived as the original Islam of the 600s. Such reasoning can lead to diametrically opposed interpretations.

Iman – Faith and taking action to prove one's faith in Islam.

Jahiliyya/Jahili – Ignorance, referring to the period of time before the rise of Islam and those societies (Jahili) who still do not accept Islam, whether they be Jews and Christians as *dhimmis*, atheists or polytheists.

Jihad – Islamic Holy War. "Defensive Jihad" is to recapture all lands ever held by Islam at any point in history and "Offensive Jihad" is to capture remaining lands worldwide.

Jinn – Genies, a type of spirit, supernatural creature. Can be good or bad, Muslim or non-Muslim. They are neither angels nor humans.

Jizya – A special tax to be paid by *dhimmis* (Jews and Christians) so these communities may retain their special status. Failure to pay may lead to forced conversion to Islam or execution. Often used to fund Jihad. Found in Koran 9:29.

Karaj – Islamic land tax.

Kitman – Deception and lying to advance the cause of Islam. Similar to *taqiyya*. Emphasis as deception.

Kuffar – An infidel or non-believer in Islam. Can be a Jew or Christian *dhimmi,* but the term is often reserved for "total" non-believers such as Hindus, Buddhists and pagans.

Low Intensity Conflict/LIC – Limited use of military force against a guerrilla and/or terrorist type adversary representing a non-state or semi-state actor often working out of civilian areas. Also can be considered a "counter insurgency." What is commonly called the "Second Intifada" is considered by this author much more as an LIC and not a "popular uprising" as often portrayed in the media.

Messiah/Messianism – The Messiah is the Savior expected to bring in the End of Days, Judgment Day and/or a Day of Resurrection. The world ceases to exist as we know it. Differing interpretations as to what actions need to be taken and what will be after the Messianic arrival. For Jews and Sunni Muslims, the Messiah has not arrived. In Christianity, Jesus of Nazareth is the Messiah who died for the sins of humanity, was rejected and will return. In Shiite Islam, the Messiah is the 12th Imam who arrived and will return.

Mujahideen – Islamic Freedom Fighters. Contains same root as Jihad: j-h-d.

Mushrikun – Idolaters, polytheists, pagans or anyone denying the Oneness of Allah.

Nakba – From the Palestinian Arab perspective, the "catastrophe" of the 1948 War whereby Israel gained independence. The Palestinians expected success but ended the war without a state, having their lands divided by Israel, Egypt (Gaza) and Jordan (West Bank).

Orientalism – The study of the East by Western scholars. Somewhat of a derogatory term indicating such intellectuals view their subjects as objects and invoke a superior and patronizing attitude in their analysis. In the case of the Middle East, Edward Said makes this accusation against Western scholars researching the Arab/Muslim world.

Ottomanism – Ideal held in the last years of the Ottoman Turkish rule to unify its divergent nationalities and religions through equal rights and a common future for all within its territories. The attempt to hold the empire together failed and the Turkish State developed in its stead after WWI. Today some speak of "neo-Ottomanism," but this is defined more as an attempt by modern day Turkey to spread its influence and build alliances throughout the Middle East.

Qibla – The prayer direction in Islam. Originally, until 624, the *quibla* was Jerusalem. Since then, Muslims pray in the direction of Mecca.

Replacement Theology – Theological understanding whereby Christianity replaced Judaism. In seventh century Arabia, Islam was meant to replace both Judaism and Christianity as Divinely revealed religion.

Salafi/Salafist – The ultra-orthodox extreme right wing literalist interpretation in Sunni Islam. Purists whose more fanatical offshoots advocate a never ending Jihad. Examples of such movements are Al-Qaeda, the Islamic State (ISIS/ISIL) and affiliates. The Muslim Brotherhood including Hamas considers itself Salafist, but more modernist in application of Islamic principles.

Sgula – Virtuous. Referring to the Israelites, the covenant and the connection of the Israelite/Jews to the Promised Land or Land of Israel.

Shaheed – In Islam, a martyr.

Sharia Law – Islamic law as determined by the Koran and for most Muslims the Hadith as well. The law is given to interpretation by Islamic jurists.

Shiite – Makes up somewhat over 10 percent of the Islamic population globally. Iran is the center of Shiism. The Koran and certain Hadith collections are the foundations of the belief. Shiites believe Ali, Mohammed's son-in-law and cousin should have been the caliph to replace the Prophet upon his death. Caliphs were to be from the family line of Mohammed.

Shirk – Idolatry or polytheism, in opposition to monotheism.

Sunni – Orthodox or mainstream Islam comprising close to 90 percent of Muslims worldwide. Accepts that caliphs not from the direct line of Mohammed would serve as the replacement leadership after the Prophet's death. The Koran and Hadith are the basis in this Islamic understanding.

Surgun – Forced population resettlement or exile under the Ottoman Empire. Usually enforced on *dhimmi* communities, but Muslims were also transferred to fulfill government policy objectives.

Tahadiya – Calm, usually instituted after conflict and before a *hudna*.

Tanzimat – In Turkish, literally "reorganization." Refers to the modernization reforms developed in the Ottoman Empire 1839-76. Included reforms in government, taxation, the military, equal rights, etc. For the purposes of this book, the 1856 cancellation of the *dhimma* status and accompanying measures bringing equality to minorities were the most significant reorganizations.

Taqiyya – Deception and lying to advance the cause of Islam. Similar to *kitman*. Emphasis on lying.

Ulama – Religious scholars in the clerical hierarchy of Islam.

Ummah – The Islamic nation, community or people.

Wahhabism – Reactionary movement in Sunni Islam originating in the Najd region of the eastern Arabian Peninsula (including oil producing Persian Gulf area) during the eighteenth century. Was the ideological basis for the

Sunni uprising and invasion into Iraq in the nineteenth century and is the foundational ideology for much of the Muslim Brotherhood and especially the Islamic State (ISIS or ISIL). Wahhabism is basic to Saudi religious understandings today.

Waqf – Endowed Islamic lands. Any land ever captured by Muslims belongs to Islam forever, these lands being captured in the name of Allah.

Zionism – Jewish nationalism and the activity to establish a Jewish State in the Land of Israel in the modern period beginning in the early 1880s and formulated politically in the late 1890s by Theodor Herzl and the Zionist Congresses.

Documents

Allon Plan – A territorial compromise plan developed by Yigal Allon after the 1967 Six Day War. Israel would annex the eastern part of the West Bank along the Jordan River as a territorial and military buffer against an attack from the east (Iraq and Jordan) and retain the areas surrounding Jerusalem. Basis of Labor party settlement policies 1967-77. The remaining territories were to be negotiated back to Jordan, including Gaza with a road connecting between the coast to the West Bank. All Palestinian/Jordanian areas were to be demilitarized. The Jordan River as Israel's "security border" is still retained by Israel's defense establishment today.

Bush "Road Map" – Developed by President George W. Bush in 2003. Called for a negotiated settlement between Israel and the Palestinians during the Second Intifada or Low Intensity Conflict. Envisions a halt to hostilities and settlement activities, a temporary Palestinian State established and finally a permanent status agreement arranged handling borders, security, Jerusalem, refugees and all other outstanding issues. Objective of mutual recognition between Israel and a Palestinian State co-existing leading to "End of Conflict" scenario.

Camp David Accords (1978-79) – Two accords negotiated by Israel and Egypt. Peace agreement between Israel and Egypt whereby Israel withdraws from all of Sinai while full recognition and diplomatic relations are established between the two nations. The Framework for Peace in the Middle East outlines the need for peace between Israel and its Arab neighbors while advocating an interim agreement for Palestinian autonomy.

Charter of Omar – Legal stipulations and restrictions as relates to the *dhimmi* second class status of Christians and Jews. Initiated by Omar II in the early 700s.

Clinton Parameters – Framework for permanent status peace agreement between Israel and the Palestinians presented by US President Clinton in December 2000. Israel was to withdraw fully from Gaza, 95 percent of the West Bank and split Jerusalem with the Palestinians. Joint security to be arranged and Palestinian refugees to move to the Palestinian state in the making. Compensation instead of right of return along the guidelines of UN Res. 194 clause 11 (December 1948). Shared sovereignty in Temple Mount area of Jerusalem's Old City. Accepted by Israel, rejected by Palestinians.

Declaration of Principles/DOP – See Oslo Accords

Hadith – Among Islamic scriptures, second in importance to the Koran. A collection of narratives relating to the Prophet Mohammed as recounted by witnesses, passed down through the ages and recorded by scribes. Contributes to the basis of Sharia Law.

***The Hamas Covenant*/HC (1988)** – Also known as the "Charter of Allah." Statement of ideals, values, objectives and purpose of Hamas. Islamist, Jihadi, antisemitic, anti-Western and anti-secular document of faith and action.

Koran – Primary Islamic Scriptures. Holy book for Muslims. Origins of Sharia (Islamic) law.

Mein Kampf – Literally "My Struggle." Written by Adolf Hitler explaining his worldview as pertained to his racist ideals and the future glory of Germany. His hatred of the Jews was made vividly clear to all.

Oslo Accords – Signed in 1993 in Washington between Israeli Prime Minister Yitzhak Rabin and PLO Chairman Yasir Arafat. Negotiations began in Oslo, Norway between Israeli and Palestinian academics and lower level diplomats. Included was the initial Declaration of Principles (DOP) where Israelis and Palestinians were to work toward conflict resolution. Considered a "land for peace" arrangement, in essence the Palestinians pursued sovereignty while Israel sought security. Negotiations and partial agreements continued until the year 2000, but ended with the Second Intifada or Low Intensity Conflict 2000-04.

The Palestinian National Charter/PNC (1968) – Document representing the beliefs, objectives and actions to be taken by the Palestine Liberation Organization (PLO). Secular purpose, objectives and call to action of Palestinian Arab nationalism.

Partition Plan – UN Resolution 181 of November 29, 1947, which called for a two-state solution in the Palestine Mandate to include Jewish and Arab States living side by side as two political entities and one economy. Jerusalem and Bethlehem were to be internationalized. The Partition Plan was accepted by the Zionists and rejected by the Arabs.

The Protocols of the Elders of Zion – Czarist police forgery from 1903, which purported to prove a Jewish conspiracy to conquer the world. The supposed "plot" dates back to King Solomon and culminates with Zionist leaders in the early twentieth century. Also referred to as "*The Protocols of the Learned Elders of Zion.*"

Saudi Peace Plan (2002) – Saudi/Arab League initiative to end the conflict between Israel and the Palestinians. Israel to withdraw to the 1967 borders while Palestinian "refugee return" appeared imminent as a foundation of the proposal and not particularly compensation (UN Res. 194, Dec. 1948). Palestinian State in West Bank and Gaza with East Jerusalem as its capital. Arab world offered full peace. Israel rejected the plan, Palestinian Authority accepted, Hamas rejected.

Tanakh – Hebrew Scriptures or Old Testament. Holy book of the Jewish People. Origins of Halacha or Jewish religious law, which is defined by the 613 commandments found in the *Torah*, or Five Books of Moses, the first five books of the *Tanakh* or Hebrew Scriptures.

Wye Accords – Signed in 1998 by Israeli Prime Minister Benyamin Netanyahu and PA Chairman Yasir Arafat. Overall, Israel was to relinquish another 13-14 percent of the West Bank to Palestinian control, and the Palestinians were to bolster security and change *The Palestinian National Charter* eliminating calls for Israel's destruction. Though signed, the Wye Accords were not implemented.

Places

Al-Aksa Mosque and Dome of the Rock Domain – Al-Aksa is the "Outer" Mosque or Temple in Jerusalem, as recounted in Chapter 17 "The Night Journey" in the Koran. Inferred is that the Prophet Mohammed journeyed from Mecca to Jerusalem. It is clearly stated that this Temple compound belonged to the Israelites in the past and was destroyed because of their sins. The golden Dome of the Rock stands over the ruins of both of the Jewish Temples.

Cave of the Machpela/Ibrahimi Mosque – A holy site in Hebron where tradition holds that Abraham, Isaac, Jacob, Sarah, Rebecca and Leah are buried. Many Jews believe Adam and Eve are buried here as well. Today the holy site is divided between Jews and Muslims.

Gaza Strip – A narrow landmass bordering Sinai to its southwest, Israel to the east and north, and the Mediterranean Sea to the west. In the 1947 UN Partition Plan, the Gaza Strip was allotted as part of the western Negev to the Palestinian Arab State-to-be.

Taken by Egypt in the 1948 War, then captured by Israel and returned to Egypt in 1956, and taken again by Israel in 1967. Part of the Palestinian Authority as of 1994, but overrun by Hamas in 2007. Since then the Gaza Strip has been a semi-independent non-recognized mini-state ruled by Hamas.

Israel, Land of – In Jewish tradition, the "Promised Land." Minimal and most accepted geographical contours are from the Jordan River to the Mediterranean Sea. Generally, the *Tanakh* or Hebrew Scriptures speak of from "Dan to Beersheva" while other understandings add lands to the east of the Jordan River and in the west to the "brook of Egypt" usually interpreted as the El Arish river. Extreme maximalists speak of from the Nile to the Euphrates.

Israel, State of – Political nation state entity representing the culmination of Jewish nationalism as expressed thru the Zionist movement efforts in the Land of Israel. International recognition of armistice lines borders (1949-67). Officially comprises 78 percent of the Palestine Mandate. Has partial military control in the West Bank and fully withdrew from Gaza in 2005. Retains the Golan Heights since 1967.

Jerusalem – Holy city to Jews, Christians and Muslims. In the 1947 UN Partition Plan Jerusalem was designated as an international zone along with Bethlehem. West Jerusalem was the capital of Israel since 1948. During the 1948 War, the eastern side, including the Old City and holy sites, were captured by Jordan. After 1967 Israel unilaterally "unified" the city as its "eternal" capital. Fatah and the Palestinian Authority claim the eastern side as their capital while Hamas demands Jerusalem as a Muslim domain.

Jewish National Home – Promised to the Zionist organization by Great Britain in the Balfour Declaration (1917) during WWI. Implemented through the British administered Palestine Mandate as approved by the League of Nations in the 1920s. Did not grant the Jews a state, but allowed the framework for Zionist development necessary before the establishment of the State of Israel.

Knesset – Israel's parliament containing 120 members elected by proportional voting. For example if a political faction wins 10 percent of the vote, they receive 12 seats.

Palestine – Derived from the word "Philistine," the Romans renamed ancient Judea in the aftermath of the Bar Kokhva Revolt (132-135 CE). The region from the Jordan River to the Mediterranean is claimed by both the Jewish and Palestinian national movements. Hamas claims this land in the name of the Islamic *waqf*, or lands previously captured by Islam.

Palestine Mandate – Administered by the British 1920-48. Originally included both the region from the Jordan River to the Mediterranean Sea, the Negev Desert, and what is today the Hashemite Kingdom of Jordan. By 1922 would only include the region from the Jordan River to the Mediterranean.

Palestine, State of – As approved by the United Nations General Assembly since November 2012 to include the West Bank and Gaza as an independent Palestinian State existing side by side with Israel. Not approved by the Security Council due to a US veto and demand that details for conflict resolution be resolved in direct talks between the two parties. Palestine has UN "non-member observer state" status.

Temple Mount/Noble Sanctuary (Haram al-Sharif) – In Judaism, Mt. Moriah, traditionally where Abraham went to sacrifice his son Isaac and was halted by the angel. Also site of the ancient Israelite-Judean (Jewish) Temples destroyed in antiquity. Known as Haram al-Sharif or Noble Sanctuary in Islam. Houses the Al-Aksa Mosque and Dome of the Rock.

West Bank – Part of the landmass west of the Jordan River allocated to the Palestinian Arab State in the 1947 UN Partition Plan. Captured by Jordan in 1948 and annexed in 1950. Captured by Israel in the 1967 War and jointly administered by Israel and the Palestinian Authority since 1994. Seat of the Palestinian Authority (Ramallah). Approximately 98 percent of the population (Areas A and B) is de facto ruled by the secular Fatah faction led by Mahmoud Abbas.

Conflicts (chronologically)

1948 War – Israel's War of Independence, Palestinian Arab "Nakba" or catastrophe. Arab initiative taken against Jews beginning November 30, 1947. Jewish counter attack only succeeded in April-May 1948. Five Arab armies invaded (Egypt, Jordan, Syria, Lebanon and Iraq plus contingents from other Arab countries and irregular forces) upon declaration of independence by Israel. Israeli counter offensive and eventual outcome led to the State of Israel, but no Palestinian Arab polity. Gaza taken by Egypt, West Bank by Jordan, and other areas by Israel. Armistice lines determined by negotiations between the Arab States and Israel in 1949.

1967 War – Threatened invasion by Egypt, Jordan and Syria led to Israeli pre-emptive attack. Battles lasted six days. Israel captured Sinai and Gaza (Egypt), West Bank and East Jerusalem (Jordan) and the Golan Heights (Syria).

1973 War – Also known as the Yom Kippur War or in the Arab world as the Ramadan War. Joint Egyptian-Syrian surprise attack against Israel, initially successful. Israeli counter attack determined battlefield outcome in Israel's favor, but there were massive Israeli diplomatic failures in the aftermath of the war. Led to turning point in Israel-Arab relations, eventual peace treaty with Egypt and disengagement of forces arrangement with Syria.

First War in Lebanon (1982) – Israeli initiative to eliminate PLO security threat emanating from south Lebanon. Yasir Arafat and the PLO were banished. The conflict was a factor in the rise of Hezbollah who replaced the PLO as Israel's adversary. Full Israeli withdrawal to the international border in May 2000.

First Intifada (1987-1991) – Considered popular Palestinian uprising against Israel in the West Bank and Gaza. Took Israel and PLO by surprise and became the background for official establishment of Hamas. Israel said to have regained control. Was the springboard for Oslo Accords between Israel and Palestinians. Hamas gained inroads, commencing the organization's rise to prominence.

Second Intifada/Low Intensity Conflict/LIC (2000-04/5) – Broke out at the end of September 2000 after the failure of negotiations at Camp David that summer between Israeli Prime Minister Ehud Barak and PA Chairman Yasir Arafat. Overall directed conflict, first by Arafat and then in conjunction with Hamas. Organized terror attacks against Israeli civilians and security forces. Israeli Army (IDF) engaged in counter insurgency, anti-terror response. Virtual recapture of West Bank by Israel (after Oslo agreements) and near collapse of the Palestinian Authority.

Second War in Lebanon (2006) – July/August conflict began with Hezbollah attack on Israel's northern border, several soldiers killed and two abducted, which later died of wounds. Quick escalation led to 4000 Hezbollah rockets fired into Israel and some 12,000 Israeli air strikes. Israeli ground operations into south Lebanon were only partially successful.

Cast Lead (December 2008-January 2009) – First conflict between Israel and Gaza, lasting three weeks. Short (Qassam) and medium (katusha) range Hamas rockets versus Israeli artillery, air strikes and ground assault. Resulted in UN Goldstone Report condemning Israel.

Pillar of Defense (November 2012) – Clash between Israel and Palestinians in Gaza, lasting eight days. Palestinian rockets versus Israeli air assault. No ground invasion by Israel.

Protective Edge (July-August 2014) – Clash between Israel and Palestinians in Gaza, lasting 50 days. Israeli air attacks and ground invasion. Hamas rockets landed as far as central Israel. Hamas weakened militarily but retained and even increased popular Palestinian support. Thousands of Israeli air and artillery strikes.

People Groups and Organizations

al-Qaeda – Fanatical Jihadi terrorist organization established by Osama bin Laden, who was killed by American forces in May 2011. Ideology of Islamic world conquest, spawned the Islamic State (ISIS/ISIL). Responsible for attacks throughout the Middle East and Africa beginning in the late 1990s and the 9/11 assaults on the US.

Armenians – Christian Orthodox Church national group living in northeast Ottoman Empire, geographically the Caucus mountains around Mt. Ararat until WWI. One and a half million killed by the Ottomans in WWI, virtually none left in modern Turkey. Independent state on the other side of the Turkish northeast border as a result of the collapse of the Soviet Union.

Assyrian Christians (includes Nestorians and Chaldeans) – Eastern Churches suffered massacres totaling into the 100,000s at the hands of Turks, Kurds, Circassians and Arabs during and just after WWI. Domicile was and still is for the few remaining in southeast Turkey, northern Iraq and Syria of today. Assyrians suffered massacres and expulsions in recent years, especially in Iraq by Jihadi and Islamic State types.

Democratic Front for the Liberation of Palestine (DFLP) – representing extreme left wing secular Maoist views. Organized by a Christian Palestinian Naif Hawatmah in 1969, broke from the PFLP (see below) and a member organization in the PLO.

Fatah – Palestinian guerilla and terror organization established by Yasir Arafat in 1965 to destroy the State of Israel and replace it with a Palestinian Arab State. Evolved politically and became the largest faction in the PLO. Fatah is the political leadership of the PA.

Hamas – Palestinian Muslim Brotherhood. Follows Sharia law and challenges the secular Palestinian organizations such as the PLO/Fatah. Victorious in the Palestinian Authority elections of 2006. Rule in Gaza after banishing the Fatah dominated PA in 2007. Involved in terrorist activity against Israel seeking destruction of the Jewish State.

Israel Defense Forces/IDF – The Israeli Army.

IHH – Turkish NGO and Islamist organization whose official mandate is to provide humanitarian relief. In May 2010 the IHH sent the Mavi Marmara flotilla to Gaza and was intercepted by the Israeli navy due to suspicions that military contraband was being supplied to Hamas.

Islamic Jihad – Smaller, more fanatical terrorist organization established 1981. Break off from the Palestinian Muslim Brotherhood (eventually Hamas) and remains a separate military but non-political faction to this day. In early 1980s accused Hamas of not taking armed action to destroy Israel. Credited with forcing Hamas into physical Jihadi activity. Almost entirely a guerilla terrorist operation, but with educational wings the organization is only Jihadi and not designed to be a political faction.

Islamic Resistance Movement – See Hamas. Arabic acronym for "Hamas"

Islamic State – Also known as ISIS (Islamic State in Iraq and Syria) or ISIL (Islamic State in Iraq and the Levant). Most fanatical Islamic "state" in existence. Follows the strictest Salafist and al-Qaeda interpretations of Sharia law. No recognition of state boundaries, fully enforces the *dhimma* status on Christians and discriminatory laws against women. Supporters challenge state authorities throughout the Middle East including Hamas in Gaza. Occupies eastern Syria and western Iraq. Known for media coverage of its own brutality.

League of Nations – International organization established after WWI and similar in intent and function to the UN. Failed to halt hostilities leading to WWII.

Labor Party – Mainstream Israeli left of center political party, moderate on foreign policy. Favors mixed economy on domestic issues.

Likud – Mainstream Israeli right of center political party, hawkish on foreign policy. Capitalist on domestic issues.

Muslim Brotherhood – Islamic fundamentalist organization established by Hasan al-Banna in Egypt in 1928, at first to resist the British. Is an Islamist activist organization opposing the secular Arab nation state, European influences and the State of Israel. Capitalized on the uprisings of 2011 making inroads in Iraq, Syria and Libya through Islamic State type activism and in Tunisia and Egypt through elections. Was entrenched previously in Sudan and Gaza (a mixture of elections and military force).

National Unity Government/NUG – A National Unity Government occurs when two major opposing sides (generally left vs. right, or religious vs. secular) form a government together because neither side can gain a majority and ensure regime stability. The example in Israel is when Labor (left) and the Likud (right) formed a coalition. In Palestinian politics it occurs when Fatah/PLO (secular) and Hamas (religious) worked together in a similar manner.

Ottoman Empire – The last great empire in the Islamic World. At its height in the 1500s–1600s, it spread from North Africa to the Balkans, as far east as Iraq, and south to Yemen in the Arab Middle East. Lost territory and power throughout the nineteenth century, especially in the Balkans. Final destruction by the Entente Powers (Britain in particular) during WWI. Replaced at its center by the modern Turkish State. Today, the Middle East, North Africa and the Balkans are ruled by independent nations.

Palestinian Authority/PA – Quasi-state regime established by the Palestinians under Yasir Arafat in 1994 as a result of the Oslo Accords. Established as Israel's negotiating partner to arrive at a permanent status agreement between Israel and the Palestinians. Officially, the PA agrees to a two-state solution.

Palestine Liberation Organization/PLO – Established in 1964 as a framework for a Palestinian Arab government in exile. By 1969 was unified with the Fatah, PFLP and DFLP armed factions in an overall umbrella group under Yasir Arafat. The PLO remains the largest Palestinian secular organization and considered internationally as the "sole legitimate representative of the Palestinian People." With the establishment of the Palestinian Authority in 1994 and the rise of Hamas the claim is called into question.

Popular Front for the Liberation of Palestine/PFLP – Represents Marxist-Leninism secular and anti-clerical positions. Established in 1967, and is a member of the PLO. Led by the Christian Palestinian George Habash.

Supreme Muslim Council/SMC – The ruling body responsible for Muslim affairs in the British administered Palestine Mandate. Established in 1921, its most dominant figure was Haj Amin el-Husseini who was forced to flee abroad after the outbreak of the 1936 Arab Revolt.

United National Leadership of the Uprising/UNLU – Considered generally moderate local Arab leadership in the West Bank and Gaza that organized the First Intifada at the outset in December 1987. Responsible for demonstrations and tax strikes. Was eventually overtaken by PLO/Fatah and Hamas activism.

United Nations High Commissioner for Refugees/UNHCR – General UN commission for refugees worldwide. A person does not inherit refugee status.

United Nations Relief Works Agency/UNRWA – Established December 1949 to aid refugees from the 1948 War and the later 1967 conflict. UNRWA officials are 99 percent Palestinian Arab and through their mandate also render assistance to the refugee descendants. Presently 5 million Palestinian Arabs claim refugee status.

Bibliography

Bibliography appears online at
www.HamasJihad.com

Books

Abitbol, Michel, *The Jews of North Africa During the Second World War*, Wayne State University Press, Detroit, Michigan, 1989.

Ahroni, Reuben, *Yemenite Jewry*, Indiana University Press, Bloomington, Indiana, USA, 1986.

Alexander, Yona, *Palestinian Secular Terrorism*, Transnational Publishers, Ardsley, New York, USA, 2003.

Antonius, George, *The Arab Awakening*, Capricorn Books, New York, USA, (1946) 1965.

al-Banna, Hasan, *The Five Tracts of Hasan al-Banna*, translated by Charles Wendell, University of California Press, 1978.

Azzam, Abdullah, *Defence of the Muslim Lands,* Internet version in *Religiscope, www.religioscope.com/info/.../azzam_defence.htm.*

Azzam, Abdullah, *Join the Caravan,* Internet version in *Religiscope, www.religioscope.com/.../azzam_caravan.htm.*

Baron, Salo W., *The Russian Jew Under Tsars and Soviets*, Schocken Books, New York, 1987.

Bat Ye'or, *Islam and Dhimmitude Where Civilizations Collide,* translated from the French by Miriam Kochan and David Littman, Fairleigh Dickinson University Press, Teaneck NJ, USA, 2002.

Bat Ye'or, *The Dhimmi, Jews and Christians Under Islam,* translated from the French by David Maisel and Paul Fenton, Fairleigh Dickinson University, Associated University Presses, NJ., USA, 1985.

Ben Dov, Meir, *Jerusalem, Man and Stone,* Modan Publishing House, Tel Aviv, Israel, 1990.

Benvenisti, Meron and Khayat, Shlomo, *The West Bank and Gaza Atlas,* The Jerusalem Post, Jerusalem, 1988.

Bostom, Andrew, (ed.) *The Legacy of Jihad, Islamic Holy War and the Fate of Non-Muslims,* Prometheus Books, Amherst, New York, USA, 2005.

Bostom, Andrew, *The Legacy of Islamic Antisemitism,* Prometheus Books, Amherst, New York, USA, 2008.

Brinton, Crane, *The Anatomy of Revolution,* Vintage Books, New York, USA, Expanded Edition, 1965.

Chehab, Zaki, *Inside Hamas,* Nation Books, New York, N.Y., 2007.

Chouraqui, Andre N., *Between East and West, A History of the Jews of North Africa,* Temple Books, Atheneum, New York, 1973.

Dawidowicz, Lucy, *The War Against the Jews 1933 - 1945*, Bantam Books, New York, NY, USA, 1976.

Dershowitz, Alan, *The Case for Israel,* John Wiley and Sons Inc., Hoboken, NJ, 2005.

Eisenhower, Dwight D., *Crusade in Europe*, Doubleday, Garden City, New York, 1948.

Eldar, Shlomi, *Getting to Know Hamas,* Keter Publishers (Hebrew), Israel, 2012.

Enderlin, Charles, *Shattered Dreams,* Other Press, New York, USA, 2002.

Gilbert, Martin, *In Ishmael's House, A History of Jews in Muslim Lands,* Yale University Press, New Haven and London, 2010.

Goitein, S. D., *Arabs and Jews, Their Contacts Though the Ages,* Schocken Books, New York, USA, 1974.

Gunning, Jeroen, *Hamas in Politics: Democracy, Religion, Violence,* Hurst & Company, London, 2007.

Gurr, Ted, *Why Men Rebel,* Woodrow Wilson School of International Affairs, Princeton University Press, New Jersey, 1970.

Hanus, George D., *The Compendium,* Gravitas Media, Chicago, USA, 2002.

Harkabi, Yehoshafat, *Palestinians and Israel,* Keter Publishing House, Jerusalem, 1974.

Harris, Sam, *The End of Faith,* W.W. Norton, New York, USA, 2004.

Herzog, Chaim, *The Arab-Israeli Wars,* Steimatzky's Agency Ltd., Tel Aviv, Israel, 1984.

Hirszowicz, Lukasz, *The Third Reich and the Arab East,* Rutledge and K. Paul, London, 1966.

Hillel, Shlomo, *Operation Babylon,* translated by Ina Friedman from the Hebrew, Fantana/Collins, Glasgow, Great Britain, 1988.

Hitler, Adolf, *Mein Kampf,* originally published 1925/26, accessed online at *Hitler Historical Museum,* http://hitler.org/writings/Mein-Kampf.

Hourani, Albert, *A History of the Arab Peoples,* Fabar and Fabar, London, England, 1991.

Howard, Michael, *War and the Liberal Conscience,* Oxford University Press, England, 1978.

Hroub, Khalid, *Hamas Political Thought and Practice,* Institute for Palestine Studies, Washington DC, 2002.

Huntington, Samuel P., *The Clash of Civilizations and the Remaking of the World Order,* Simon and Schuster, New York, USA (original 1996), 2003.

Jensen, Michael Irving, *The Political Ideology of Hamas: A Grass Roots Perspective,* translated from the Danish by Sally Laird, I.B. Tauris, London and New York, 2009.

Kadourie, Elie and Haim, Sylvia, eds., *Zionism and Arabism in Palestine and Israel,* F. Cass, London, 1982.

Khomeini, Ruhallah, *Islam and Revolution I,* translated and annotated by Hamid Algar, Mizan Press, Berkley CA, USA, 1981.

Kohen, Eli, *A History of the Turkish Jews and Sephardim: Memories of a Past Golden Age,* University Press of America, Maryland, 2007.

Kurzman, Dan, *Genesis 1948: The First Arab Israeli War,* Sefer Ve Sefel Publishing, Jerusalem, Israel, (original copyright 1970), 2005.

Landau, Jacob, *Jews, Arabs and Turks,* Magnum Press, Jerusalem, 1993.

Laqueur, Walter Zeev, *A History of Zionism,* Schoken Books, New York, NY, USA, 1976.

Laqueur, Zev and Rubin, Barry, eds., *The Israel-Arab Reader,* Penguin Books, New York, NY, 1995.

Lebel, Jennie, *The Mufti of Jerusalem, Haj Amin el-Husseini and National Socialism,* Cigoja Stampa Publishers, English translation, Paul Munch, Belgrade, 2007.

Levitt, Matthew, *Hamas: Politics, Charity and Terrorism,* Yale University Press, New Haven and London, 2006.

Lew, Uzrad, *Inside Arafat's Pocket,* (Hebrew) Kinneret, Zmora-Bitan, Dvir, Israel, 2005.

Lewis, Bernard, *Semites and Anti-Semites,* WW Norton and Co., New York and London, 1999.

Lewis, Bernard, *The Crisis of Islam, Holy War and Unholy Terror,* Phoenix Paperback, London, UK, 2004.

Lewis, Bernard, *The Arabs in History,* Harper and Row Publishers, New York, 1966.

Lewis, Bernard, *What Went Wrong?* Harper Perennial, New York, USA, 2002.

Maududi, Sayyid Abul A'la, *West versus Islam,* translated by S. Waqar Ahmad Gardezi and Abdul Waheed Khan, Markaz Maktaba Islami Publishers, New Delhi, India, 2005.

Manji, Irshid, *The Problem With Islam Today,* St. Martin's Griffin Press, New York, USA, 2003.

Mishal, Shaul and Sela, Avraham, *The Palestinian Hamas, Vision, Violence and Coexistence,* Columbia University Press, New York, USA, 2000.

Morris, Benny, *Righteous Victims,* Vintage Books, New York, USA, 2001, p. 572 and Mishal and Sela.

Muslih, Mohammed, *The Origins of Palestinian Nationalism,* Institute for Palestinian Studies, Washington DC, 1989.

Nettler, Ronald, *Past Trials and Present Tribulations: A Muslim Fundamentalist's View of the Jews,* Vidal Sasson International Study of Antisemitism, Hebrew University, Jerusalem, Israel, 1987.

Newby, Gordon Darnell, *A History of the Jews in Arabia, From Ancient Times to Their Eclipse Under Islam,* University of South Carolina Press, 1988.

Porath, Yehoshua, *The Emergence of the Palestinian National Movement 1918-1929,* Frank Cass, London, 1974.

Porath, Yehoshua, *The Palestine Arab National Movement 1929-1939*, Frank Cass, London, 1977.

Parkes, James, *The Conflict of the Church and the Synagogue*, A Temple Book, Atheneum, New York, 1969.

Peters, Joan, *From Time Immemorial, the Origins of the Arab Jewish Conflict Over Palestine*, JKAP Publications, USA 1984.

Patterson, David, *A Genealogy of Evil, Anti-Semitism from Nazism to Islamic Jihad*, Cambridge University Press, New York, NY, USA, 2010.

Rapoport, Louis, *Stalin's War Against the Jews, The Doctor's Plot and the Soviet Solution,* The Free Press, New York, 1990.

Qutb, Sayyid, *Milestones,* Internet version from *Studies in Islam and the Middle East: SIME Journal* - USA: SIME journal, http://majalla.org/books/2005qutb-nilestone, - retrieved January 5, 2010.

Rejwan, Nissim, *The Jews of Iraq, 3000 Years of History and Culture,* Weidenfeld and Nicolson, London, England, 1985.

Ross, Dennis, *The Missing Peace*, Farrar, Straus and Giroux, New York, USA, 2005.

Rubin, Barry and Rubin Judith Colp, *Yasir Arafat, A Political Biography,* Oxford University Press, New York, 2005.

Sachar, Howard, *A History of Israel From the Rise of Zionism to Our Time,* Alfred A. Knopf, New York, 2007.

Said, Edward W., *Orientalism,* Vintage Books, New York, USA, 1978, Afterward, 1994, Twenty-Fifth Anniversary Edition, 2003.

Safrai, Zeev, *The Galilee in the Time of the Mishna and the Talmud,* (Hebrew), Ministry of Education and Culture, Israel, 1985.

Sharan, Shlomo, and Bukay, David, *Crossovers Anti-Zionism and Anti-Semitism*, Transaction Publishers, New Brunswick, New Jersey, 2010.

Schiff, Ze'ev, and Ya'ari Ehud, *Intifada,* translated by Ina Friedman, Simon and Schuster, New York, 1990.

Shirer, William, *The Rise and Fall of the Third Reich,* Simon and Schuster, New York, 1960.

Shor, Natan, *Toldot Tzfat* - History of Safed, (Hebrew) Dvir Co. and Am Oved Publishers, Israel, 1983.

Taggar, Yehuda, *The Mufti of Jerusalem and Palestine Arab Politics 1930-1937,* University of London, London, 1973.

Taheri, Amir, *Holy Terror, The Inside Story of Islamic Terrorism,* Sphere Books Limited, Great Britain, 1987.

Tamimi, Azzam, *Hamas, A History from Within,* Olive Branch Press, Northampton, Mass, USA, 2007.

Thompson, Jason, *A History of Egypt,* American University in Cairo Press, Cairo, Egypt, 2008.

Weizmann, Chaim, *Trial and Error,* Hamilton LTD, London, 1949.

Yousef, Mosab Hassan (with Ron Brackin), *Son of Hamas,* Tyndale House Publishers, USA, 2010.

Texts and Documentation

The Jerusalem Bible, (Old Testament or Hebrew Scriptures), English text revised and edited by Harold Fisch, Koren Publishers, Ltd., Jerusalem, 1969.

The Koran, translated by N.J. Dawood, Penguin Classics, Penguin Books, New York, 1977.

Latrun Armor Museum, Israel
    Exhibit on Jewish Soldiers in World War II

Memri - The Middle East Media Research Institute - www.memri.org
  Hamas Covenant, translation by Memri. Special Dispatch - No. 1092, February 14, 2006.

Memri clips - www.memritv.org
  Gaza Speech by Hamas PM Ismail Haniyeh, Dec. 14, 2011
  We are a Nation of Jihad and Martyrdom, by Hamas PM Ismail Haniyeh, November 15, 2010

Israel's Foreign Relations, Selected Documents, 1947-1974, Medzini, Meron, ed., Ministry for Foreign Affairs, Vol. I, Isratypeset, Jerusalem, Israel, 1976.

Mideast Web Historical Documents - www.mideastweb.org
  Fayyad, Salam, "Ending the Occupation, Establishing the State," August 26, 2009
  Agreement Between Emir Faisal and Dr. Weizmann, January 3, 1919

The Palestinian National Charter in Kadi, Leila S. (ed.), "*Basic Political Documents of the Armed Palestinian Resistance Movement*," Palestine Research Centre, Beirut, December 1969.

The Protocols of the Learned Elders of Zion
  www.solargeneral.org/wp-content/uploads/library/protocols-of-zion.pdf

United Nations - www.un.org

Flotilla Incident
  www.un.org/News/dh/infocus/middle.../Gaza_Flotilla_Panel_Report.pdf
  Report on the Secretary-General's Panel of Inquiry of the 31 May Flotilla Incident

United Nations Human Rights Office of the High Commissioner
  www.ohchr.org

United Nations Fact Finding Mission on the Gaza Conflict
  Convention relating to the Status of Refugees to Palestinian Refugees (1951)

Yad VaShem Holocaust Memorial and Museum: Exhibits
 Testimonies of North African Jews
 Book Burning (short film)

Encyclopedias/Dictionaries/Factbooks
Printed and Online

Britannica - www.britannica.com
 Totalitarianism

CIA World Book (CIA Factbook)
 www.cia.gov/library/publications/the-world-factbook

Encyclopedia.com - www.encyclopedia.com
 Morrison-Grady Plan (1946)

Encyclopedia Judaica, Keter Publishing House, Jerusalem, LTD, P.O.B. 7145, Jerusalem, Israel, 1972.
 Articles:
  Omar, Covenant of
  Persia
  Iraq

Geneva Accord - www.geneva-accord.org
 Joint Israeli-Palestinian Poll - Harry S. Truman Institute for the Advancement of Peace

Israel Ministry of Foreign Affairs - www.mfa.gov.il
 UN Resolution 242
 Israel-PLO Recognition-Exchange of Letters Between PM Rabin and Chairman Arafat- Sept. 9, 1993
 Camp David Accords
 Declaration of Principles
 Terrorist Bombing at Hebrew University Cafeteria
 Exchange of Letters Between PM Sharon and President Bush
 Main Terrorist Attacks Carried Out at Gaza Strip Crossings
 Address by PM Netanyahu at Bar Ilan University

Merriam Webster - www.merriam-webster.com
   Tolerance

The New Hebrew Dictionary, Even Shoshan, Abraham, "Kiryat-Sefer" Ltd., Jerusalem, Israel, 1992.

Theodora - www.theodora.com/wfb/abc_world_fact_book.html
   Gaza Strip Economy 2011

Wikipedia the free encyclopedia, https://en.wikipedia.org.
   To access, Google "Wikipedia" and the name of entry below in quotes.
   (Example, to find Azzam, Abdullah, Search for: Wikipedia "Azzam Abdullah")

   Articles:
   2014 Israel Gaza Conflict
   Azzam, Abdullah
   Assyrian Genocide
   Casualties of Gaza War
   Dawah
   East Jerusalem
   Fatah-Hamas Conflict
   Battle of Gaza (2007)
   Fida'i
   Gaza War
   Greco-Turkish War
   Inter-Services Intelligence in Afghanistan
   Islamization of Gaza Strip
   Karine A. Affair
   Jews in the Byzantine Empire
   Operation Pillar of Defense
   Palestinian General Election 1996
   Palestinian General Election 2006
   Palestinian Refugee
   Qaradawi, Yusuf
   Refugee
   Toynbee, Arnold
   Valley of Peace Initiative

Wikisource - www.wikisource.org
   Constitution of Palestine

YIVO Encyclopedia of Jews in Eastern Europe - www.yivoencyclopedia.org
   World War I

Media
Includes print, online, and TV
To access, Google URL and article name. (For example, search for: www.al-monitor.com "Egypt Floods Gaza's Smuggling Tunnels")

Al-Arabiya
   https://english.alarabiya.net
   Abbas backs Egyptian crackdown on Gaza tunnels, Dec. 12, 2014

Al-Monitor: The Pulse of the Middle East
   www.al-monitor.com
   Egypt Floods Gaza's Smuggling Tunnels, Feb. 19, 2013

Associated Press
   www.ap.org
   Palestinian Authority Admits: Warfare was Planned, March 4, 2001

CBN News
   www.cbnnews.com
   Wage, John, "Hamas Leaders Enforce Sharia Law in Gaza Strip" Oct. 12, 2009

Haaretz
   www.haaretz.com
   Haaretz, "Word of the Day/Hamas the Terror Movement that didn't do its Homework," no date
   Stern, Yoav and Issacharoff, Avi, "Major Hezbollah Attack in Sinai Thwarted," April 19, 2009
   Hasson, Nir "Demographic Debate Continues // How Many Palestinians Actually Live in the West Bank?" Oct. 1, 2014

Iris (Information Regarding Israel's Security)
   www.iris.org.il
   Arafat in Johannesburg (Arafat's Speech in Johannesburg)

Israel TV (Channel 10)
   www.nana10.co.il
   London and Kirshenbaum, Interview with Uri Lubrani, June 6, 2010
   Interview with Zvi Yehezkeli, Dec. 9, 2013

Jerusalem Post
   www.jpost.com
   Abu Toameh, Khaled and Keinon, Herb, "Fatah Official says Two-State Solution is Over," Oct. 12, 2010

   Abu Toameh, Khaled, "Palestinian Population in W. Bank, Gaza, about 4.5 Million," Nov. 7, 2013

Jewish Chronicle Online
   www.thejc.com
   "Qaradawi Predicts a Muslim Apocalypse," May 2008.

Jihad Watch, by Robert Spencer
   www.jihadwatch.org
   Hamas Parliament Votes for Sharia in Gaza

The National World
   www.thenational.ae/world
   Egypt's military says it has destroyed 1370 Gaza smuggling tunnels, March 12, 2014

The New York Times
   www.nytimes.com
   Fathi, Nazila "Wipe Israel Off the Map" Oct. 27, 2005
   Bernard, Avishai, "The Israel Peace Plan That Could Still Be," Feb. 7, 2011
   Akram, Fares, "At Rally for Hamas for Hamas Celebration and Vows," Dec. 14, 2011

Reuters
　www.reuters.com
　Al-Mughrabri, Nidal, "Egypt tunnel blockade takes toll on Gaza business, Dec. 9, 2011

Speigel
　www.spiegel.de/international
　Musharbash, Yassin, "Surprising Study on Terrorism: Al-Qaeda Kills Eight Times More Muslims than Non-Muslims," Dec. 3, 2009

USA Today
　www.usatoday.com
　President Obama's Cairo Speech Text

Washington Post
　www.washingtonpost.com
　Goldstone, Richard, "Reconsidering the Goldstone Report on Israel and War Crimes," April 1, 2011

Washington Times
　www.washingtontimes.com
　Armstrong, William, "The Endless Wars of Islam," Sept. 12, 2010.
　Constantine, Tom, "Eighty-One Percent Support Islamic State and Its Barbaric Acts," May 28, 2015

Articles

It is the intent of this author that the reader also become familiar with the various journals, periodicals and organizations listed below and not only with the specifics articles or entries as mentioned in the endnotes. Should one desire to access an online specific work mentioned, the reader is to enter the URL through Google and then type in the name of the article requested, placed inside quotation marks. **Note: Articles are listed alphabetically by name of publication when appropriate.**

Al-Bab
　www.al-bab.com
　The Arab Peace Initiative 2002

Aldawaba
www.aldawaba.com
World Bank: Read GDP Growth in West Bank and Gaza in 2008 2%, June 4, 2009.

The American Interest
www.the-american-interest.com
Herf, Jeffrey, "A Pro-Hamas Left Emerges," August 26, 2014.

American Rhetoric
www.americanrhetoric.com
George W. Bush Rose Garden Speech on Israel-Palestine

Answering Islam
www.answering-islam.org
Arlandson, James M., "Did Allah Transform Jews into Apes and Pigs?"

Association for Civil Rights in Israel
www.acri.org.il/en
East Jerusalem by the Numbers

Bogdanor, Paul
www.paulbogdanor.com
Understanding the Arab-Israeli Conflict

Brookings Institute
www.brookingsinstitute.edu
Reidel, Bruce, "Pakistan, Taliban and the Afghan Quagmire," August 24, 2013.

Budnitskii, Oleg
Russian Jews Between the Reds and the Whites 1917- 1920, *University of Pennsylvania Press,* www.upenn.edu/pennpress

Chodoff, Elliot
Theory of Delayance, unpublished, 2010

Daniel Pipes Middle East Forum
    www.danielpipes.org
    [UNRWA] The Refugee Curse, August 19, 2003

Emerson, Steve
    Abdullah Assam: The Man Before Osama Bin Laden, www.iacsp.com/itobli3.html

Foreign Affairs
    www.foreignaffairs.com/articles
    Allon, Yigal, Israel: the Case for Defensible Borders, Fall 1976

Freethought Nation
    www.freethoughtnation.com
    What does the Koran say about women?

Gatestone Institute
    www.gatestoneinstitute.org
    Zahran, Mudar, Is Jordan's King Losing Control Over the Bedouin? July 20, 2011

German Jewish Soldiers
    www.germanjewishsoldiers.com/epilogue/php
    Epilogue: Relevant Commentaries on the Lists

George W. Bush White House Archives
    www.georgewbush-whitehouse.archives.gov
    President discusses roadmap for peace in the Middle East, March 14, 2003

Global Jihad
    www.globaljihad.net
    Israel seized weapons ship 'Victoria', March 22, 2011

Islam Watch
    www.islamwatch.org
    Bostom, Andrew, Antisemitism in Islam

Jerusalem Center for Public Affairs
www.jcpa.org
Cook, David, Anti-Semitic Themes in Muslim Apocalyptic and Jihadi Literature
Lapidot, Ruth, Legal Aspects of the Palestinian Refugee Question

Jewish Virtual Library (Bard, Mitchell)
www.jewishvirtuallibrary.org
Myths and Facts Online, Arab/Muslim Attitudes Towards Israel
Bornstein-Makovetsky, Leah, Suleiman I

Jews of Egypt Foundation
www.jewsofegyptfoundation.com
History

Kadouri, Elie and Haim, Sylvia, eds., *Zionism and Arabism in Palestine and Israel*, F. Cass, London, 1982.
Lachman, Shai, "Arab Rebellion and Terrorism in Palestine 1929-1939: The Case of Sheikh Izz a-Din al-Qassam and His Movement."
Mayer, Thomas, "The Military Force of Islam: The Society of the Muslim Brethren and the Palestine Question, 1945-48."

Long War Journal
www.longwarjournal.org
Mayer, Alexander, Muslims Account for 85% of Casualties in Al-Qaeda Attacks, December 9, 2009

Lyrics Translate
www.lyricstranslate.com/en
Palestinian National Anthem - Fida'i

Middle East Forum
www.me-forum.org
Bukay, David, Peace or Jihad? Abrogation in Islam, in Middle East Quarterly, Fall 2007
Karsh, Ephraim, Obama's Middle East Delusions, Middle East Quarterly, Winter, 2016
Stillwell, Cinnamon, Academia on San Bernardino Attack: No Jihad Here, December 11, 2015

Mideast on Target
www.me-ontarget.org
Blogs are listed in order of appearance in the book. All blogs written by Yisrael Ne'eman unless otherwise indicated. To access, Google www.me-ontarget.org plus name of article. (For example, to find the article titled "Israel Forces a Hudna," search for: www.me-ontarget.org "Israel Forces a Hudna")

Blogs:
  Arafat and Netanyahu face the Sharon Evacuation Plan, Feb. 3, 2004
  Israel Forces a Hudna, June 24, 2003
  Dismantling the Terror Infrastructure-the Only Issue, July 28, 2003
  Middle East Cycle of Violence is Linear, Aug. 22, 2003
  Hudna Cease-Fire or Disarmament? May 11, 2003
  Hudna Lunacy, June 23, 2003
  Israel Forces a Hudna, June 24, 2003
  The Islamic Movement Considers a Hudna, July 1, 2003
  Dismantling the Terror Infrastructure, July 28, 2003
  Hudna Collapse, Aug. 9, 2003
  Removing Sheikh Yasin, March 22, 2004
  The Schadeh Elimination, July 25, 2002
  After Schadeh: The Morality Debate, July 26, 2002
  Likud Referendum, May 2, 2004
  Hamas Victory: Following the Trend, Jan. 28, 2006
  Palestinian Islamic Revolution-On the March, June 15, 2007
  A Time of Reckoning, July 14, 2006
  Seeking Total Victory, July 16, 2006
  Cross-Cultural Misunderstandings, July 28, 2006
  Hezbollah-Israel War: Comments, Aug. 4, 2006
  Hezbollah Victory, Sept. 3, 2006
  Hamas Dilemma, Nov. 25, 2007
  Gaza Humanitarian Issues in the Service of Islamists, Jan. 29, 2008
  Closing in on Hamas? March 5, 2008
  Coexistence or the Erez Crossroads as an Enemy, March 8, 2004
  Increasing Palestinian Suffering for Islamist Gains, April 10, 2008
  Karni Attack Directed at Palestinians, Not Israel, Jan. 15, 2005
  Lessons of War - by Elliot Chodoff, Jan. 5, 2009
  Egypt vs. Hamas (and the Moslem Brotherhood), Jan. 8, 2009
  Hamas Perspectives and Options, Jan. 10, 2009
  Phase 3: Israel's Military and Diplomatic Options, Jan. 14, 2009

Demanded: End Game Scenario, Jan. 16, 2009
Hamas War and Cease-Fire, Jan. 19, 2009
The EU Contains Obama's Initiative? Jan. 21, 2009
Anti-Semitism and the Liberal/Extreme, Feb. 1, 2009
Obama's Emerging Permanent Status Agreement, July 29, 2009
The Flotilla: Turkish Move to Lead the Muslim World, June 4, 2010
The New Ottomanism Taps the Palestinian Venue, Sept. 6, 2011
Some Gaza Conclusions, Aug. 25, 2014
Arab World Heroes, July 31, 2006

Militant Islam Monitor
www.militantislammonitor.org
Hitler Predicted Holocaust as Early as January 30, 1939

My Jewish Learning
www.myjewishlearning.com
Barnavi, Eli, World War I and the Jews (*Jewish History 1914-1948*)

Nizkor Project
www.nizkor.org
Heinsohn, Gunnar, Why Auschwitz? Shofar ATP Archive File

Nairaland Forum
www.nairaland.com
Taqiyya and Kitman: Are Muslims Permitted to Lie?

No Beliefs.com for Free Thinkers
www.nobeliefs.com
Walker, Jim, Hitler's religious beliefs and fanaticism

Palestine Facts
www.palestinefacts.org
Church of the Nativity - 2002
Palestine National Anthem - Fida'i

Palestine Media Watch
www.palwatch.org
Hamas: Ruling West Bank it could destroy Israel with speed that no one can imagine, Oct 5, 2014
"The Protocols of the Elders of Zion," (usage by Palestinian media)

Palestine Information with Provenance
www.cosmos.ucc.ie/cs1064/jabowen/IPSC/php/home.php
Arab League recognizes PLO as sole legitimate representative of the Palestinian People

Peace For Our Time?
www.peaceforourtime.org.uk
What if the Jews Lost any War
Why Did Yasser Arafat Sign the Oslo Accord?

Pew Research
www.pewresearch.org
In Nations with Significant Muslim Populations Much Disdain for ISIS

Qaradawi, Yusuf, "[UNTITLED]" in Bostom, Andrew, (ed.), *The Legacy of Jihad: Islamic Holy War and the Fate of Non-Muslims*, Prometheus Books, Amherst, New York, USA, 2005.

Qazi, Farhana
www.farhanaqazi.com/72-virgins-in-heaven-fact-or-fiction
72 Virgins in Heaven: Fact or Fiction

Qutb, Sayyid, "Our Struggle with the Jews," in Nettler, Ronald, *Past Trials and Present Tribulations: A Muslim Fundamentalist's view of the Jews*, Oxford, Pergamon Press, 1987.

Religion of Peace
www.thereligionofpeace.com
The Politically Incorrect Truth About Islam Lying (Taqiyya and Kitman)
A Tiny Minority of Extremists?

Rutgers University - Newark Campus of Arts and Sciences
   www.ncas.rutgers.edu
   The Assyrian Genocide 1914-23 and 1933 to the Present

Simple to Remember.com Judaism Online
   www.simpletoremember.com
   Quotes from Adolf Hitler

Vajda, Georges, "Jews and Muslims according to the Hadith" in Bostom, Andrew, (ed.), *The Legacy of Islamic Antisemitism,* Prometheus Books, Amherst, New York, USA, 2008.

Washington Institute for Near East Policy
   www.washingtoninstitute.org
   The Karine A-Affair: A Strategic Watershed in the Middle East?

Index

A

Abbas II, King of Persia 41
Abbas, Mahmoud/Abu Mazen 199, 212, 214, 220, 249, 411, 424, 449
Abdullah II, King 95, 239, 433
Abrogation 106, 149, 340, 377–379, 390, 395, 430
   Clause in Koran 149, 340, 377
   Conflict Resolution 17, 109, 111, 375–390
   Reverse Abrogation 17, 378, 399, 402, 430
Abyssinia/Abyssinian 25
Accusation Reversal 62, 362
Adrisi, Hian al- 190
Afghanistan 11, 88, 98, 102–107, 110, 115, 169–170, 192, 233, 255, 286, 293, 403–404, 422–425, 433
   Azzam, Abdullah 169–170
   Taliban 286, 399, 425
Africa 80, 143, 192, 232, 239, 287, 293, 451, 453
   Horn of 255
Ahmedinejad, Pres. Mahmoud 142, 280, 307, 401, 424
Al-Aksa Martyr Brigades 206
Al-Aksa Mosque/Dome of the Rock
   Domain 57, 138, 149, 177, 186, 189–190, 196, 201–202, 206–211, 278, 281, 336, 387, 447, 449
   1967 Six Day War 303
   Husseini, Haj Amin el- 125–128, 137, 202
   Intifada 202
   Khomeini 141
   Visitation Rights 336
Aleppo 47
Ali, Mohammed 60
Aliya 18, 118, 136, 438
Allah
   Actions in the name of
      Jihad/War/Death/Killing 39, 64, 75, 80, 90–92, 111, 269, 332
      Martyrdom 64, 75, 77, 91, 102, 128, 220, 226–227, 259, 267, 285
      Obedience 79–80, 89–90
   Christians, attitudes toward 32–37, 81–83, 389–395
   Hamas Covenant 12, 18, 111, 257–261, 264–265, 267–271, 273–279, 281–283, 285, 294–297, 301, 304–305, 309, 312–317
   Jews
      Broken Covenant with/punished by 16, 24–41, 28, 34, 82, 108
      Follow Covenant with 380–390
      Reward/Favor of 28–29
   Koran, Mentioned in the 27–34, 28, 82, 88, 101, 380–394
Allon Plan 167, 178, 444
Almohads 39, 44
Almoravids 45
al-Mujamma' al-Islami 166
al-Qaeda 13, 73, 200, 235, 240, 246, 403, 415, 422–424, 431, 451, 452
Al-Sha'm 272, 276, 314, 318
America
   Arab/Islamic View of 12, 19, 83, 141, 147, 343, 399, 417
   Destruction of 417, 422–423
   Israeli/Palestinian Relations 293, 422
   Jewish Emigrants 119
   Qutb, Sayyid 79, 83–86

Amir, Yigal 184
Andrews, Lewis 129
Anglo-American Committee 133
Anglo-Palestine Bank 121
Annapolis 223–224
Anthropocentric/Anthropocentrism 59, 62, 80, 91, 108–109, 190, 350, 409, 418, 438
Antisemitism/Antisemitic 411, 416, 418, 430, 437, 438, 445
  Anti-Zionism 147, 149
  Czarist 125, 148, 256, 306, 362–371, 395
    Influence on Palestinian Jihad 355–374
  Eastern European 13, 15
  Eastern Orthodox 53, 55–59, 147–148
  European 59, 61, 118–119, 147, 148, 232
  Haj Amin 131
  Hamas 106, 395
  Hamas Covenant 169, 232, 238, 255–261, 267–269, 275–278, 288–289, 292–295, 303, 306, 310–313, 315–316, 319, 362–371
  Islamic Traditional 11, 13, 15, 21, 23–70, 146, 154
  Jihadi 62–64, 72, 154, 170, 179, 256
  Nazi 16, 62, 117–118, 126, 154, 306, 362–371, 395
  North America 19
  Palestinian 55–59, 118, 124, 145–146, 149, 169
    Denial of 143, 158
    Palestinian National Charter 323, 339–341
  Qutb 15, 79, 92, 96
Antonius, George 57, 68
Apostate 95, 276
Aqnin, Ibn 41
Aqta, Rabbi Sulayman al- 43
Arab al-Na'im 19–20
Arab Higher Committee 128–129, 135, 435

Arabian Peninsula 73, 93, 97, 105, 123, 278, 389, 417, 435, 444
  Cleansing from Non-Muslims 35
  Early Islam 23
  Jewish Presence 15, 23–26
Arab League 136–137, 431, 446
Arab Liberation Army/ALA 48, 135
Arabs
  in the Holy Land 49
  Jewish Relations with 26, 41, 46
  Nationalist, definition of 55
  View/Treatment of Israel/Jews 45, 48–50, 58
Arab Spring 12, 87, 114, 200, 238, 406, 438
Arafat, Suha 149
Arafat, Yasir 16, 20, 95, 104, 109, 118, 128, 130, 151–155, 161, 164, 169, 170, 176, 194, 199, 201, 207, 221, 230, 241, 247, 250, 298, 323, 347, 353, 410–411, 419, 433, 434, 446, 447, 450, 451, 453
  At the UN 143–145
  Low Intensity Conflict/Al-Aksa/Second Intifada 201–215
  Palestinian Authority/Oslo Accords 181–193
  PLO 142–145, 148–155, 161–166, 169–180
Armenia/Armenians 53–56, 122–123, 451
  Genocide 54
Army of Islam 222
Art/Artwork 263, 286–287, 320, 328
Ashdod 228, 234
Assad, Hafiz el- 140
Assyrian Christians/Nestorians/Chaldeans 56, 451
Ataturk, Mustafa Kemal 54, 369, 419
Avariz 38, 439
Aviram 28
Ayyash, Yahya 191
Ayyub, Sayyid 41
Aziz, Colonel Abd al- 133
Aziz, Omar ibn Abdel (Omar II) 37

Index

Azzam, Abdullah 15, 71, 73, 79,
    98–115, 136–137, 154, 165,
    169, 194, 216, 248, 255, 301,
    304, 320, 328, 330, 334, 433
    Ideals/Theology 98–110

B

Baath/Baathist/Baathism 114, 138,
    140, 163, 321, 333, 423
Badr, Battle of 26
Baghdadi, Abdul Bakr al-/Ibrahim
    Awwad Ibrahim 281, 433
Bakr, Caliph Abu 26
Balfour Declaration 57–58, 117, 123,
    294, 327, 339, 368, 371, 448
Balkans 12, 52–53, 78, 120, 356, 393,
    453
Banna, Hasan al- 15, 71–72, 78–79,
    89, 107, 112, 117, 132, 157,
    255, 259, 280–281, 305,
    318–320, 360, 433, 453
    Ideals/Theology 71–89
    Tracts by
        Between Yesterday and Today 73,
            112, 257
        On Jihad 72–73, 112, 434
Banu an-Nadir 26
Banu Aus 25
Banu Qaylah 25
Banu Qaynuqa 26
Banu Qurayza 26
Barak, Ehud 20, 149, 188, 196, 199,
    202, 205, 224, 433, 434, 450
Barghouti, Bilal 212
Barghouti, Marwan 189, 206, 208,
    213
Bar Kokhva 24, 49, 448
Bashir, Omar al- 192, 422
Bat Ye'or 54, 64–68, 145, 158,
    352–354
Baybars 305, 316, 319
Bedouin 19–20, 50, 224, 239–240,
    252

Beersheva 201, 228, 428–429, 447
Begin, Menachem 167, 169, 434
Beirut 52, 301, 353
Beit Hanoun 228
Beit Lahiya 228
Bin Laden, Osama 13, 98, 115, 169,
    255, 399, 417, 451
Black Hand 117, 127–128, 415, 436
Blood Libel 55, 158, 289
Bolshevik/Bolshevism 263, 287, 357,
    361, 361–362, 369
Brinton, Crane 406
Britain 52–57, 80, 84, 103, 122–123,
    127, 129, 131, 155, 356, 359,
    372, 404, 437, 448, 453
Bukhara 103
Bulgaria 103, 400
Burma 103
Bush, George W. 204, 211, 444
Bush "Road Map" 212, 223, 444
Byzantine Empire 37, 39, 65, 80, 272,
    307, 389, 423
Byzantines 25, 103

C

Cairo 46, 68, 98, 124, 132, 132–133,
    137–141, 150, 224, 232, 240,
    251, 360, 413–415, 422, 426,
    431
Camp David Accords 172, 181–182,
    195, 310–311, 347, 434, 437,
    445
Capitalism 11, 35, 81, 83, 107, 262,
    369, 417
Capitulation Treaties 53, 60, 118
Cast Lead 21, 200, 222, 225, 227–228,
    231–233, 239, 288, 291, 412,
    414, 450
Cave of the Machpela/Ibrahimi
    Mosque 50, 167, 183, 447
Chad 103
Charter of Omar 15, 108, 130, 219,
    266, 308, 319, 389, 445

Dhimma Status 23, 51
Implementation from the Middle Ages to Nineteenth Century 42–45
Stipulations/Early Implementation 35–43
Chehab, Zaki 222
China 80, 293, 400, 404
Chodoff, Elliot 432
Chouraqui, Andre 44
Christendom 62, 92, 389–390, 421
  Clash with Islam 52–54, 107, 375
Christians 11, 17, 80, 82, 84, 94, 104–105, 107–108, 110, 115, 117, 121, 123, 129, 133, 142–146, 192, 197, 204, 232, 255–258, 304, 307–308, 315–316, 324, 330, 336, 341, 370, 377–379, 394–395, 401, 403, 436, 439, 441, 445, 448, 451, 452
  Anti-Semitism 56–59
  As Dhimmis 26, 33, 36–40, 52–53, 216
  Clash with Islam 298
  Hamas Covenant 257–258, 266, 275–276, 293, 297, 319
  In the Koran
    Compared to Jews 35
    Condemnation of 32–36
  Islamic View of 36, 40, 56
    Early Days 44
  Jihad Against 75–78, 80–84, 105–110, 371
  Persecution/Massacres 129, 192
  Persecution/Massacres of 53–56, 61
  Positive Attitudes Toward 389–393
  Pro-Jewish/Zionist 293
  Secular Arab Nationalism 55–59, 133, 145–146
Church of the Holy Sepulcher 55
Clinton, Bill 149, 186, 199, 422
Clinton Parameters 247, 445
Committee of Union and Progress 119

Communism 11, 35, 81, 96–97, 107, 121, 262, 287, 298, 359, 366, 369, 401, 419–420
  Communist East 81, 294–295, 343, 367, 423
Concentration Camps 46, 359
Conscience 49, 83, 370, 403
Constantinople 50, 73, 103, 307
Constitution of Medina 25
Constitution of Palestine (2003) 344–345, 352–353
Covenant, Israelite/Jewish 28–30, 32, 34, 48–49, 56, 146–147, 304, 340, 380–381, 393
Crusaders 11, 49, 80, 107, 273, 279, 281, 300, 304–305, 314, 316, 319, 417–418
Czarism 361

D

Dahlan, Mohammed 187
Dajjal 36, 41, 48, 62, 439
Damascus 52, 55, 60, 121, 128, 133, 140, 187, 212, 218, 235, 239, 241, 301
Dar al-Harb 77, 85, 91, 108, 423, 439
Dar al-Islam 77, 85, 86, 91, 108, 423, 439
Darby, John Nelson 118
Darwin, Charles 61
Darwish, Abdullah Nimr 168
Davos World Economic Forum 205
Dawa/Dawah 161, 164, 166, 173–174, 179, 192, 195, 200, 206, 225–227, 246, 439
Day of Resurrection 25, 30, 90, 271–273, 312–314, 380–384, 441
Dayton, Keith 221, 406, 412, 434
Declaration of Intervention 136–137
Declaration of Principles/DOP 181–185, 196, 445, 446
Dehamshe, Abdulmalik 309
de Lagarde, Paul 61

Index

Democratic Front for the Liberation of Palestine/DFLP 104, 175, 329, 451, 453
Deuteronomy, Book of 387–388
Dhimmi/Dhimma 15, 17, 117–123, 130, 135–136, 141, 144, 146, 153, 155, 179, 216, 219, 238, 258, 266, 268, 276, 297, 307–308, 317, 319, 327, 330, 336, 340, 351–352, 398–399, 418, 422, 439, 441, 443, 445, 452
    After Ottoman Empire ends 52–55
    Charter of Omar 31, 35–49
    Ideologues 75, 77, 78, 94, 96, 115, 117
    Jewish Defiance of 48, 118–120
    Post WWII 46
    Various Muslim Countries views of 42–45
Dibani, Sheikh Abdul-Hamid Attiyah al- 48
Din, Deha a- 121
Diocentric/Diocentrism 11, 59, 62, 79, 80, 86, 109, 269, 351, 379, 409, 418, 439
Dor Alon 225
Doust, Muhsen Rafiq 219
Dual Sovereignty 219, 227, 410, 428

E

Eban, Abba 140, 141
Effendi 120, 120–121, 124, 126, 128, 148, 439
Egypt
    1948 War of Independence 97–98, 131–135
    1967 Six Day War 138–141
    Gaza Tunnels 223–224, 227, 232, 243–245
    Muslim Brotherhood 72, 74, 79, 85, 117
Eichmann, Adolf 359
Eisenhower, Dwight D. 45–46, 67

Eldar, Shlomi 217, 236
End of Conflict 18, 109, 188, 193, 218, 428, 444
End of Days 34, 40, 93, 167, 266, 268, 337, 385, 441
Enlightenment 21, 59–61, 80–81, 256, 280, 301, 339, 370, 398, 419–420, 430
Erasement Theology 64, 339, 352, 439
Erdogan, Recep Tayyip 234, 242, 296, 425–426, 434
Erez Crossing 225
Eritria 103
Eshchar 19
Europe/European
    Conflict with Ottomans 52–55
    Eastern 52, 119, 126, 130, 158, 287, 356, 359, 361–362
    Imperialism in the Middle East 52
    Media 293
    Western 119, 127, 298, 400
Exodus, Book of 27–28, 65, 93, 146, 380–381, 383, 385, 389, 402, 435
Ezra, the scribe 27, 33, 35, 68, 82, 387

F

Falastin 119
Falouji, Imad 202
Fard Ayn 99, 103, 106, 439
Fard Kifaya 99, 440
Far East 86, 255, 320, 395, 401, 419
Farouk, King 72, 137, 161, 418
Fascism 109, 256, 263, 280, 285, 287, 355
Fatah
    1960s to 1970s 165–166
    1987 to 1993 Intifada 169–181
    Hamas Challenge of (LIC/Second Intifada) 201–215
    Lebanon 1982 168, 170
    Palestinian National Charter 323–351

Fatahland 221
Fatwa 98, 115, 150, 190
Fay 40, 440
Fayyad, Salam 221, 235–236, 250, 424
Fellah/Fellahin 120, 440
Fez 45, 67
First Intifada (1987-1991) 289, 450, 454
First War in Lebanon (1982) 437, 450
First Zionist Congress 119, 434
Fitnah 99–100, 440
France 47, 52, 55, 59–61, 73, 80, 293, 307, 393, 404
France/French 45–46, 54–56, 59–61, 64, 68, 158, 210, 294, 353, 361, 398, 400, 405, 430
    Christians 54–56
    In the Middle East 60–61, 122–123, 147, 281, 371
    Jews 45–46
    Revolution 59–60, 292, 365, 408
    Vichy 45–46, 356, 358, 400
Freemasons 283–284, 292–293, 302, 362, 365–366
Free Officers 85, 107, 109, 130, 163, 436
Futuwwa (Youth Movement) 132

G

Galilee 19–20, 49, 67, 129, 135, 165, 189, 221
Gaza/Gaza Strip
    Egyptian Rule of 161–165
    Gaza and Jericho First 151
    Gaza City 152, 171, 228
    Hamas Capture of 199
    Israeli Occupation 165–172, 180–182
Gedaliah of Siemiatyce 50
Germany 54, 58, 72, 122, 135, 141, 256, 262, 269, 285, 287, 290, 319, 361–362, 401, 404, 435, 445

Germans 45, 54, 62, 131, 356–358, 370, 400
Ghoshah, Abdullah 49
Giado 46
Goebbels 97
Goldstein, Barukh 183
Goldstone Report 230–231, 251, 450
Gospel/Gospels 32, 65, 315, 391–393, 402, 439
Great Depression 126
Greco-Turkish War 123, 156
Greece 80, 83, 124, 146, 235
Green Line 183
Gunning, Jeroen 242
Gurr, Ted 194, 408

H

Habash, George 57, 104, 453
Hadith 23, 27, 35–36, 62, 65, 68, 95, 102, 114–115, 131, 152, 255, 257, 268, 276, 282, 305, 318, 360, 443, 445
Haifa 57, 124, 127–129, 133, 209–210, 428
Haifa Congress of Palestinian Arabs 124
Hakim, Caliph al- 39, 44
Halabia, Dr. Abu- 150, 190
Halacha 446
Halevy, Yehuda 52
Halutz, Dan 237
Hamadan 42
Hamad, Fathi 228
Hamastan 221
Hamed, Ibrahim 212
Hamid II, Sultan Abdul 95, 119
Hanafi 75, 440
Handschar Division 359
Haniyeh, Ismail
    Cast Lead and Aftermath 227–242
    Hamas Prime Minister 200, 216–217

Harith, Zaynab Bint al- 62
Harkabi, Yehoshafat 323
Hawatmeh, Nayef 57
Heaven 27, 75, 85, 102, 114, 172, 259, 273–274, 279, 281, 388, 406–407, 440
Hebron 49–50, 67, 118, 125, 167, 169, 183–184, 186, 191, 213, 244, 447
Hejira 25, 79, 93, 440
Hell 31, 33–36, 81, 100, 312, 315, 387, 390, 393
Herzl, Theodor 119, 363, 434, 444
Hess, Moses 118
Hezbollah
　Alliance with Hamas 179, 181, 211, 219–222
Hitler, Adolf 154, 292, 374, 398, 445
Holocaust 52, 58, 63, 67, 141, 154, 159, 294, 307, 320, 352, 362, 372, 400, 435
Holy Basin 204
Holy Sepulcher, Church of the 55
Holy War 12, 40, 72–76, 78, 80, 90, 92, 98–99, 103, 106, 109, 112, 142, 158–159, 264, 312, 319, 332, 351, 377, 395, 439, 441
Homosexuals 101
Houri/Houris 102, 440
Hroub, Khalid 170, 191
Hudaybia
　Battle of 26
　Treaty/Pact/Hunda of 105, 111, 191
Hudna
　As Defined by Islamic Ideologues 63, 77–78, 90–92, 105–106, 111–112
　As Used by Hamas 179–180, 191, 213, 216, 221, 224, 228, 238, 243–244, 416, 418
Huntington, Samuel 419
Husseini, Abdul Qader el- 135
Husseini, Amin el- 434
Husseini, Muza Kazem el- 57

Hussein, King 138, 418, 433, 434
Hussein, Saddam 176–177, 190, 207, 220, 289, 299, 423
Hussein, Sharif 123, 371

I

Iberian Peninsula 12, 52, 78, 307, 377
Idolatry/Idolaters 33, 39, 73, 77, 80, 87, 104, 261–262, 315, 371, 393, 399, 417, 442, 443
IHH 200, 234, 452
Ijtihad 86, 300, 394, 440
Iman 98, 440
India 12, 80, 96, 293, 358, 377
Infidels 33–34, 38, 76–78, 90, 99, 163, 269, 275, 312, 316, 398, 440
Iran
　Ally of Hezbollah 187–188, 220
　Destroy Israel 296
　Destruction of Israel 135, 141–142
　Jews in 39, 42
　Nuclear Program 400–401
　Supports Hamas and PLO 187, 192, 211, 218–220, 232, 242
　War with Iraq 170, 220, 307
Iranian Revolution 167–170, 220
Iraq
　Jews 43–47
　Supports Palestinians (Intifada 1987-93) 177, 207
　War of Independence (1948) 48, 135
　War with Iran 170, 220, 307
　WWII/Farhud 130–131, 135, 356–357
Islam and Judaism 131, 359–360, 373
Islamic Awakening 12, 17, 87, 200, 233, 235, 238, 296, 406, 416, 423, 425, 428–429, 438
Islamic Jihad 168–169, 173, 175, 178–179, 183, 193, 200, 207, 210, 226, 292, 316, 452

Islamic State/ISIS/ISIL 13, 64, 68, 88, 200, 255–257, 280–281, 287, 308, 351, 377, 398–399, 403, 415–417, 432, 433, 442, 444, 451, 452, 453
   Jihadi Influences Shared with Hamas 13, 73, 246, 267, 272, 296, 424–429
Israel, Children/People of 28, 108, 259, 339, 380–384, 389–390
   Reverse Abrogation 380–390
Israel Defense Forces/IDF 167, 180, 204–208, 211–212, 223, 228–231, 237, 243, 245, 427–428, 437, 450, 452
Israel, Land of 15, 17, 18, 24, 41, 43, 58, 62, 158, 180, 252, 339–341, 381, 384, 389, 393, 418–419, 434, 437, 439, 442, 444, 448
   Belongs to Israelites/Jews 380–390
   Greater 167, 186
   Hamas Covenant 271–273, 314–315
   Ottoman rule 49–54
   PLO denies 143–148, 154–155, 205, 324, 339–341
   Zionism 117–119
Israel, People of 259, 339, 382
Israel, State of 13, 15–16, 21, 47–48, 62, 78, 143–148, 155, 161–163, 182, 201, 307, 326, 360, 389, 434, 436, 437, 448, 449, 451, 453
   Immigration to 39, 43, 47
   Islamists/Hamas Battles Against 64, 93, 96–97, 103, 107–108, 147, 151–152, 163, 173, 185, 217–218, 222, 273, 299, 304, 330–334, 340, 351, 376, 401
   Peace 148, 201
   PLO battles against 337–338, 344, 351
   Wars 48, 64, 97, 117, 144, 267, 294

J

Jaabari, Ahmed 200, 221, 236, 238, 413, 435
Jahiliyya/Jahili 79–92, 107–108, 111, 301, 402, 417, 441
Japan 80, 293, 400–401
Jensen, Michael 218, 249
Jerusalem
   Attacks/Bombings 133–135, 184, 186, 188, 190, 209–210, 213
   British 56–57, 117, 124–127
   Negotiations Over 182–183, 202–205, 233, 237
   Ottoman 50, 55, 118–122
Jewish National Fund 273
Jewish National Home 57, 58, 62, 117, 124–127, 130–131, 162, 339, 357, 358, 371, 448
Jihad
   Army 372
   Defensive 76, 78, 99, 102, 108–111, 139, 237, 272, 274, 278, 303
   Offensive 76, 78, 106, 110–112, 426
   Of the Heart 107
Jihad Army 135
Jinn 264, 441
Jizya 26, 33, 35, 37–38, 40, 43, 45, 65–66, 75, 82, 90–91, 98, 108, 371, 389, 439, 441
Jordan
   Hamas in 219
   Muslim Brotherhood 164–165, 239
   Peace Agreement with Israel (1994) 310–311
   Six Day War (1967) 165–166
   War of Independence (1948) 48, 134, 372
   West Bank 135–138, 163, 165, 175
Joshua 28, 381, 385
Judaism/Jews 442, 449
   Arabia 24–27

Demonization of 148, 158, 258, 295
Dhimmi Status 31–36
Hadith 36–37
Haj Amin el-Husseini 359–360
Hamas Covenant 306–307, 315–316
Islamic Acceptance of 376–379
Judeo-Islamic Relationship History and Theology 24–28
Koran 27–35
Palestinian National Charter 339–342, 351
Soldiers 362–363, 373

K

Kadima 223, 436
Karaj 38, 40, 441
Karbala 60, 307
Karine A' (ship) 211, 219
Kashmir 103, 192
Kastel 135, 157
Kawasmeh, Abdullah 213
Kaybar 26, 35, 62, 359
Kerem Shalom/Rafiah Crossing 223
Kfar Darom 134
Khamenei, Ali 401
Khomeini, Ruhollah (Ayatollah) 71, 435
Khomeinist 11, 87, 141, 161, 167, 187, 192, 232, 242, 281, 400
Kilani, Rashid Ali al- 356, 358
Kiryat Gat 228
Kitman 193, 197, 207–208, 218, 441, 443
Knesset 206, 225, 309, 448
Kolhozy 148
Korach 28
Koran 445
Kuffar 99–100, 103–105, 441
Kuwait 177, 187, 423

L

Labadi, Majid 175
Labib, Mahmud 132
Labor Party 434, 453
Lane, Edward 44
Last Day 25, 30, 33, 37, 82
League of Nations 55, 57, 125, 294, 339, 371–372, 448, 452
Lebanon
  First Lebanon War (1982) 168–169
  Hamas Expulsion to 168, 170
  Hezbollah 168–181, 422
  Second Lebanon War (2006) 237
  War of Independence (1948) 48, 135
Levant 52–53, 56, 60, 256, 272, 452
Levitt, Matthew 173
Lewis, Bernard 147, 375, 422
Libya 13, 46–48, 67, 88, 141, 232, 242, 255, 308, 406, 423, 438, 453
Likud 150, 166–169, 178, 184–186, 189, 202, 205–206, 211, 214, 223, 244, 249, 434, 436, 437
Low Intensity Conflict/LIC/Second Intifada
  History of 201–215
Lubrani, Uri 42
Luther, Martin 61, 158
Luttke, Moritz 44

M

MacMahon Correspondence 117, 123
Madrid Conference 177
Magreb 44
Makadme, Ibrahim 187, 199
Makarem Society 133
Mandatory Police 128
Manji, Irshad 394–395, 402, 417
Mao/Maoist 12, 280
Marmara Flotilla 200
Maronite Catholic 54

Marrakesh 45
Marranos 41
Martyr/Martyrdom 64, 75, 91, 102, 107, 128, 220, 227–228, 259, 267, 285, 440, 443
Mashal, Khaled 106, 435
Maududi, Sayyid Abul A'la 71, 146
Mauza 43
Mecca 25–26, 60, 79, 85, 105, 111, 123, 149, 164, 216, 221, 277–278, 281, 371, 387, 417, 426, 440, 442, 447
Mecca Agreement 216
Media
  Accusation of Jewish Control 62, 105, 283–284, 363–364, 368
  Arab/Muslim 283
  Palestinian 153, 189, 200, 205, 225–226, 230–232, 279–280, 283–284, 293, 305
Medina 24–26, 39, 60, 79, 91, 93–95, 99, 123, 149, 277–278, 359, 389, 426, 440
Mein Kampf
  Influences Hamas Covenant 255, 261, 292, 306, 311, 319, 363–369
Meshhed 42
Messiah/Messianism 33, 36, 43, 48, 82, 121, 152, 390–393, 409, 441
Milestones 79, 85, 92, 110, 112, 263, 431, 437
Misgav 309
Mohammed, Prophet/Messenger
  Early Islam 23–40, 62, 93–95
  In Hamas Covenant 269, 277–278, 308, 314–315
Money Lending 30
Mongols 304–305, 316, 319, 417–418
Morocco 45–46, 66–67, 247, 431
Morsi, Mohammed 13, 238, 240, 243, 406, 413–414, 435, 437
Moses
  in the Koran 27–28
  Israelites/Exodus 381–387
Mubarak, Hosni 219, 239, 293, 412

Mudayris, Ibrahim 153
Mujahideen/Freedom Fighters 100, 103, 442
Mukata'a 208, 210–211
Multi-culturalism 396
Murad III 50
Mushrikun 104, 110, 442
Muslih, Muhammed 122
Muslim Brotherhood
  Egyptian 72, 130–135, 239–243, 360, 415, 423
  Gaza and West Bank (1948-67) 162–165
  Gaza and West Bank (1987) 165–172
  Gaza and West Bank (1987-93) 172–180
  Hamas defined as 263
Muslim Christian Association 58

N

Nahal Oz 225
Najah University, al- 191, 227
Najjada (Youth Movement) 132
Najjar, Mustapha Mohammed 219
Nakba 163, 188, 442, 449
Napoleon 59–60
Nashashibi Family 120, 122, 128–129, 356
Nasrallah, Hasan 229–230, 436
Nasser, Gamal Abdul 47–48, 79, 85, 95, 102–103, 107, 109, 114, 130, 138–142, 161–167, 333, 355, 417–418, 436, 437
Nasserism 114, 138
National Unity Government/NUG
  Israel 178, 206, 211, 214
  Palestinian 216, 218, 220–221, 243, 249, 410–416, 424–425, 430
Nazi/Nazism
  Ideals Incorporated by Islamists 61–64

Impact on Palestinian
Nationalism 125–127, 130–131,
135
In Hamas Covenant 255–256,
287, 290, 292, 308–309, 311–312,
355, 372
In The Hamas Covenant 146
Netanyahu, Benyamin 184–188, 223,
232, 241, 244, 247, 251, 434,
436, 437, 447
Netanyahu-Lapid Government 244
Nettler, Ronald 65, 93, 113–114, 320
New Christians 41
New Muslims 41
New York Times 54, 142, 158, 293,
432
Nigeria 103, 255
Night Journey 93, 204, 281, 320,
385–387, 447
NILI 122, 156
North Africa 23, 39, 44–46, 52,
65–67, 73, 141, 307, 358
453
Nuwas, Yusuf Dhu 24

O

Obama, Barack 422
Olmert, Ehud 217, 222–224, 237,
428, 434, 436
Open Bridges (policy) 165
Orientalism 108, 115, 300, 321, 376,
398, 402, 442
Orr Commission 202
Oslo Accords 18–20, 118, 148, 150,
150–151, 155, 178–181,
187–191, 217, 241, 299, 323,
347, 411, 429, 433, 436, 437,
445, 446, 450, 453
Ottoman Empire 24, 52–55, 60, 67,
118–119, 122–123, 155, 307,
361, 375, 417, 443, 451, 453
Ottomanism 120, 122–123, 252, 442

Ottomans 51, 53–54, 117–120, 123,
451
Owadally, Mohamad Yasin 41

P

Pagan/Paganism 11, 25–26, 34–35,
121, 191, 286–287, 382, 417
Pakistan 88, 98, 255, 293, 399, 404,
422–423
Palestine, Land of 49, 58, 119, 121,
136–140, 169, 179–180, 201,
226, 260, 271–272
In the Hamas Covenant 260, 263,
265–268, 270–281, 298, 301, 303,
310–311, 314–317
Palestinian National
Charter 323–353
Palestine Liberation Organization/
PLO
Adopts Hamas Stereotype of
Jews 143–155
Development of 164–172
First Intifada (1987-1993) 172–
180
Low Intensity Conflict/Second
Intifada (2000-04) 199, 201–215,
220, 350, 412, 424
Palestine Mandate 47, 56–57, 98,
124–126, 130–132, 161–162,
339, 372, 411, 438, 446, 448,
454
Palestine National Liberation Movement/PNLM 164
Palestine, State of 18, 129, 138,
163–164, 170, 203–204, 212,
218, 224, 232–233, 235, 240,
250, 327, 336, 345, 350, 426,
428, 444, 446, 449
Palestinian Authority/PA
Cast Lead 227–234
Hamas Challenge of 235–247
Loss to Hamas 215–227
Low Intensity Conflict/Second
Al-Aksa Intifada 201–215

Oslo Period 181–194
Palestinian National
 Charter 344–345, 349–350
Palestinian Civil Conflict of 2007 220
Palestinianism 145–146, 155, 352
Palestinian Legislative Elections of
 1995 and 2006 249
Palestinian National Charter, The/
 PNC
 Comparison with Hamas
 Covenant 323–353
Palestinian Reform and Development
 Plan (2008-2010) 223
Palmer Report 234
Pan-Arab/Pan-Arabism 139, 333–335
Pan-Ottomanism 120
Pan-Turkism 120
Paris 91, 287, 399, 403, 426
Partition Plan 47, 54, 132–133,
 136–137, 144, 157, 326, 338,
 343, 372, 446, 447, 448, 449
Pasha, Azzam 136–137
Pasha, Da'ud 43
Pasha, Djemal 121
Pasha, Mehmed Sharif Rauf 119
Peace 129, 139, 157, 159, 166, 172,
 181–183, 193–196, 205, 224,
 247–252, 310–311, 373, 402,
 425, 432, 445, 446
Peel Commission 128
People of the Book 27–28, 31–34,
 38, 40, 77, 82, 84, 89–91, 94,
 99, 104, 107–109, 258, 266,
 276, 319, 340, 378, 380, 383,
 389–391, 395, 439
Peres, Shimon 18, 180, 184, 206, 224,
 236, 247, 436
Persia 39, 42, 66, 73, 80, 83, 103
Philippines 80, 103
Pillar of Defense 239, 241, 252,
 413–415, 435, 451
Pogroms 42, 45–47, 125–126, 372
Popular Front for the Liberation of
 Palestine/PFLP 104, 167, 175,
 329, 451, 453

Popular Resistance Committees 222
Population Statistics
 Gaza 229, 245
  People Killed in Gaza 166, 224,
   228, 229, 245–246, 289
 Hamas Supporters 13, 191, 229,
  245–246, 256
 Jews 48, 136, 256
 Muslims 13, 256, 421, 443
 Palestinians 169–170, 184, 192,
  223, 239, 289, 309
Portugal 80, 307
Protective Edge 13, 200, 244, 246–
 247, 291, 414–415, 427, 451
Protocols of the Learned Elders of
 Zion, The
 Compared to Hamas
 Covenant 363–370

Q

Qaradawi, Yusuf al- 63, 71, 153–154,
 159, 248, 362, 436
Qassam Brigades, Al- 212
Qassamites 128–131
Qassam, Izz a-Din al- 58, 117, 126–
 128, 154, 156, 162, 200, 222,
 237, 267, 412–415, 435, 436
Qawuqji, Fawzi 135, 372
Qibla 25–26, 164, 277, 442
Qutb, Sayyid
 Anti-Semitic/Anti-Israel 92–98
 Ideals, Theology 79–92

R

Rabin, Yitzhak 18, 148, 159, 178–180,
 183–184, 195, 247, 433, 436,
 437, 446
Rachel's Tomb 50
Rafiah 223–224, 227, 243, 243–244,
 413
 Rafiah Crossing 223
Rajoub, Jibril 189

Ramallah 135, 175, 208–212, 224, 414, 438, 449
Ramat Rachel 134
Rambam (Maimonides/Moshe ben Maimon) 41, 157
Rantisi, Abdul Aziz al- 171, 213
Rayashi, Rim Salih al- 226
Rechabites 26
Refugees
   Arab 19, 187–188, 203–205
   Jewish 145
Relative Deprivation/RD 408–409, 430
Replacement Theology 58, 64, 340, 352, 403, 442
Reverse Abrogation 17, 378, 399, 402, 430
Revolution, Stages of 405–409
Robinson, Mary 231
Rome 24, 80, 83, 103, 146, 387
Roosevelt 46
Royal Fusilier Battalions 121
Russia, Russian 12, 59, 79, 122, 147, 216, 319, 363, 395, 404
   Russian Orthodox 147

S

Sabri, Ikrima 63
Sa'd 26
Sadat, Anwar 142, 166–167, 182, 310, 355, 418, 434, 437
Sa'di, Farhan al- 128
Safed 49–50, 61, 118, 125
Said, Edward 57, 113, 115, 300, 321, 376, 402, 442
Salafi/Salafist 11, 200, 235, 240, 296, 413, 429, 442, 452
Samaria 127, 129, 167, 186, 196, 208
San Bernardino 399, 403
Saudi Arabia 216, 307, 417
Saudi Peace Plan 446
Sderot 224, 228, 429

Second War in Lebanon 2006 222, 436, 450
Sgula 383, 442
Shafi'ite 76
Shaftesbury, Lord 118
Shaheed 91, 102, 214, 227, 285, 440, 443
Shahian, Abd el-Aziz 149
Shalit, Gilad 200, 222, 228, 236, 435
Shamir, Yitzchak 167, 178, 437
Shanab, Abu 218
Sharia Law
   Azzam 105–107
   Clash with Secularism 54–55, 58
   Gaza 222–225
   Hamas Covenant 272, 275, 300–301, 307–308
   Islamic Awakening 417–420, 423–424
   Qutb 79–80, 84–87, 95
Sharon, Ariel 189, 202, 205–211, 214–215, 247, 249, 437
Shehadeh, Salah 168, 171, 176, 213
Shiite 42, 60, 135, 142, 169, 181, 188, 219–220, 234–235, 242, 255, 257, 287, 308, 401, 424, 435, 436, 441, 443
Shikaki, Fathi al- 168, 169
Shikaki, Khalil 191
Shiraz 42
Shirk 80, 82, 84–85, 100, 262, 443
Shomron, Dan 180
Shukri, Hasan 128
Sinai
   Exodus/Covenant 27, 93, 380–384, 387
   Peace with Egypt 311
   Six Day War (1967) 48, 138–140
   Tunnels 200, 219, 239–241, 243, 413
   War of Independence (1948) 133–134
Sisi, Abdul Fatah a- 200, 240, 243, 293, 406, 413–414, 437
Skanderberg Division 359

Soviet Union 109, 178, 295, 343, 356, 362, 400, 439, 451
Spain 38, 41, 52, 73, 80, 103, 307, 375, 393
Special Night Squads/SNS 129
Stalin 280–281, 362, 373–374
Steinetz, Yuval 225
Sukkari, Ahmed 132
Sulieman (Sultan) 50
Summer Rain Operation 222
Sunni 12, 135, 142, 169, 219, 232, 234–235, 241–242, 255, 296, 308, 318, 414, 424, 426, 441, 442, 443, 444
Supreme Muslim Council/SMC 124, 126–127, 129, 454
Sur Baher 134
Surgun 50, 443
Syria
  Jews in 47, 141
  Six Day War (1967) 48, 138–141
  Support for Hamas 220, 235
  War of Independence (1948) 48, 372

T

Taba Talks 204
Tahadiya 224, 443
Taheri, Amir 141
Talahme, Saleh 212
Taliban 11, 88, 233, 286, 377, 399, 404, 424–425
Tamimi, Azzam 216
Tamimi, Taysir al- 63
Tamiya, Ahmed ibn 417
Tanakh 16, 27, 65, 339, 385, 387, 389, 402, 435, 439, 446, 447
Tanzimat 119, 443
Taqiyya 193, 197, 207–208, 218, 441, 443
Tax, Taxes 26, 28, 30, 33, 35, 37–40, 43, 45, 50, 60, 65–66, 82, 90–91, 98, 108, 220, 244, 371, 389, 407, 439, 441, 454
Tehran 42, 220, 241–242, 296, 400, 401, 404
Tel Aviv 97, 122, 134, 184, 190, 201, 209–210, 228, 240, 308, 320, 428–429
Temple Mount Faithful 177
Temple Mount/Noble Sanctuary/Haram al-Sharif 48, 57, 68, 124, 127, 138, 167, 177, 189–190, 194, 202, 204–205, 209, 303–304, 336, 356, 445, 449
Third Temple 125, 177, 196
Tiberias 49, 118, 129, 210
Toameh, Khaled Abu 233
Toledano, Nissim 178
Torah 25, 29, 36, 44, 65, 93, 153, 315, 381–392, 402, 439
Totalitarian/Totalitarianism 109, 263, 280, 287, 370, 403
Transjordan 57, 123, 356
Trench, Battle of the 26, 99, 389
Tunisia, Tunisian 11, 13, 192, 242, 419, 423, 438, 453
Turkey 54, 61, 67, 121, 124, 232, 234, 242–243, 255, 293, 421, 425–426, 434, 442, 451
Tzvi, Shabtai 43

U

Ubayy, Abd Allah b. 95
Uganda 103
Ulama 279, 443
Ummah 102, 443
Umm el-Fahm 309
United National Leadership of the Uprising/UNLU 175–176, 454
United Nations/UN
  Palestinian Independence 240
  Res. 181 137, 145, 158, 163, 181, 203, 326–327, 338, 372, 446
  Res. 194 163, 203, 327

Res. 242 145, 158, 181
Res. 338 181
"Zionism is Racism" 144–145
United Nations/UN High Commissioner for Refugees/UNHCR 203, 454
United Nations/UN Relief Works Agency/UNRWA 158, 201, 203, 326, 353, 454
United Palestinian [Muslim] Brotherhood Organization 165
United States/US
  North Africa (WWII) 45
  Oslo Accords and as Intermediary 202, 204, 206, 213–217, 240–241
  Qutb's Writings 79, 81
Uthman (Caliph) 95

V

Vatican 61
Victoria (ship) 219
Vienna 52, 375, 393

W

Wahhabism/Wahhabist 26, 60, 64, 68, 417, 444
Waqf
  as defined by Islamic Ideologues 76–78, 89, 99, 110
  Hamas Covenant 270–274
  Land of Israel/Palestine 124–125, 136–137, 139, 146, 201, 204, 224
  Mosque Lands 303
Wars
  1948 War 16, 95, 97, 144, 147, 161–162, 355, 361, 442, 447, 448, 449, 454
  1967 War 98, 139, 162, 167, 449
  1973 War (Yom Kippur War, Ramadan War) 181, 449

1982 First War in Lebanon 437, 450
2006 Second War in Lebanon 222, 436, 450
War of Independence 132, 308, 449
World War 1 (WWI) 53–56, 121–123, 361–362, 371
World War 2 (WWII)
  Haj Amin el-Husseini 130–131, 355–361
  Hamas Covenant 294, 368
  Jewish Fighters 362
Weizmann, Chaim 156, 363, 371, 374, 437
West Bank
  First Intifada (1987-93) 175–180
  Hamas Support in 406, 412–416
  Israeli rule (including First Intifada) 165–167, 170–172, 176–178
  Jordanian rule 135–137
  Low Intensity Conflict/Second / Al-Aksa Intifada 208–215
  Oslo Accords 184, 186–188, 190–192, 203–205
  Palestinian Authority/Israeli Rule 220–224, 235–236, 239–244
  Six Day War (1967) 141, 163
West, Capitalist 81, 294, 295, 343, 366–367
Western Wall 50, 125, 177, 185, 202, 204, 208
White Paper of 1931 126, 129, 131
Wingate, Orde 129
Women
  Hamas Covenant 283–286
  Human Shields 214, 225
  In the Koran 88
  Islamic Ideologues' Commentary On 76–77, 83–84, 87–88, 96, 99, 102, 105, 109
Wye Accords 148, 186–188, 216, 323, 352, 436, 447

Y

Ya'bad 128
Yahya (Imam) 43, 66
Yasin, Ahmed 71, 161, 165, 168–169, 171, 176, 187, 261, 296, 438
Yemen 24, 35, 39, 42–43, 60, 66, 141, 242, 255, 307, 308, 400, 423, 438, 453
Young Turks 54, 119–120
Young, W.T. 51
Yousef, Hassan 171, 175, 181, 184–185, 189, 206, 209, 438
Yousef, Mosab 175, 184, 209, 438

Z

Zahar, Mahmoud al- 180, 217–218, 221, 246
Zanzibar 103
Zionism 93, 96–97
Zionism, anti- 21, 46, 117–118, 147, 149, 283, 302, 310, 317, 329–330, 338–343, 351
Zionist Organization 120–121
Zion Mule Corps 121
Zion/Zionism
  Arafat 143–149
  British Palestine Mandate 57–58, 124, 135–136
  Dhimmi Status Violation 46, 56, 58–59
  Hamas Covenant 267, 277, 302, 310–313
  Ottomans 117–120
  Palestinian National Charter 325, 329, 331, 335, 341–343

About the Author

Yisrael Ne'eman

Yisrael Ne'eman is a historian. He received a BA in Middle Eastern Studies from Rutgers University in New Jersey, and an MA in Modern Jewish History from the University of Haifa in Israel. His area of expertise includes Middle East Conflict, modern Jewish History, and Israel.

Yisrael began as a lecturer at the University of Haifa International School in 1991. He teaches *Contemporary Israel: History and Society* during the year, and leads a field study program *Israel: A Mosaic* during the summer. He teaches *Jewish History, Ancient History/Archeology* and *The Development of Middle East Nationalism* at the Technion—Israel Institute of Technology, and teaches similar topics at the Givat Haviva Jewish-Arab Peace Center.

In 1992, Yisrael completed the National Tour Guide course. Over the years, he has lectured and led tours for various educational organizations throughout Israel. Many of his outings are academic touring experiences, which can be approved for use as university credits. Currently, Yisrael works with his son, Alon, guiding family, group, and academic tours throughout the country.

From 1989-2012 he partnered with Elliot Chodoff in *Hamartzim Educational Services* as a lecturer and guide. Together they operated the *Mideast on Target* website (www.me-ontarget.org), which provided commentary on Israel and the Middle East from 2002-2015. The two colleagues continue to work together on developing and implementing educational ventures including lectures and seminars in Israel and abroad.

Yisrael moved to Israel (made aliyah) from New Jersey in 1977 and worked at Kibbutz Elrom, a collective farm in the Golan Heights. The next year, he was drafted and served as a combat medic from 1978 to 1980. Later, as a reservist, he was the senior non-commissioned officer of a battalion aid/evacuation station. During the 1980s, he served on six tours of active duty in the First Lebanon War. Afterward, he spent his reserve service in border patrol along the Lebanese and Golan frontiers, and was discharged in 2001.

About the Author

Following the path of early Labor ideologues, he joined the working class and learned a trade as a stainless steel craftsman. By late 1984, Yisrael moved his family to the Galilean town of Ma'alot seeking out the pioneering challenge of developing the barren hilltops of northern Israel. While in Ma'alot, he taught metal work and English as a second language to newly arrived Ethiopian teenagers.

In early 1988, Yisrael, his wife Susan and their children joined the Eshchar community village in central Galilee. The town's ideology advocates cooperation between Jewish secular and religious lifestyles. Yisrael volunteered on the local council for twelve years, ten of them as chairman. He helped facilitate the arrival and integration of another 100 families into the town.

Furthermore, he worked with the Bedouin tribe of Arab Al-Na'im to establish municipal borders and build cooperation to ensure development for both communities. The communities agreed upon a guiding principle of "good neighborly relations" as opposed to a minimalist policy of "coexistence."

Hamas Jihad: Antisemitism, Islamic World Conquest and Manipulation of Palestinian Nationalism is Yisrael's first book.

Contact Yisrael online at:
Author@HamasJihad.com or www.HamasJihad.com.

published by

White Hart Publications

a division of
Our Written Lives, LLC
in San Antonio, Texas

www.OurWrittenLives.com

www.ingramcontent.com/pod-product-compliance
Lightning Source LLC
Chambersburg PA
CBHW071723080526
44588CB00013B/1881